Biosafety and Ecological Assessment of Genetically Engineered and Edited Crops

Biosafety and Ecological Assessment of Genetically Engineered and Edited Crops

Editors

Wei Wei

Charles Neal Stewart, Jr.

MDPI

Basel • Beijing • Wuhan • Barcelona • Belgrade • Novi Sad • Cluj • Manchester

Editors
Wei Wei
Institute of Botany, Chinese
Academy of Sciences
Beijing, China

Charles Neal Stewart, Jr.
University of Tennessee
Knoxville, TN, USA

Editorial Office
MDPI
St. Alban-Anlage 66
4052 Basel, Switzerland

This is a reprint of articles from the Special Issue published online in the open access journal *Plants* (ISSN 2223-7747) (available at: https://www.mdpi.com/journal/plants/special_issues/Biosafety_Edited_Crops).

For citation purposes, cite each article independently as indicated on the article page online and as indicated below:

Lastname, A.A.; Lastname, B.B. Article Title. *Journal Name* **Year**, *Volume Number*, Page Range.

ISBN 978-3-0365-8592-5 (Hbk)
ISBN 978-3-0365-8593-2 (PDF)
doi.org/10.3390/books978-3-0365-8593-2

Contents

About the Editors

Wei Wei

Wei Wei received a PhD degree in Botany from the Institute of Botany, Chinese Academy of Sciences. He is currently an Associate Professor at this institution. His research interests include biodiversity conservation, and biosafety assessment of biotechnology, as well as plant-environment interaction. He has more than 20 years' experience in risk assessment and governance policy study of genetically engineered (GE) crops. He has recently studied the way forward to utilize past experiences with GE crops for evaluating those organisms developed by newly emerged breeding tools, such as gene drives and gene editing. He completed eight related projects in this area as a principal investigator and participated in more than 10 others. He has authored and co-authored more than 50 scientific papers in English as well as 50 papers in Chinese. He served as an expert member of the ad hoc technical expert group on risk assessment and management of the United Nations' Convention on Biological Diversity from 2008 to 2016, and an expert member of the International Union for Conservation of Nature (IUCN) Task Force on synthetic biology and biodiversity conservation from 2018 to 2021.

Charles Neal Stewart, Jr.

Neal Stewart, Ph.D, is the Ivan Racheff Chaired Professor in plant molecular genetics, at the University of Tennessee-Knoxville, where he has been on faculty since 2002. In 2018, he and Scott Lenaghan co-founded the Center for Agricultural Synthetic Biology, where they are co-directors. Neal Stewart took a PhD from Virginia Tech in biology. He was elected as a AAAS Fellow in 2015 and from 2014-2016 served on the National Academies committee responsible for publishing "Genetically Engineered Crops: Experiences and Prospects" in 2016. He is an author of 340 journal articles, 12 books, and an inventor on 10 issued patents.

plants

MDPI

Editorial

Biosafety and Ecological Assessment of Genetically Engineered and Edited Crops

Wei Wei [1,*] and Charles Neal Stewart, Jr. [2,*]

1 State Key Laboratory of Vegetation and Environmental Change, Institute of Botany, Chinese Academy of Sciences, Beijing 100093, China
2 Department of Plant Sciences and Center for Agricultural Synthetic Biology, 112 Plant Biotechnology Building, University of Tennessee, Knoxville, TN 37996, USA
* Correspondence: weiwei@ibcas.ac.cn (W.W.); nealstewart@utk.edu (C.N.S.J.)

Nearly three decades have passed since the first commercial cultivation of genetically engineered (GE) crops. Even prior to commercialization, there were studies testing the biosafety and ecological risks of the release of GE plants. While we have learned much, and the National Academies of Sciences, Engineering, and Medicine Committee on genetically engineered crops found no substantiated evidence showing foods from GE crops were less safe than foods from non-GE crops [1], concerns and controversial views remain. The vast majority of GE crops cultivated by area are annual row crops that are either resistant to herbicide or produce pesticide or of stacked traits of both herbicide resistance and pesticide production. We have gained much experience growing these crops [1]. Although people are optimistic about the environmental and economic benefits conferred by the adoption of GE crops, there have been some predicted risks that have been actualized. For example, secondary insect pests have replaced the niche of the primary pests after the wide adoption of resistant plants that target and suppress populations of primary insect pests, e.g., the study on GE cotton by Lu et al. [2]. Genetic contamination either by pollen or seed flow in native maize varieties of Mexico has been confirmed [3].

Genes isolated from *Bacillus thuringiensis* (*Bt*) are widely used in the first generation of GE crops and to express various insecticidal Bt proteins in host plants to protect them from insect damage. When insect pests evolved resistance to Bt proteins, new Bt proteins were produced in host plants for pest control [4]. Studies have been performed to evaluate the efficacy of those new Bt strains, or a combination of them, against insect pests. There are two research papers in this Special Issue, titled "Biosafety and Ecological Assessment of Genetically Engineered and Edited Crops", addressing this concern [5,6].

Once GE plants are released into the environment, they interact with various factors, including those involving food chains and competition at multiple levels (gene, individual, population, community, and ecosystem) through trophic connections, nutrient cycles and energy flows, as well as biogeochemical cycles, in contact with abiotic/biotic elements of soils, water, and above- and underground ecosystems in the receiving environment [7–13]. For instance, GE plants as primary producers play important roles in the receiving environment to convert light energy or chemical energy into organic compounds, which are used as food for other organisms in natural ecosystems. Herbivorous insect pests feed on GE plants and are then preyed upon and/or parasitized by predators and/or parasitoids, e.g., the work of Wei et al. [14] and Guan et al. [15]. When the plants die and decay, decomposers convert those decaying materials (and other wastes, including dead animals) into inorganic materials in soils that support a new cycle commenced by the growth of new primary producers, such as plants. Plant compounds could also be exuded from the roots into soils and may affect soil organisms [16]. The engineered genes and their products (such as proteins) in GE plants could accumulate at or transfer through different trophic levels and actively participate in natural processes (cycles) in the receiving ecosystems and could cause unintended effects to the exposed organisms (Figure 1).

Citation: Wei, W.; Stewart, C.N., Jr. Biosafety and Ecological Assessment of Genetically Engineered and Edited Crops. *Plants* **2023**, *12*, 2551. https://doi.org/10.3390/plants12132551

Received: 27 June 2023
Accepted: 29 June 2023
Published: 5 July 2023

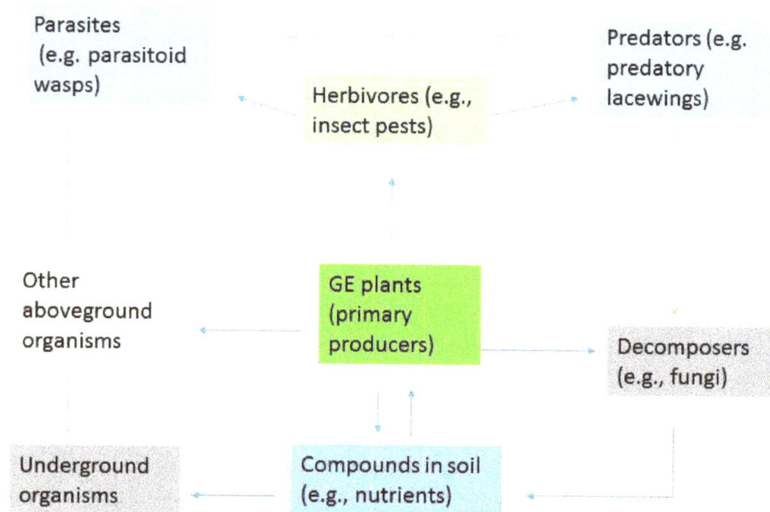

Figure 1. An example of the interaction of genetically engineered (GE) plants with organisms in released environment.

This Special Issue includes four research papers on the impacts of Bt crops on the arthropod community [15,17–19] and two research reports on the effects on soil microbiome [20,21] in the lab and in the field. Generally, no overall significant change was observed in the field studies as the environmental conditions and plant growth stage were likely stronger effects than the engineered status of crop cultivars. Two review papers [16,22] analyzed and discussed the current progress of the impacts of GE crops on soil microbiota.

Similarly, a feeding study in this Special Issue showed the absence of adverse effects of a drought-tolerant GE wheat line to experimental rats compared to its non-GE parent crop [23]. "Omics" technologies are proposed to quantify the differences between GE and non-GE foods to inform regulation strategies [24]. A proteomic case study presented in this Special Issue suggested that no shared change occurred between the two GE oilseed rape (*Brassica napus*) lines transformed by the *Bt Cry1Ac* gene [25]. However, in GE corn, the transformation of the *Bt Cry1Ab* gene may affect plant defenses with plant hormones [26], which could suggest that a potential change in metabolomics could be important [24]. Another important aspect of the biosafety concern for the release of GE crops is transgene escape through pollen or seed flow, which may increase or reduce the fitness of the gene flow recipient plants [27]. One review paper included in this Special Issue proposed and discussed potential approaches to bioconfine transgene flow [28].

While controversial debates on those adverse impacts of the first generation of GE plants using transgenic technology continue and some of those concerns remain unresolved, new breeding tools such as gene editing have been developed and widely employed in scientific research for quantity and quality/nutrition improvement in agriculture and food production [29]. Regarding the fast development of gene-edited crops, this Special Issue also published a review paper on the regulation perspectives of these novel crops [29]. In some countries, this kind of breeding tool may be exempt from the sorts of regulation imposed on GE crops when genetic manipulation does not involve or result in the presence of transgenes. Although there are still concerns regarding the release of gene-edited crops, it is unfortunate that no experimental studies have been reported yet on the biosafety and ecological consequences of the edited crops. However, we believe that the commercially adopted edited crops can benefit from lessons learned from the first generation of engineered crops. Holistic approaches may be helpful to evaluate both the benefits and risks of those GE crops in the view of sustainable agriculture.

In summary, the papers collected in this Special Issue addressed some crucial aspects of the interaction of GE crops with organisms in the environment. Although there is no report on the experimental evaluation of the application of novel breeding tools, such as gene editing, previous works with genetically engineered crops may provide valuable experiences for new gene-edited plants.

Author Contributions: All authors contributed equally. All authors have read and agreed to the published version of the manuscript.

Acknowledgments: We sincerely thank all the authors for contributing their papers to this Special Issue, titled "Biosafety and Ecological Assessment of Genetically Engineered and Edited Crops". On their behalf, we extend appreciation to all the reviewers and two Academic Editors of the journal who generously gave their time and energy to ensure the expected high scientific standards were met. Finally, we thank the Managing Editor and the editorial office of the journal for their professional assistance for the completion of this Special Issue.

Conflicts of Interest: The authors declare no conflict of interest.

References

1. National Academies of Sciences, Engineering, and Medicine (NASEM). *Genetically Engineered Crops: Experiences and Prospects*; The National Academies Press: Washington, DC, USA, 2016; p. 606. [CrossRef]
2. Lu, Y.-H.; Wu, K.-M.; Jiang, Y.-Y.; Xia, B.; Li, P.; Feng, H.-Q.; Wyckhuys, K.A.G.; Guo, Y.-Y. Mirid bug outbreaks in multiple crops correlated with wide-scale adoption of Bt cotton in China. *Science* **2010**, *328*, 1151–1154. [CrossRef]
3. Ceniceros-Ojeda, E.A.; Hayano-Kanashiro, C.; Martínez, O.; Reyes-Valdés, M.H.; Hernández-Godinez, F.; Pons-Hernández, J.L.; Simpson, J. Large scale sampling of Mexican maize landraces for the presence of transgenes. *Transgenic Res.* **2023**. [CrossRef]
4. Tabashnik, B.E.; Fabrick, J.A.; Carrière, Y. Global patterns of insect resistance to transgenic Bt crops: The first 25 years. *J. Econ. Entomol.* **2023**, *116*, 297–309. [CrossRef]
5. Li, G.-P.; Ji, T.-J.; Zhao, S.-Y.; Feng, H.-Q.; Wu, K.-M. High-dose assessment of transgenic insect-resistant maize events against major Lepidopteran pests in China. *Plants* **2022**, *11*, 3125. [CrossRef] [PubMed]
6. Skoková Habuštová, O.; Svobodová, Z.; Kodrík, D.; Sehnal, F. Cry3Aa toxin is not suitable to control Lepidopteran pest *Spodoptera littoralis* (Boisd.). *Plants* **2022**, *11*, 1312. [CrossRef]
7. Trevors, J.T.; Kuikman, P.; Watson, B. Transgenic plants and biogeochemical cycles. *Mol. Ecol.* **1994**, *3*, 57–64. [CrossRef]
8. Andow, D.A.; Hilbeck, A. Science-based risk assessment for nontarget effects of transgenic crops. *BioScience* **2004**, *54*, 637–649. [CrossRef]
9. Kolseth, A.-K.; D'Hertefeldt, T.; Emmerich, M.; Forabosco, F.; Marklund, S.; Cheeke, T.E.; Hallin, S.; Weih, M. Influence of genetically modified organisms on agro-ecosystem processes. *Agric. Ecosyst. Environ.* **2015**, *214*, 96–106. [CrossRef]
10. Guan, Z.-J.; Lu, S.-B.; Huo, Y.-L.; Guan, Z.-P.; Liu, B.; Wei, W. Do genetically modified plants adversely affect soil microbial communities? *Agric. Ecosyst. Environ.* **2016**, *235*, 289–305. [CrossRef]
11. Mandal, A.; Sarkar, B.; Owens, G.; Thakur, J.K.; Manna, M.C.; Niazi, N.K.; Jayaraman, S.; Patra, A.K. Impact of genetically modified crops on rhizosphere microorganisms and processes: A review focusing on Bt cotton. *Appl. Soil Ecol.* **2020**, *148*, 103492. [CrossRef]
12. Schiemann, J.; Dietz-Pfeilstetter, A.; Hartung, F.; Kohl, C.; Romeis, J.; Sprink, T. Risk assessment and regulation of plants modified by modern biotechniques: Current status and future challenges. *Annu. Rev. Plant Biol.* **2019**, *70*, 699–726. [CrossRef]
13. Schulz, R.; Bub, S.; Petschick, L.L.; Stehle, S.; Wolfram, J. Applied pesticide toxicity shifts toward plants and invertebrates, even in GM crops. *Science* **2021**, *372*, 81–84. [CrossRef]
14. Wei, W.; Schuler, T.H.; Clark, S.J.; Stewart, C.N., Jr.; Poppy, G.M. Movement of transgenic plant-expressed Bt Cry1Ac proteins through high trophic levels. *J. Appl. Entomol.* **2008**, *132*, 1–11. [CrossRef]
15. Guan, Z.-J.; Zhou, Q.-J.; Shi, H.; Tang, Z.-X.; Liu, B.; Wei, W. Effect of transgenic cotton expressing Bt Cry1Ac or Cry1Ab/Ac toxins on lacewing larvae mediated by herbivorous insect pests. *Plants* **2022**, *11*, 2755. [CrossRef]
16. Lebedev, V.; Lebedeva, T.; Tikhonova, E.; Shestibratov, K. Assessing impacts of transgenic plants on soil using functional indicators: Twenty years of research and perspectives. *Plants* **2022**, *11*, 2439. [CrossRef]
17. Ren, Z.-T.; Yang, M.-Z.; He, H.-P.; Ma, Y.-J.; Zhou, Y.-J.; Liu, B.; Xue, K. Transgenic maize has insignificant effects on the diversity of arthropods: A 3-Year study. *Plants* **2022**, *11*, 2254. [CrossRef] [PubMed]
18. Yang, Y.; Chen, Y.; Xue, J.-B.; Wang, Y.-Y.; Song, X.-Y.; Li, Y.-H. Impact of transgenic Cry1Ab/2Aj maize on abundance of non-target arthropods in the field. *Plants* **2022**, *11*, 2520. [CrossRef] [PubMed]
19. Yin, J.-Q.; Wang, D.-M.; Liang, J.-G.; Song, X.-Y. Negligible impact of drought-resistant genetically modified maize on arthropod community structure observed in a 2-year field investigation. *Plants* **2022**, *11*, 1092. [CrossRef]
20. Xu, X.-H.; Liu, X.; Li, F.; Hao, C.-F.; Sun, H.-W.; Yang, S.-K.; Jiao, Y.; Lu, X.-B. Impact of insect-resistant transgenic maize 2a-7 on diversity and dynamics of bacterial communities in rhizosphere soil. *Plants* **2023**, *12*, 2046. [CrossRef]

21. Yang, S.-K.; Liu, X.; Xu, X.-H.; Sun, H.-W.; Li, F.; Hao, C.-F.; Lu, X.-B. Effects of insect-resistant maize 2A-7 expressing mCry1Ab and mCry2Ab on the soil ecosystem. *Plants* **2022**, *11*, 2218. [CrossRef] [PubMed]
22. Li, Y.-J.; Wang, C.; Ge, L.; Hu, C.; Wu, G.-G.; Sun, Y.; Song, L.-L.; Wu, X.; Pan, A.-H.; Xu, Q.-Q.; et al. Environmental behaviors of *Bacillus thuringiensis* (Bt) insecticidal proteins and their effects on microbial ecology. *Plants* **2022**, *11*, 1212. [CrossRef] [PubMed]
23. Tian, J.; Ke, X.-H.; Yuan, Y.; Yang, W.-X.; Tang, X.-Q.; Pei, L.-J.; Fan, J.; Zhuo, Q.; Yang, X.-G.; Liu, J.-F.; et al. Subchronic toxicity of GmDREB3 gene modified wheat in the third generation wistar rats. *Plants* **2022**, *11*, 1823. [CrossRef] [PubMed]
24. Gould, F.; Amasino, R.M.; Brossard, D.; Buell, C.R.; Dixon, R.A.; Falck-Zepeda, J.B.; Gallo, M.A.; Giller, K.E.; Glenna, L.L.; Griffin, T.; et al. Toward product-based regulation of crops. *Science* **2022**, *377*, 1051–1053. [CrossRef] [PubMed]
25. Guan, Z.-J.; Zheng, M.; Tang, Z.-X.; Wei, W.; Stewart, C.N., Jr. Proteomic analysis of Bt cry1Ac transgenic oilseed rape (*Brassica napus* L.). *Plants* **2023**, *12*, 2319. [CrossRef]
26. Feng, Y.-J.; Wang, X.-Y.; Du, T.-T.; Shu, Y.-H.; Tan, F.-X.; Wang, J.-W. Effects of exogenous salicylic acid application to aboveground part on the defense responses in Bt (*Bacillus thuringiensis*) and Non-Bt corn (*Zea mays* L.) seedlings. *Plants* **2022**, *11*, 2162. [CrossRef] [PubMed]
27. Wei, W.; Stewart, C.N., Jr. (Eds.) *Gene Flow: Monitoring, Modeling and Mitigation*; CABI: Wallingford, UK, 2021; p. 169.
28. Stockdale, J.N.; Millwood, R.J. Transgene bioconfinement: Don't flow there. *Plants* **2023**, *12*, 1099. [CrossRef] [PubMed]
29. Jones, M.G.K.; Fosu-Nyarko, J.; Iqbal, S.; Adeel, M.; Romero-Aldemita, R.; Arujanan, M.; Kasai, M.; Wei, X.; Prasetya, B.; Nugroho, S.; et al. Enabling trade in gene-edited produce in Asia and Australasia: The Developing Regulatory Landscape and Future Perspectives. *Plants* **2022**, *11*, 2538. [CrossRef]

plants

Article

Cry3Aa Toxin Is Not Suitable to Control Lepidopteran Pest *Spodoptera littoralis* (Boisd.)

Oxana Skoková Habuštová [1,*], Zdeňka Svobodová [1], Dalibor Kodrík [1,2] and František Sehnal [1,†]

1 Institute of Entomology, Biology Centre, Czech Academy of Sciences, 370 05 České Budějovice, Czech Republic; svobodova@entu.cas.cz (Z.S.); kodrik@entu.cas.cz (D.K.)
2 Faculty of Science, University of South Bohemia, 370 05 České Budějovice, Czech Republic
* Correspondence: habustova@entu.cas.cz
† Deceased author.

Abstract: The toxicity of the *Bacillus thuringiensis* (Bt) toxin Cry3Aa—originally used against the main potato pest, the Colorado potato beetle, *Leptinotarsa decemlineata*—was verified on this species and then evaluated against the Egyptian armyworm, *Spodoptera littoralis*, which is a pest of several economically important plants. Larvae of *S. littoralis* were fed a semi-artificial diet supplemented either with a recombinant or with a natural Bt toxin Cry3Aa and with the genetically engineered (GE) potato of variety Superior NewLeaf (SNL) expressing Cry3Aa. Cry3Aa concentration in the diet and the content in the leaves were verified via ELISA (enzyme-linked immunosorbent assay) and HPLC (high-performance liquid chromatography) during and at the end of the experiments. The biological effectiveness of the coleopteran-specific Cry3Aa with previous reports of activity against *S. littoralis* was tested on five different populations of *S. littoralis* larvae by monitoring 13 parameters involving development from penultimate instar, weight, the efficiency of food conversion to biomass, ability to reproduce, and mortality. Although some occasional differences occurred between the Cry3Aa treatments and control, any key deleterious effects on *S. littoralis* in this study were not confirmed. We concluded that the Cry3Aa toxin appears to be non-toxic to *S. littoralis*, and its practical application against this pest is unsuitable.

Keywords: *Spodoptera littoralis*; *Leptinotarsa decemlineata*; recombinant Cry3Aa; natural Cry3Aa; Superior NewLeaf; integrated pest management; biological control

Citation: Skoková Habuštová, O.; Svobodová, Z.; Kodrík, D.; Sehnal, F. Cry3Aa Toxin Is Not Suitable to Control Lepidopteran Pest *Spodoptera littoralis* (Boisd.). *Plants* **2022**, *11*, 1312. https://doi.org/10.3390/plants11101312

Academic Editors: Wei Wei and C. Neal Stewart, Jr.

Received: 19 April 2022
Accepted: 13 May 2022
Published: 15 May 2022

1. Introduction

One of the environmentally friendly methods used to reduce insect pest populations is the practical utilisation of the insecticidal crystal protein (Cry) that occurs naturally in the soil bacterium *Bacillus thuringiensis* (Ber.) (Bt). Cry toxins are usually applied via spraying or in genetically engineered (GE) plants. Cry toxins kill host cells and thus allow Bt germination in dead arthropods. Cry toxins are intestinal pore-forming δ-endotoxins that, after activation by host proteases in the midgut, interact with receptors on the midgut epithelium. For example, in Lepidoptera, aminopeptidase N (APN) receptors, cadherin-like receptors, and ATP binding cassette (ABC) protein family function as toxin receptors for Cry1A. They are involved in the cleavage of the amino-terminal end, including the helix, and the formation of a pre-pore oligomer of Cry toxin, which leads to membrane insertion and pore formation. The pore formation results in osmotic cell lysis or else activation of the oncotic cell death pathway [1,2]. Because of their interaction with greatly diversified receptors, Cry toxins are highly specific to certain species of the insect orders Lepidoptera, Coleoptera, Hymenoptera, Diptera, Orthoptera, and Mallophaga, and also to Nematoda, Acari, and Protozoa [3]. However, some Cry toxins have an expanded spectrum of action to two or more taxonomic categories. For example, Cry1B is one of those that present a remarkable activity against larvae of Lepidoptera, Diptera, and Coleoptera [4].

Cry1 act on lepidopteran pests, and therefore Cry1Ab suppresses *Spodoptera littoralis* (Boisd.) (Lepidoptera: Noctuidae) [5–7]. On the other hand, Cry3 toxins are specific to coleopteran species [8–10]; Cry3Aa is used against potato pest *Leptinotarsa decemlineata* (Say) (Coleoptera: Chrysomelidae). Interestingly, in some previous experiments, a certain cross-activity of the Cry toxins among the insect orders was recorded; for example, the Cry3Aa was found to affect non-target lepidopterans, namely, the early instars of *Acherontia atropos* (L.), *Manduca sexta* (L.) (both Lepidoptera: Sphingidae), and *Autographa gamma* (L.) (Lepidoptera: Noctuidae) [11]. Later, it was reported that the Cry3Aa toxin also reduced larval growth of *S. littoralis* when fed a Cry3Aa-expressing potato, and larval growth, pupal size, and adult fecundity when fed Cry3Aa in a semi-artificial diet [12–14]. Therefore, we decided to review and extend the data on the effect of the Cry3Aa toxin on five different populations of *S. littoralis*, two of which should have high sensitivity to insecticides.

It is well known that *S. littoralis* is a polyphagous and economically important pest of many cultivated plants in the Mediterranean region [15]. It is an A2 quarantine pest in the European Union (EU) with occasional occurrence in Central Europe, where its permanent existence is not yet possible. However, climate change could alter its distribution, and Central European potatoes and other crops may be in danger by this novel pest [16]. New agents against this pest applicable in integrated pest management (IPM) and organic farming would be appreciated by farmers because pressure to utilise sustainable agriculture practices is considerable worldwide.

We used natural and recombinant Cry3Aa toxins applied in a semi-artificial diet and the GE potato Superior NewLeaf™ (MONSANTO Technology LLC, St. Louis, MO, USA), which expresses Cry3Aa. This potato is resistant to *L. decemlineata* with a simultaneous absence of effects on beneficial arthropods such as lady beetles and carabid beetles in laboratory and field studies [17–19].

The main objective of the present study was to investigate and extend the existing data on the efficacy of various forms of Cry3Aa (recombinant, natural, and expressed in GE potato) on the pest *S. littoralis* with possible implications in IPM.

2. Results

2.1. Cry3Aa Content

The relative amount of Cry3Aa in the working solutions was determined by semiquantitative RP HPLC (Figure 1). The results suggested a higher (about 7.5-fold) level of the recombinant protein than the natural protein in the corresponding solutions. The results obtained using the RP HPLC were supported by ELISA. These showed that the working solutions of recombinant and natural Cry3Aa contained 3.418 μg/mL and 279 μg/mL protein, respectively. These amounts were stable and constant until the end of the experiments. The diet used in the experiments contained an amount of Cry3Aa that was based on the concentrations determined by ELISA. Further, the content of the Cry3Aa in potato leaves used for bioassays 1 and 3 ranged from 1.31 to 1.96 μg/g Cry3Aa of fresh weight.

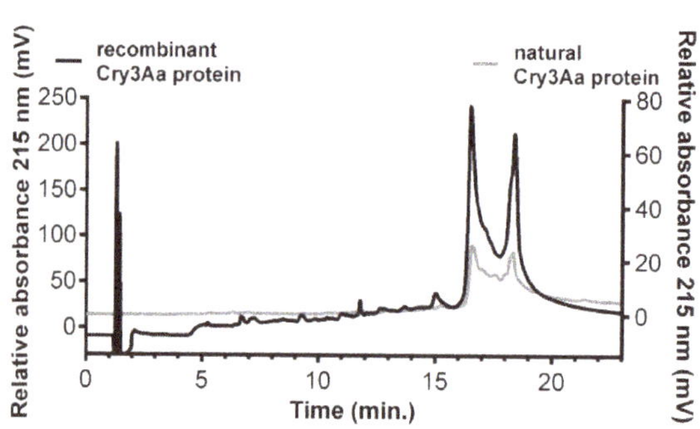

Figure 1. The RP HPLC elution profiles of recombinant Cry3Aa (left y-axis) and natural Cry3Aa (right y-axis) solutions (200 μL).

2.2. Effect of Cry3Aa on Survival of *L. decemlineata*

The results showed that the effect of natural Cry3Aa was more pronounced than effect of recombinant Cry3Aa (Figure 2A,B). Thus, lower concentrations of natural Cry3Aa were utilised to determine the LC_{50} and LC_{90}. LC_{50} was determined to be 1.8 µg/g (95% confidence limits: 0.78–2.74) and 0.1 µg/g (95% confidence limits: 0.03–0.23) for recombinant and natural Cry3Aa, respectively. LC_{90} was calculated to be 8.1 µg/g (95% confidence limits: 4.91–33.68) and 1.2 µg/g (95% confidence limits: 0.56–4.41) for recombinant and natural Cry3Aa, respectively. The values of LC_{50} and LC_{90} for the natural Cry3Aa was about 18 and 6.8 times lower than their recombinant forms, respectively. The effectiveness of Cry3Aa toxins was different for various concentrations (Log-rank test: recombinant Cry3Aa: $\chi2 = 105.3$, df = 6, $p < 0.0001$; natural Cry3Aa: $\chi2 = 282.9$, df = 9, $p \leq 0.0001$, results of post hoc tests in Table S1A). The effect of Cry3Aa expressed in leaves of GE potato SNL was also evident from the second day. Survival curves were significantly different between GE potato SNL and control (Log-rank test: $\chi2 = 93.9$, df = 1, $p \leq 0.0001$). On the fifth day, more than 90% of *L. decemlineata* on leaves of GE potato SNL were dead (Figure 2C). Compared with that, we estimated from Figure 2A,B that recombinant and natural form in same concentrations caused approximately 20% and 38% mortality, respectively. On the basis of the results of bioassay 1 (LC_{90}), in bioassay 2, we worked with a concentration of 8 µg/g of Cry3Aa in a semi-artificial diet.

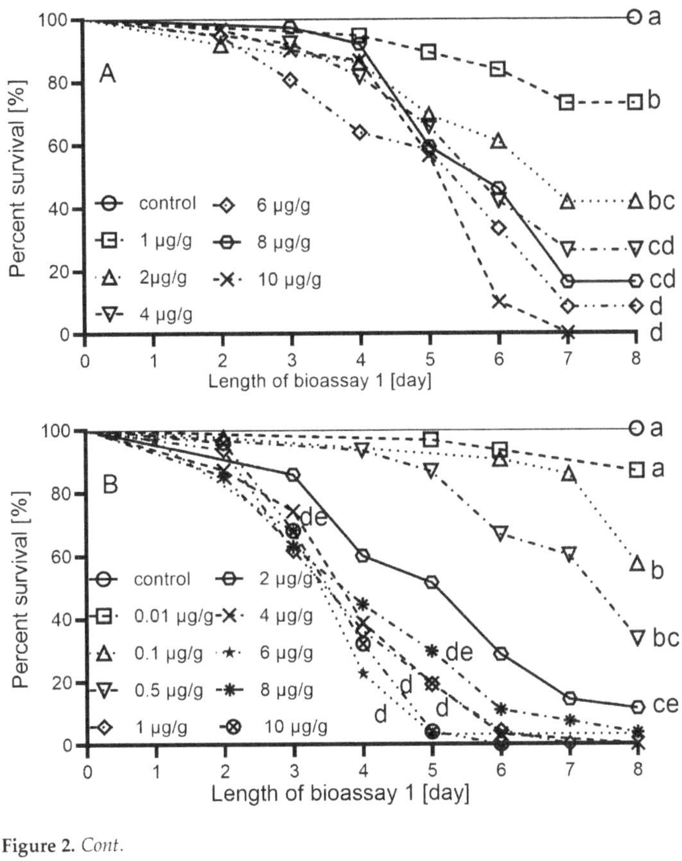

Figure 2. *Cont.*

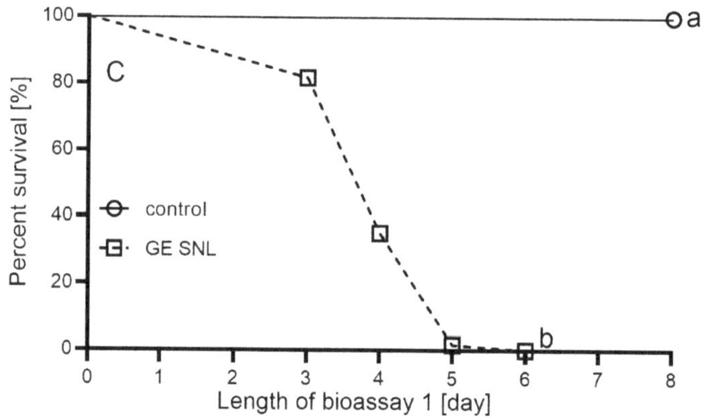

Figure 2. Survival of *L. decemlineata* larvae on the semi-artificial diet with different concentrations of recombinant (**A**) and natural (**B**) Cry3Aa, and on the leaves of GE potato SNL plants expressing Cry3Aa (**C**) in bioassay 1. The same letters denote non-significant differences, while different letters denote statistically significant differences in trend of survival between treatments. The values of statistical tests are available in Table S1A.

2.3. Effect of Cry3Aa on Larval and Pupal Mortality of S. littoralis

In bioassays 2 and 3, certain differences in larval and pupal mortalities between Cry3Aa-treated *S. littoralis* and corresponding controls were recorded (Tables 1 and 2). However, the difference was mostly insignificant. In the NRC population within bioassay 2 (Tables 1 and S1B), the overall test ($p = 0.046$ for larvae, and $p = 0.036$ for pupae) indicated statistical significance; however, post hoc tests did not reveal any specific difference between treatments. Further, in bioassay 3 within the SF population, pupal mortality was significantly higher (about 2.4-fold) in the control compared to GE potato SNL feeding ($p = 0.011$) (Tables 2 and S1C).

Table 1. Examined parameters (mean ± SD) of three *S. littoralis* populations in bioassay 2 with 8 µg/g recombinant and natural Cry3Aa in a semi-artificial diet. Different letters (in bold) denote statistically significant differences, while letters are not assigned when statistical difference was not found. Statistical comparison was performed for recombinant and natural Cry3Aa together. The values of statistical tests are available in Table S1B.

Population	Examined Parameter	Control	Recombinant Cry3Aa	Natural Cry3Aa
NRC	Initial larval weight (mg)	59.1 ± 3.3	60.3 ± 4.0	59.2 ± 3.5
	ECI (%) [1]	24.2 ± 1.7	36.5 ± 1.9	31.8 ± 11.3
	Weight increment (mg)	950.0 ± 176.8 **b**	1077.7 ± 149.5 **a**	945.8 ± 153.1 **b**
	Maximal body weight (mg)	1009.1 ± 176.4 **b**	1137.9 ± 148.7 **a**	1004.9 ± 152.9 **b**
	Length of fifth instar (days)	2.9 ± 0.4	2.9 ± 0.3	2.9 ± 0.3
	Length of sixth instar (days)	2.9 ± 0.4	2.8 ± 0.4	2.9 ± 0.4
	Length of prepupal stage (days)	2.6 ± 0.6	2.7 ± 0.5	2.5 ± 0.5
	Larval mortality (%)	0	0	7.5
	Pupal weight (mg)	363.4 ± 52.5 **b**	386.9 ± 45.9 **a**	366.2 ± 40.4 **b**
	Length of pupal stage (days)	8.7 ± 0.7 **b**	10.2 ± 0.7 **a**	8.5 ± 0.5 **c**
	Pupal mortality (%)	0	0	8.1
	No. of laid eggs per female per day	281.2 ± 125.6	294.7 ± 121.3	274.6 ± 127.8
	No. of hatched eggs per female per day	136.1 ± 79.4	140.2 ± 127.9	147.1 ± 108.3
	Hatching rate per female per day (%)	43.8 ± 17.0	36.9 ± 30.4	50.2 ± 22.0

Table 1. *Cont.*

Population	Examined Parameter	Control	Recombinant Cry3Aa	Natural Cry3Aa
SE	Initial larval weight (mg)	58.6 ± 3.6	59.7 ± 3.2	58.0 ± 2.8
	ECI (%)	25.0 ± 7.5	28.7 ± 1.8	20.2 ± 3.8
	Weight increment (mg)	825.6 ± 104.8 **a**	800.1 ± 114.6 **a**	709.7 ± 177.0 **b**
	Maximal body weight (mg)	884.3 ± 103.4 **a**	859.8 ± 114.0 **a**	767.7 ± 176.4 **b**
	Length of fifth instar (days)	2.6 ± 0.5 **a**	2.2 ±0.4 **b**	2.6 ± 0.5 **a**
	Length of sixth instar (days)	2.9 ± 0.4	3.1 ± 0.5	3.1 ± 0.5
	Length of prepupal stage (days)	2.5 ± 0.5	2.4 ± 0.6	2.7 ± 0.8
	Larval mortality (%)	20.0	32.5	32.5
	Pupal weight (mg)	327.2 ± 42.3 **a**	303.9 ± 43.8 **ab**	299.0 ± 52.9 **b**
	Length of pupal stage (days)	8.5 ± 0.7	8.2 ± 0.9	8.3 ± 0.8
	Pupal mortality (%)	6.3	0	3.7
	No. of laid eggs per female per day	298.9 ± 132.5	328.8 ± 148	252.2 ± 123.9
	No. of hatched eggs per female per day	256.9 ± 150.3	292.2 ± 170.8	214.3 ± 135.5
	Hatching rate per female per day (%)	75.8 ± 29.5	81.8 ± 23	72.9 ± 34.9
SF	Initial larval weight (mg)	59.9 ± 3.1	60.2 ± 3.5	59.3 ± 3.3
	ECI (%)	21.4 ± 2.8	22.2 ± 3.4	23.2 ± 1.8
	Weight increment (mg)	861.8 ± 211.9	822.9 ± 131.3	811.1 ± 138.0
	Maximal body weight (mg)	921.7 ± 212.2	883.3 ± 131.2	870.4 ± 138.7
	Length of fifth instar (days)	2.3 ± 0.5	2.2 ± 0.4	2.2 ± 0.4
	Length of sixth instar (days)	3.1 ± 0.4	3.1 ± 0.5	3.0 ± 0.4
	Length of prepupal stage (days)	2.7 ± 0.5	2.9 ± 0.5	2.8 ± 0.5
	Larval mortality (%)	17.5	15.0	25.0
	Pupal weight (mg)	330.5 ± 46.1	326.6 ± 39.1	317.4 ± 41.7
	Length of pupal stage (days)	8.2 ± 0.8 **ab**	8.1 ± 0.7 **b**	8.6 ± 1.3 **a**
	Pupal mortality (%)	15.2	11.8	3.3
	No. of laid eggs per female per day	245.8 ± 152.1	334.4 ± 180.6	259.0 ± 133.2
	No. of hatched eggs per female per day	157.1 ± 118.9	227.3 ± 179.6	165.9 ± 120.2
	Hatching rate per female per day (%)	48.5 ± 30.1	61.0 ± 29.8	52.4 ± 32.7

[1] ECI: efficiency of food conversion to biomass.

Table 2. Examined parameters (mean ± SD) of five *S. littoralis* populations in bioassay 3 with Cry3Aa expressed in GE potato SNL. Different letters (in bold) denote statistically significant differences, while letters were not assigned when statistical difference was not found. The values of statistical tests are available in Table S2C. Abbreviation in the table—for details, see Section 5, Materials and Methods.

Population	Examined Parameter	Control	GE Potato SNL
N	Initial larval weight (mg)	75.7 ± 2.4	74.7 ± 2.5
	ECI (%) [1]	24.1 ± 12.1	23.1 ± 5.1
	Weight increment (mg)	964.6 ± 141.5	933.3 ± 170.5
	Maximal body weight (mg)	1040.3 ± 141.6	1008.0 ± 170.5
	Length of fifth instar (days)	2.4 ± 0.5	2.2 ± 0.4
	Length of sixth instar (days)	3.5 ± 0.6	3.4 ± 0.6
	Length of prepupal stage (days)	1.7 ± 0.6	1.9 ± 0.6
	Larval mortality (%)	33.3	43.3
	Pupal weight (mg)	341.3 ± 30.2	326.1 ± 43.3
	Length of pupal stage (days)	10.4 ± 0.7	10.4 ± 0.6
	Pupal mortality (%)	0 [2]	0
	No. of laid eggs per female per day	217.3 ± 96.6	260.0 ± 142.5
	No. of hatched eggs per female per day	144.4 ± 130.3	150.6 ± 147.6
	Hatching rate per female per day (%)	60.8 ± 21.5	46.9 ± 29.6

Table 2. *Cont.*

Population	Examined Parameter	Control	GE Potato SNL
NRC	Initial larval weight (mg)	72.7 ± 2.4	71.7 ± 1.8
	ECI (%)	24.0 ± 5.7	33.1 ± 16.1
	Weight increment (mg)	872.7 ± 125.6	913.6 ± 109.4
	Maximal body weight (mg)	945.4 ± 125.2	985.3 ± 110.0
	Length of fifth instar (days)	2.0 ± 0^2	2.0 ± 0
	Length of sixth instar (days)	3.4 ± 0.6 **a**	3.0 ± 0 **b**
	Length of prepupal stage (days)	2.6 ± 0.5	2.4 ± 0.5
	Larval mortality (%)	36.7	53.3
	Pupal weight (mg)	333.3 ± 17.8	340.1 ± 25.5
	Length of pupal stage (days)	8.2 ± 0.4 **b**	9.4 ± 0.7 **a**
	Pupal mortality (%)	0^2	0
	No. of laid eggs per female per day	297.5 ± 70.4	376.2 ± 152.7
	No. of hatched eggs per female per day	232.8 ± 94.3	262.7 ± 159.1
	Hatching rate per female per day (%)	74.5 ± 20.2	67.9 ± 31.2
CU	Initial larval weight (mg)	74.5 ± 3.3	73.9 ± 3.3
	ECI (%)	17.9 ± 4.2	19.8 ± 4.7
	Weight increment (mg)	703.5 ± 204.7	740.6 ± 138.9
	Maximal body weight (mg)	778.0 ± 203.5	814.5 ± 138.0
	Length of fifth instar (days)	2.0 ± 0.2	2.1 ± 0.3
	Length of sixth instar (days)	3.0 ± 0.2	2.9 ± 0.4
	Length of prepupal stage (days)	1.7 ± 0.5	1.9 ± 0.3
	Larval mortality (%)	40.0	46.7
	Pupal weight (mg)	287.4 ± 39.9	288.3 ± 42.6
	Length of pupal stage (days)	8.1 ± 0.9	7.6 ± 0.8
	Pupal mortality (%)	0^2	0
	No. of laid eggs per female per day	348.6 ± 108.3	300.4 ± 73.1
	No. of hatched eggs per female per day	321.9 ± 110.1	217.1 ± 78.0
	Hatching rate per female per day (%)	91.8 ± 4.0	71.5 ± 14.9
SE	Initial larval weight (mg)	60.8 ± 6.8	61.6 ± 6.6
	ECI (%)	24.2 ± 4.0	25.6 ± 4.1
	Weight increment (mg)	868.3 ± 145.	912.2 ± 148.4
	Maximal body weight (mg)	929.1 ± 142.8	973.8 ± 149.9
	Length of fifth instar (days)	2.3 ± 0.7	2.3 ± 0.5
	Length of sixth instar (days)	3.8 ± 0.8	3.8 ± 0.5
	Length of prepupal stage (days)	1.4 ± 0.5	1.6 ± 0.5
	Larval mortality (%)	6.7	16.7
	Pupal weight (mg)	291.0 ± 52.2	298.8 ± 47.9
	Length of pupal stage (days)	8.8 ± 0.8 **b**	9.3 ± 0.9 **a**
	Pupal mortality (%)	3.6	8.3
	No. of laid eggs per female per day	322.9 ± 134.2	317.8 ± 76.9
	No. of hatched eggs per female per day	194.1 ± 94.6	161.7 ± 60.2
	Hatching rate per female per day (%)	60.2 ± 16.4	51.0 ± 17.7
SF	Initial larval weight (mg)	73.6 ± 3.8	74.3 ± 3.9
	ECI (%)	16.7 ± 3.6	21.0 ± 7.7
	Weight increment (mg)	769.9 ± 324.3	973.8 ± 294.2
	Maximal body weight (mg)	843.5 ± 325.1	1048.1 ± 293.0
	Length of fifth instar (days)	1.4 ± 0.5 **a**	1.1 ± 0.3 **b**
	Length of sixth instar (days)	3.8 ± 0.9	4.4 ± 1.0
	Length of prepupal stage (days)	1.3 ± 0.4 **b**	1.4 ± 0.5 **a**
	Larval mortality (%)	6.7	3.3
	Pupal weight (mg)	295.0 ± 81.1	351.1 ± 92.2

Table 2. *Cont.*

Population	Examined Parameter	Control	GE Potato SNL
	Length of pupal stage (days)	10.7 ± 0.8 **a**	10.1 ± 0.8 **b**
	Pupal mortality (%)	57.1 **a**	24.1 **b**
	No. of laid eggs per female per day	332.9 ± 203.7	255.7 ± 146.7
	No. of hatched eggs per female per day	114.1 ± 145.4	150.9 ± 149.0
	Hatching rate per female per day (%)	32.2 ± 36.4	41.4 ± 39.8

[1] ECI: efficiency of food conversion to biomass. [2] Statistical comparison is impossible because of no variability in data. N—population of *S. littoralis* reared in our laboratory; NRC—population of *S. littoralis* obtained from National Research Centre, Egypt; CU—population of *S. littoralis* obtained from Cairo University, Egypt; SE—sensitive population of *S. littoralis* obtained from Egypt; SF—sensitive population of *S. littoralis* obtained from France.

2.4. Sublethal Effects of Cry3Aa on S. littoralis

Within bioassay 2 in the NRC population, recombinant Cry3Aa treatment caused an increase in body weight and a difference in maximal body weight (1.1 times for both) in comparison with those in the natural Cry3Aa and control treatments ($p < 0.001$), but the body weight gain was similar between treatments (Tables 1 and S1A, Figure 3A). In the SE population, recombinant Cry3Aa treatment produced a higher weight increment ($p = 0.029$) and maximal body weight ($p = 0.025$) than natural Cry3Aa. Both parameters were 1.1 times higher than those in the natural Cry3Aa treatment, which was also 1.2 times lower than in the control ($p = 0.002$) for both parameters. Similarly, the body weight gain was highest in the control, followed by the recombinant and natural Cry3Aa treatments (Figure 3B). The length of the fifth instar in the SE population fed recombinant Cry3Aa was 1.2 times shorter than those fed natural Cry3Aa, and the control ($p = 0.003$). In the NRC population, the pupal weight of the recombinant Cry3Aa treatment was 1.1 times higher than in the other treatments ($p = 0.017$ compared with the control, and $p = 0.047$ compared with the natural Cry3Aa treatment), and the length of the pupal stage was 1.2 times longer than the other two treatments (both tests: $p < 0.001$). Moreover, the length of the pupal stage in the natural Cry3Aa treatment was 1.2 times longer than natural Cry3Aa ($p = 0.045$). In the SE population, the pupal weight of the control was 1.1 times higher than in both the natural and recombinant Cry3Aa treatments ($p = 0.035$). In the SF population, the length of the pupal stage was 1.1 times longer in the natural than in the recombinant Cry3Aa treatments ($p = 0.033$). Other results of the statistical comparison were not significantly different (Tables 1 and S1B, Figure 3C). The highest hatching rate was found in the SE population, followed by the SF and NRC populations.

Although the curves for body weight gain in bioassay 3 looked similar for both treatments, in the NRC population, body weight increased more intensively at the end of sixth instar in the GE potato SNL treatment than in the control (Figure 4B). The opposite trend at the end of the sixth instar was recorded in the SF population (Figure 4E). In the SE population, the body weight of the GE potato SNL treatment increased slower than in the control, but maximal body weight was higher in the GE potato SNL treatment than in the control (Figure 4D). In the NRC populations, the sixth instar was 1.1 times longer in the control than those in the GE potato SNL ($p = 0.031$). Similarly, in the SF population, the fifth instar was 1.3 times longer in the control than those in the GE potato SNL ($p = 0.029$), and the length of the prepupal stage of the GE potato SNL treatment was 1.1 times longer than in the control ($p < 0.001$). On the contrary, the pupal stage was 1.1 times longer in the control ($p = 0.018$). In the NRC and SE populations, the pupal stage was longer in the GE potato SNL treatment than in the control by 1.1 times ($p < 0.001$, both). Other results of the statistical comparison were not significantly different (Tables 2 and S1C, Figure 4). The highest hatching rate was identified in the CU population, followed by the NRC, SE, N, and SF populations.

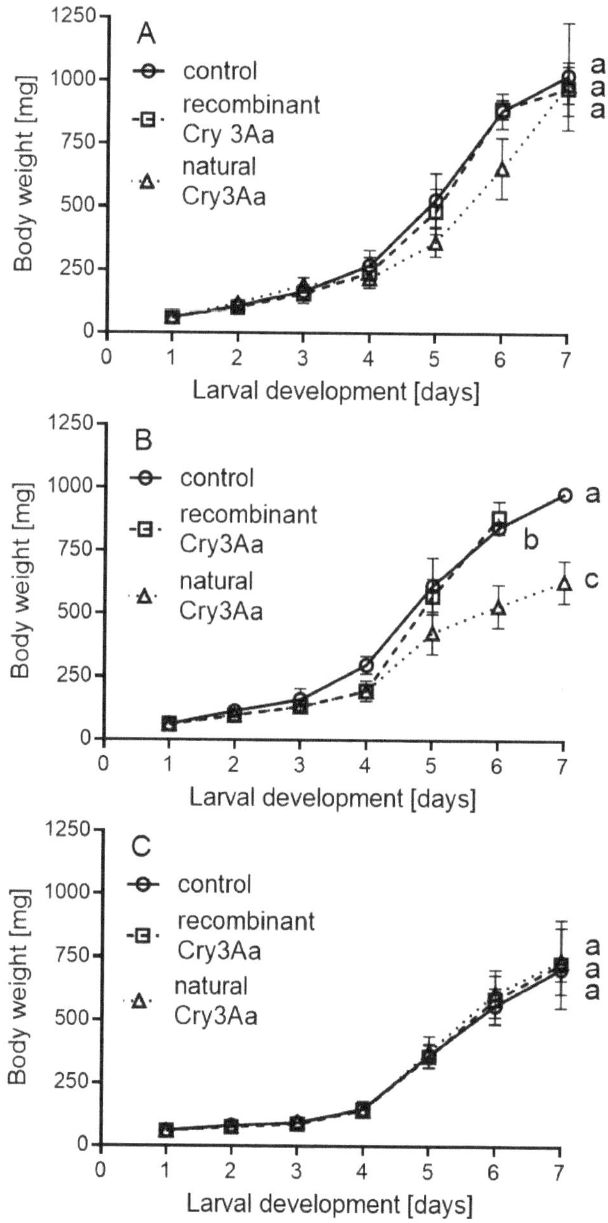

Figure 3. The body weight gain of *S. littoralis* (mean ± SD) larvae in NRC (**A**), SE (**B**), and SF (**C**) populations in bioassay 2 with 8 µg/g recombinant and natural Cry3Aa in a semi-artificial diet. The same letters denote non-significant differences, while different letters denote statistically significant differences. The values of statistical tests are available in Table S2B. N—population of *S. littoralis* reared in our laboratory; NRC—population of *S. littoralis* obtained from National Research Centre, Egypt; CU—population of *S. littoralis* obtained from Cairo University, Egypt; SE—sensitive population of *S. littoralis* obtained from Egypt; SF—sensitive population of *S. littoralis* obtained from France.

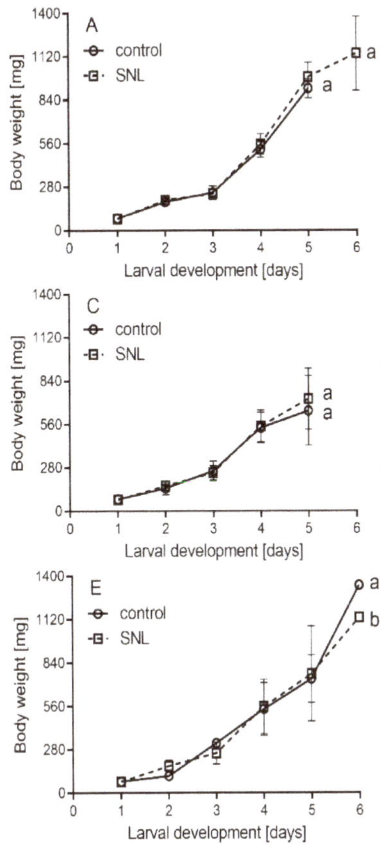

Figure 4. The body weight gain of *S. littoralis* (mean ± SD) larvae in N (**A**), NRC (**B**), CU (**C**), SE (**D**), and SF (**E**) populations in bioassay 3 with GE potato SNL expressing Cry3Aa and control potato Superior. The same letters denote non-significant differences, while different letters denote statistically significant differences. The values of statistical tests are available in Table S2C. N-population of *S. littoralis* reared in our laboratory; NRC—population of *S. littoralis* obtained from National Research Centre, Egypt; CU—population of *S. littoralis* obtained from Cairo University, Egypt; SE—sensitive population of *S. littoralis* obtained from Egypt; SF—sensitive population of *S. littoralis* obtained from France.

3. Discussion

In this study, we sought to characterise the effect of a Cry3Aa toxin on the lepidopteran pest *S. littoralis*. Three types of the Cry3Aa toxin were tested—recombinant, natural, and that expressed by GE potato SNL.

3.1. Bioassay 1—Effect of Cry3Aa on L. decemlineata

We verified the efficacy of three forms of Cry3Aa on first instar *L. decemlineata* larvae. The results confirmed the high effectiveness of tested toxins. Interestingly, the recombinant and natural Cry3Aa toxins differed in their efficacy. Cry3Aa expressed in GE potato SNL showed the highest efficiency. The distinction could be caused by any difference in protein three-dimensional arrangement given by organisms in which they were synthetised. Moreover, the recombinant and natural Cry3Aa in solution could be less efficient because they could be subject to varying degrees of degradation in comparison with Cry3Aa in GE potato SNL leaves that was permanently synthetised in leaves [11,20]. However, variation in LC_{50s} of Cry3Aa has already been subject of research and extensive discussion [21,22]. Generally, our mean LC_{50} values are approximately 10 times lower than mean value obtained by Robertson et al. [22], but they are in the range of values they determined. The results of bioassay 1 showed that all examined forms of Cry3Aa toxin were active and effective against *L. decemlineata*. These results are not surprising, because the efficacy of the Cry3Aa toxin against *L. decemlineata* is generally known, e.g., [17,23–25], and Cry3Aa toxin is used in agricultural practice in the form of spray (organic farming) or incorporated in GE potatoes [26].

3.2. Bioassay 2—Effect of Cry3Aa in Semi-Artificial Diet on S. littoralis

First, we analysed the effect of Cry3Aa applied in a semi-artificial diet. We selected Cry3Aa toxin concentration on the basis of its efficacy on *L. decemlineata* in bioassay 1. We assumed lower activity on *S. littoralis*. We tested penultimate larval instar of two insecticide-sensitive populations, SE and SF, and one common population, NRC. Significant differences in several parameters were found. These results suggested certain Cry3Aa activities. Haider and Ellar [27] suggested that the specific endotoxin-binding receptors in the gut were not necessarily a precondition of a toxic effect. Thus, Lepidoptera eventually would not need a suitable receptor, but suitable protease would be necessary to activate Cry3Aa [11]. Furthermore, Hussein et al. [12] claimed that Cry3Aa ingested by the *S. littoralis* larvae could cause reparable injuries in their midgut epithelium, as was described for *Manduca sexta* after ingestion of suspended crystal endotoxin from *B. thuringiensis* ssp. kurstaki HD1 [28]. This phenomenon could influence the food intake and reduced food consumption, which causes smaller biomass increments and longer development, but manifestation of sublethal effects is limited after switching to non-toxic food [28]. This phenomenon may explain the longer pupal stage for the NRC and SF population in recombinant and natural Cry3Aa treatment, respectively. Nevertheless, the fifth instar of recombinant Cry3Aa treatment in the SE population was shorter than in other treatments. We recorded higher values of body weight parameters in the NRC population for recombinant Cry3Aa in comparison with other treatments, but conversely lower values of body weight parameters in SE population for natural Cry3Aa in comparison with other two treatments. These results suggest that this phenomenon is not fully applicable to our results. Furthermore, it seems that it is possible that pupal weight is not affected even in lepidopteran species sensitive to consumed Bt toxin, although mortality and prolonged development was recorded [29]. Nevertheless, reports about heavier pupae are also available [30].

We used two different Cry3Aa toxins. Although, the different methods of their preparation might play a crucial role in Bt toxin efficiency [11], in contrast to the effect on *L. decemlineata*, we did not determine any constant difference in observed parameters that would imply dissimilarity in Cry3Aa toxicity. However, there is another phenomenon that can substantially affect the efficacy of Cry toxin—the age of the tested insect [11]. The high susceptibility of the first larval instars of *S. littoralis* to lepidopteran-specific Cry toxins and decline in its effectiveness in following larval development have been demonstrated several times [11,31]. Whilst this has been satisfactorily explained [32], there are also reports of adverse effects of lepidopteran-specific Cry toxins at all stages of *S. littoralis* development [33–36].

Results from bioassay 2 showed that the concentration of 8 μg of Bt Cry3Aa toxin per gram of the semi-artificial diet, which is approximately five times higher than in GE potato SNL leaves, did not cause any evident and uniform effect on observed parameters of *S. littoralis*.

3.3. Bioassay 3—Effect of Cry3Aa Expressed in GE Potato SNL on S. littoralis

In bioassay 3, we tested the effect of Cry3Aa toxin expressed in GE potato SNL on selected populations of *S. littoralis* (sensitive SE and SF populations, and N, NRC, and CU long-term laboratory populations).

Primarily, the results showed occasional significant differences in length of larval, prepupal, and pupal stages and mortality and body weight between SNL potato and the corresponding control. Nevertheless, the actual differences were not dramatic. Other parameters including female fecundity and hatchability of progeny were not affected by the Cry3Aa treatment. In addition, as in bioassay 2, all results showed no clear tendency to indicate positive or negative effects of the Cry3Aa, because the differences were recorded in both directions (higher/lower, longer/shorter) for GE potato SNL and control. Anyhow, in this assay, we tested several populations of *S. littoralis* and received a sometimes significant but generally low effect of the Cry3Aa toxin. Thus, we assume we can generalise our results to other *S. littoralis* populations and conclude that GE potato SNL is not significantly resistant to *S. littoralis*.

3.4. Larval and Pupal Mortality (Survival) of S. littoralis in Cry3Aa Treatments

Mortality is a basic aspect for assessing the deleterious effect of any toxin. Differences in mortality levels after the Cry3Aa treatments within the populations were not significant. Thus, it is evident that mortality was not dependent on the applied insecticidal Cry3Aa toxin but was likely affected by any other parameter(s). It is interesting to note that mortality in bioassay 3, where the GE potato SNL was used, was in the most cases higher than in bioassay 2, where only a semi-artificial diet was used. We can speculate that switch of diet may play a role (basic cultures were kept on a semi-artificial diet); this is supported by relatively high mortalities in controls (see Table 2). Nevertheless, it is peculiar that the sensitive SF population showed higher mortality (for pupae, even significantly higher) in controls than that in treated insects within bioassay 3; we are at present unable to offer any satisfactory explanation for this, but some connection with the diet switch cannot be excluded. In contrast, *S. littoralis* is a polyphagous species and should tolerate wider spectrum of diets; therefore, we decided to start our experiments immediately after the populations were delivered to our laboratory. Additionally, we wanted to preserve their features and not affect them by breeding in our conditions. Furthermore, it is usual that tested species are exposed to unusual food sources in the assessment of Cry toxins and GE crops expressing Cry toxins without becoming accustomed to new food, e.g., [19,37], but it is important to separate the effect of Cry toxin and nutritional stress [38].

3.5. Sublethal Effects of Cry3Aa on Different S. littoralis Populations

No differences were found in four of the tested parameters after the Cry3Aa treatments. Nevertheless, there were significant differences in eight recorded parameters, namely, in weight increment, body weight gain, maximal body weight, pupal weight, length of the fifth and sixth instar, and length of prepupal and pupal stage between both Cry3Aa treatments and the controls in both bioassays. However, as mentioned above for mortality, the actual differences of studied parameters were just slight—both positive and negative—and thus it is impossible to specify any constant effect of Cry3Aa from these results. In contrast, we can speculate that some of these differences might be explained as a consequence of Cry3Aa ingestion, which could cause reparable effect of midgut epithelium and slowdown in development [29,30]. We cannot exclude the effect of the food switch or natural variability of tested individuals.

Natural variation is a numerical difference in response that is detected each time a bioassay is repeated with one genetic group (population in our case) either within a single generation or population [21]. As a result of natural variation, responses of a tested group at any one time will therefore never be the same as responses of another group tested either at the same or different time [22]. Robertson et al. [21] and recently also Chen et al. [38] demonstrated that once variation for cohorts or generations are assessed, realistic conclusions about values outside the range of natural variation can be drawn. For this reason, in any study of population sensitivity, responses of any species must be estimated with unselected cohorts within a population or for several generations [21] as we did.

4. Conclusions

Our study did not show a deleterious effect of Cry3Aa on the pest *S. littoralis*. We explained the observed differences in the parameters between the Cry3Aa and control treatments primarily as a result of food switch and natural variation. Thus, according to our results, Cry3Aa toxin is not suitable for the control of *S. littoralis* populations in any form, and therefore we do not recommend using it as a natural insecticide against *S. littoralis* in IPM and organic farming.

5. Materials and Methods

5.1. Culture of Leptinotarsa Decemlineata

The adults and larvae of *L. decemlineata* were collected from the potato plants in the vicinity of Biology Centre, Czech Academy of Sciences, České Budějovice, Czech Republic (48.97417 N, 14.44867 E), in several consecutive series. The collected *L. decemlineata* was placed inside the fine mesh cage (100 × 50 × 50 cm). Culture was maintained in controlled greenhouse conditions (25 °C, 75% relative humidity, photoperiod of 16 h of light/8 h of dark). Culture was supplemented daily with fresh potato plants of the variety Magda. Potato plants of the variety Magda were obtained in the form of tubers and tissue cultures (tiny plants) from the Potato Research Institute, Havlíčkův Brod, Czech Republic. Plants were grown in a pot with a diameter of 21 cm and a volume of 4 L, watered regularly, and kept in the same conditions as a *L. decemlineata* culture.

5.2. Cultures of Spodoptera Littoralis

Five populations of *S. littoralis* were obtained from different localities and were kept in different conditions before the experiments started.

Population N: Larvae were collected in the vicinity of Cairo, Egypt, and kept in the National Research Centre, Giza, Egypt, for many years. This population was obtained by our laboratory several years ago and kept on a Manduca–Heliothis Premix diet (Stonefly Industries Inc., Bryan, TX, USA).

Population NRC: The population was obtained from the National Research Centre, Giza, Egypt, where it was kept for many years on the castor *Ricinus communis* (Euphorbiaceae) leaves with occasional feeding of some generations on the agar-bean semi-artificial diet (see below). To maintain the culture in our laboratory, the Manduca–Heliothis Premix diet was used.

Population CU: This population was obtained from the Faculty of Agriculture, Cairo University, Egypt, where it was kept on the *R. communis* leaves. Culture was kept on the Manduca–Heliothis Premix diet in our laboratory.

Population SE: This population sensitive to insecticides was received from the Central Agricultural Pesticides Laboratory, Agricultural Research Centre, Giza, Egypt, where it was maintained on the agar-bean semi-artificial diet (see below) for many years. To increase the vigour of the progeny, one generation per year was fed *R. communis* leaves. In our laboratory, the same agar-bean semi-artificial diet was used.

Population SF: The sensitive population was obtained from the French National Institute for Agricultural Research (INRA), Versailles, France, where they were reared on a diet based on soya powder and maize bran (pinole) with antibiotics (see below). In our laboratory, larvae were kept on the same diet.

In our laboratory, all *S. littoralis* cultures were kept at 25 °C at a photoperiod of 16:8 h, and they were fed ad libitum. Experiments were carried out with the first generation of larvae that were delivered to our laboratory.

5.3. Semi-Artificial Diets

The recipe for semi-artificial agar diet for *L. decemlineata* is available in S1: D. The recipe for semi-artificial agar bean diet for *S. littoralis* is given in S1: E. The recipe for soy powder and corn bran diet for *S. littoralis* is described in S1: F. The Manduca–Heliothis Premix diet was prepared from commercially available powder (Stonefly Industries Inc.,

TX, USA) according to the instructions in the manual, but potassium bicarbonate buffer pH 10.6 was used instead of water. The use of buffer shifted diet pH from 5.1 (prepared with distilled water) to 8.2, which is more favourable for Cry3Aa stability. Diets were stored in the refrigerator (4 °C) for up to one month.

5.4. Origin of Cry3Aa Toxins

The recombinant Cry3Aa crystals produced in *Escherichia coli* was provided by MONSANTO Technology LLC. The crystals were dissolved in 0.1 M potassium bicarbonate buffer (pH 10.6) to prepare working solution, centrifuged, stored in a refrigerator, and used within two weeks.

The purified natural Cry3Aa crystals from *B. thuringiensis* ssp. tenebrionis were provided by Igor A. Zalunin (Scientific Research Institute for Genetics and Selection of Industrial Microorganisms, Moscow, Russia, [39]). Working solutions were prepared with 0.05 M potassium bicarbonate buffer (pH 10.6, 0.001 M EDTA), centrifuged, stored in a refrigerator, and used within two weeks.

GE potato variety Superior NewLeaf™ (SNL) expressing Cry3Aa was obtained from MONSANTO Technology LLC, St. Louis, MO, USA, The GE potato SNL plants and their near-isogenic unmodified variety Superior were grown according to standard techniques [14].

5.5. Quantification of Cry3Aa Toxins

The relative amount of Cry3Aa in working solutions was verified by reversed-phase high-performance liquid chromatography (RP HPLC). Both Cry3Aa toxins (recombinant and natural) were dissolved in 0.11% TFA (trifluoroacetic acid) and analysed on the RP HPLC system by Clarity software (Data Apex version 8.0) with a Waters 2487 UV detector (wavelength 215 nm), using a Chromolith Performance RP-18e column 150−4.6 mm (Merck), solutions A and B (A—0.11% TFA in water; B—0.1% TFA in 60% acetonitrile), and a flow rate of 1 mL/min. The relative titre of the toxins was estimated from the areas of the corresponding HPLC peaks.

In another series of experiments, the levels of recombinant and natural Cry3Aa in working solutions, in a semi-artificial diet, and potato plants were checked by using the commercial enzyme-linked immunosorbent assay (ELISA) PathoScreen Complete Kit PSA 05900/0288 Bt-Cry3A (Agdia-Biofords, Evry Cedex, France) at the time of diet preparation and the end of storage of the working solutions. The assay was performed according to the manufacturer's protocol. A positive control, supplied with an ELISA kit, was used to construct a standard curve with a twofold dilution series ranging from 0.16 to 20 ng/mL for potato leaves and stock solution, and 1.25 to 160 ng/mL for semi-artificial diet. The sensitivity threshold of the assay was 0.16 ng Cry3Aa per 1 g of fresh plant tissue and per 1 mL of both stock solution and 1.25 ng Cry3Aa per 0.1 g of semi-artificial diet. Absorbance was determined using an ELISA reader (Spectra MAX 340 PC, Molecular Devices, LLC., Sunnyvale, CA, USA) at 630 nm.

5.6. Bioassays

Bioassay 1—*L. decemlineata*: We verified the efficacy of tested Cry3Aa toxins on the larvae of susceptible coleopteran *L. decemlineata*. Freshly laid eggs were used. The eggs were transferred one by one from the potato leaf by a needle and entomological forceps and dipped individually for 1 s in 0.1% formaldehyde. Excess formaldehyde was removed by touching a filter paper, and eggs were transferred onto a sterile wet filter paper in a sterile glass/plastic Petri dish and incubated at 25 °C and a photoperiod of 16:8 h until larvae hatching. Mobile larvae not older than 30 h were put into a 48-well titre plate on a semi-artificial diet (S1: D) with different concentrations of Cry 3Aa toxin to investigate 50 and 90% lethal concentration (LC_{50} and LC_{90}). In the case of testing effect of Cry3Aa expressed in GE potato SNL leaves, freshly hatched larvae were placed into a 48-well titre plate on the cut-out disk of control and GE potato SNL leaves. Potato disks were underlaid with moistened filter paper. The plates were tightly closed with a food foil (Saran wrap), punctured 3 times over each well with an insect pin (size 00), covered with a provided plastic lid, and kept at 25 °C and a photoperiod of 16:8 h. Mortality was recorded daily. The bioassay 1 was terminated in 8 days. For the exact number of larvae per treatment, see S2: G.

Bioassay 2—*S. littoralis* on a semi-artificial diet: Freshly moulted fifth (penultimate) instar larvae were selected from the NRC, SE, and SF populations. Larvae were divided into three treatments: a semi-artificial diet with natural Cry3Aa, a semi-artificial diet with recombinant Cry3Aa, and a control diet. Both recombinant and natural Cry3Aa toxins were administered in the Manduca–Heliothis Premix diet at a final concentration of 8 μg/g Cry3Aa in the diet. Larvae were kept separately, each in a Petri dish (9 cm in diameter); for exact number of larvae per treatment, see S2: H. Each experiment was repeated three times. Pupae were sexed and kept separately in plastic cups (4.5 cm diameter, 0.18 l volume) filled with two layers consisting of a 2 cm layer of fine sawdust and a 5 cm soil layer. Cups were sealed by netting until adult eclosion. The adults (1 ± 0.5 days old) were randomly paired, one of each sex, transferred into paper cylinders (10 cm high, 9 cm diameter), sealed on both sides with a Petri dish lid, and provided with the 10% honey solution (without added toxin). The experiment was terminated 10 days after the start of egg laying. The following parameters were monitored daily in each population: initial larval weight; body weight gain; ECI (efficiency of conversion of ingested materials; weight gain/(ingested diet—vapor) ∗ 100); maximum body weight; pupal weight; fifth instar length; sixth instar length; larval mortality; prepupal and pupal stage length; pupal mortality; number (no.) of laid eggs per female per day; no. of hatched eggs per female per day; and hatching rate (no. of laid/hatched eggs) per female per day. Larvae in bioassay 2 were maintained at the same temperature and light conditions as the stock cultures.

Bioassay 3—*S. littoralis*: Freshly moulted *S. littoralis* larvae of the fifth (penultimate) instar of populations N, NRC, CU, SE, and SF were divided into two treatments: feeding GE potato SNL and control with isogenic Superior that does not produce toxin. Larvae were placed individually in plastic cups (9 cm top diameter, 0.5 l volume), covered with netting, and fed daily fresh leaves placed in a small tube containing water; squares of cotton pads and aluminium foil prevented water leakage. For the exact number of larvae per treatment, see S2: I. The remaining procedures and monitoring parameters were the same as in bioassay 2.

5.7. Data Analysis

Bioassay 1: Log-rank (Mantel–Cox) tests with Bonferroni correction of the significance level of post hoc tests [40] were calculated to analyse the difference between survival curves of *L. decemlineata* on a diet with different concentrations of Cry3Aa. Probit analysis was applied for LC_{50} and LC_{90} calculations.

Bioassay 2 and **3**: Analysis of covariance (ANCOVA) was used to eliminate the effect of sex included as a covariate in the analysis. ANCOVA was used to evaluate the data of initial larval weight, weight gain, and maximum body weight; fifth and sixth instar length; prepupal and pupal stage length; and pupal weight. In bioassay 2, where three treatments were compared (control, recombinant Cry3Aa, natural Cry3Aa), Tukey's post hoc test followed significant tests to specify the results (between which treatments the difference was found). One-way ANOVA was used for ECI, the number of laid and hatched eggs per female per day. The chi-squared test was used for larval and pupal mortality. The chi-squared test for trend was used for the body weight gain during development. In bioassay 2, Bonferroni correction of significance level was applied in the chi-squared test and the chi-squared test for trend.

Data were analysed using PoloPlus (probit analysis, LeOra Software, Robertson et al. [41]), STATISTICA 8 for Windows (ANOVA, ANCOVA, StatSoft Inc., Tulsa, OK, USA) [42], and GraphPad Prism 5 (Log-rank test, chi-squared test, chi-squared test for trend, GraphPad Software Inc.) [43]. If not stated otherwise, a two-sided α-value of 5% was used to determine the level of significance. F-values were accompanied by degrees of freedom and degrees of freedom of the error (within-population degrees of freedom). On the basis of the Cochran C, Hartley, and Bartlett statistic, homogeneity of variances was confirmed, and normal approximation was applied. Chi-squared values were accompanied by degrees of freedom. Graphs were constructed in GraphPad Prism 5. Mean values were presented with standard deviation (mean ± SD).

Supplementary Materials: The following supporting information can be downloaded at https://www.mdpi.com/article/10.3390/plants11101312/s1, **Table S1: A.** Results of Log-rank (Mantel–Cox) test with Bonferroni correction of α level of difference between survival curves of *L. decemlineata* in bioassay 1 with recombinant and natural Cry3Aa in artificial diet and with Cry3Aa expressed in GM potato SNL. Statistically significant *p*-values are highlighted in bold. **Table S1B.** Result of statistical tests of three *S. littoralis* populations in bioassay 2 with 8 μg/g recombinant and natural Cry3Aa in artificial diet. Statistically significant *p*-values are highlighted in bold (Bonferroni correction: $\alpha = 0.017$ in chi-squared test and chi-squared test for trend). The type of test used for each parameter and abbreviations of *S. littoralis* populations are described in Section 5, Materials and Methods. **Table S1C.** Result of statistical tests of five *S. littoralis* populations in bioassay 3 with Cry3Aa expressed in GM potato SNL. Statistically significant *p*-values are highlighted in bold. The type of test used for each parameter and abbreviations of *S. littoralis* populations are described in Section 5, Materials and Methods. **S1D.** Preparation of semi-artificial diet for *L. decemlineata* larvae (description). **S1E.** Preparation of a semi-artificial agar bean diet for *S. littoralis* larvae (description). **S1F.** Preparation of a semi-artificial soy powder and corn bran diet for *S. littoralis* larvae (description). **S2G.** Original raw data of the bioassay 1 including the exact number of *L. decemlineata* used for statistical analyses. **S2H.** Original raw data of the bioassay 2 including the exact number of *S. littoralis* used for statistical analyses. **S2I.** Original raw data of the bioassay 3 including the exact number of *S. littoralis* used for statistical analyses.

Author Contributions: Conceptualisation, O.S.H. and F.S.; methodology, O.S.H. and F.S.; software, Z.S.; validation, O.S.H., Z.S., D.K. and F.S.; formal analysis, Z.S.; investigation, O.S.H. and Z.S.; resources, O.S.H.; data curation, Z.S.; writing—original draft preparation, O.S.H.; writing—review and editing, O.S.H., Z.S., D.K. and F.S.; visualisation, Z.S.; supervision, F.S.; project administration, O.S.H.; funding acquisition, O.S.H. and D.K. All authors have read and agreed to the published version of the manuscript.

Funding: This research was funded by NAZV QK1910270 (as an introduction of the standard methodology performing on new species) and by the Operational Program Integrated Infrastructure within the project Sustainable smart farming systems taking into account the future challenges (ITMS 313011W112), co-financed by the European Regional Development Fund. Plant material was provided by the Potato Research Institute Havlíčkův Brod, Ltd., Czech Republic, supported by the project of the Ministry of Agriculture of the Czech Republic, E-97/01-3160-0200. None of the funders had any role in the study design, data collection, and analysis; decision to publish; or preparation of the manuscript. The APC was funded by the Czech Academy of Sciences (RVO:60077344).

Institutional Review Board Statement: Not applicable.

Informed Consent Statement: Not applicable.

Data Availability Statement: All published data are available within the article and Supplementary Materials.

Acknowledgments: The authors are grateful to MONSANTO Technology LLC, St. Louis, MO, USA, especially Sylvia Fernandez for her support with experiment performance, Igor A. Zalunin for providing Cry3Aa proteins, bachelor student Jana Husáková for assisting with the conduction of some of the experiments, and Radka Tanzer Fabiánová for technical assistance.

Conflicts of Interest: The authors declare no conflict of interest. The funders had no role in the design of the study; in the collection, analyses, or interpretation of data; in the writing of the manuscript; or in the decision to publish the results.

References

1. Bravo, A.; Gomez, I.; Porta, H.; Garcia-Gomez, B.I.; Rodriguez-Almazan, C.; Pardo, L.; Soberon, M. Evolution of *Bacillus thuringiensis* Cry toxins insecticidal activity. *Microb. Biotechnol.* **2013**, *6*, 17–26. [CrossRef]
2. Jurat-Fuentes, J.L.; Crickmore, N. Specificity determinants for Cry insecticidal proteins: Insights from their mode of action. *J. Invertebr. Pathol.* **2017**, *142*, 5–10. [CrossRef] [PubMed]
3. Ye, W.; Zhu, L.; Liu, Y.; Crickmore, N.; Peng, D.; Ruan, L.; Sun, M. Mining new crystal protein genes from *Bacillus thuringiensis* on the basis of mixed plasmid-enriched genome sequencing and a computational pipeline. *Appl. Microbiol. Biotechnol.* **2012**, *78*, 4795–4801. [CrossRef] [PubMed]
4. Van Frankenhuyzen, K. Cross-order and cross-phylum activity of *Bacillus thuringiensis* pesticidal proteins. *J. Invertebr. Pathol.* **2013**, *114*, 76–85. [CrossRef] [PubMed]
5. Vojtech, E.; Meissle, M.; Poppy, G.M. Effects of Bt maize on the herbivore *Spodoptera littoralis* (Lepidoptera: Noctuidae) and the parasitoid *Cotesia marginiventris* (Hymenoptera: Braconidae). *Transgenic Res.* **2005**, *14*, 133–144. [CrossRef] [PubMed]
6. Alfazairy, A.A.; El-Ahwany, A.M.; Mohamed, E.A.; Zaghloul, H.A.; El-Helow, E.R. Microbial control of the cotton leafworm *Spodoptera littoralis* (Boisd.) by Egyptian *Bacillus thuringiensis* isolates. *Folia Microbiol.* **2013**, *58*, 155–162. [CrossRef]
7. BenFarhat-Touzri, D.; Saadaoui, M.; Abdelkefi-Mesrati, L.; Saadaoui, I.; Azzouz, H.; Tounsi, S. Histopathological effects and determination of the putative receptor of *Bacillus thuringiensis* Cry1Da toxin in *Spodoptera littoralis* midgut. *J. Invertebr. Pathol.* **2013**, *112*, 142–145. [CrossRef]
8. Krieg, A. *Bacillus Thuringiensis—Ein Mikrobielles Insektizid. Grundlagen und Anwendung (Acta Phytomedica, Heft 10)*; DM 58,00; Paul Parey: Berlin/Hamburg, Germany, 1986; 191p, ISBN 3-489-60826-7. [CrossRef]
9. Keller, B.; Langerbruch, G.A. Control of coleopteran pests by *Bacillus thuringiensis*. In *Bacillus Thuringiensis, an Environmental Biopesticide: Theory and Practice*; Entwistle, P.F., Cory, J.S., Bailey, M.J., Higgs, S., Eds.; The Quarterly Review of Biology; Wiley: Chichester, UK, 1993; Volume 69, pp. 171–191. [CrossRef]
10. Knowles, B.H. Mechanism of action of *Bacillus thuringiensis* insecticidal δ-endotoxins. *Adv. Insect Physiol.* **1994**, *24*, 275–308. [CrossRef]
11. Deml, R.; Meise, T.; Dettner, K. Effects of *Bacillus thuringiensis* δ-endotoxins on food utilization, growth, and survival of selected phytophagous insects. *J. Appl. Entomol.* **1999**, *123*, 55–64. [CrossRef]
12. Hussein, H.M.; Habuštová, O.; Sehnal, F. Beetle-specific *Bacillus thuringiensis* Cry3Aa toxin reduces larval growth and curbs reproduction in *Spodoptera littoralis* (Boisd.). *Pest Manag. Sci.* **2005**, *61*, 1186–1192. [CrossRef]
13. Hussein, H.M.; Sehnal, F.; Habuštová, O. Bt-potatoes resistant to Colorado potato beetle affect the performance of Egyptian armyworm (*Spodoptera littoralis*). *Acta Fytotech. Et Zootech.* **2005**, *8*, 38–41.
14. Hussein, H.M.; Habuštová, O.; Turanli, F.; Sehnal, F. Potato Expressing beetle-specific *Bacillus thuringiensis* Cry3Aa toxin reduces performance of a moth. *J. Chem. Ecol.* **2006**, *32*, 1–13. [CrossRef] [PubMed]
15. European and Mediterranean Plant Protection Organization [OEPP/EPPO]. Diagnostic protocol for *Spodoptera littoralis, Spodoptera litura, Spodoptera frugiperda, Spodoptera eridania. OEPP/EPPO Bull.* **2015**, *34*, 257–270.
16. Lopez-Vaamonde, C. Species accounts of 100 of the most invasive alien species in Europe. In *Handbook of Alien Species in Europe*; Springer: Dordrecht, The Netherlands, 2008; Volume 3, pp. 269–374. [CrossRef]
17. Reed, G.L.; Jensen, A.S.; Riebe, J.; Head, G.; Duan, J.J. Transgenic Bt potato and conventional insecticides for Colorado potato beetle management: Comparative efficacy and non-target impacts. *Entomol. Exp. Et Appl.* **2001**, *100*, 89–100. [CrossRef]
18. Duncan, D.; Hammond, D.; Zalewski, J.; Cudnohufsky, J.; Kaniewski, W.; Thornton, M.; Bookout, J.; Lavrik, P.; Rogan, G.; Feldman-Riebe, J. Field performance of transgenic potato, with resistance to Colorado potato beetle and viruses. *HortScience* **2002**, *37*, 275–276. [CrossRef]
19. Ferry, N.; Mulligan, E.A.; Majerus, M.E.N.; Gatehouse, A.M.R. Bitrophic and tritrophic effects of Bt Cry3A transgenic potato on beneficial, non-target, beetles. *Transgenic Res.* **2007**, *16*, 795–812. [CrossRef]
20. Latham, J.R.; Love, M.; Hilbeck, A. The distinct properties of natural and GM cry insecticidal proteins. *Biotechnol. Genet. Eng. Rev.* **2017**, *33*, 62–96. [CrossRef]
21. Robertson, J.L.; Preisler, H.K.; Ng, S.S.; Hickle, L.A.; Gelernter, W.D. Natural variation: A complicating factor in bioassay with chemical and microbial pesticides. *J. Econ. Entomol.* **1995**, *88*, 1–10. [CrossRef]
22. Robertson, J.L.; Russell, R.M.; Preisler, H.K.; Savin, N.E. *Bioassays with Arthropods*, 2nd ed.; CRC Press: Boca Raton, FL, USA; Taylor and Francis Group: Boca Raton, FL, USA, 2007; 199p.
23. Perlak, F.J.; Stone, T.B.; Muskopf, Y.M.; Petersen, L.J.; Parker, G.B.; McPherson, S.A.; Wyman, J.; Love, S.; Reed, G.; Biever, D.; et al. Genetically improved potatoes: Protection from damage by Colorado potato beetles. *Plant Mol. Biol.* **1993**, *22*, 313–321. [CrossRef]
24. Whalon, M.E.; Wierenga, J.M. *Bacillus thuringiensis* resistant Colorado potato beetle and transgenic plants: Some operational and ecological implications for deployment. *Biocontrol Sci. Technol.* **1994**, *4*, 555–561. [CrossRef]
25. Grafius, E.J.; Douches, D.S. The present and future role of insect-resistant genetically modified potato cultivars in IPM. In *Integration of Insect-Resistant Genetically Modified Crops within IPM Programs*, 1st ed.; Springer: Dordrecht, The Netherlands, 2008; pp. 195–221.
26. International Service for the Acquisition of Agri-Biotech Applications [ISAAA]. Global Status of Commercialized Biotech/GM Crops ISAAA Brief No. 55, ISAAA, Ithaca, USA. 2019. Available online: https://www.isaaa.org/resources/publications/briefs/55/ (accessed on 1 April 2022).

27. Haider, M.Z.; Ellar, D.J. Mechanism of action of *Bacillus thuringiensis* insecticidal δ-endotoxin: Interaction with phospholipid vesicles. *Biochim. Et Biophys. Acta (BBA) Biomenbrane* **1989**, *978*, 216–222. [CrossRef]
28. Spies, A.G.; Spence, K.D. Effect of sublethal *Bacillus thuringiensis* crystal endotoxin treatment on the larval midgut of a moth, *Manduca sexta*—SEM study. *Tissue Cell* **1985**, *17*, 379–394. [CrossRef]
29. Peacock, J.W.; Schweitzer, D.F.; Carter, J.L.; Dubois, N.R. Laboratory assessment of the effects of *Bacillus thuringiensis* on native Lepidoptera. *Environ. Entomol.* **1998**, *27*, 450–457. [CrossRef]
30. Moreau, G.; Bauce, E. Developmental polymorphism: A major factor for understanding sublethal effects of *Bacillus thuringiensis*. *Entomol. Exp. Appl.* **2001**, *98*, 133–140. [CrossRef]
31. Sneh, B.; Schuster, S.; Broza, M. Insecticidal activity of *Bacillus thuringiensis* strains against the Egyptian cotton leafworm, *Spodoptera littoralis* Boisd. (Lep. Noctuidae). *Entomophaga* **1981**, *26*, 179–190. [CrossRef]
32. Keller, M.; Sneh, B.; Strizhov, N.; Prudovsky, E.; Regev, A.; Koncz, C.; Schell, J.; Zilberstein, A. Digestion of δ-endotoxin by gut proteases may explain reduced sensitivity of advanced instar larvae of *Spodoptera littoralis* to CryIC. *Insect Biochem. Mol. Biol.* **1996**, *26*, 365–373. [CrossRef]
33. Mohamed, S.A.; Badr, N.A.; El-Hafez, A.A. Efficacy of two formulations of pathogenic bacteria *Bacillus thuringiensis* against the first instar larvae of *Spodoptera littoralis* (Boisd.) and *Agrotis ipsilon* (Han.) (Lepidoptera: Noctuidae). *Egypt. J. Agric. Res.* **2000**, *78*, 1025–1040.
34. Porcar, M.; Caballero, P. Molecular and insecticidal characterization of a *Bacillus thuringiensis* strain isolated during a natural epizootic. *J. Appl. Microbiol.* **2000**, *89*, 309–316. [CrossRef]
35. Romeilah, M.; Abdel-Megeed, M.A. The role of certain bacterial preparations (*Bacillus thuringiensis*) in controlling the cotton leafworm *Spodoptera littoralis* (Boisd.). *Egypt. J. Agric. Res.* **2000**, *78*, 1877–1887.
36. Dutton, A.; Klein, H.; Romeis, J.; Bigler, F. Prey-mediated effects of *Bacillus thuringiensis* spray on the predator *Chrysoperla carnea* in maize. *Biol. Control* **2003**, *26*, 209–215. [CrossRef]
37. Svobodová, Z.; Shu, Y.; Skoková Habuštová, O.; Romeis, J.; Meissle, M. Stacked Bt maize and arthropod predators—Exposure to insecticidal Cry proteins and potential hazards. *Proceeding R. Soc. B Biol. Sci.* **2017**, *284*, 20170440. [CrossRef] [PubMed]
38. Chen, Y.; Romeis, J.; Meissle, M. Performance of *Daphnia magna* on flour, leaves, and pollen from different maize lines: Implications for risk assessment of genetically engineered crops. *Ecotoxicol. Environ. Saf.* **2021**, *212*, 111967. [CrossRef] [PubMed]
39. Chestukhina, G.G.; Kostina, L.I.; Zalunin, I.A.; Kotova, T.S.; Katruka, S.P.; Kuznetsov, Y.S.; Stepanov, V.M. Proteins of *Bacillus thuringiensis* delta-endotoxin crystals. *Biokhimiia* **1977**, *42*, 1660–1667. [PubMed]
40. Dunn, O.J. Multiple comparisons among means. *J. Am. Stat. Assoc.* **1961**, *56*, 52–64. Available online: http://www.jstor.org/stable/2282330 (accessed on 28 March 2022). [CrossRef]
41. Robertson, J.L.; Preisler, H.K.; Russell, R.M. *PoloPlus Probit and Logit Analysis USER'S GUIDE*; CRC Press: Boca Raton, FL, USA, 2002–2003; p. 36.
42. StatSoft Inc. *Statistica Electronic Manual*; StatSoft Inc.: Tulsa, OK, USA, 2015. Available online: http://documentation.statsoft.com/STATISTICAHelp.aspx?path=common/AboutSTATISTICA/ElectronicManualIndex (accessed on 8 March 2018).
43. GraphPad Software Inc. *GraphPad Prism 5.0 User 's Guide*; GraphPad Software Inc.: San Diego, CA, USA, 2007. Available online: https://www.graphpad.com/guides/prism/7/user-guide/index.htm?citing_graphpad_prism.htm (accessed on 15 May 2020).

plants

MDPI

Article

High-Dose Assessment of Transgenic Insect-Resistant Maize Events against Major Lepidopteran Pests in China

Guoping Li [1], Tingjie Ji [1], Shengyuan Zhao [2], Hongqiang Feng [1] and Kongming Wu [2,*]

[1] Key Laboratory of Integrated Pest Management on Crops in Southern Part of Northern China, Ministry of Agriculture and Rural Affairs, Institute of Plant Protection, Henan Academy of Agricultural Sciences, Zhengzhou 450002, China

[2] State Key Laboratory for Biology of Plant Diseases and Insect Pests, Institute of Plant Protection, Chinese Academy of Agricultural Sciences, Beijing 100193, China

* Correspondence: wukongming@caas.cn

Abstract: Lepidopteran pests present a key problem for maize production in China. In order to develop a new strategy for the pest control, the Chinese government has issued safety certificates for insect-resistant transgenic maize, but whether these transformation events can achieve high dose levels to major target pests is still unclear. In this paper, the transformation events of DBN9936 (Bt-Cry1Ab), DBN9936 × DBN9501 (Bt-Cry1Ab + Vip3A), Ruifeng 125 (Bt-Cry1Ab/Cry2Aj), and MIR162 (Bt-Vip3A) were planted in the Huang-huai-hai summer corn region of China to evaluate the lethal effects on major lepidopteran pests, *Spodoptera frugiperda, Helicoverpa armigera, Ostrinia furnacalis, Conogethes punctiferalis, Mythimna separata, Leucania loreyi*, and *Athetis lepigone*, using an artificial diet containing lyophilized Bt maize tissue at a concentration representing a 25-fold dilution of tissue. The results showed that the corrected mortalities of DBN9936 (Bt-Cry1Ab), DBN9936 × DBN9501 (Bt-Cry1Ab + Vip3A), Ruifeng 125 (Bt-Cry1Ab/Cry2Aj), and MIR162 (Bt-Vip3A) to the seven pests were in the ranges 53.80~100%, 62.98~100%, 57.09~100%, and 41.02~100%, respectively. In summary, the events of DBN9936, DBN9936 × DBN9501, and MIR162 reached high dose levels to *S. frugiperda*. DBN9936 × DBN9501 only at the R1 stage reached a high dose level to *H. armigera*. DBN9936, DBN9936 × DBN9501, and Ruifeng 125, at most growth stages, reached high dose levels to *O. furnacalis*, and these three events at some stages also reached high dose levels to *A. lepigone*. Ruifeng 125 presented a high dose level only to *C. punctiferalis*. However, no transformations reached high dose levels to either *M. separata* or *L. loreyi*. This study provides a support for the breeding of high-dose varieties to different target pests, the combined application of multiple genes and the commercial regional planting of insect-resistant transgenic maize in China.

Keywords: transgenic insect-resistant maize; bioassay; target pests; high dose

Citation: Li, G.; Ji, T.; Zhao, S.; Feng, H.; Wu, K. High-Dose Assessment of Transgenic Insect-Resistant Maize Events against Major Lepidopteran Pests in China. *Plants* **2022**, *11*, 3125. https://doi.org/10.3390/plants11223125

Academic Editors: C. Neal Stewart, Jr. and Wei Wei

Received: 2 November 2022
Accepted: 12 November 2022
Published: 16 November 2022

1. Introduction

Genetically modified insect-resistant corn was commercially grown in the United States in 1996 and quickly spread to major corn-producing countries such as Brazil. As a result, pests such as *Ostrinia nublilalis* and *Spodoptera frugiperda* were effectively controlled [1,2], achieving significant economic, social, and ecological benefits [3–5]. However, one main threat to sustainable cultivation of transgenic insect-resistant maize is the resistance of target pests. Thus the "high dose/refuge" strategy is crucial for pest resistance management [6]. Its theoretical basis is as follows: (1) insect-resistant crops should express a high dose of insecticidal protein, which can kill almost all resistant heterozygous individuals RS or all sensitive individuals SS; (2) the initial frequency of resistance genes in the target pest population is very low; and (3) adults from resistant crop plots and refuges are randomly mated in the field [6]. Among them, insect-resistant crops expressing high-dose insecticidal proteins, that is, the breeding of insect-resistant varieties with high-dose effects on target pests, is not only the most effective control of field pests but also the most significant key to the effective implementation of resistance management strategy.

Field resistance cases of *S. frugiperda* to the TC1507 event expressing Cry1F and the MON810 event expressing Cry1Ab corn have been reported in Puerto Rico, Argentina, and Brazil [7–9]. *Busseola fusca* in South Africa and *Helicoverpa zea* in the United States evolved resistance to MON810 [10,11]. *H. zea* in the United States also presented resistance to MON89034 maize expressing Cry1A.105 + Cry2Ab [12]. A review of the global development and application history of transgenic insect-resistant crops showed that all successful cases strictly implemented the "high dose + refuge" strategy, while others did not [13,14]. Especially when some varieties do not reach high doses, such as the first generation of TC1507 maize expressing single Cry1F, in Puerto Rico, Brazil, and other regions where refuge measures were not in place, resistance problems can easily develop [15]. Thus, the key to resistance management is "source control", that is, Bt crops planted in the pest occurrence area should have high dose expression of the target pests in the area. High dose expression of Bt crops means that the amount of Bt insecticidal protein expressed by Bt crops can kill 100% of sensitive homozygous individuals (SS) and 95% of sensitive heterozygous individuals (RS) in the target pest population [6], which can be achieved by expressing one or a combination of Bt proteins [16]. It is difficult to directly test RS heterozygous individuals with Bt crops because it is difficult to obtain resistant populations before the application of Bt crops for registration. Therefore, this quantitative index cannot be accurately measured before the development of resistance. Internationally, the dose with an expression level $\geq 25 \times LC_{99.9}$ is usually considered as the standard of operationally high dose, that is, the dose that is 25 times higher than the concentration to kill sensitive larvae [17].

Currently, there are five methods to determine whether transformation events reach high dose levels to target pests [18,19]. (1) The expression level of transformation events to be registered should be determined by enzyme-linked immunosorbent assay (ELISA) or other more reliable techniques, and the expression level should be less than 25 times that of commercially grown varieties before bioassay of registered transformation events. (2) Serial dilution and bioassay were performed on lyophilized Bt crop tissues by artificial diet, and non-Bt crop tissues were used as control. (3) In the common pest occurrence area, a large number of investigations on the occurrence of pests on the plants to be tested for transformation event were conducted to ensure that the expression level of transformation event reached $LD_{99.99}$ or higher, so as to ensure that at least 95% of heterozygous Sr could be killed. (4) A method similar to Method 3 but using controlled infestation with the LD_{50} value of the laboratory pest population had an LD_{50} value similar to that of the field pest population. (5) An instar with a large target pest was found, and the LD_{50} of this instar was 25 times higher than that of the newly hatched larvae. Larvae from this instar were then tested on Bt crops to determine whether 95% or more of the older larvae were killed. A combination of two of these methods is generally recommended to determine that the transformation event has reached a high dose level [20].

Although Bt transgenic insect-resistant maize has not been commercially grown in China, several transformation events have obtained the national production safety certificates. These include the insect resistant and herbicide tolerant maize DBN9936 with *cry1Ab* and *epsps* genes; Ruifeng 125 with *cry1Ab/cry2Aj* and *G10evo-epsps* genes; DBN9501 with *vip3Aa19* and *pat* genes; Zhedaruifeng 8 with *cry1Ab* and *cry2Ab* genes; DBN9936 × DBN9501 (DBN3601T) with *cry1Ab, vip3Aa19*, pat, and *mepsps* genes; ND207 with *mcry1Ab* and *mcry2Ab* genes; Bt11 × GA21 with *cry1Ab, pat*, and *mepsps* genes; and Bt11 × MIR162 × GA21 with *cry1Ab, pat, vip3Aa20*, and *mepsps* genes (http://www.moa.gov.cn/ztzl/zjyqwgz/spxx/ accessed on 15 October 2022). Previous studies have shown that DBN9936 [21], DBN3601 [22,23], DBN9936, DBN9501, DBN9936 × DBN9501, Bt11 × MIR162, Ruifeng125 [24–26], and other transformation events showed high control effects against the invasive pest *S. frugiperda* and other lepidopteran pests in the laboratory and field, showing good commercialization prospects. In 2021, the Ministry of Agriculture and Rural Affairs of China carried out the pilot work of the commercialization of insect-resistant transgenic maize, and the effect was obvious. It not only effectively prevented and controlled the damage of lepidopteran pests and improved the yield and quality of maize but also reduced the application of insecticides and protected the environment, which indicated that the commercialization of insect-resistant transgenic maize in China is promising.

Maize is the grain crop with the largest sown area in China. The annual planting area is 4.13×10^7 hm^2, and the total output is 2.6×10^8 t (National Bureau of Statistics, http://data.stas.gov.cn/ accessed on 10 September 2022). It is divided into six production areas, namely North spring corn region, Huang-huai-hai summer corn region, Southwest hilly corn region, South hilly corn region, Northwest inland corn region, and Qingzang plateau corn region [27]. There are great differences in planting area, farming system, and insect species in different maize planting areas. The area of spring sown corn in the North spring corn region accounts for 35% of the total area of corn planted in China. *O. furnacalis* and *Mythimna separata* are the main pests in this area. The Southwest hilly corn region and South hilly corn region accounted for 20%, which are the main planting areas of autumn and winter maize in China, dominated mainly by the newly invaded *S. frugiperda*. *O. furnacalis* is the main pest in Northwest inland corn region, accounting for 4%. The summer maize area in the Huang-huai-hai summer corn region is the largest maize production area in China, accounting for 40% of the planting area. *Helicoverpa armigera*, *O. furnacalis*, *Conogethes punctiferalis*, *Athetis lepigone*, *M. separata*, *Leucania loreyi*, and *S. frugiperda* are common maize pests in this area [15,28]. Therefore, according to the occurrence characteristics of the main pests in different maize growing regions, it is very necessary to breed and plant maize transformation events with high dose expression to the main pests in these areas.

Although we have previously determined that several transformation events have strong insect resistance to *S. frugiperda*, it is not clear whether they have reached high dose levels to *S. frugiperda* and other pests. This is not conducive to seed research and development companies to carry out targeted breeding of insect-resistant transgenic maize varieties and resistance management departments to decide whether to commercialize planting of different transformation in different areas. So to explore whether the transformation events in China Huang-huai-hai summer corn region mainly reached high dose level to lepidopteran pests, an artificial diet containing lyophilized tissues of Bt corn at a 25-fold dilution bioassay was used for research. The study contributes to developing Bt protein expression levels and high dose assessments for several main target pests in China's existing insect-resistant maize transformation events, which provide technical and theoretical support for the regional layout and resistance management of transgenic insect-resistant maize commercial planting in China.

2. Results

Analysis of variance showed that the corrected mortality rates after 7 d and 14 d were related mainly to insect species, different transformation events, and different tissues of transformation events ($p = 0.000$). The corrected mortality rates between 7 d and 14 d were significantly different ($T = -19.704$, $df = 323$, $p = 0.000$), and there was a significant positive correlation ($r = 0.891$, $p = 0.000$). In addition to the direct lethal effect of Bt protein on pests, it has an inhibitory effect on the growth and development of pests so they cannot complete the life cycle. In order to better and more accurately reflect the lethal effect of transformation events on various pests, the corrected mortality rate after 14 days was used as the index to express, as discussed in the following sections.

2.1. Corrected Mortalities of DBN9936 (Bt-Cry1Ab) to the Seven Pests

The corrected mortality rates of newly hatched larvae of seven lepidopteran pests on an artificial diet containing lyophilized tissues of DBN9936 (Bt-Cry1Ab) at a 25-fold dilution relative to isogenic negative control are shown in Figure 1, and the corrected mortality rates of 7 d and 14 d after feeding were significantly different ($T = -9.995$, *df* = 80, p = 0.000). The 14 d corrected mortality rates of *S. frugiperda* in DBN9936 (Bt-Cry1Ab) at the V6–V8, V12, VT, and R1 stages were 93.43~100%, which were significantly higher than that of R4 (73.72% ± 1.98%) (Figure 1A, $p < 0.05$). The corrected mortality rate of *H. armigera* was 84.79% ± 1.66% in DBN9936 at the V6–V8 stages, which was significantly higher than 58.05~72.70% at other stages (Figure 1B, $p < 0.05$). The corrected mortality rate of *O. furnacalis* in DBN9936 (Bt-Cry1Ab) at the V6–V8, V12, VT, and R1 stages were 100%, which were significantly higher than 94.70% ± 0.76% at the R4 stage (Figure 1C, $p < 0.05$). The corrected mortality rates of *C. punctiferalis* in DBN9936 (Bt-Cry1Ab) at V6–V8 and V12 were 93.59% ± 0.82% and 94.46% ± 0.89%, respectively, which were significantly higher than 88.57% ± 2.03% at R4 (Figure 1D, $p < 0.05$). The corrected mortality rates of *M. separata* of DBN9936 (Bt-Cry1Ab) at V6–V8 and V12 were 87.31% ± 6.43% and 88.35% ± 2.59%, respectively, which were significantly higher than 53.80% ± 3.54% at the R4 stage (Figure 1E, $p < 0.05$). The corrected mortality rates of *L. loreyi* larvae of DBN9936 (Bt-Cry1Ab) at V6–V8, V12, and R4 were 80.06% ± 4.41%, 81.53% ± 3.95%, and 80.13% ± 1.71%, respectively, and the difference was not significant among these stages (Figure 1F, $p > 0.05$). The 14 d corrected mortality rate of *A. lepigone* reached 100% in DBN9936 (Bt-Cry1Ab) at stages V6–V8, V12, and R4, and there was no significant difference (Figure 1G, $p > 0.05$).

2.2. Corrected Mortalities of DBN9936 × DBN9501 (Bt-Cry1Ab + Vip3A) to the Seven Pests

The corrected mortality rates of newly hatched larvae of seven lepidopteran pests fed with an artificial diet containing lyophilized tissues of DBN9936 × DBN9501 (Bt-Cry1Ab + Vip3A) at a 25-fold dilution relative to isogenic negative control are shown in Figure 2, and the corrected mortality rates at 7 d and 14 d of feeding were significantly different ($T = -11.633$, *df* = 80, p = 0.000). The 14 d corrected mortality rates of *S. frugiperda* of DBN9936 × DBN9501 (Bt-Cry1Ab + Vip3A) at the VT, R1, and R4 stages were 100%, which were significantly higher than 89.36% ± 3.98% and 87.18% ± 0.84% at the V6–V8 and V12 stages, respectively (Figure 2A, $p < 0.05$). The 14 d corrected mortality rate of *H. armigera* of DBN9936 × DBN9501 (Bt-Cry1Ab + Vip3A) at the R1 stage was 100%, which was significantly higher than that of 32.82~82.51% at other stages (Figure 2B, $p < 0.05$). The corrected mortality rates of *O. furnacalis* of DBN9936 × DBN9501 (Bt-Cry1Ab + Vip3A) at V6–V8, V12, VT, and R1 were 100%, which was significantly higher than at the R4 stage with 92.02% ± 2.20% (Figure 2C, $p < 0.05$). The corrected mortality rates of DBN9936 × DBN9501 (Bt-Cry1Ab + Vip3A) at V6–V8, V12, and R4 to *C. punctiferalis* were 84.28% ± 2.37%, 85.92% ± 2.39%, and 88.18% ± 1.36%, respectively, and there was no significant difference among the three stages (Figure 2D, $p > 0.05$). The corrected mortality rates of DBN9936 × DBN9501 (Bt-Cry1Ab + Vip3A) at V6–V8, V12, and R4 to *M. separata* were 64.44% ± 2.61%, 62.98% ± 4.11%, and 66.11% ± 2.29%, respectively, and there was no significant difference among the three stages (Figure 2E, $p > 0.05$). The corrected mortality rates of DBN9936 × DBN9501 (Bt-Cry1Ab + Vip3A) at the V6–V8, V12, and R4 stages to *L. loreyi* were 73.01% ± 3.19%, 63.94% ± 3.98%, and 69.79% ± 2.36%, with no significant difference (Figure 2F, $p > 0.05$). The 14 d corrected mortality rate of DBN9936 × DBN9501 (Bt-Cry1Ab + Vip3A) at the V12 stage to *A. lepigone* was 100%, which was significantly higher than 88.80% ± 1.56% at the V6–V8 stage and 86.30% ± 3.07 at the R4 stage (Figure 2G, $p < 0.05$).

2.3. Corrected Mortalities of Ruifeng 125 (Bt-Cry1Ab/Cry2Aj) to the Seven Pests

The corrected mortality rates of newly hatched larvae of seven lepidopteran pests fed with an artificial diet containing lyophilized tissues of Ruifeng 125 (Bt-Cry1Ab/Cry2Aj) at a 25-fold dilution relative to isogenic negative control are shown in Figure 3, and the corrected mortality rates at 7 d and 14 d after feeding were significantly different ($T = -10.704$, *df* = 80, p = 0.000). Ruifeng 125 (Bt-Cry1Ab/Cry2Aj) at the V12 stage had the highest corrected mortality rate for *S. frugiperda*, reaching 96.60% ± 0.93%, which was significantly higher than 78.15—89.13% at other stages (Figure 3A, $p < 0.05$). The corrected mortality rate for *H. armigera* of Bt-(Cry1Ab/Cry2Aj) at V6–V8 was 95.35% ± 2.69, which was significantly higher than that of other stages with 57.09~67.26% (Figure 3B, $p < 0.05$). Ruifeng 125 (Bt-Cry1Ab/Cry2Aj) at five growth stages had no significant difference in *O. furnacalis* with 100% corrected mortality (Figure 3C, $p > 0.05$). The corrected mortality rates for *C. punctiferalis* of Ruifeng 125 (Bt-Cry1Ab/Cry2Aj) at V6–V8 and V12 were 100%, which were higher than that of 92.10% ± 2.13% in the R4 stage (Figure 3D, $p < 0.05$). The corrected mortality rates of Ruifeng 125 (Bt-Cry1Ab/Cry2Aj) at the V6–V8 and V12 stages to *M. separata* were 76.87% ± 6.42% and 69.55% ± 4.29%, respectively, which were higher than 59.99% ± 2.13% in the R4 stage (Figure 3E, $p < 0.05$). The corrected mortality rate for *L. loreyi* of Ruifeng 125 (Bt-Cry1Ab/Cry2Aj) at V6–V8 was 88.90% ± 0.67%, which was significantly higher than 75.94% ± 1.73% and 65.92% ± 3.43% in the V12 and R4 stages, respectively (Figure 3F, $p < 0.05$). The 14 d corrected mortality rate of for *A. lepigone* Ruifeng 125 (Bt-Cry1Ab/Cry2Aj) at the V12 stage was 100%, which was significantly higher than 94.39% ± 2.19% and 82.13% ± 0.60% at the V6–V8 stage and R4 stage, respectively (Figure 3G, $p < 0.05$).

2.4. Corrected Mortalities of MIR162 (Bt-Vip3A) to the Seven Pests

The corrected mortalities of seven lepidopteran pests on an artificial diet containing lyophilized tissues of MIR162 (Bt-Vip3A) at a 25-fold dilution relative to isogenic negative control are shown in Figure 4. The corrected mortality rates on 7 d and 14 d after feeding were significantly different ($T = -12.182$, $df = 80$, $p = 0.000$). The corrected mortality rates of *S. frugiperda* of MIR162 (Bt-Vip3A) at V6–V8, VT, R1, and R4 were 100%, which were significantly higher than that at V12 (78.81% ± 1.63%) (Figure 4A, $p < 0.05$). The corrected mortality rate of *H. armigera* in MIR162 (Bt-Vip3A) in the R4 stage was 87.20% ± 0.75%, which was significantly higher than that of other stages (41.02~79.51%) (Figure 4B, $p < 0.05$). The corrected mortality rates of *O. furnacalis* in MIR162 (Bt-Vip3A) at the VT and R1 stages were 100%, which were significantly higher than 67.23% ± 8.64%, 80.74% ± 3.30%, and 89.61% ± 0.76% at the V6–V8, V12, and R4 stages, respectively (Figure 4C, $p < 0.05$). The corrected mortality rate of *C. punctiferalis* in MIR162 (Bt-Vip3A) at the R4 stage was 89.04% ± 3.28%, which was significantly higher than that of V6–V8 and V12, 72.26% ± 2.54% and 85.75% ± 2.63%, respectively (Figure 4D, $p < 0.05$). The corrected mortality rates to *M. separata* of MIR162 (Bt-Vip3A) at the V6–V8, V12, and R4 stages were 66.41% ± 4.70%, 79.51% ± 1.67%, and 68.46% ± 4.55%, respectively, and there was no significant difference among the three stages (Figure 4E, $p > 0.05$). The corrected mortality rates of MIR162 (Bt-Vip3A) at V6–V8, V12, and R4 to *L. loreyi* larvae were 75.83% ± 0.83%, 74.71% ± 5.23%, 79.89% ± 3.35%, respectively, and there was no significant difference among the three stages (Figure 4F, $p > 0.05$). The corrected mortality rate of *A. lepigone* in MIR162 (Bt-Vip3A) at the V12 stage was 96.66% ± 1.67%, which was significantly higher than 77.94% ± 1.58% at the V6–V8 stage and 85.46% ± 1.49% at the R4 stage (Figure 4G, $p < 0.05$).

2.5. Average Corrected Mortalities of Four Transformation Events to the Seven Pests

DBN9936 (Bt-Cry1Ab) had a significant effect on the average corrected mortality of seven insect species ($F = 21.166$, $df = 6, 81$, $p = 0.000$). The 14 d average corrected mortality rates of *A. lepigone*, *O. furnacalis*, *S. frugiperda*, and *C. punctiferalis* were 100%, 98.94% ± 0.58%, 92.65% ± 2.73%, and 92.21% ± 1.14%, respectively. This was significantly higher than the average corrected mortality of *L. loreyi* (80.57% ± 1.73%), *M. separata* (76.48% ± 6.10%), and *H. armigera* (68.68% ± 2.81%) (Figure 5, $p < 0.05$). The order of lethal effect of DBN9936 (Bt-Cry1Ab) to seven insect species was: *A. lepigone*, *O. furnacalis*, *S. frugiperda*, *C. punctiferalis* > *L. loreyi* ≥ *M. separata*, *H. armigera*.

DBN9936 × DBN9501 (Bt-Cry1Ab + Vip3A) also had a significant effect on the average corrected mortality of seven pests ($F = 17.004$, $df = 6, 81$, $p = 0.000$). The 14 d average corrected mortality rates of *O. furnacalis*, *S. frugiperda*, and *A. lepigone* were 98.40% ± 0.93%, 95.31% ± 1.69%, and 91.70% ± 2.33%, respectively. This was significantly higher than *C.*

punctiferalis, L. loreyi, H. armigera, and *M. separata* with 86.13% ± 1.19%, 68.91% ± 2.10%, 67.99% ± 6.54%, and 64.51% ± 1.62%, respectively (Figure 5, $p < 0.05$). The order of lethal effect of DBN9936 × DBN9501 (Bt-Cry1Ab + Vip3A) to seven insect species was: *O. furnacalis, S. frugiperda, A. lepigone* ≥ *C. punctiferalis* > *L. loreyi, H. armigera, M. separata*.

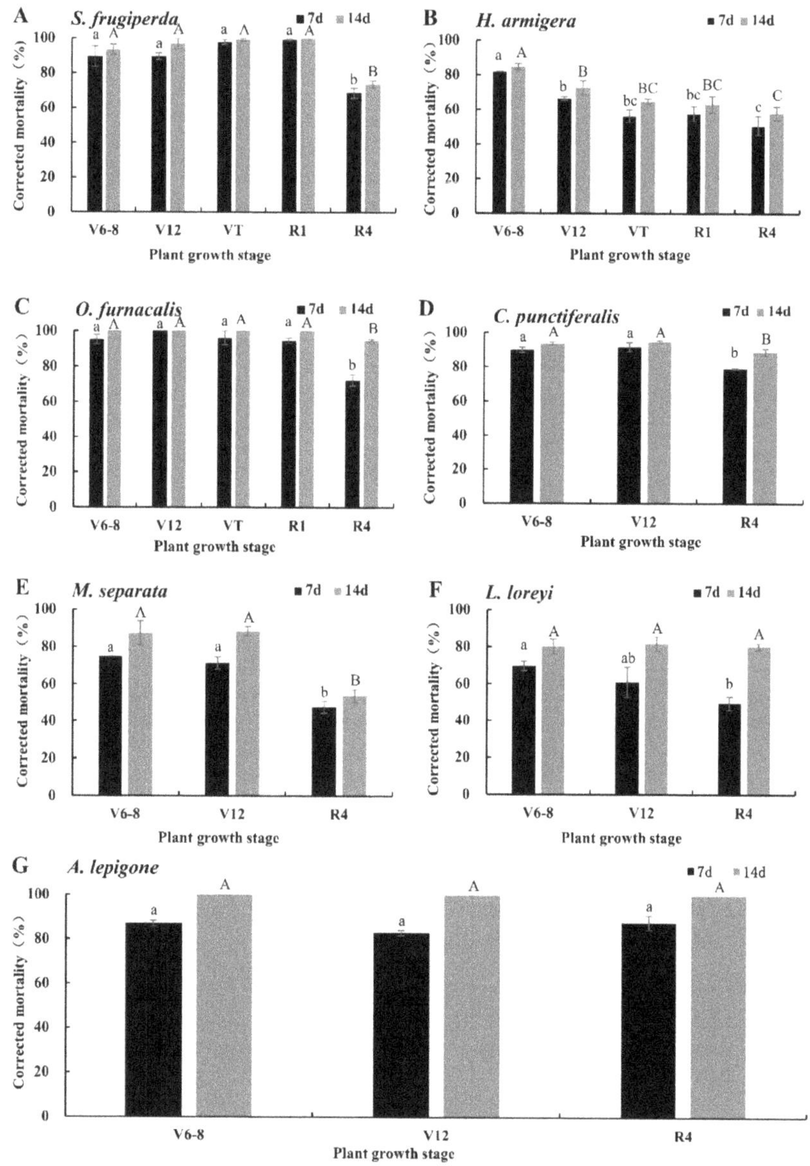

Figure 1. Corrected mortalities of neonates for seven lepidopteran pests on artificial diet containing lyophilized different tissues of DBN9936 corn (expressing Cry1Ab) in different growth stages at a 25-fold dilution relative to the isogenic negative control. Values represent means ± SE. Different lowercase and uppercase letters above black and gray bars indicate significant difference for the same treatment time by Duncan's multiple range test ($p < 0.05$). (**A**) *S. frugiperda*; (**B**) *H. armigera*; (**C**) *O. furnacalis*; (**D**) *C. punctiferalis*; (**E**) *M. separata*; (**F**) *L. loreyi*; and (**G**) *A. lepigone*.

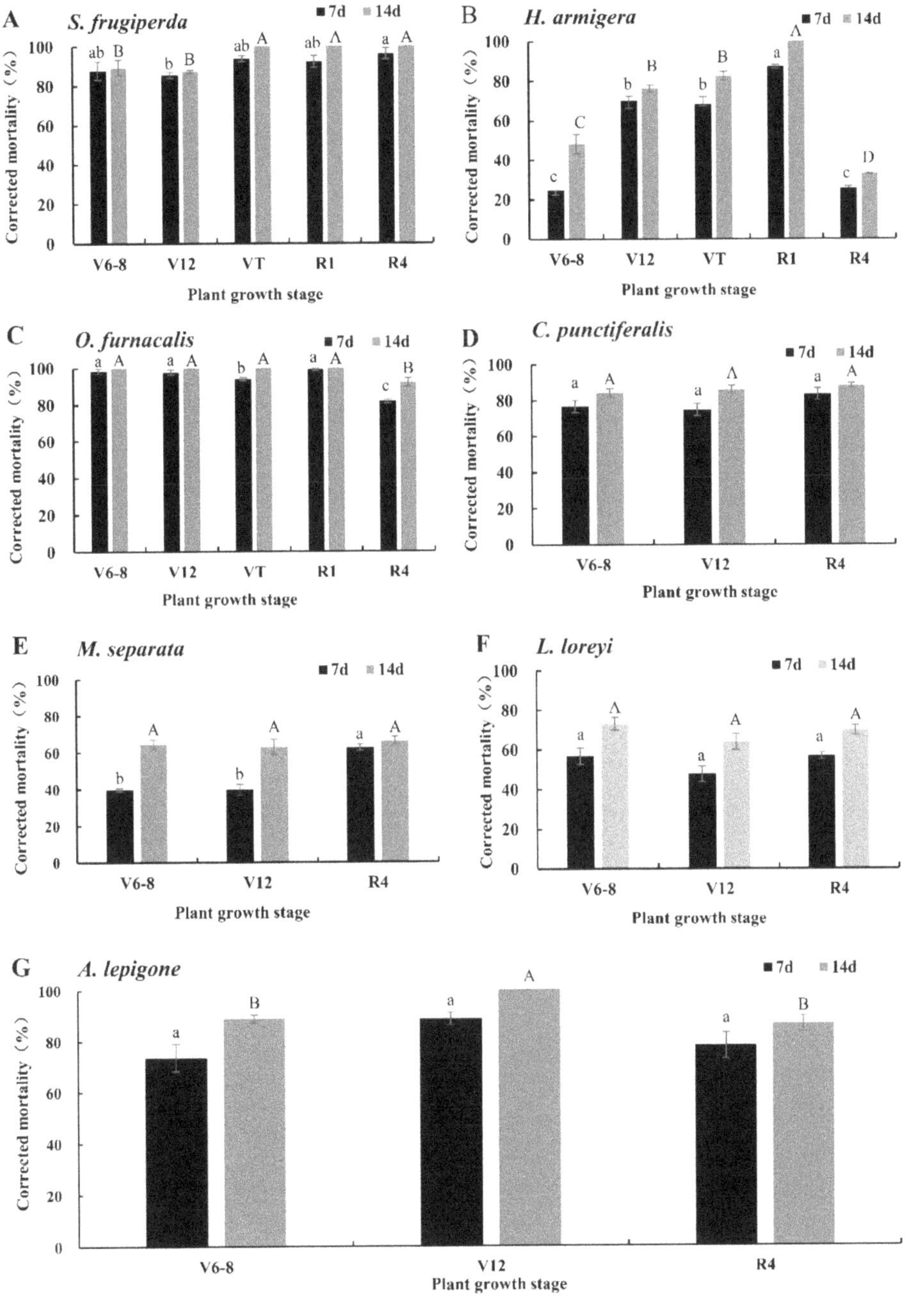

Figure 2. Corrected mortalities of neonates for seven lepidopteran pests on artificial diet containing lyophilized different tissues of DBN9936 × DBN9501 corn (expressing Cry1Ab and Vip3Aa) in different growth stages at a 25-fold dilution relative to the isogenic negative control. Values represent means ± SE. Different lowercase and uppercase letters above black and gray bars indicate significant difference for the same treatment time by Duncan's multiple range test ($p < 0.05$). (**A**) *S. frugiperda*; (**B**) *H. armigera*; (**C**) *O. furnacalis*; (**D**) *C. punctiferalis*; (**E**) *M. separata*; (**F**) *L. loreyi*; and (**G**) *A. lepigone*.

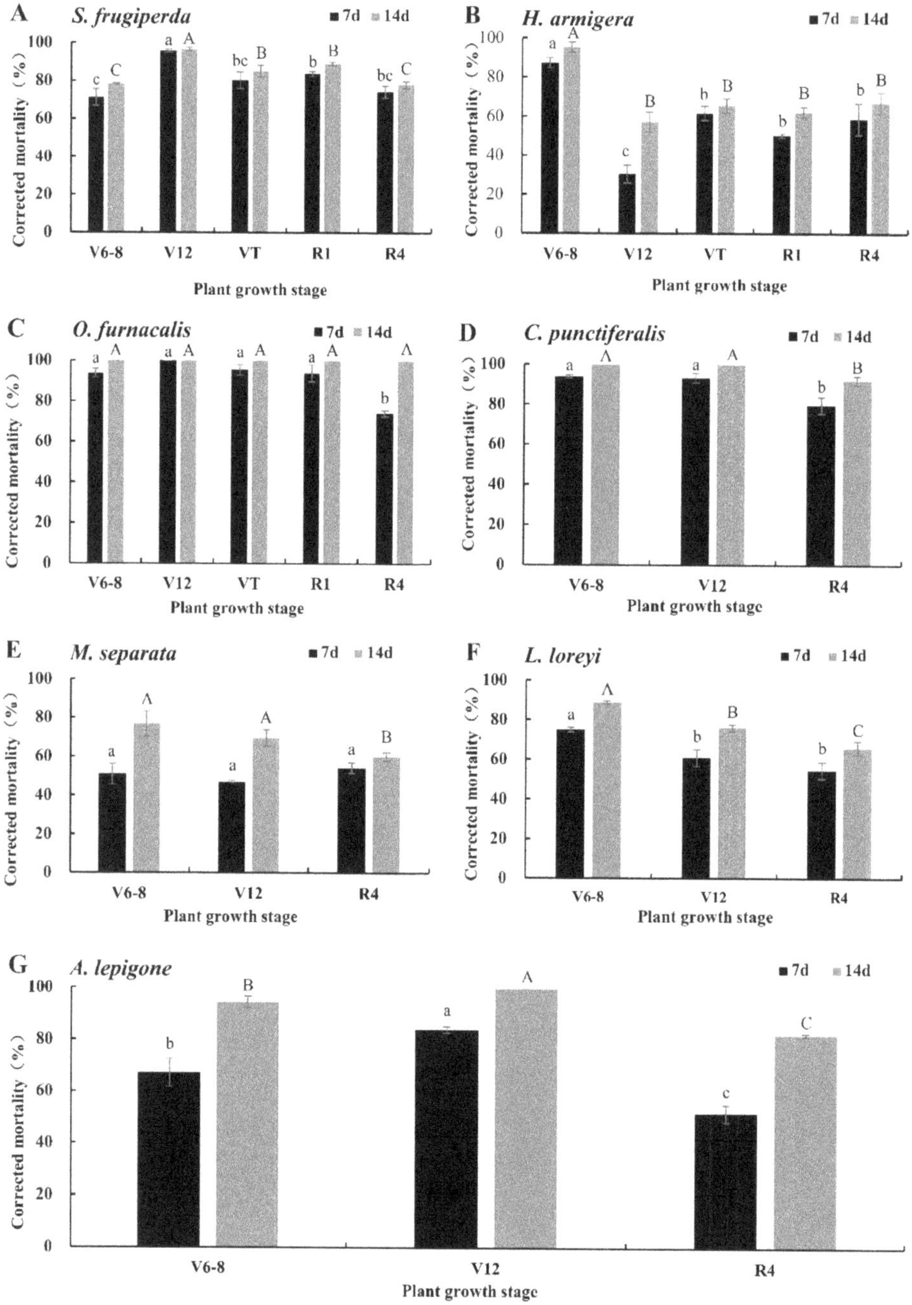

Figure 3. Corrected mortalities of neonates for seven lepidopteran pests on artificial diet containing lyophilized different tissues of Ruifeng 125 corn (expressing Cry1Ab/Cry2Aj) in different growth stages at a 25-fold dilution relative to the isogenic negative control. Values represent means ± SE. Different lowercase and uppercase letters above black and gray bars indicate significant difference for the same treatment time by Duncan's multiple range test ($p < 0.05$). (**A**) *S. frugiperda*; (**B**) *H. armigera*; (**C**) *O. furnacalis*; (**D**) *C. punctiferalis*; (**E**) *M. separata*; (**F**) *L. loreyi*; and (**G**) *A. lepigone*.

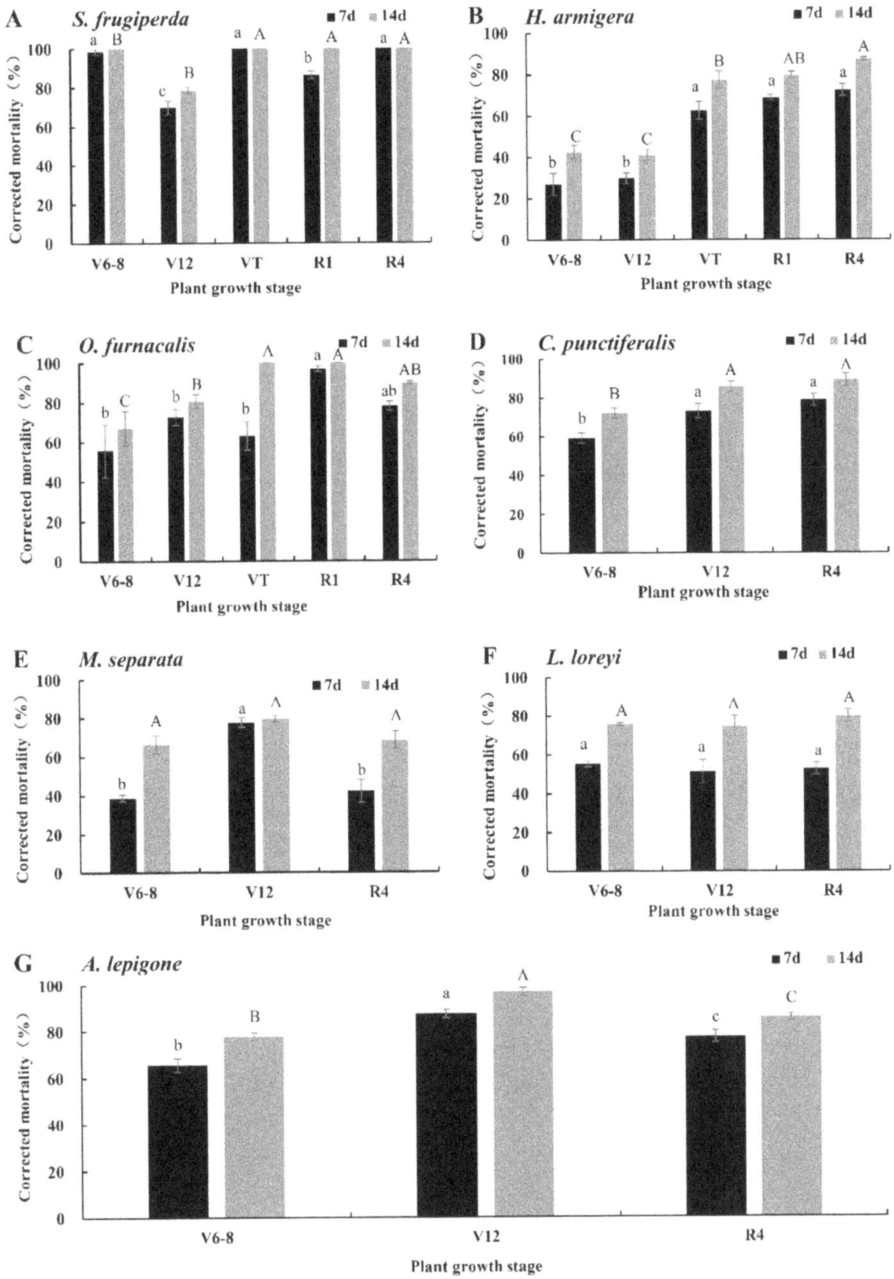

Figure 4. Corrected mortalities of neonates for seven lepidopteran pests on artificial diet containing lyophilized different tissues of MIR162 corn (expressing Vip3Aa) in different growth stages at a 25-fold dilution relative to the isogenic negative control. Values represent means ± SE. Different lowercase and uppercase letters above black and gray bars indicate significant difference for the same treatment time by Duncan's multiple range test ($p < 0.05$). (**A**) *S. frugiperda;* (**B**) *H. armigera;* (**C**) *O. furnacalis;* (**D**) *C. punctiferalis;* (**E**) *M. separata;* (**F**) *L. loreyi;* and (**G**) *A. lepigone.*

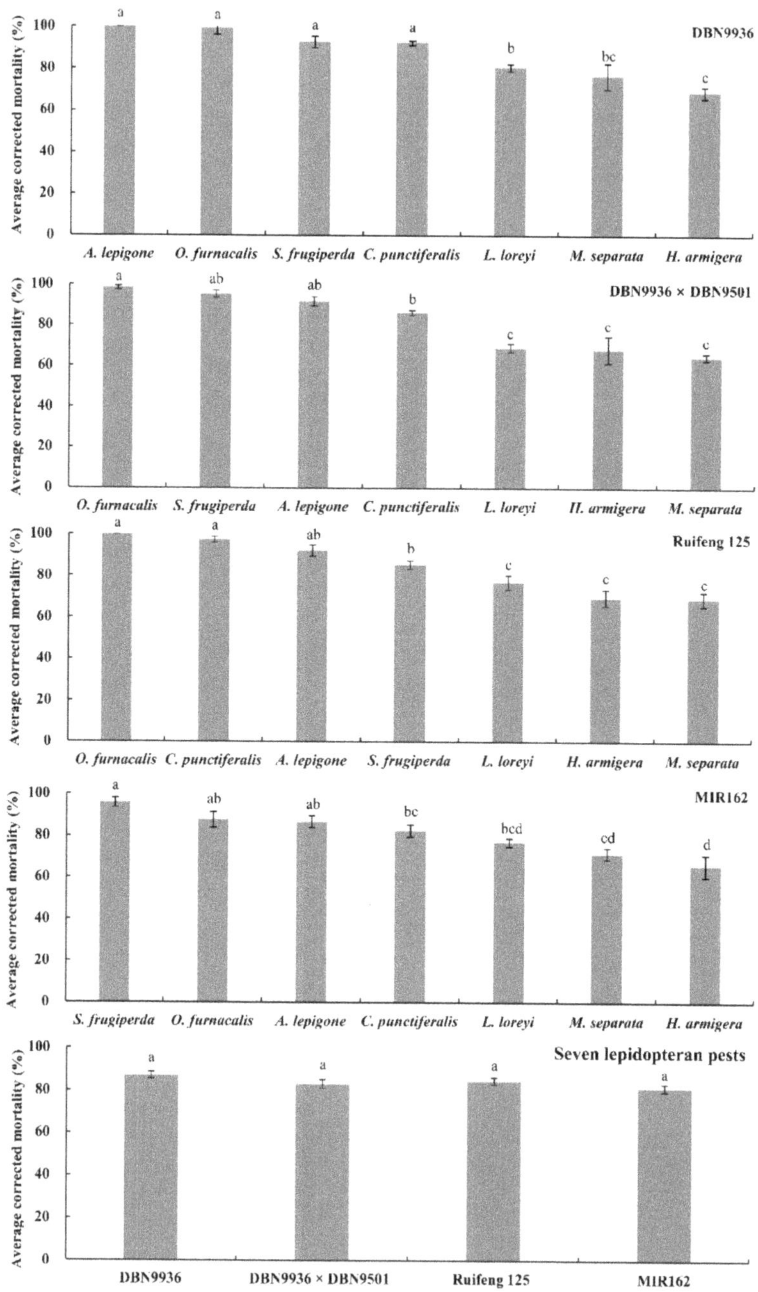

Figure 5. Average corrected mortalities of neonates for seven lepidopteran pests on artificial diet containing lyophilized different tissues of insect-resistant maize transformation in different growth stages at a 25-fold dilution relative to the isogenic negative control. Values represent means ± SE. Different lowercase letters above bars indicate significant difference by Duncan's multiple range test ($p < 0.05$).

Ruifeng 125 (Bt-Cry1Ab/Cry2Aj) also had a significant effect on the average corrected mortality of seven pests ($F = 10.419$, $df = 6, 81$, $p = 0.000$). The average corrected mortality rates of 14 d were 100%, 97.73% ± 1.45%, and 92.17% ± 2.72% for *O. furnacalis*, *C. punctiferalis*, and *A. lepigone*, respectively. It was significantly higher than the average corrected mortality of *S. frugiperda* (85.44% ± 1.98%), *L. loreyi* (76.92% ± 3.51%), *H. armigera* (69.51% ± 3.89%), and *M. separata* (68.80% ± 3.36%) (Figure 5, $p < 0.05$). The order of lethal effect of Ruifeng 125 (Bt-Cry1Ab/Cry2Aj) to seven insect species was: *O. furnacalis*, *C. punctiferalis*, *A. lepigone* ≥ *S. frugiperda* > *L. loreyi*, *H. armigera*, *M. separata*.

MIR162 (Bt-Vip3A) had a significant effect on the average corrected mortality in 14 d of seven insect species ($F = 17.505$, $df = 6, 81$, $p = 0.000$). The average corrected mortality of *S. frugiperda*, *O. furnacalis*, *A. lepigone*, and *C. punctiferalis* was 95.76% ± 2.28%, 87.52% ± 3.68%, 86.69% ± 2.83%, and 82.35% ± 2.93%, respectively. This was significantly higher than 76.81% ± 1.97% and 71.46% ± 2.82% for *L. loreyi* and *M. separata*, respectively, and 65.53% ± 5.37% for *H. armigera* (Figure 5, $p < 0.05$). The order of lethal effect of MIR162 (Bt-Vip3A) to seven insect species was: *S. frugiperda* ≥ *O. furnacalis*, *A. lepigone*, *C. punctiferalis*, *L. loreyi* ≥ *M. separata*, *H. armigera*.

In general, the lethal effects of DBN9936 (Bt-Cry1Ab), DBN9936 × DBN9501 (Bt-Cry1Ab + Vip3A), Ruifeng 125 (Bt-Cry1Ab/Cry2Aj), and MIR162(Bt-Vip3A) to seven lepidopteran pests were 87.00% ± 1.62%, 83.05% ± 1.99%, 84.46% ± 1.67%, and 81.33% ± 1.79%, respectively, and there was no significant difference among them ($F = 1.828$, df = 3, 323, $p = 0.142$, Figure 5).

3. Discussion

The insecticidal effect of insect-resistant transgenic maize exogenous Bt depends on its insecticidal protein expression [29,30]. Previous studies have confirmed that Bt protein expression in different regions, different transgenic crops, and different growth stages of the same transgenic crop has significant spatiotemporal variation [21,31–34]. For example, the expression level of Cry1Ab in DBN9936 was significantly lower in Xinxiang, Langfang, and Harbin than in Wuhan and Shenyang [21]. Similarly, the expression level of Cry1Ab in MON810 maize differed 20-fold, on average, in different regions [35]. This variation may expose local target pests to low and medium dose levels, which not only affects their field control effectiveness but also increases their survival rate due to exposure to sublethal doses, accelerating the evolution of resistance [36]. Therefore, it is of great significance for field planting layout and resistance management techniques for specific transformation events to determine whether insect-resistant transgenic crops achieve high dose levels to major pests in the local area.

Artificial diet containing lyophilized tissues of Bt crop events at 25-fold dilution bioassay is one of the most commonly used high-dose assays. We used this method to determine the high-dose levels of four transformation events to different lepidopteran pests at different stages. The results showed that two transformation events, DBN9936 × DBN9501 (Bt-Cry1Ab + Vip3A) and MIR162 (Bt-Vip3A), reached high dose levels to *S. frugiperda*, and DBN9936 (Bt-Cry1Ab) approached high dose levels. This is consistent with the results of this study [37]. The lethal sensitivity of *S. frugiperda* population to five Bt proteins in Yunnan was Vip3Aa > Cry1Ab > Cry1F > Cry2Ab > Cry1Ac. Therefore, planting insect-resistant maize expressing Cry1Ab, Vip3Aa, or superimposed Cry1Ab + Vip3Aa can meet the requirement of high dose of *S. frugiperda*. The corrected mortality of DBN9936 (Bt-Cry1Ab), DBN9936 × DBN9501 (Bt-Cry1Ab + Vip3A), and Ruifeng 125 (Bt-Cry1Ab/Cry2Aj) to *O. furnacalis* was more than 99.99%, while MIR162 (Bt-Vip3A) reached 100% only at the R1 and R4 stages. Our previous laboratory bioassay showed that *O. furnacalis* exhibited a high sensitivity to Cry1Ab with a LC_{50} value of 2.11 (1.64–2.19) ng/cm^2, while its sensitivity to Vip3A was low with a value 328.44 (183.99–660.54) ng/cm^2 [38], which is a 155-fold difference. Studies have shown that Vip3A has little or no activity on *O. nubilalis*, which may be caused by the two different species. Therefore, planting Cry1Ab-based multi-gene superimposed pest resistant maize is suggested to meet the high dose demand of *O. furnacalis*.

In the previous study, the sensitivity of *C. punctiferalis* to different Bt proteins was similar to that of *O. furnacalis* [38]. However, in this study, only Ruifeng 125 (Bt-Cry1Ab/Cry2Aj) was exposed to high dose levels under the determination of 25-fold dilution concentration. The reason for this result needs to be studied further.

DBN9936 × DBN9501 (Bt-Cry1Ab + Vip3A) was close to the high dose level to *H. armigera*, while DBN9936 (Bt-Cry1Ab), Ruifeng 125 (Bt-Cry1Ab/Cry2Aj), and MIR162 (Bt-Vip3A) did not reach the high dose level to *H. armigera*. Compared with other pests, *H. armigera* was the most sensitive to Cry2Ab [38]. It is suggested that multi-gene superimposed insect-resistant maize with Cry1Ab + Vip3A and Cry2Ab + Vip3A can meet the demand of high dose of *H. armigera*. The four transformation events did not reach high dose levels to *M. separata* and *L. loreyi*, which is consistent with our study that their sensitivity to Cry1Ab and Vip3A is lower than that of *H. armigera* and *O. furnacalis* [38]. Therefore, multigene superposition is more essential for *M. separata* control to achieve high dose levels. High dose levels of DBN9936 (Bt-Cry1Ab) were reached to *A. lepigone*, and near high dose levels of DBN9936 × DBN9501 (Bt-Cry1Ab + Vip3A) and Ruifeng 125 (Bt-Cry1Ab/Cry2Aj) were reached to this pest as well. *A. lepigone* feeds mainly on the base of maize stems, but in this paper, leaf, silk, and grain were used to determine the high dose levels of the four transformation events, which may not accurately reflect the actual situation.

"High dose/refuge" strategy is an effective measure to ensure sustainable application of insect-resistant transgenic crops [13,39–41]. Learning from foreign transgenic insect-resistant corn and resistance of target insect pest management experience and lessons, on the basis of the biological characteristics of major pests in maize growing areas of China, such as regional occurrence, host infestation and migration and dispersal, we proposed a planting layout of transgenic insect-resistant maize with "zoning layout and source control" and a high-dose/refuge resistance management strategy suitable for China's national conditions, and effectively implemented the refuge strategy [15]. In terms of the layout of transgenic insect-resistant maize, the autumn and winter corn in Southwest hilly and South hilly corn region in China are the annual breeding areas of *S. frugiperda* and *M. separata*, the concentrated landing sites of imported populations from abroad, and the important insect sources in the Huang-huai-Hai summer corn region and North spring corn region of China. Therefore, in order to reduce the occurrence of *S. frugiperda* and *M. separata* in the source area, Bt maize varieties should be planted to efficiently control *S. frugiperda* and *M. separata* and meet the requirements of high dose. The Huang-huai-hai summer corn region is the main occurrence area of *H. armigera*, *O. furnacalis*, *C. punctiferalis*, and *A. lepigone*. Bt maize varieties are planted to meet the requirements of high-dose control of *H. armigera*, *O. furnacalis*, *C. punctiferalis* and *A. lepigone*. In the Northern spring corn region, cultivars with high dose requirements for *O. furnacalis* are planted.

According to the results of high-dose determination of different transformation events against different pests, the research and application of transgenic insect-resistant maize in China should focus on the following points. For areas dominated by a single pest of *S. frugiperda*, Bt maize events such as Cry1Ab + Vip3Aa, expressing Vip3Aa or superimposed with Cry1Ab, should be cultivated. In the area where *S. frugiperda* and *M. separata* co-occur, the combined application of Cry1Ab + Vip3Aa, Cry1F, and Cry2Ab can also achieve an effective high dose level for *M. separata* [38]. For the co-occurrence areas of *H. armigera*, *O. furnacalis*, *C. punctiferalis*, and *A. lepigone*, multi-gene insect-resistant maize varieties containing mainly Cry1Ab + Cry2Ab or combined with Vip3A should be cultivated to achieve the goal of controlling multiple target pests. For the areas where the prevention and control of *O. furnacalis* is the main goal, we should focus on breeding varieties containing Cry1Ab and Cry1Ab + Cry2Aj, which can effectively control the damage of *O. furnacalis*. At the same time, according to the occurrence and damage characteristics of pests, the type and size of refuge should be formulated for each event in each area. Because China has commercially planted Bt-Cry1Ac Cotton since 1997 [39], one concern arises for cross-resistant pest development with Bt corns. This is not an important issue because more than 80 percent of cotton is grown in western China, which is a non-corn-producing area.

For mixed cotton- and maize-growing areas in eastern China, the planting of Bt-Cry1A maize with cross-resistance together with the cotton should be avoided in consideration of resistance management.

Clarifying the relationship between the dose expression of Bt-gene-resistant crop toxin protein and the pest response to it is one of the important aspects of the target pest resistance management work, which is conducive to the establishment of resistance management measures, such as the establishment and size of structural refuge or seed mixed refuge [42,43]. In the United States and Canada, according to the standards proposed by the USEPA, the only transformation with a clear evaluation of whether the transformation meets the high dose standard for a few pests are the following: Bt11 and MON810 expressing Cry1Ab did not reach high dose levels to *H. zea* and *S. frugiperda* but showed high dose levels to *O. nubilalis* [44]. Cry1A.105 and Cry2Ab in MON89034 single protein did not reach high doses to *H. zea* and *O. nubilalis* and was unknown to *S. frugiperda*. However, MON89034 showed better field resistance to these three pests than MON810 [45]. TC1507 expressing Cry1F was not high dose to *H. zea* but was high dose to *O. nubilalis* and was unknown to *S. frugiperda* [44]. MIR162 expressing Vip3A did not reach high dose to *H. zea* and *O. nubilalis* but reached high dose to *S. frugiperda* [46]. As for other pests, such as *Agrotis ipsilon*, *Diatraea grandiosella*, *D. saccharalis*, and *S. exigiua*, corresponding high-dose data are lacking, although they are active [36]. Most of the studies focused on insect resistance after laboratory and field planting, so they could not give clear high-dose results, and it was expressed as close to high dose [47], low dose [48], moderate dose [49], and other designations. Among the few results of the above definitive evaluation, high-dose expression assay was used by 25-fold dilution of lyophilized tissues of transgenic insect-resistant crops. This method—the transgenic insect-resistant crops lyophilized tissues 25 times dilution method—is easy to implement and also the most direct support as it reached 25 times higher doses of the evaluation method. We used this method before China's commercial cultivation of different transformations not only to the *S. frugiperda*, *H. armigera*, and *O. nubilalis* main pests (such as the high dose assessment) but also carried it out for other four insect pests, namely *M. separata*, *L. loreyi*, *A. lepigone*, and *C. punctiferalis*. The high-dose levels of different transformations to major maize pests in China were comprehensively and systematically evaluated, which is of great significance for guiding the breeding and regional commercial planting of insect-resistant transgenic maize varieties in China.

In this study, the high-dose levels of four transformation events on seven major lepidopteran pests in China were measured, and the dose–response relationships between different tissues of different transformation events and different pests were preliminarily clarified, which provided a basis for the breeding and application of transgenic maize varieties and the establishment of refuges in China. However, as a result of different regions and varieties of genetically modified crops, Bt protein expression at different stages of genetically modified crops have significant differences in characteristics of space and time. For the future, further study is needed on the harm of the pests on corn as well as feeding characteristics in different regions of China's corn belt. Resistance monitoring is the basis for resistance management of the target pests. The susceptible baselines and resistance allele frequencies of the major target pests in different ecological regions should be established before commercialization, and a regular program for resistance monitoring should be conducted after commercialization.

4. Materials and Methods

4.1. The Transformation Events of Insect-Resistant Transgenic Maize

DBN9936 transformation event maize (Bt-Cry1Ab), DBN9936 × DBN9501 event (Bt-Cry1Ab + Vip3A), and isogenic negative control (Nonghua 106) were provided by Beijing DaBeiNong Biotechnology Co., Ltd. (Beijing, China). Ruifeng 125 transformation event (Bt-Cry1Ab/Cry2Aj) and its isogenic negative control (Hongshuo 899) were provided by Hangzhou Ruifeng Biotechnology Co., Ltd. (Hangzhou, China). MIR162 transformation event (Bt-Vip3Aa) and its isogenic negative control (Xianda 901) were provided by Syngenta Biotechnology (China) Co., Ltd. (Beijing, China). MIR162 transformation event (Bt-Vip3Aa) has been commercially grown in the United States, Brazil, and other countries. All the above maize varieties were planted in the transgenic maize field (35°13′ N, 113°42′ E) of the Modern Agricultural Science and Technology Base of Henan Academy of Agricultural Sciences located in the Huang-huai-hai summer maize region on 25 June 2020, with an area of 200 m^2 in each plot. Plant spacing was 28 cm, row spacing was 60 cm, and spacing between plots was 1.5 m, repeated three times; routine water and fertilizer management were implemented.

When maize plants grew to the V6–V8, V12, VT, R1, and R4 stages, tissue samples of leaves, tassels, silk, and grains were taken as the criterion, as shown in Table 1. After each sampling, the Cry1Ab or Vip3A protein expression was confirmed by dipstick tests (AA0331-LS for testing Cry1Ab and AA1632-LS for testing Vip3A, Shanghai Youlong Biotechnology Co., Ltd. (Shanghai, China)), and samples were stored on dry ice and stored in −20 °C freezer within 2 to 4 h. The fresh tissue was ground into a fine powder by a low-temperature pulverization mixer (Robot coupe model R10 V.V.SV, speed 3000 rpm) and then dried by a freeze dryer (Ningbo Xinzhi Biotechnology Co., Ltd. (Ningbo, China), Scientz-12nd) at −50 °C for 24 h. After drying, they were divided into 50 mL centrifuge tubes and stored in the refrigerator at −80 °C until use.

Table 1. Maize at different growth stages and tissue sampling requirements.

Maize Growth Stage	Tissue	Specific Sampling Requirements
V6–V8 (6–8 leaves have visible collars)	Leaf	The youngest leaf that emerged and was at least 20 cm in length was cut from the leaf tip.
V12 (12 leaves have visible collars)	Leaf	The youngest leaf that emerged and was at least 20 cm in length was cut from the leaf tip.
VT (The lowest branch of the tassel is visible, but silks are not)	Tassel	One tassel was extracted from each corn plant
R1 (Silk is visible)	Silk	The ear was bagged, and the ear with the bag was removed from the plant and moved to a pollen-free environment, and the silk were cut from the ear.
R4 (Kernel contents are pasty as starch accumulates)	Grain	Thirty young grains were collected from each ear.

4.2. Collection and Culture of Insect Species

The susceptible strain of *S. frugiperda* was collected in a maize field of Mengmao Town, Ruili, Dehong Prefecture, Yunnan Province, China, in January 2019. The collected insect species were mainly the 3rd to 5th instar larvae. The population was sensitive to Bt after laboratory biological tests [37]. *H. armigera, O. furnacalis, A. lepigone*, and *C. punctiferalis* were collected from Modern Agricultural Science and Technology Base of Henan Academy of Agricultural Sciences (Yuanyang County, Henan Province) from 2015 to 2016. *H. armigera* and *O. furnacalis* were collected from conventional corn ears and *C. punctiferalis* was collected from both corn and sorghum. *A. lepigone* was captured by light trap. *M. separata* population was collected from the maize field in Lingbao County, Henan Province in 2016, and the larvae were collected as 4–5 instar larvae. The *L. loreyi* larvae were collected from the spring maize field in Ganan Town, Pingqiao District, Xinyang City, Henan Province in 2019.

S. frugiperda, H. armigera, O. furnacalis, M. separata, L. loreyi, and *A. lepigone* larvae feed formula in laboratory, artificial ingredients with corn flour, soybean meal, wheat germ and bran, casein as the main ingredient [50] and *C. punctiferalis* feed with chestnut powder and corn flour were acquired, and the formula is shown in [51]. Adults were fed 5–10% honey water to supplement nutrition and water in the cage (40 × 30 × 25 cm^3). *O. furnacalis* eggs were collected using wax paper, *S. frugiperda*, *H. armigera*, *C. punctiferalis*, and *A. lepigone* eggs were collected using white medical gauze. *M. separata* and *L. loreyi* eggs were collected using nylon rope.

All larvae and adults of the above species were incubated in an incubator with temperature of (27 ± 1) °C, humidity of 60% ± 10%, and photoperiod of L/D = 16 h/8 h. No chemical insecticides or Bt insecticidal proteins were exposed in the feeding process of the test insects. The specific information of the insect source is shown in Table 2.

Table 2. Source information of tested insects.

Insect Species	Date	Collecting Location	Insect Number Collected	Insect Stage	Host Plants
S. frugiperda	January 2019	Ruili City, Yunnan Province (23°58′ N, 97°48′ E)	150–200	3–5 instar larvae	Maize
H. armigera	August 2016	Yuanyang County, Henan Province (35°13′ N, 113°42′ E)	150–200	4–5 instar larvae	Maize
O. furnacalis	May to June 2015	Yuanyang County, Henan Province (35°13′ N, 113°42′ E)	150–200	4–5 instar larvae	Maize
C. punctiferalis	September 2016	Yuanyang County, Henan Province (35°13′ N, 113°42′ E)	250–300	3–5 instar larvae	Maize and sorghum
M. separata	August 2016	Lingbao County, Henan Province (34°36′ N, 110°48′ E)	300–350	4–5 instar larvae	Maize
L. loreyi	May 2019	Xinyang City, Henan Province (32°17′ N, 114°01′ E)	100–150	4–5 instar larvae	Maize
A. lepigone	May to June 2016	Yuanyang County, Henan Province (35°13′ N, 113°42′ E)	100–120	Adults	Captured by light trap

4.3. High-Dose Bioassays

Dilution and bioassay of lyophilized Bt maize tissues were performed using artificial diet high-dose assays [52,53]. When artificial diets cooled ca. 50 °C, 0 g and 20 g of transgenic insect-resistant maize tissue lyophilized powder and 20 g and 0 g of isotype control maize tissue lyophilized powder were mixed into 480 g artificial diet and stirred evenly, and the concentrations of 0 (0 times) and 4% (25 times) were finally formed. After the artificial diets solidified, 1 m^3 artificial diet portions were put into 128-well culture plates (diameter of each well 16 mm; height: 13 mm), and one neonate larvae (0–24 h old) of each species was added to each well using a fine brush, and then covered with a breathable plastic mucous membrane to prevent larvae from escaping. The bioassays were repeated three times for each species, forty-eight larvae in each concentration for a total of 144 larvae. All culture plates were incubated at 27 ± 1 °C, 60–70% relative humidity, and 16 h/8 h light. The number of dead larvae was recorded at 7 d and 14 d, respectively. The larvae that could not move normally were considered dead, and the larvae that did not reach the second instar after 7 d and 14 d were also considered dead.

4.4. Statistics and Analysis

Equations (1) and (2) were used to calculate the mortality rate and corrected mortality rate of several pests in different stages of different maize transformation events. If the corrected mortality rate reached 100% under the treatment of 4% concentration (25 times dilution concentration), it indicated that the Bt protein content of the maize transformation event in this stage reached the requirement of high dose for the pests.

$$\text{Mortality (\%)} = \text{Number of dead insects after treatment/total number of insects tested before treatment} \times 100\% \quad (1)$$

$$\text{Corrected mortality (\%)} = (\text{treatment group mortality} - \text{control group mortality})/(100 - \text{control group mortality}) \times 100\% \quad (2)$$

The differences were analyzed by one-way analysis of variance. Duncan's new complex range method was used for significance test. SPSS 20.0 software was used for statistical analysis of the test data.

5. Conclusions

Herein, we tested the high-dose levels of four transformation events on seven major lepidopteran pests in China. Different transformations at different growth stages showed different dose–mortality relationships for different pests. The three events of DBN9936 (R1), DBN9936 × DBN9501 (VT, R1, R4), and MIR162 (V6–V8, VT, R1, R4) reached high dose level to *S. frugiperda*. DBN9936 × DBN9501 (R1) reached a high dose level to *H. armigera*. DBN9936 (V6–V8, V12, VT, R1), DBN9936 × DBN9501 (V6–V8, V12, VT, R1), and Ruifeng 125 (V6–V8, V12, VT, R1, R4) reached high dose levels to *O. furnacalis*. DBN9936 (V6–V8, V12, R4), DBN9936 × DBN9501 (V12), and Ruifeng 125 (V12) reached high dose levels to *A. lepigone*. Ruifeng 125 (V6–V8, V12) reached a high dose level to *C. punctiferalis*. No transformations reached high dose levels to either *M. separata* or *L. loreyi*. The results of this study can provide support for the breeding of high-dose varieties for different target pests, the combined application of multiple genes, and the commercial regional planting of insect-resistant transgenic maize in China.

Author Contributions: Conceptualization, K.W.; methodology, G.L. and T.J.; data analysis, G.L., T.J. and S.Z.; resources, K.W.; writing—original draft preparation, G.L.; writing—review and editing, K.W. and H.F. All authors have read and agreed to the published version of the manuscript.

Funding: This study was funded by National Modern Agricultural Industry Technology System Construction Fund of China (CARS-02) and Science and Technology Innovation Team of Henan Academy of Agricultural Sciences (2022TD13).

Acknowledgments: We thank Beijing DaBeiNong Biotechnology Co., Ltd., Hangzhou Ruifeng Biotechnology Co., Ltd., and Syngenta for providing materials of Bt corn and advice on field planting.

Conflicts of Interest: The authors declare no conflict of interest.

References

1. Koziel, M.G.; Beland, G.L.; Bowman, C.; Carozzi, N.B.; Crenshaw, R.; Crossland, L.; Dawson, J.; Desai, N.; Hill, M.; Kaell, S.; et al. Field performance of elite transgenic maize plants expressing an insecticidal protein derived from *Bacillus thuringinesis*. *Nat. Biotechnol.* **1993**, *11*, 194–200. [CrossRef]
2. Mason, C.E.; Rice, M.E.; Calvin, D.D.; Van Duyn, J.W.; Showers, W.B.; Hutchinson, W.D.; Witkowski, J.F.; Higgins, R.A.; Onstad, D.W.; Dively, G.P. *European Corn Borer Ecology and Management*; North Central Regional Extension Publication, No.327; Iowa State University: Ames, IA, USA, 1996; p. 57.
3. Hutchison, W.D.; Burkness, E.C.; Mitchell, P.D.; Moon, R.D.; Leslie, T.W.; Fleischer, S.J.; Abrahamson, M.; Hamilton, K.L.; Steffey, K.L.; Gray, M.E.; et al. Area wide suppression of European corn borer with Bt maize reaps savings to non-Bt maize growers. *Science* **2010**, *330*, 222–225. [CrossRef] [PubMed]
4. Dively, G.P.; Venugopal, P.D.; Bean, D.; Whalen, J.; Holmstrom, K.; Kuhar, T.P.; Doughty, H.B.; Patton, T.; Cissel, W.; Hutchison, W.D. Regional pest suppression associated with widespread Bt maize adoption benefits vegetable growers. *Proc. Natl. Acad. Sci. USA* **2018**, *115*, 3320–3325. [CrossRef] [PubMed]
5. ISAAA. *Global Status of Commercialized Biotech/GM Crops in 2019: Biotech Crops Drive Socio-Economic Development and Sustainable Environment in the New Frontier*; ISAAA: Ithaca, NY, USA, 2019.
6. Gould, F. Sustainability of transgenic insecticidal cultivars: Integrating pest genetics and ecology. *Annu. Rev. Entomol.* **1998**, *43*, 701–726. [CrossRef] [PubMed]
7. Storer, N.P.; Babcock, J.M.; Schlenz, M.; Meade, T.; Thompson, G.D.; Bing, J.W.; Huckaba, R.M. Discovery and characterization of field resistance to Bt maize: *Spodoptera frugiperda* (Lepidoptera: Noctuidae) in Puerto Rico. *J. Econ. Entomol.* **2010**, *103*, 1031–1038. [CrossRef]
8. Chandrasena, D.I.; Signorini, A.M.; Abratti, G.; Storer, N.P.; Olaciregui, M.L.; Alves, A.P.; Pilcher, C.D. Characterization of field-evolved resistance to *Bacillus thuringiensis*-derived Cry1F δ-endotoxin in *Spodoptera frugiperda* populations from Argentina. *Pest Manag. Sci.* **2018**, *74*, 746–754. [CrossRef]
9. Omoto, C.; Bernardi, O.; Salmeron, E.; Sorgatto, R.J.; Dourado, P.M.; Crivellari, A.; Carvalho, R.A.; Willse, A.; Martinelli, S.; Head, G.P. Field-evolved resistance to Cry1Ab maize by *Spodoptera frugiperda* in Brazil. *Pest Manag. Sci.* **2016**, *72*, 1727–1736. [CrossRef]
10. Van Rensburg, J.B.J. First report of field resistance by the stem borer, *Busseola fusca* (Fuller) to Bt-transgenic maize. *S. Afr. J. Plant Soil* **2007**, *24*, 147–151. [CrossRef]

11. Dively, G.P.; Venugopal, P.D.; Finkenbinder, C. Field-evolved resistance in corn earworm to Cry proteins expressed by transgenic sweet corn. *PLoS ONE* **2016**, *11*, e0169115. [CrossRef]
12. Kaur, G.; Guo, J.; Brown, S.; Head, G.P.; Price, P.A.; Paula-Moraes, S.; Ni, X.; Dimase, M.; Huang, F. Field-evolved resistance of *Helicoverpa zea* (Boddie) to transgenic maize expressing pyramided Cry1A.105/ Cry2Ab2 proteins in northeast Louisiana, the United States. *J. Invertebr. Pathol.* **2019**, *163*, 11–20. [CrossRef]
13. Tabashnik, B.E.; Carrière, Y. Surge in insect resistance to transgenic crops and prospects for sustainability. *Nat. Biotechnol.* **2017**, *35*, 926–935. [CrossRef] [PubMed]
14. Tabashnik, B.E.; Carrière, Y. Global patterns of resistance to Bt crops highlighting pink bollworm in the United States, China and India. *J. Econ. Entomol.* **2019**, *112*, 2513–2523. [CrossRef] [PubMed]
15. Li, G.P.; Wu, K.M. Commercial strategy of transgenic insect-resistant maize in China. *J. Plant Prot.* **2022**, *49*, 17–32. (In Chinese)
16. Carrière, Y.; Crickmore, N.; Tabashnik, B.E. Optimizing pyramided transgenic Bt crops for sustainable pest management. *Nat. Biotechnol.* **2015**, *33*, 161–168. [CrossRef] [PubMed]
17. U.S. Environmental Protection Agency (EPA). *The Environmental Protection Agency's White Paper on Bt Plant-Pesticide Resistance Management*; EPA/739/S-98/001; Biopesticides and Pollution Prevention Division: Washington, DC, USA, 1998. Available online: http://www.epa.gov/EPA-PEST/1998/January/Day-14/paper.pdf (accessed on 9 September 2021).
18. FIFRA Scientific Advisory Panel. *Transmittal of the Final Report of the FIFRA Scientific Advisory Panel Subpanel on Bacillus thuringiensis (Bt) Plant-Pesticides and Resistance Management, Meeting Held on February 9 and 10, 1998*; US Environmental Protection Agency: Washington, DC, USA, 1998; p. 59.
19. FIFRA Scientific Advisory Panel. *Bt Plant-Pesticides Risk and Benefit Assessments*; Report No. 2000-07; US Environmental Protection Agency: Washington, DC, USA, 2001; p. 78.
20. U.S. Environmental Protection Agency (EPA). *Terms and Conditions for Bt Corn Registrations 30 Sept 2010*; Office of Pesticide Programs: Washington, DC, USA, 2010. Available online: http://www3.epa.gov/pesticides/chem_search/reg_actions/pip/bt-corn-terms-conditions.pdf (accessed on 16 May 2022).
21. Liang, J.G.; Zhang, D.D.; Li, D.Y.; Zhao, S.Y.; Wang, C.Y.; Xiao, Y.T.; Xu, D.; Yang, Y.Z.; Li, G.P.; Wang, L.L.; et al. Expression profiles of Cry1Ab protein and its insecticidal efficacy against the invasive fall armyworm for Chinese domestic GM maize DBN9936. *J. Integr. Agric.* **2021**, *20*, 792–803. [CrossRef]
22. Zhang, D.D.; Wu, K.M. The bioassay of Chinese domestic Bt-Cry1Ab and Bt-(Cry1Ab+Vip3Aa) maize against the fall armyworm, *Spodoptera frugiperda*. *Plant Prot.* **2019**, *45*, 54–60. (In Chinese)
23. He, L.M.; Zhao, S.Y.; Gao, X.W.; Wu, K.M. Ovipositional responses of *Spodoptera frugiperda* on host plants provide a basis for using Bt-transgenic maize as trap crop in China. *J. Integr. Agric.* **2020**, *19*, 2–12. [CrossRef]
24. Wang, W.W.; Zhang, D.D.; Zhao, S.Y.; Wu, K.M. Susceptibilities of the Invasive Fall Armyworm (*Spodoptera frugiperda*) to the Insecticidal Proteins of Bt maize in China. *Toxins* **2022**, *14*, 507. [CrossRef]
25. Yang, X.M.; Zhao, S.Y.; Liu, B.; Gao, Y.; Hu, C.X.; Li, W.J.; Yang, Y.Z.; Li, G.P.; Wang, L.L.; Yang, X.Q.; et al. Bt maize can provide non-chemical pest control and enhance food safety in China. *Plant Biotechnol. J.* **2022**. [CrossRef]
26. Zhao, S.Y.; Yang, X.M.; Liu, D.Z.; Sun, X.X.; Li, G.P.; Wu, K.M. Performance of the domestic Bt-corn event expressing pyramided Cry1Ab and Vip3Aa19 against the invasive *Spodoptera frugiperda* (J. E. Smith) in China. *Pest Manag. Sci.* **2022**. [CrossRef]
27. Tong, P.Y. *Maize Plant District in China*; Chinese Agricultural Science and Technology Press: Beijing, China, 1992. (In Chinese)
28. Wang, Z.Y.; Wang, X.M. Current status and management strategies for corn pests and diseases in China. *Plant Prot.* **2019**, *45*, 1–11. (In Chinese)
29. Erasmus, R.; Pieters, R.; Du, P.H.; Hilbeck, A.; Trtikova, M.; Erasmus, A.; Van den Berg, J. Introgression of a cry1Ab transgene into open pollinated maize and its effect on Cry protein concentration and target pest survival. *PLoS ONE* **2019**, *14*, e0226476. [CrossRef] [PubMed]
30. Wang, F.; Peng, S.B.; Cui, K.H.; Nie, L.X.; Huang, J.L. Field performance of *Bt* transgenic crops: A review. *Aust. J. Crop. Sci.* **2014**, *8*, 18–26.
31. Ramirez-Romero, R.; Desneux, N.; Chaufaux, J.; Kaiser, L. Bt-maize effects on biological parameters of the non-target aphid *Sitobion avenae* (Homoptera: Aphididae) and Cry1Ab toxin detection. *Pestic. Biochem. Physiol.* **2008**, *91*, 110–115. [CrossRef]
32. Bakhsh, A.; Shahzad, K.; Husnain, T. Variation in the spatio-temporal expression of insecticidal genes in cotton. *Czech J. Genet. Plant Breed* **2011**, *47*, 1–9. [CrossRef]
33. Székács, A.; Weiss, G.; Quist, D.; Takács, E.; Darvas, B.; Meier, M.; Swain, T.; Hilbeck, A. Inter-laboratory comparison of Cry1Ab toxin quantification in MON 810 maize by enzyme-immunoassay. *Food Agric. Immunol.* **2012**, *23*, 99–121. [CrossRef]
34. Trtikova, M.; Wikmark, O.G.; Zemp, N.; Widmer, A.; Hilbeck, A. Transgene expression and Bt protein content in transgenic Bt maize (MON810) under optimal and stressful environmental conditions. *PLoS ONE* **2015**, *10*, e0123011. [CrossRef]
35. Székács, A.; Lauber, É.; Takács, E.; Darvas, B. Detection of Cry1Ab toxin in the leaves of MON 810 transgenic maize. *Anal. Bioanal. Chem.* **2010**, *396*, 2203–2211. [CrossRef]
36. Reisig, D.D.; DiFonzo, C.; Dively, G.; Farhan, Y.; Gore, J.; Smith, J. Best Management practices to delay the evolution of Bt resistance in lepidopteran pests without high susceptibility to Bt toxins in North America. *J. Econ. Entomol.* **2022**, *115*, 26–36. [CrossRef]
37. Li, G.P.; Ji, T.J.; Sun, X.X.; Jiang, Y.Y.; Wu, K.M.; Feng, H.Q. Susceptibility evaluation of invaded *Spodoptera frugiperda* population in Yunnan province to five Bt proteins. *Plant Prot.* **2019**, *45*, 15–20. (In Chinese)

38. Li, G.P.; Feng, H.Q.; Ji, T.J.; Huang, J.R.; Tian, C.H. What type of Bt corn is suitable for a region with diverse lepidopteran pests: A laboratory evaluation. *GM Crops Food* **2021**, *12*, 115–124. [CrossRef] [PubMed]
39. Wu, K.M. Monitoring and management strategy for *Helicoverpa armigera* resistance to Bt cotton in China. *J. Invertebr. Pathol.* **2007**, *95*, 220–223. [CrossRef] [PubMed]
40. Wan, P.; Xu, D.; Cong, S.; Jiang, Y.; Huang, Y.; Wang, J.; Wu, H.; Wang, L.; Wu, K.; Carrière, Y.; et al. Hybridizing transgenic Bt cotton with non-Bt cotton counters resistance in pink bollworm. *Proc. Natl. Acad. Sci. USA* **2017**, *114*, 5413–5418. [CrossRef]
41. Huang, F.N.; Andow, D.A.; Buschman, L.L. Success of the high-dose/refuge resistance management strategy after 15 years of Bt crop use in North America. *Entomol. Exp. Appl.* **2011**, *140*, 1–16. [CrossRef]
42. Caprio, M.A.; Kurtz, R.; Catchot, A.; Kerns, D.; Reisig, D.; Gore, J.; Reay-Jones, F.P.F. The corn–cotton agroecosystem in the midsouthern United States: What insecticidal event pyramids should be used in each crop to extend Vip3A durability. *J. Econ. Entomol. 2019. 112*, 2894–2906.
43. DiFonzo, C. Handy Bt Trait Table for U.S. Corn Production. 2021. Available online: https://www.texasinsects.org/bt-corn-trait-table.html (accessed on 12 September 2021).
44. U.S. Environmental Protection Agency (EPA). *Biopesticides Registration Action Document Cry1Ab and Cry1F Bacillus thuringiensis (Bt) Corn Plant-Incorporated Protectants;* U.S. Environmental Protection Agency Office of Pesticide Programs Biopesticides and Pollution Prevention Division: Washington, DC, USA, 2010. Available online: https://www3.epa.gov/pesticides/chem_search/reg_actions/pip/cry1f-cry1ab-brad.pdf (accessed on 16 December 2021).
45. U.S. Environmental Protection Agency (EPA). *Biopesticide Registration Action Document: Bacillus thuringiensis Cry1A.105 and Cry2Ab2 Insecticidal Proteins and the Genetic Material Necessary for Their Production in Corn [PC Codes 006515 (Cry2Ab2), 006514 (Cry1A.105)];* U.S. Environmental Protection Agency Office of Pesticide Programs Biopesticides and Pollution Prevention Division: Washington, DC, USA, 2010. Available online: https://www3.epa.gov/pesticides/chem_search/reg_actions/pip/mon-89034-brad.pdf (accessed on 16 December 2021).
46. U.S. Environmental Protection Agency (EPA). *Biopesticide Registration Action Document Bacillus thuringiensis Vip3Aa20 Insecticidal Protein and the Genetic Material Necessary for Its Production (via Elements of Vector pNOV1300) in Event MIR162 Maize (OECD Unique Identifier: SYN-IR162-4) PC Code: 006599;* U.S. Environmental Protection Agency Office of Pesticide Programs Biopesticides and Pollution Prevention Division: Washington, DC, USA, 2010. Available online: https://www3.epa.gov/pesticides/chem_search/reg_actions/registration/decision_PC-006599_3-Apr-09.pdf (accessed on 16 December 2021).
47. Burkness, E.C.; Dively, G.; Patton, T.; Morey, A.C.; Hutchison, W.D. Novel Vip3A *Bacillus thuringiensis* (Bt) maize approaches high dose efficacy against *Helicoverpa zea* (Lepidoptera: Noctuidae) under field conditions: Implications for resistance management. *GM Crops.* **2010**, *1*, 337–343. [CrossRef]
48. Sousa, F.F.; Mendes, S.M.; Santos-Amaya, O.F.; Araújo, O.G.; Oliveira, E.E.; Pereira, E.J. Life-history traits of *Spodoptera frugiperda* populations exposed to low-dose Bt maize. *PLoS ONE* **2016**, *11*, e0156608. [CrossRef]
49. Horner, T.A.; Dively, G.P.; Herbert, D.A. Development, survival and fitness performance of *Helicoverpa zea* (Lepidoptera: Noctuidae) in MON810 Bt field corn. *J. Econ. Entomol.* **2003**, *96*, 914–924. [CrossRef]
50. Li, G.P.; Huang, J.R.; Ji, T.J.; Tian, C.H.; Zhong, J.; Feng, H.Y.; Feng, H.Q. The Artificial Diet for Rearing Ostrinia furnacalis and Its Application. China Patent CN108813114A, 30 July 2021.
51. Li, G.P.; Ji, T.J.; Huang, J.R.; Tian, C.H.; Zhong, J.; Feng, H.Y.; Feng, H.Q. The Artificial Diet for Rearing Conogethes punctiferalis and Its Application. China Patent CN108740611A, 29 October 2021.
52. Macrae, T.C.; Baur, M.E.; Boethel, D.J.; Fitzpatrick, B.J.; GAO, A.G.; Gamundi, J.C.; Harrison, L.A.; Kabuye, V.T.; Mcpherson, R.M.; Miklos, J.A.; et al. Laboratory and field evaluations of transgenic soybean exhibiting high-dose expression of a synthetic *Bacillus thuringiensis* Cry1AGene for control of Lepidoptera. *J. Econ. Entomol.* **2005**, *98*, 577–587. [CrossRef]
53. Bernardi, O.; Malvestiti, G.S.; Dourado, P.M.; Wladecir, S.O.; Martinelli, S.; Berger, G.U.; Head, G.P.; Omoto, C. Assessment of the high-dose concept and level of control provided by MON87701 × MON89788 soybean against *Anticarsia gemmatalis* and *Pseudoplusia includens* (Lepidoptera: Noctuidae) in Brazil. *Pest Manag. Sci.* **2012**, *68*, 1083–1109. [CrossRef]

plants

MDPI

Article

Effect of Transgenic Cotton Expressing Bt Cry1Ac or Cry1Ab/Ac Toxins on Lacewing Larvae Mediated by Herbivorous Insect Pests

Zheng-Jun Guan [1,2,†], Qiu-Ju Zhou [3,†], Hong Shi [1,†,‡], Zhi-Xi Tang [1], Biao Liu [4,*] and Wei Wei [1,*]

[1] State Key Laboratory of Vegetation and Climate Change, Institute of Botany, Chinese Academy of Sciences, Beijing 100093, China
[2] Department of Life Sciences, Yuncheng University, Yuncheng 044000, China
[3] Institutes of Science and Development, Chinese Academy of Sciences, Beijing 100190, China
[4] Nanjing Institute of Environmental Sciences, Ministry of Ecology and Environment of P. R. China, Nanjing 210042, China
* Correspondence: liubiao@nies.org (B.L.); weiwei@ibcas.ac.cn (W.W.)
† These authors contributed equally to this work.
‡ Current address: Central Airlines, Haikou 571100, China.

Abstract: A simple food chain (plant, insect pests, and predatory arthropods) in an agro-ecosystem was set up here as a model system to elucidate the potential effect of transgenic *Bacillus thuringiensis* (Bt) cotton on non-target organisms. The system included transgenic/non-transgenic cotton, neonate larvae of three herbivorous insects (*Spodoptera exigua*, *Helicoverpa armigera*, and *S. litura*), and predatory lacewing larvae (*Chrysopa* spp.), which represent the first, second, and third trophic levels, respectively. The results showed that transgenic treatments and different densities of prey had significant effects on both body-weight gain of neonate herbivorous larvae and the number of prey captured by lacewing larvae, respectively. It was found that Bt toxin could persist at the third trophic level in lacewing larvae. The diet mixture bioassay showed that body-weight gain of lacewing larvae was significantly affected by various treatments, especially at lower concentrations of plant-expressed Bt toxin in the diet mixture, which caused significant decreases in body-weight gain. In contrast, synthetic Bt toxin at higher concentrations in the diet did not show this effect. Thus, we inferred that Bt toxin indirectly affected the growth of the lacewings and the lacewings may not be susceptible to Bt toxin or are able to metabolize it.

Keywords: beet armyworm; cotton bollworm; insect predator; metabolism; non-target effect; trophic level

Citation: Guan, Z.-J.; Zhou, Q.-J.; Shi, H.; Tang, Z.-X.; Liu, B.; Wei, W. Effect of Transgenic Cotton Expressing Bt Cry1Ac or Cry1Ab/Ac Toxins on Lacewing Larvae Mediated by Herbivorous Insect Pests. *Plants* **2022**, *11*, 2755. https://doi.org/10.3390/plants11202755

Academic Editor: Fermín Morales

Received: 10 July 2022
Accepted: 13 October 2022
Published: 18 October 2022

1. Introduction

The ecological safety of transgenic crops has been a focus of scientific research and public debate [1–3]. Specifically, the effects of transgenic crops with *Bacillus thuringiensis* (Bt) gene on non-target organisms are of great concern in ecological risk assessment [4]. Bt cotton can significantly reduce damage by insect pests [5]; however, Bt toxin expression and insect resistance in transgenic cotton vary in different organs or tissues during different growth periods [6–8]. Among the target pests of Bt cotton, cotton bollworm (*Helicoverpa armigera*; CBW) is a main focus in scientific reports. CBW is a dominating insect pest for cotton in fields and its damage has caused heavy losses to agricultural production [9]. With the use of Bt cotton, it was reported that the population of CBW had greatly decreased during the growing season [10]; however, Bt cotton did not offer efficient control of CBW in the late growing season [6,11]. Evolved resistance to Bt cotton has been reported in cotton fields [12,13].

Previous studies indicated that Bt cotton can significantly reduce the growth of target pest populations, including body weight, developmental age, pupation, eclosion, behavior, and survival rate [14–16]. It was suggested that Bt cotton can provide better control of CBW populations from the first generation to the third generation, but the effect on the fourth generation was relatively poor [17]. The lethality of Bt cotton to CBW is positively correlated with the expression level of Bt insecticidal toxin [16,18]. In addition, the population boom of secondary insect pests has become a critical problem in cotton fields since the commercialized planting of Bt cotton [18–20]. Beet armyworm (*Spodoptera exigua*; BAW) is a secondary insect pest in cotton fields and its tolerance to Bt toxin could be generated through laboratory selection [21,22]. Although BAW is not a target pest of Bt cotton, it could become an important pest and much more difficult to control, especially given the widespread application of Bt cotton in recent years [23]. This pest is especially not susceptible to Bt toxin expressed in transgenic cotton and the food source of BAW larvae varies [24]. Resistance of BAW larvae to Bt insecticides has been reported previously [21]. Another pest of the same genus, the cluster caterpillar (*S. litura*; CLC), was also found to be less susceptible to Bt toxin [25].

Whether transgenic plants affect natural insect predators has been a hot topic in research and is controversial. Hilbeck and Otto [26] suggested that Bt toxin could negatively affect the growth of predatory lacewing larvae (*Chrysopa cornea*) (Neuroptera: Chrysopidae). Dutton et al. [27] argued that transgenic plants would affect the growth of Lepidopteran insects and, thus, affect the prey quality for lacewings. On the other hand, Bt toxin, expressed in Bt cotton, can be transmitted to higher trophic levels through a non-target pest (e.g., the cotton aphid) and may alter the biology and behavior of a predatory ladybug [28]. In contrast, Tian et al. [29] and Romeis et al. [30] confirmed that Bt toxin did not pose a hazard to the green lacewing, whereas the larvae of both Lepidoptera and cabbage looper (*Trichoplusia ni*) fed on Bt cotton and maize leaves that were exposed to green lacewing larvae (*C. rufilabris*). It was also argued that Bt toxins have no detrimental or adverse effects on natural enemies [31–33]. Bt toxin concentrations decreased considerably from one trophic level to the next in the food web, e.g., Wei et al. [34], except for spider mites [35]. However, those studies were performed without considering the effects of insecticides sprays. Lu et al. [36] demonstrated that bio-control services would benefit from the widespread adoption of Bt cotton, as it was assumed that transgenic Bt cotton would boost the amount of prey consumed by lacewing larvae due to the reduced application of insecticides sprays.

Therefore, there is interest in understanding the relationship between the amounts of non-target insects preyed upon by lacewing and the effects of the Bt plant. In this study, we tried to investigate the effect of Bt cotton on insect pests and the preying ability of lacewing larvae in the presence of transgenic plants, to trace the transmission of Bt toxins through trophic levels and to assay the effect of Bt cotton on the predator mediated by herbivorous insect pests.

2. Results

2.1. Number and Weight of Prey Consumed by Lacewing Larvae

In bioassay 1, transgenic or non-transgenic cotton treatments ($F_{1, 50} = 15.71$, $p = 0.0002$) and different densities of prey ($F_{1, 50} = 110.82$, $p < 0.0001$) had a significant effect on the number of prey captured by lacewing larvae (*C. formosa*); both transgenic cotton treatment of prey and an increase in prey density could boost the number of prey captured by lacewing larvae. In experiment 1, where the density of prey was 10 larvae per cup, the number of prey consumed by lacewing larvae for transgenic Bt and non-transgenic cotton-treated prey was 5.4 ± 0.53 and 4.3 ± 0.59 individuals, respectively. In experiment 2, where the density of BAW larvae was 20 individuals, the number of prey captured by lacewing larvae was 14.3 ± 0.60 and 10.0 ± 0.95 for Bt and non-Bt treatments, respectively. Although on different experimental days, the recorded number of prey consumed by the predator varied ($F_{4, 50} = 12.79$, $p < 0.0001$), no interaction existed between days and treatments.

In bioassay 2, the number of prey captured by *C. formosa* larvae was different between the two experiments ($F_{1, 40} = 5.00$, $p = 0.031$) and the two treatments ($F_{1, 40} = 30.46$, $p < 0.0001$) and among different days ($F_{4, 40} = 12.91$, $p < 0.0001$). In experiment 1, Bt cotton treatment enhanced the number of prey captured by lacewing larvae from 8.4 ± 0.81 to 11.4 ± 0.81, whereas in experiment 2, the number of prey captured by lacewing larvae for Bt cotton treatment (14.7 ± 0.74) was higher than that for non-Bt-cotton-treated prey (8.7 ± 0.87). In the second experiment, neonate CBW larvae fed actively on cotton disks that were placed at the bottom of the cup after switching of prey from CLC; thus, they were assumed to have been treated by Bt and non-Bt cotton leaves. The change in prey did not alter the number of prey captured by lacewing larvae ($F_{1, 20}$ = 1.31, p = 0.27) in experiment 2.

The results of bioassay 3 are shown in Figure 1. The body weights of neonate BAW larvae were significantly lower for transgenic Bt cotton treatment than non-Bt cotton treatment ($F_{1, 16}$ = 134.73, p < 0.0001, Figure 1a) after the same feeding duration of 24 h on cotton leaves in experiment 1. Remarkably, body weights were not different for BAW larvae feeding on transgenic cotton for 24 h or feeding on non-transgenic cotton for 12 h in experiment 2. The number of BAW prey consumed by lacewing larvae (*C. sinica*) was significantly higher when body weight of prey was low, but the numbers were not different than when body weight of prey was similar (Figure 1b) between Bt and non-Bt cotton treatments. This suggested that Bt cotton treatment could reduce the body weight of prey compared to control and stimulate lacewings to prey on more individuals ($F_{1, 40}$ = 11.68, p = 0.0013).

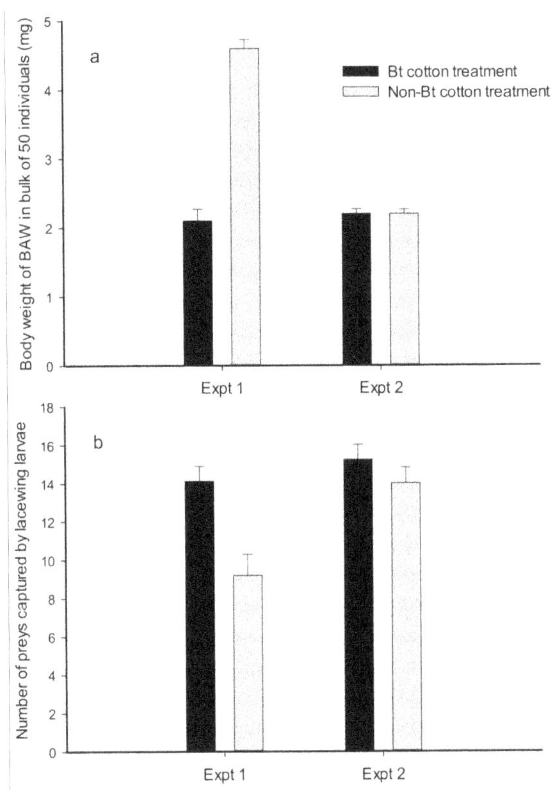

Figure 1. Body weight (**a**) and number (**b**) of neonate beet armyworm (BAW) caterpillars captured by *Chrysopa sinica* larvae. Experiment 1 (Expt 1) included neonate BAW larvae treated for 24 h with transgenic cotton 'Zhong-30' leaves and non-transgenic parent 'Zhong-16' leaves for Bt and non-Bt treatments, respectively. Experiment 2 (Expt 2) included neonate BAW larvae treated for 24 h with transgenic cotton 'Zhong-30' leaves and for 12 h with non-transgenic cotton 'Zhong-16' leaves for Bt and non-Bt treatments, respectively.

2.2. Persistence of Cry1Ac Toxin at Trophic Levels

Three samples from the 20 transgenic cotton plants (Zhong-30) contained very low concentrations of Bt toxin and, thus, were not used to feed neonate BAW larvae. The average Bt toxin concentration in the remaining 17 cotton plants was 744.6 ± 62.65 ng/g fresh weight (FW), whereas none was detected in Zhong-16. The average Bt toxin content accumulated in the bodies of neonate BAW larvae was 321.7 ± 7.33 ng/g FW (n = 3), which was nearly half of Bt toxin concentration in leaves. Only 4.5 ± 1.93 ng/g FW (n = 8) of Bt toxin persisted at the third trophic level in lacewing larvae when measured right after feeding on BAW larvae was stopped. Bt toxin concentration between these three trophic levels was significantly different ($\chi^2 = 20.32$, df = 2, $p < 0.001$). Bt toxin content in lacewing bodies declined (Figure 2) during starvation, up to 24 h but the difference was not significant. This implied a great variation in the metabolism of Bt toxin in lacewing larvae.

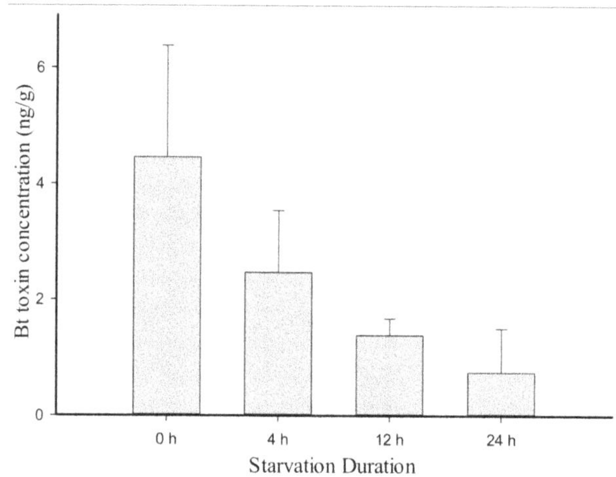

Figure 2. Changes in Bt toxin content in lacewing larvae (*Chrysopa sinica*) bodies during starvation.

2.3. Effect of Bt Toxin Residual in Larval Body on the Growth of Lacewing Larvae

No Bt toxin was detected in the leaf samples of Simian-3 (data not shown) and relatively high concentrations of Bt toxins were measured (Table 1) in the third expanded leaves of GK-19 and the results varied among the four times of Bt toxin accumulations ($F_{3,47} = 5.98$, $p = 0.003$). Approximately 0.77~19.48 ng/g Bt toxin was detected in the larval powder of BAW that fed on Bt cotton leaves (Table 1), whereas Bt toxin was not detectable in the larval powder of BAW larvae feeding on Simian-3 (data not shown). Thus, no Bt toxin was presented in the diet mixture for "treatment B" in the following feeding test with lacewing larvae. Bt toxins present in larval powder of BAW decreased 80~90% compared to the leaves. Owing to the possible degradation of Bt toxin stored in the freezer, its concentration was very low at the two earlier accumulations. This caused a very low content in the diet mixture fed to lacewing larvae (Table 1).

Body-weight gain of lacewing larvae was significantly affected by various treatments ($F_{3,322} = 72.34$, $p < 0.0001$). The average weight gain of lacewing larvae in the four synthetic diet treatments ranged from 0.40 to 1.96 mg, whereas the aphid controls ranged from approximately 1.74 to 9.96 mg during the 8 days of the bioassay; the former was obviously lower than the latter (Figure 3). Various treatment durations (i.e., the first, second, third, and fourth 2-day durations) had a significant effect on weight gain of lacewing larvae ($F_{3,322} = 180.98$, $p < 0.0001$), and they also had an interaction with the various treatments ($F_{9,322} = 8.54$, $p < 0.0001$). Overall, both "treatment A" (feeding synthetic diet plus larval powder of BAW fed on transgenic cotton GK-19) and "treatment B" (feeding on synthetic diet plus larval powder of BAW fed on non-transgenic cotton Simian-3) were significantly different from each other and from other treatments (C and D), whereas no difference was detected between treatment C (treated with synthetic diet plus synthetic Bt pro-toxin) and D (treated with blank synthetic diet only). This result suggested a 5 ng/g concentration of synthetic Bt pro-toxin did not affect the body-weight gain of lacewing larvae, but Bt toxin of lower concentration (Table 1) that was isolated from Bt cotton leaves and was presented in the larval powder could have a negative effect on lacewing larvae. On the fourth and eighth day, the recorded variation between treatments had the same trend with the overall significance, whereas no difference was found among various treatments on the second day, probably due to very low Bt toxin content in the feed. The difference between treatments B, C, and D only bordered on significance.

Table 1. Bt toxin contents in third expanded leaves from the plant apex of GK-19 at the six-leaf stage, in larval powder of beet armyworm (BAW) during the four times of Bt toxin accumulations and in the diet mixture feeding to lacewing larvae.

	Bt Toxin Content in GK-19 Leaves (n = 12) (ng/g Fresh Weight)	Bt Toxin Content in the Powder of Larvae Fed on GK-19 Cotton Leaves (ng/g Fresh Weight)	Bt Toxin Content in Diet Mixture (ng/g Diet Weight)
1	77.4 ± 10.47	0.77	0.004
2	77.8 ± 18.84	1.48	0.016
3	155.5 ± 25.86	19.48	0.418
4	61.1 ± 8.32	11.72	0.251

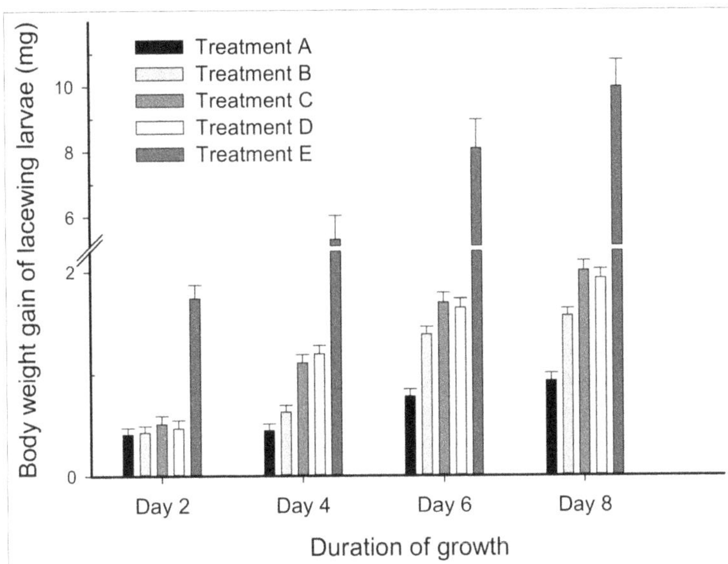

Figure 3. Body-weight gain of lacewing larvae (*Chrysopa sinica*) treated with a synthetic diet plus larval powder of beet armyworm (BAW) fed on transgenic cotton (GK-19) (Treatment A), synthetic diet plus larval powder of BAW fed on non-transgenic cotton (Simian-3) (Treatment B), synthetic diet plus synthetic Bt pro-toxin (Treatment C), blank synthetic diet (Treatment D), and aphids (Treatment E), respectively.

In contrast to the significant variation in body-weight gain among the various treatments, different treatments did not result in differences in the growth duration from the first instar to the pupae stage or in the actual weight of the lacewing pupae. Treatment A resulted in the longest growth duration of 16.0 ± 0.3 d, while the aphid control (treatment E) resulted in the shortest growth period of 12.8 ± 0.2 d. The pupal weight of the five treatments ranged from 10.4 ± 0.5 to 12.7 ± 0.3 mg, where the weight of the aphid-treated pupae was the lowest since they took the least amount of time (less than 13 days) to reach pupation. Treatments A and C had similar and the lowest pupation rates (57% for A, 55% for C), whereas B and D had the highest rates (73% for B, 70% for D); the aphid-treated larvae (E) exhibited an intermediate rate of pupation (65%). The number of lacewing adults that emerged from pupae was recorded, but their individual dates of eclosion were not included. Although less pupae were obtained for the five treatments, the rate of eclosion was not low; in total, 15, 22, 11, 15, and 12 (male and female) adults emerged from the A, B, C, D, and E treatments, respectively (Table 2).

Table 2. Pupation and eclosion of lacewing in the five diet mixture treatments: (A) synthetic diet plus larval powder of beet armyworm (BAW) fed on transgenic cotton (GK-19); (B) synthetic diet plus larval powder of BAW fed on non-transgenic cotton (Simian-3); (C) synthetic diet plus synthetic Bt pro-toxin; (D) blank synthetic diet; and (E) aphid feed.

Treatment	No. of Treated Larvae	No. of Pupae	Eclosion Rate (%)	No. of Adult Males	No. of Adult Females
A	30	17	88.2	10	5
B	30	22	100	10	12
C	20	11	100	7	4
D	20	15	100	7	8
E	20	13	92.3	5	7

3. Discussion

While both transgenic Bt and non-transgenic cottons contain secondary metabolites that may adversely affect the growth of insect pest and the integration of foreign Bt gene might alter the metabolite profile in cotton [37,38], our results here showed that the Bt cotton could still provide better control over herbivorous insect pests (e.g., BAW) compared to non-Bt cotton. This study showed that there was a reduced body-weight gain for prey subjected to the Bt cotton treatment and that the number of prey (including CBW, BAW, and CLC) captured by lacewing larvae in transgenic treatments was significantly larger than non-transgenic treatments after 24 h of feeding. However, a feeding duration of 12 h did not result in difference in weight gain; therefore, the number of BAWs captured by lacewing larvae was not affected. The results implied that different feeding durations likely affected the development of prey and eventually changed the number of prey consumed by lacewing larvae. The prey showed growth retardation, weight loss, and small body size after feeding on Bt cotton leaves; thus, the predators had to consume a high number of prey to obtain adequate nutrients [27]. Therefore, the increased number of prey consumed by lacewing larvae might result from the indirect effects of Bt toxin on the nutritional quality of prey, rather than its direct effects.

It was found that the developmental periods of synthetic diet-treated lacewing larvae were longer than the aphid-treated one. Although the four different synthetic diets showed no difference in the developmental period of the lacewing pupae, the growth period of lacewing larvae in treatment A (synthetic diet plus larval powder of BAW fed on Bt cotton leaves) was the longest, i.e., pupation of the lacewing larvae was delayed and its growth cycle was extended. Both treatments A and C, with artificial diets containing either plant-expressed or artificially synthesized Bt toxin, had lower pupation rates than the other treatments. It was inferred that Bt toxin might have a certain influence on the transformation of larvae into adult lacewings. This specific mechanism of effect needs further study. In addition, the plant-expressed Bt toxin resulted in lower body-weight gain, which was not observed in the synthesized Bt toxin treatment. This might be due to the differences in potential insecticidal activity or protein structure between plant-expressed Bt toxin and synthesized Bt pro-toxin. Although it seems that the predatory function of predators could benefit from reduced growth and weight gain (i.e., smaller body size) of insect pest preys in the field, this advantage could become invalid if predators (e.g., lacewing) are directly or indirectly affected while consuming prey feeding on cotton plants expressing Bt toxins.

In our previous study, we found that Bt toxin residue in insect bodies and feces still had insecticidal ability and could affect other organisms exposed [15]. In this present study, Bt toxin residues at the second and third trophic levels were measured and proved. Although the Bt toxin that persisted in lacewing larvae was low, the finding illustrated that the Bt toxin can pass through the trophic levels through the food chain [34,39], which could cause unintended impacts on the natural ecosystem. The result is consistent with Zhang et al. [28], which indicated that Bt toxin expressed in transgenic cotton can be transmitted to higher trophic level into the coccinellid predator *Propylaea japonica* through a non-target pest insect. Furthermore, in our bioassays, as the starvation time prolonged, Bt toxin content in lacewing bodies declined. During its whole growing period, the lacewing could remove waste from the body only before it begins to pupate [40]. Therefore, it was assumed that Bt toxin may be metabolized inside the lacewing body. This phenomenon was also reported as Bt toxin moves through the food chain involving Bt canola (*Brassica napus*), diamondback moth (*Plutella xylostella*), and lacewing larvae (*C. carnea*) [34]. Although BAW is not the main prey of some predators, such as lacewings, the activity of residual Bt toxin that remained in and/or excreted from their bodies and persisted in the environment may influence susceptible organisms [41,42]. Our results here showed that the body-weight gain as well as pupation rate of lacewings could be indirectly affected by Bt toxin expressed in transgenic plants. Nevertheless, this study did not confirm that Bt toxin has no direct toxicity on predatory enemies of insect pests that feed on Bt plants. Direct toxicity is probably not the only factor that has a lethal effect on natural enemies; other growth indicators should also be considered when performing non-target risk assessment of Bt plants [30,43].

Considering the research results presented here, the indirect effects of Bt cotton may be the main factor affecting lacewings. Although Hassanpour et al. [44] showed that lacewing larvae could serve as potential bio-control agents against CBW, Lepidopteran insects are not high-quality prey for lacewings; lacewing larvae had more stunted growth and higher mortality after feeding on Lepidopteran insect larvae compared to other prey species [43,45,46]. In a feeding choice test, lacewing larvae tended to feed on aphids in the field, other than young larvae of Lepidopteran insects [47]. When being attacked by lacewing larvae, young larvae of Lepidopteran insects would fight back; therefore, the lacewings were more likely to become injured and experience higher mortality [26,27]; however, this phenomenon was not observed in the present study due to the limited number of neonate prey larvae used in each assay.

The results indicated that Bt plants might affect the growth and nutritional quality of herbivorous insects. Owing to low-nutrient foods, the lacewing increased prey numbers. At the end, no significant differences in the various growth indicators were observed between transgenic Bt diet treatment and non-Bt treatment [48]. If the differences in nutrition status among the treatments were eliminated, the amount of variation in prey number of lacewings could disappear. Therefore, the quality and quantity of the prey may be only indirect factors affecting growth of the lacewings. Whether Bt toxin has direct toxic effects on the lacewing (e.g., on body-weight gain and pupation rate) still needs to be further examined.

In general, successful suppression of the targeted insect pests in Bt cotton fields was demonstrated. However, the ecological niches of the suppressed targeted insect pests at the second trophic level could be replaced by secondary or non-targeted insect pests, which could be due to the reduction in broad-spectrum pesticides sprays [19], as well as the reduced presence of bio-control services by natural enemies (including arthropod predators), which resulted from direct or indirect impacts by Bt plants and accumulated Bt toxins in their prey. Impact on the arthropod predators and other natural enemies could affect the fundamental bases of IPM.

Bio-control is the cornerstone of integrated pest management (IPM) and natural pest control methods are normally preferred before using the pesticide options. IPM had called on the strategic integration of multiple control tactics [49], while limiting the dependence on one single technology. By many years of experience using genetically engineered crops, including herbicide-tolerant and/or Bt insecticidal traits, resistance inevitably evolved over time [50,51]. Studies demonstrated the benefits of genetically engineered crops expressing Bt toxins; however, managing resistance evolution of insect pests requires the development of novel resistant crops expressing new and multiple Bt toxins or other resistant traits. Therefore, it is hard to define whether the genetically engineered traits are natural host plant resistances or a series of chemical pesticides upgrading. In addition, while the traditional synthetic sprays of Bt pro-toxin insecticides need to be activated by binding to receptors in the midgut epithelium of the herbivorous larvae to become lethal, Bt toxins directly expressed in plant hosts are already in the activated form and might pose adverse impacts to arthropod natural enemies, even at a very low concentration without the presence of receptors in their digestive tract. Although this unintended effect needs to be confirmed in further studies, the potential adverse impacts on natural bio-control agents could add risk of failure to the IPM strategy, especially if there are additional action modes of Bt toxins on natural bio-control agents [26]. Host plant resistance, including traditional breeding and genetic engineering, could remain a critical component of IPM in the near future [49]; however, diverse approaches and control strategies should be employed for sustainable agricultural production.

4. Materials and Methods

4.1. Plant and Insect Materials

The plant materials included a Bt cotton (*Gossypium hirsutum*) variety 'Zhong-30' containing a *Cry1Ac* gene isolated from *Bacillus thuringiensis* and its non-transgenic parent 'Zhong-16'; a Bt cotton hybrid 'GK-19' containing a recombinant *cry1Ab/1Ac* gene [52] and its conventional and maternal parent 'Simian-3'. They were planted in a greenhouse at a temperature of 23 ± 2 °C and had light conditions set to a photoperiod of 16L:8D. The third expanded leaves from the apex at the six-leaf developmental stage were sampled for further bioassays. In addition, seeds of broad bean (*Vicia faba*) were sown in humid sand and their seedlings were used to feed aphids (*Aphis craccivora* Koch), the prey of lacewing.

The insect materials, including cotton bollworm (CBW), beet armyworm (BAW), and cluster caterpillar (CLC), and their synthetic diets were purchased from Jiyuan Baiyun Industrial Co. Ltd. (Jiyuan, China). The insects were reared in the insectary at the Institute of Botany at the Chinese Academy of Sciences (IBCAS) (Beijing, China) and the second generation was obtained for the study. The adult lacewings (*Chrysopa formosa*) were captured in the Botanic Garden of IBCAS and reared in the insectary to obtain lacewing larvae for bioassays in this study. Chinese green lacewing larvae (*C. sinica*) were kindly provided by Dr. Fan Zhang at the Plant Protection and Environmental Protection Research Institute, Beijing Agriculture and Forestry Academy of Sciences. All insect materials were kept and reared and the bioassays were conducted in the insectary of IBCAS under appropriate rearing conditions (temperature 25 ± 2 °C, humidity 50 ± 5% and light 16 h: 8 h). Aphids were reared on the seedlings of broad bean and fed to lacewings regularly during the experiment where applicable.

4.2. Bioassays on Prey Consumption by Lacewing Larvae and Cry1Ac Toxin Transfer through Trophic Levels

4.2.1. Prey Consumption Bioassays

Three bioassays were designed to investigate the effect of Bt cotton on the number of caterpillars consumed by predatory lacewing larvae (Figure 4 and Appendix A Table A1). Each bioassay contained two experiments. Lacewing larvae of *C. formosa* were used as predator in both bioassay 1 and 2, while *C. sinica* was used in bioassay 3. Both bioassays used transgenic Bt cotton Zhong-30 and non-transgenic cotton Zhong-16 to feed to herbivorous insect larvae (Figure 4).

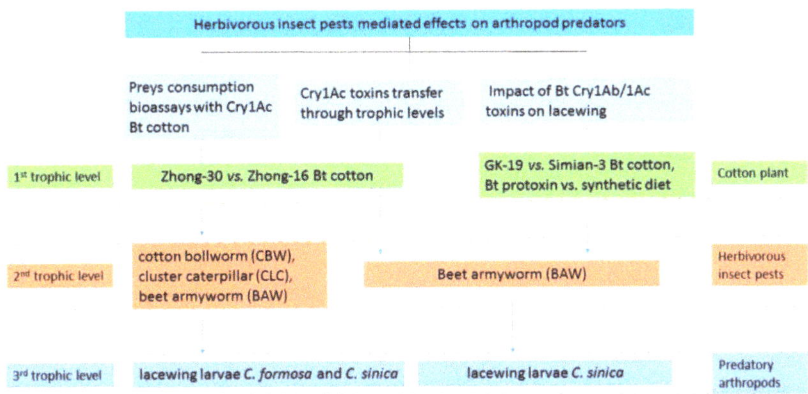

Figure 4. Experimental designs. Bt cotton 'Zhong-30' expressing Cry1Ac toxin and its non-transgenic counterpart 'Zhong-16' were used to study preys consumption bioassays and to measure toxin transfer through trophic levels, while 'GK-19' expressing Cry1Ab/1Ac fusion toxin and its non-transgenic counterpart 'Simian-3' were used to study the impact of toxin on lacewing.

In bioassay 1, neonate BAW larvae were fed by leaves of transgenic Bt cotton Zhong-30 and non-transgenic Bt cotton Zhong-16 for 24 h and then they were fed to the third instar larvae of *C. formosa* in plastic cups with permeable covers on the top and 1% agarose at the bottom. The first experiment contained eight and nine replicates for the Bt cotton-treated and non-Bt cotton-treated prey, respectively. Each replicate had 10 neonate BAW larvae and one lacewing larva. The second experiment contained seven and four replicates for the transgenic-treated and non-transgenic-treated prey, respectively. In each replicate of the second experiment, one lacewing larva was placed with 20 neonate BAW larvae, on either a transgenic or non-transgenic cotton leaf disk.

In bioassay 2, neonate CLC and CBW larvae were allowed to feed for 24 h on the third expanded leaves from the apex of Zhong-30 and Zhong-16, respectively, before they were fed to the third instar larvae of *C. formosa*. In experiment 1, 20 neonate CBW larvae were fed to one lacewing larva in each of the five replicates of both treatments. Experiment 2 contained five and six replicates for non-transgenic and transgenic treatments, respectively. During the first 2 days, 20 individuals of treated CLC larvae were fed to a third instar lacewing larva in each replicate. During the remaining 3 days, the predator was switched to untreated neonate CBW larvae due to a shortage of CLC larvae.

In bioassay 3, neonate BAW larvae were allowed to feed for 12–24 h on the third expanded leaves from the apex of Zhong-30 and Zhong-16, respectively. In experiment 1 of this assay, the larvae were fed for 24 h on both Bt and non-Bt cotton and for 12 h on Zhong-16 in experiment 2. These treated larvae were weighed in a bulk of 50 individuals nine times for each treatment of each experiment and then fed to a third instar larva of *C. sinica* at a density of 20 prey items to one predator. Each treatment of Zhong-16 and Zhong-30 had five replicates in both experiments.

The number of prey captured by lacewings was recorded in each of 5 days of observation for the three bioassays and the density was maintained at the initial levels by adding an appropriate number of caterpillars.

4.2.2. Detecting Cry1Ac Toxin Transfer through Trophic Levels

It was suggested that Bt toxin content in the leaves should be assayed in a rearing bioassay [15]. This bioassay intended to study the movement of Bt toxin through a simple food chain (cotton-BAW-lacewing) (Figure 4). Leaf disks of the third expanded leaves were sampled from each of the 20 transgenic Bt cotton plants (Zhong-30) and their Bt concentrations were measured using ELISA. Leaves of non-transgenic cotton plants (Zhong-

16) were used as a control. These cotton leaves were used to feed neonate BAW larvae for 24 h; three samples of 50 larvae feeding on 'Zhong-30' were randomly sampled to detect Bt toxin accumulation. A sample of 50 larvae that fed on non-Bt cotton was used as a control. These larvae that fed on Bt cotton leaves were transferred to 32 plastic cups with covers; each cup contained 20 larvae and a third instar lacewing larva. One cup contained BAW larvae feeding on non-Bt cotton leaves and a lacewing larva were used as a negative control. The density of BAW larvae in each cup was maintained by adding an appropriate number of larvae. After 3 days of feeding, lacewing larvae were collected and divided into four groups. The first group of eight lacewing larvae was killed immediately by placing into liquid nitrogen individually; the remaining predators were put in cleaned Petri dishes without any food. The second group of ten lacewing larvae was collected after 4 h of starvation. Seven lacewing larvae in each of the third group and the fourth group were collected individually after 12 h and 24 h of starvation, respectively. All lacewing larvae were stored at −80 °C, separately, for further analysis using ELISA Kits (Agdia Inc., Elkhart, IN, USA) to measure the presence of Bt toxin [34].

4.3. Bioassays on the Effect of Cry1Ab/1Ac Toxins Residual on Lacewing Larvae

Bt cotton 'GK-19' expressing Cry1Ab/1Ac fusion toxin and its non-transgenic counterpart 'Simian-3' were used in this experiment (Figure 4). The third expanded leaves of GK-19 at the six-leaf stage were fed to the second instar BAW larvae for 24 h to accumulate Bt toxin at the second trophic levels. Simian-3-fed BAW larvae were used as controls. Bt toxin concentration was measured in the leaves of both GK-19 and Simian-3 using ELISA test that were fed to BAW larvae. Bt toxin in larval body powders was accumulated for bioassay as described in Shi et al. [15]. Ten of the second instar larvae of BAW were added to a Petri dish for each of the twelve and four replicates for GK-19 and Simian-3, respectively. The larvae were collected after 24 h of feeding and stored separately between GK-19 treatment and Simian-3 treatment at −80 °C for >24 h and were ground into dry powder in liquid nitrogen. Part of the larval powder was used to detect Bt toxin concentration and the others was mixed with the synthetic diet [53] of lacewings and fed to the first instar lacewing larvae. The accumulation was repeated four times to generate enough larval powder for further bioassay on the lacewing larvae. Previous research suggests that Bt toxin in synthetic diets persists for 2 days [54]; thus, the mixed diet for the lacewing larvae was re-made and renewed every 2 days and the dry powder that contained Bt toxin obtained at four different accumulations was used in subsequent order to mix with the synthetic diet, respectively.

Five treatments were set up in this feeding bioassay: (A) 30 replicates of synthetic diet feeding tests with larval powder of BAW that fed on transgenic cotton (GK-19); (B) 30 replicates of synthetic diet feeding tests with larval powder of BAW that fed on non-transgenic cotton (Simian-3); (C) 20 replicates of synthetic diet feeding tests with synthetic Bt pro-toxin (1 mg/mL), and the final concentration of Bt pro-toxin in this treatment was 5 ng/g, which was equal to the reported Bt toxin concentration in larval bodies feeding on GK-19 [15]; (D) 20 replicates of blank synthetic diet feeding tests of lacewings; (E) 20 replicates of aphid feeding tests of lacewing larvae. Each replicate contained one lacewing larva and a certain amount of synthetic diet mixture or aphids. The last treatment was used as a control only and not considered for statistical analysis.

A piece of squared sponge (1 cm × 1 cm × 1 cm) was placed inside each plastic container (diameter × height = 11 cm × 8.5 cm) with a permeable cover to feed one first instar lacewing larva for each replicate after absorbing enough diet mixture in liquid in each treatment. Body-weight gain of lacewing larvae was recorded every 2 days. The synthetic diet was replaced by aphids after 8 days of bioassay until lacewing larvae pupated. Survival rate of lacewing larvae was also recorded during the bioassay, as well as the rate of pupation, weight of the pupae, and the developmental days of each larval stage. The eclosion rate of lacewing pupae, the sex ratio and oviposition rate of lacewing adults, and hatching rate of lacewing eggs were monitored and recorded to study the possible effects of the diet treatments.

4.4. Data Analysis

Statistical analyses were conducted using SPSS 16.0 software. One-way analysis of variance (ANOVA) was used to analyze the variation in Bt toxin content in leaves of transgenic cotton, lacewing bodies and diet mixture, and the number and weight of prey consumed by lacewing larvae. Chi-square test was conducted to test the difference in Bt toxin content at various trophic levels. The GLM (general linear model) procedure was used to analyze the effects of Bt toxin residual on lacewing. Differences in means between treatments were compared using a least significant difference (LSD) test.

Author Contributions: Conceptualization, W.W.; formal analysis, Q.-J.Z., H.S. and Z.-J.G.; investigation, Q.-J.Z., H.S. and Z.-X.T.; writing, Z.-J.G., Q.-J.Z. and W.W.; supervision, W.W. and B.L.; funding acquisition, B.L. and Z.-J.G. All authors have read and agreed to the published version of the manuscript.

Funding: This research was funded by the China National Special Transgenic Project "Monitoring and controlling technology for ecological risks in nature, grant number: 2016ZX08012005" and the Natural Science Foundation of Shanxi Province, China, grant number: 201701D121089".

Data Availability Statement: The data presented in this study are available on request from the corresponding author.

Acknowledgments: We thank the three anonymous reviewers whose comments/suggestions helped improve and clarify this manuscript.

Conflicts of Interest: The authors declare no conflict of interest. The funders had no role in the design of the study; in the collection, analyses, or interpretation of data; in the writing of the manuscript, or in the decision to publish the results.

Appendix A

Table A1. Bioassays to investigate the effect of Bt cotton on the number of caterpillars consumed by predatory lacewing larvae.

	Expt. No.	Prey (Neonate Larvae)	Predator (3rd Instar Larva)	Prey: Predator Ratio	Predator Feeding Replicates		Preys Pre-Treated Hours	
					Tr.	Non-Tr.	Tr. (Zhong-30)	Non-Tr. (Zhong-16)
Bioassay 1	Expt. 1 Expt. 2	BAW	Chrysopa formosa	10:1 20:1	8 7	9 4	24	24
Bioassay 2	Expt. 1 Expt. 2	CBW CLC+CBW	Chrysopa formosa	20:1	5 6	5 5	24	24
Bioassay 3	Expt. 1 Expt. 2	BAW	Chrysopa sinica	20:1	5	5	24 24	24 12

BAW (beet armyworm): *Spodoptera exigua*; CBW (cotton bollworm): *Helicoverpa armigera*; CLC (cluster caterpillar): *Spodoptera litura*. 'Zhong-30' and 'Zhong-16' is the transgenic line and the non-transgenic parent cotton, respectively. 'Tr.' and 'Non-Tr' is the abbreviation for 'transgenic' and 'non-transgenic', respectively.

References

1. Guan, Z.-J.; Lu, S.-B.; Huo, Y.-L.; Guan, Z.-P.; Liu, B.; Wei, W. Do genetically modified plants affect adversely on soil microbial communities? *Agric. Ecosys. Environ.* **2016**, *235*, 289–305. [CrossRef]
2. Wei, W.; Stewart, C.N., Jr. (Eds.) *Gene Flow: Monitoring, Modeling and Mitigation*; CABI: Wallingford, UK, 2021; p. 169.
3. Sharmaa, P.; Singh, S.P.; Iqbal, H.M.N.; Parra-Saldivar, R.; Varjani, S.; Tong, Y.W. Genetic modifications associated with sustainability aspects for sustainable developments. *Bioengineered* **2022**, *13*, 9509–9521. [CrossRef] [PubMed]
4. Belousova, M.E.; Malovichko, Y.V.; Shikov, A.E.; Nizhnikov, A.A.; Antonets, K.S. Dissecting the Environmental Consequences of *Bacillus thuringiensis* Application for Natural Ecosystems. *Toxins* **2021**, *13*, 355. [CrossRef] [PubMed]

5. Lu, Y.-H.; Wyckhuys, K.A.G.; Yang, L.; Liu, B.; Juan Zeng, J.; Jiang, Y.-Y.; Desneux, N.; Zhang, W.; Wu, K.M. Bt cotton area contraction drives regional pest resurgence, crop loss, and pesticide use. *Plant Biotech. J.* **2022**, *20*, 390–398. [CrossRef]
6. Olsen, K.M.; Daly, J.C.; Holt, H.E.; Finnegan, E.J. Season-long variation in expression of Cry1Ac gene and efficacy of *Bacillus thuringiensis* toxin in transgenic cotton against *Helicoverpa armigera* (Lepidoptera: *Noctuidae*). *J. Econ. Entomol.* **2005**, *98*, 1007–1017. [CrossRef]
7. Liu, Z.; Eltayib, H.M.A.A.; Wu, H.; Zhou, M.-Y.; Zhang, X.; Chen, Y.; Chen, D.-H. Bt insecticidal efficacy variation and agronomic regulation in Bt cotton. *J. Cotton Res.* **2019**, *2*, 23. [CrossRef]
8. Tang, F.-Y.; Gao, X.; Peng, J.-J. The dynamics of carbohydrate and associated gene expression in the stems and roots of upland cotton (*Gossypium hirsutum* L.) during carbon remobilization. *Plant Physiol. Biochem.* **2022**, *178*, 125–136. [CrossRef]
9. Li, G.-P.; Wu, K.-M.; Gould, F.; Wang, J.-K.; Miao, J.; Gao, X.-W.; Guo, Y.-Y. Increasing tolerance to Cry1Ac cotton from cotton bollworm, *Helicoverpa armigera*, was confirmed in Bt cotton farming area of China. *Ecol. Entomol.* **2007**, *32*, 366–375. [CrossRef]
10. Wu, K.-M.; Guo, Y.-Y.; Lv, N.; Greenplate, J.T.; Deaton, R. Efficacy of transgenic cotton containing a cry1Ac gene from *Bacillus thuringiensis* against *Helicoverpa armigera* (Lepidoptera: *Noctuidae*) in northern China. *J. Econ. Entomol.* **2003**, *96*, 1322–1328. [CrossRef]
11. Wan, P.; Zhang, Y.; Wu, K.; Huang, M. Seasonal expression profiles of insecticidal protein and control efficacy against *Helicoverpa armigera* for Bt cotton in the Yangtze River valley of China. *J. Econ. Entomol.* **2005**, *98*, 195–201. [CrossRef]
12. Wu, K.-M.; Lu, Y.-H.; Feng, H.-Q.; Jiang, Y.Y.; Zhao, J.Z. Suppression of cotton bollworm in multiple crops in China in areas with Bt toxin–containing cotton. *Science* **2008**, *321*, 1676–1678. [CrossRef]
13. Zhang, H.-N.; Tian, W.; Zhao, J.; Jin, L.; Yang, J.; Liu, C.-H.; Yang, Y.-H.; Wu, S.-W.; Wu, K.-M.; Cui, J.-J.; et al. Diverse genetic basis of field-evolved resistance to Bt cotton in cotton bollworm from China. *Proc. Nat. Acad. Sci. USA* **2012**, *109*, 10275–10280. [CrossRef]
14. Men, X.-Y.; Ge, F.; Liu, X.-H.; Yardim, E.-N. Diversity of arthropod communities in transgenic Bt cotton and nontransgenic cotton agroecosystems. *Environ. Entomol.* **2003**, *32*, 270–275. [CrossRef]
15. Shi, H.; Zhang, L.-F.; Hua, B.-Z.; Mi, X.-C.; Wei, W.; Zhang, Y.-J.; Ma, K.P. Insecticidal activity of residual Bt protein at the second trophic level. *Chin. Sci. Bull.* **2006**, *51*, 946–951. [CrossRef]
16. Tabashnik, B.E.; Gassmann, A.J.; Crowder, D.W.; Carriere, Y. Insect resistance to Bt crops: Evidence versus theory. *Nat. Biotech.* **2008**, *26*, 199–202. [CrossRef]
17. Zhang, Y.-J.; Wu, K.-M.; Guo, Y.Y. On the spatio-temporal expression of the contents of Bt insecticidal protein and the resistance of Bt transgenic cotton to cotton bollworm. *Acta Phytophyl. Sin.* **2001**, *28*, 1–6.
18. Wu, K.-M.; Guo, Y.-Y.; Gao, S. Evaluation of the natural refuge function for *Helicoverpa armigera* (Lepidoptera: *Noctuidae*) within *Bacillus thuringiensis* transgenic cotton growing areas in north China. *J. Econ. Entomol.* **2002**, *95*, 832–837. [CrossRef]
19. Lu, Y.-H.; Wu, K.-M.; Jiang, Y.-Y.; Xia, B.; Li, P.; Feng, H.-Q.; Wyckhuys, K.A.; Guo, Y.-Y. Mirid bug outbreaks in multiple crops correlated with wide-scale adoption of Bt cotton in China. *Science* **2010**, *328*, 1151–1154. [CrossRef]
20. Catarino, R.; Ceddia, G.; Areal, F.J.; Park, J. The impact of secondary pests on *Bacillus thuringiensis* (Bt) crops. *Plant Biotech. J.* **2015**, *13*, 601–612. [CrossRef]
21. Tabashnik, B.E. Evolution of resistance to *Bacillus thuringiensis*. *Ann. Rev. Entomol.* **1994**, *39*, 47–79. [CrossRef]
22. Adamczyk, J.J., Jr.; Gore, J. Laboratory and field performance of cotton containing Cry1ac, Cry1f, and both Cry1ac and Cry1f (widestrike®) against beet armyworm and fall armyworm larvae (Lepidoptera: *Noctuidae*). *Fla. Entomol.* **2004**, *87*, 427–432. [CrossRef]
23. Wu, G.; Harris, M.K.; Guo, J.-Y.; Wan, F.-H. Response of multiple generations of beet armyworm, *Spodoptera exigua* (Hübner), feeding on transgenic Bt cotton. *J. Appl. Entomol.* **2009**, *133*, 90–100. [CrossRef]
24. Guo, J.-Y.; Wu, G.; Wan, F.-H. Activities of digestive and detoxification enzymes in multiple generations of beet armyworm, *Spodoptera exigua* (Hübner), in response to transgenic Bt cotton. *J. Pest Sci.* **2010**, *83*, 453–460. [CrossRef]
25. Shu, Y.-H.; Du, Y.; Chen, J.; Wei, J.-X.; Wang, J.-W. Responses of the cutworm *Spodoptera litura* (Lepidoptera: *Noctuidae*) to two Bt corn hybrids expressing Cry1Ab. *Sci. Rep.* **2017**, *7*, 41577.
26. Hilbeck, A.; Otto, M. Specificity and combinatorial effects of *Bacillus thuringiensis* Cry toxins in the context of GMO risk assessment. *Front. Environ. Sci.* **2015**, *3*, 71. [CrossRef]
27. Dutton, A.; Klein, H.; Romeis, J.; Bigler, F. Uptake of Bt-toxin by herbivores feeding on transgenic maize and consequences for the predator *Chrysoperla carnea*. *Ecol. Entomol.* **2002**, *27*, 441–447. [CrossRef]
28. Zhang, G.-F.; Wan, F.-H.; Lövei, G.L.; Liu, W.-X.; Guo, J.-Y. Transmission of Bt toxin to the predator *Propylaea japonica* (Coleoptera: *Coccinellidae*) through its aphid prey feeding on transgenic Bt cotton. *Environ. Entomol.* **2006**, *35*, 143–150. [CrossRef]
29. Tian, J.-C.; Wang, X.-P.; Long, L.-P.; Romeis, J.; Naranjo, S.E.; Hellmich, R.L.; Wang, P.; Earle, E.D.; Shelton, A.M. Bt crops producing Cry1Ac, Cry2Ab and Cry1F do not harm the green lacewing, *Chrysoperla rufilabris*. *PLoS ONE* **2012**, *8*, e60125. [CrossRef]
30. Romeis, J.; Meissle, M.; Naranjo, S.E.; Li, Y.; Bigler, F. The end of a myth—Bt (Cry1Ab) maize does not harm green lacewings. *Front. Plant Sci.* **2014**, *5*, 391. [CrossRef]
31. Yao, Y.-S.; Han, P.; Niu, C.-Y.; Dong, Y.-C.; Gao, X.-W.; Cui, J.J.; Desneux, N. Transgenic Bt cotton does not disrupt the top-down forces regulating the cotton aphid in central China. *PLoS ONE* **2016**, *11*, e0166771. [CrossRef]
32. Van Den Berg, J.; Warren, J.F.; Du Plessis, H. The potential effect of Bt maize on *Chrysoperla pudica* (Neuroptera: *Chrysopidae*). *Environ. Entomol.* **2017**, *46*, 413–417. [CrossRef] [PubMed]

33. Ali, I.; Zhang, S.; Muhammad, M.S.; Iqbal, M.; Cui, J.J. Bt proteins have no detrimental effects on larvae of the green lacewing, *Chrysopa pallens* (Rambur) (Neuroptera: *Chrysopidae*). *Neotrop. Entomol.* **2018**, *47*, 336–343. [CrossRef] [PubMed]
34. Wei, W.; Schuler, T.H.; Clark, S.J.; Stewart, C.N., Jr.; Poppy, G.M. Movement of transgenic plant-expressed Bt Cry1Ac proteins through high trophic levels. *J. Appl. Entomol.* **2008**, *132*, 1–11. [CrossRef]
35. Meissle, M.; Romeis, J. Transfer of Cry1Ac and Cry2A proteins from genetically engineered Bt cotton to herbivores and predators. *Insect Sci.* **2017**, *25*, 823–832. [CrossRef]
36. Lu, Y.-H.; Wu, K.-M.; Jiang, Y.-Y.; Guo, Y.-Y.; Desneux, N. Widespread adoption of Bt cotton and insecticide decrease promotes biocontrol services. *Nature* **2012**, *487*, 362–365. [CrossRef]
37. Hagenbucher, S.; Wäckers, F.L.; Wettstein, F.E.; Olson, D.M.; Ruberson, J.R.; Romeis, J. Pest trade-offs in technology: Reduced damage by caterpillars in Bt cotton benefits aphids. *Proc. Roy. Soc. B Biol. Sci.* **2013**, *280*, 20130042. [CrossRef]
38. Ma, H.; Zhao, M.; Wang, H.-Y.; Wang, Z.-M.; Wang, Q.; Dong, H.-Z. Comparative incidence of cotton spider mites on transgenic Bt versus conventional cotton in relation to contents of secondary metabolites. *Arthropod. Plant Interact.* **2014**, *8*, 1–7. [CrossRef]
39. Svobodova, Z.; Shu, Y.-H.; Habustova, O.S.; Romeis, J.; Meissle, M. Stacked Bt maize and arthropod predators: Exposure to insecticidal Cry proteins and potential hazards. *Proc. Roy. Soc. B Biol. Sci.* **2017**, *284*, 20170440. [CrossRef]
40. New, T.R. The biology of Chrysopidae and Hemerobiidae (*Neuroptera*), with reference to their usage as biocontrol agents: A review. *Ecol. Entomol.* **1975**, *127*, 115–140. [CrossRef]
41. Icoz, I.; Stotzky, G. Fate and effects of insect-resistant Bt crops in soil ecosystems. *Soil Biol. Biochem.* **2008**, *40*, 559–586. [CrossRef]
42. Romeis, J.; Bartsch, D.; Bigler, F.; Candofi, M.P.; Hartley, S.E.; Hellmich, R.L.; Huesing, J.E.; Jepson, P.C.; Layton, R.; Quemada, H.; et al. Assessment of risk of insect-resistant transgenic crops to nontarget arthropods. *Nat. Biotech.* **2008**, *26*, 203–208. [CrossRef]
43. Meissle, M.; Zünd, J.; Waldburger, M.; Romeis, J. Development of *Chrysoperla carnea* (Stephens) (Neuroptera: *Chrysopidae*) on pollen from Bt-transgenic and conventional maize. *Sci. Rep.* **2015**, *4*, 5900. [CrossRef]
44. Hassanpour, M.; Mohaghegh, J.; Iranipour, S.; Nouri-Ganbalani, G.; Enkegaard, A. Functional response of *Chrysoperla carnea* (Neuroptera: *Chrysopidae*) to *Helicoverpa armigera* (Lepidoptera: *Noctuidae*): Effect of prey and predator stages. *Insect Sci.* **2011**, *18*, 217–224. [CrossRef]
45. Obrycki, J.J.; Hamid, M.N.; Sajap, A.S.; Lewis, L.C. Suitability of corn insect pests for development and survival of *Chrysoperla carnea* and *Chrysopa oculata* (Neuroptera: *Chrysopidae*). *Environ. Entomol.* **1989**, *18*, 1126–1130. [CrossRef]
46. Klingen, I.; Johansen, N.S.; Hofsvang, T. The predation of *Chrysoperla carnea* (Neurop., *Chrysopidae*) on eggs and larvae of *Mamestra brassicae* (Lep., *Noctuidae*). *J. Appl. Entomol.* **1996**, *120*, 363–367. [CrossRef]
47. Meier, M.S.; Hilbeck, A. Influence of transgenic *Bacillus thuringiensis* corn-fed prey on prey preference of immature *Chrysoperla carnea* (Neuroptera: *Chrysopidae*). *Basic Appl. Ecol.* **2001**, *2*, 35–44. [CrossRef]
48. Lawo, N.C.; Wäckers, F.L.; Romeis, J. Characterizing indirect prey-quality mediated effects of a Bt crop on predatory larvae of the green lacewing, *Chrysoperla carnea*. *J. Insect Physiol.* **2010**, *56*, 1702–1710. [CrossRef]
49. Naranjo, S.E.; Hellmich, R.L.; Romeis, J.; Shelton, A.M.; Velez, A.M. The role and use of genetically engineered insect-resistant crops in integrated pest management systems. In *Integrated Management of Insect Pests: Current and Future Developments*; Kogan, M., Higley, L., Eds.; Burleigh Dodds Science Publishing: London, UK, 2019; pp. 283–340. [CrossRef]
50. Kos, M.; van Loon, J.J.A.; Dicke, M.; Louise, E.M.; Vet, L.E.M. Transgenic plants as vital components of integrated pest management. *Trends Biotechnol.* **2009**, *27*, 621–627. [CrossRef]
51. Anderson, J.A.; Ellsworth, P.C.; Faria, J.C.; Head, G.P.; Owen, M.D.K.; Pilcher, C.D.; Shelton, A.M.; Meissle, M. Genetically Engineered Crops: Importance of Diversified Integrated Pest Management for Agricultural Sustainability. *Front. Bioeng. Biotechnol.* **2019**, *7*, 24. [CrossRef]
52. Jia, S.-R.; Guo, S.-D.; An, D.-C. *Transgenic Bt Cotton*; Science Press: Beijing, China, 2001. (In Chinese)
53. Zhang, F.; Wang, S.-Q.; Luo, C.; Chen, Y.-H.; Li, F. Effects of the artificial diets and breeding means on growth and development of *Chrysopa septempunctata* Wesmael. *Plant Prot.* **2004**, *30*, 36–40, (In Chinese with English abstract).
54. Romeis, J.; Dutton, A.; Bigler, F. *Bacillus thuringiensis* toxin (Cry1Ab) has no direct effect on larvae of the green lacewing *Chrysoperla carnea* (Stephens) (Neuroptera: *Chrysopidae*). *J. Insect Physiol.* **2004**, *50*, 175–183. [CrossRef]

plants

Article

Transgenic Maize Has Insignificant Effects on the Diversity of Arthropods: A 3-Year Study

Zhentao Ren [1,2], Muzhi Yang [1,2], Haopeng He [2], Yanjie Ma [2], Yijun Zhou [2], Biao Liu [3,*] and Kun Xue [1,2,3,*]

1 Country Key Laboratory of Ecology and Environment in Minority Areas, Minzu University of China, National Ethnic Affairs Commission, Beijing 100081, China
2 College of Life and Environmental Sciences, Minzu University of China, Beijing 100081, China
3 Nanjing Institute of Environmental Sciences, Ministry of Ecology and Environment, Nanjing 210042, China
* Correspondence: liubiao@nies.org (B.L.); xuekun@muc.edu.cn (K.X.)

Abstract: In order to provide more evidence for the evaluation of the ecological risks of transgenic maize, arthropod population dynamics and biodiversity in fields planted with two kinds of transgenic maize (DBN9868, expressing the *PAT* and *EPSPS* genes, and DBN9936, expressing the *Cry1Ab* and *EPSPS* gene) were investigated by direct observation and trapping for three years. The recorded arthropod species belonged to 19 orders and 87 families, including Aphidoidea, Chrysomelidae, Coccinellidae, Chrysopidae and Araneae. The species richness, Shannon–Wiener diversity index, Pielou evenness index, dominance index and community similarity index of arthropod communities in maize fields were statistically analyzed, and the results showed that (1) the biodiversity difference of arthropod communities between transgenic maize and non-transgenic maize was smaller than that between different conventional cultivars; (2) the differences between ground-dwelling arthropod communities were less obvious than those between plant-inhabiting arthropod communities; and (3) Lepidoptera, the target pests of Bt maize, were not the dominant population in maize fields, and the dominant arthropod population in maize fields varied greatly between years and months. Combining those results, we concluded that the transgenic maize DBN9868 and DBN9936 had no significant effect on the arthropod communities in the field.

Keywords: genetically modified organisms; maize; Bt; corn; biodiversity; arthropods; biosafety

Citation: Ren, Z.; Yang, M.; He, H.; Ma, Y.; Zhou, Y.; Liu, B.; Xue, K. Transgenic Maize Has Insignificant Effects on the Diversity of Arthropods: A 3-Year Study. *Plants* **2022**, *11*, 2254. https://doi.org/10.3390/plants11172254

Academic Editors: Wei Wei and C. Neal Stewart Jr.

Received: 27 June 2022
Accepted: 26 August 2022
Published: 30 August 2022

1. Introduction

Genetically modified (GM) crops have been used commercially for more than 20 years. GM maize is one of the most adopted GM crops in the world, with 60.9 million hectares planted in 2019, which is 31% of the global planted area (https://www.isaaa.org/ (accessed on 1 May 2022)). In 2020, three biosafety certificates for GM maize were granted by the Ministry of Agriculture, PR China. Although GM crops have been commercially planted at a large scale in recent decades, their biosafety is still controversial in certain countries [1,2]. One concern about GM crops is the potential negative impact on the diversity and abundance of non-target organisms and, subsequently, on ecosystem functions [3–5]. In maize field ecosystems, non-target arthropods provide important ecological functions such as biological control, recycling of organic matter, and pollination [6,7]. Prior to biosafety certification being granted, local field trials are required to assess the potential adverse effects of GM maize on agroecosystems, including arthropod communities, relative to a non-GM control.

A growing number of studies have revealed the non-target impact of GM maize on arthropod biodiversity in the field. To date, most studies have shown that insect-resistant and herbicide-tolerant traits, either single events or stacked products, do not adversely affect biodiversity, the populations of natural enemies, or other ecologically important non-target arthropods [6,8–16]. In recent years, some long-term studies have also supported this point of view. For example, a farm-scale evaluation of the impact of Bt maize (*Cry1Ab*) on non-target arthropods through a three-year study was reported, indicating that Bt maize had no negative effect on hemipteran herbivores (leafhoppers and planthoppers) and hymenopteran parasitic wasps (mymarids) [17]. A three-year study showed that Bt maize (*Cry1Ac*) had no significant negative effects on selected non-target arthropod diversity in the field [18]. Another three-year study also showed that Bt maize (*Cry1Ab* and *Cry2Ab*) was compatible with non-target arthropods [19]. Most studies compared transgenic maize with its parental line, but non-transgenic controls of different varieties were not included in the comparison.

Evaluation of GM crops follows the principles of precaution, familiarity and substantial equivalence, which are concepts arising from the knowledge and experience gained over time [20]. They are based on a series of comparative assessment approaches that consider the GM crop in the context of appropriate non-GM control and the known variation of the crop [21]. The mean values of the key characteristics of the GM crops are assessed against the range of values of the reference varieties or from the literature in the context of known values common for the crop [22]. Those comparative assessments can reveal the potential risks of the cultivation of GM crops [23,24]. This study uses comparative assessment approaches to evaluate the effects of GM crops on arthropod communities in the field. If the mean value of the arthropod diversity of a GM crop is beyond the value range of the cultivated crops, the detected differences are then used to assess whether they could potentially alter arthropod communities.

In this study, the arthropod biodiversity in plots with four kinds of maize was investigated in Northeast China from 2015 to 2017. The maize cultivars were commercial maize XianYu335, non-transgenic maize DBN318, transgenic maize DBN9868 providing tolerance to herbicides, and transgenic maize DBN9936 providing tolerance to herbicides and resistance to lepidopteran pests. In order to evaluate the impact of transgenic maize on arthropod communities, species richness (S), Shannon–Wiener index (H), dominance index (C), Pielou evenness index (J) and community similarity index (C') of plant-inhabiting and ground-dwelling arthropods were analyzed. We had the following objectives: (1) to compare the effects of different types of transgenic maize on arthropod communities in the field; (2) to compare the effects of maize on arthropod communities on the plant and on the land; (3) to determine the dynamics of arthropod communities in maize fields in different months and years. The results will provide a statistical baseline and theoretical basis for the ecological risk assessment of transgenic crops.

2. Results

2.1. Composition of Arthropod Communities

The arthropod species in maize plots were investigated throughout the growing season in 2015, 2016 and 2017. A total of 89 families from 19 orders were identified and recorded, including insects, spiders and other arthropods (Tables S1 and S2).

A simplified food web, including the main plant-inhabiting and ground-dwelling arthropod species in maize fields, is shown in Figure 1. During all stages of the maize, *Rhopalosiphum maidis* and *Monolepta hieroglyphica* were the main plant-inhabiting herbivores. Lepidoptera herbivores rarely occurred, mainly including the Asian corn Borer (*Ostrinia furnacalis*), the cotton bollworm (*Helicoverpa armigera*) and the oriental armyworm (*Mythimna separata*). Natural enemies included predators and parasitoids. Ladybeetles (*Harmonia axyridis* and *Propylaea japonica*), minute pirate bugs (*Orius sauteri*), lacewings (*Chrysopa pallens*) and spiders (*Misumenops tricupiuatus* and *Synema globosum*) were the most frequent predators on maize. Meanwhile, parasitoids were difficult to find. The detritivores included fruit flies and psocids. Ants are omnivorous insects that protect aphid populations and repel ladybeetles on maize.

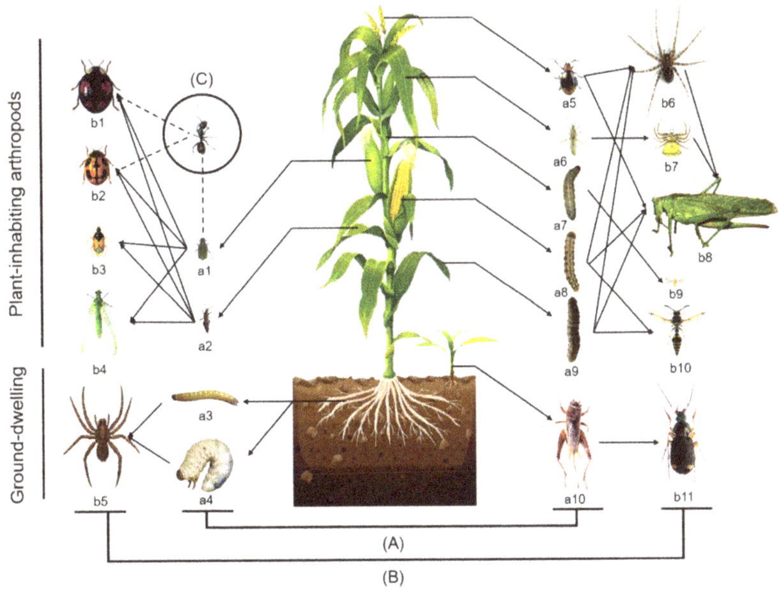

Figure 1. Important arthropod species in the food webs in maize field in Yitong, northeast China. (**A**) The herbivores included *Rhopalosiphum maidis* (**a1**), *Haplothrips aculeatus* (**a2**), *Monolepta hieroglyphica* (**a5**), *Trigonotylus ruficornis* (**a6**), *Ostrinia furnacalis* (**a7**), *Helicoverpa armigera* (**a8**), *Mythimna separata* (**a9**), eating leaves and buds, and wireworms (Coleoptera: Elateridae) (**a3**), *Anomala corpulenta* (**a4**), *Teleogryllus emma* (**a10**), eating roots and stems; (**B**) the predators and parasites were *Harmonia axyridis* (**b1**), *Propylaea japonica* (**b2**), *Orius sauteri* (**b3**), *Chrysopa pallen* (**b4**), *Pirata subpiraticus* (**b5**), funnel-web spiders (Araneida: Agelenidae) (**b6**), *Misumenops tricupiuatus* (**b7**), *Tettigonia chinensis* (**b8**), *Trichogramma ostriniae* (**b9**), potter wasps (Hymenoptera: Vespidae) (**b10**) and *Chlaenius bioculatus* (**b11**); (**C**) another related arthropod was *Camponotus japionicus*. The solid arrows indicate predation relationships; the dashed arrows indicate regulatory relationships.

The number of species of ground-dwelling arthropods collected in pitfall traps was less than that of plant-inhabiting arthropods (Figure 1). The cricket (*Teleogryllus emma*) was the most abundant ground-dwelling herbivorous species. The ground-dwelling predators included the wolf spider (*Pirata subpiraticus*) and ground beetles (*Chlaenius bioculatus* and *Dolichus halensis*). The main scavenging insects on the ground were springtails (Collembola).

The cumulative number (summed data from all sampling dates) of arthropods over the growing season was analyzed (Figure 2, Table S3). The most abundant herbivore groups were the aphids (Aphidoidea) and leaf beetles (Chrysomelidae), and important predator groups were ladybeetles (Coccinellidae), lacewings (Chrysopidae) and spiders (Araneae). The abundance of these groups varied highly between different years. For example, in 2015, the number of leaf beetles was the highest, and that of aphids was the second. However, this was reversed in 2016 and 2017, and the leaf beetle population almost disappeared in 2017.

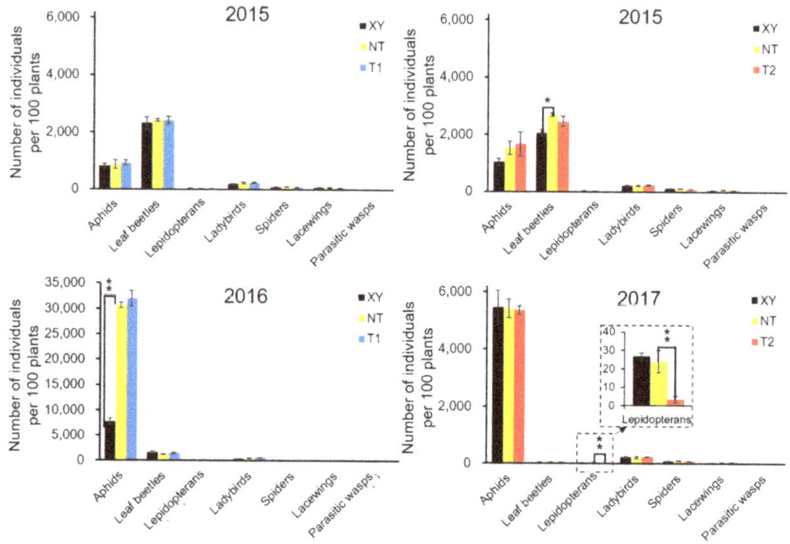

Figure 2. From 2015 to 2017, the cumulative number of main arthropod groups in the field of cultivated variety XianYu335 (XY), non-transgenic recipient variety DBN318 (NT), transgenic maize DBN9868 (T1) and DBN9936 (T2) over the growth season. The Y-axis shows the cumulative number of individuals per 100 plants during the growth period, and the X-axis shows the arthropod groups. We only compared XY with NT and NT with T1 or T2 by using a Student's *t*-test. Asterisks denote significant differences, where $p < 0.05$ (*), $p < 0.01$ (**).

A *t*-test was conducted to compare the total number of main arthropod groups in the plots of XY and NT, NT and T1, NT and T2 (Figure 2). In 2015, Chrysomelidae abundance on NT was significantly higher than that on XY ($p = 0.047$). In 2016, the Aphidoidea abundance on NT was significantly higher than that on XY ($p = 0.0001$). In 2017, Lepidopteran abundance on NT was significantly higher than that on T2 ($p = 0.005$). Other groups showed no significant difference.

2.2. The biodiversity Indices of Arthropods

Four indices were used for arthropod biodiversity analysis. The species richness (*S*) directly represents the biodiversity in an agroecosystem. The Shannon–Wiener biodiversity index (*H*) reflects species richness and community structure information, which is sensitive to changes in the rare species of a sample [25]. The Pielou evenness index (*J*) describes the relative abundance or ratio of individuals in certain species to the total number of individuals, which is often used in combination with *H* [26]. The dominance index (*C*), based on the Simpson's index, is sensitive to changes in species abundance in the sample and less sensitive to species richness [27,28].

The *S* presented similar dynamics each year, regardless of the type of maize (Figure 3). The dynamics of *S* of the arthropod community in the plots between XY and NT, NT and T1, NT and T2 were compared using a *t*-test. The *S* of plant-inhabiting arthropods on NT was significantly different from that on XY for each pair of comparisons (26 August 2015, $p = 0.047$; 23 June 2016, $p = 0.016$). On 1 September 2016, the *S* of plant-inhabiting arthropods on NT was significantly higher than that on T1 ($p = 0.013$). In 2017, on sample date 7 September ($p = 0.002$), the *S* of plant-inhabiting arthropods on NT was significantly higher than that on T2. In 2015, on the sample dates of 21 July ($p = 0.016$) and 16 September ($p = 0.047$), the *S* of plant-inhabiting arthropods on XY was significantly higher than that on

NT. There were no significant differences between the NT and XY, NT and T1, NT and T2 plots in terms of the S of ground-dwelling arthropods at each sampling time.

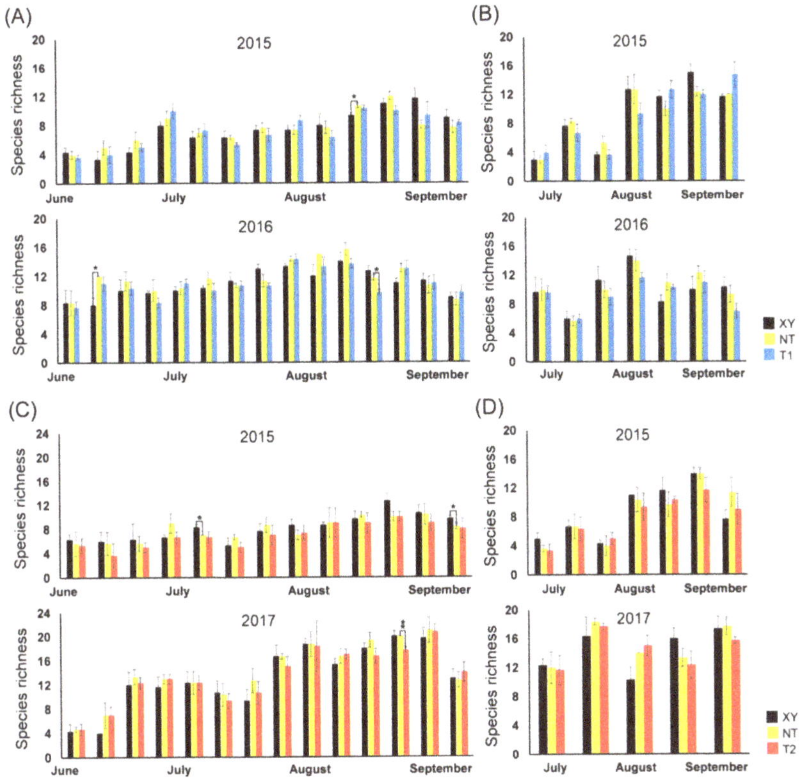

Figure 3. From 2015 to 2017, the species richness (S) dynamics of the arthropod community in the plots of cultivated variety XianYu335 (XY), non-transgenic recipient variety DBN318 (NT), transgenic maize DBN9868 (T1) and DBN9936 (T2). (**A**) The S of plant-inhabiting arthropods in XY, NT and T1; (**B**) the S of ground-dwelling arthropods in XY, NT and T1; (**C**) the S of plant-inhabiting arthropods in XY, NT and T2; (**D**) the S of ground-dwelling arthropods in XY, NT and T2. Y-axis: Mean ± SE (n = 3) of S at each sampling point. We only compared XY with NT and NT with T1 or T2 by using a Student's t-test. Asterisks denote significant differences, where $p < 0.05$ (*), $p < 0.01$ (**).

In this study, H also had similar dynamics each year, regardless of the maize varieties. There were eight significant differences between XY and NT on different sampling dates and no significant differences between NT and transgenic varieties (Figure 4).

For the Pielou evenness index, there were eight significant differences and two extremely significant differences between XY and NT in Student's t-test, while there were two significant differences between NT and T1 and NT and T2, respectively (Figure S1). The C presented similar dynamics each year, regardless of the maize varieties. For the dominance index, there were seven significant differences that only existed in comparisons between XY and NT with Student's t-test (Figure S2).

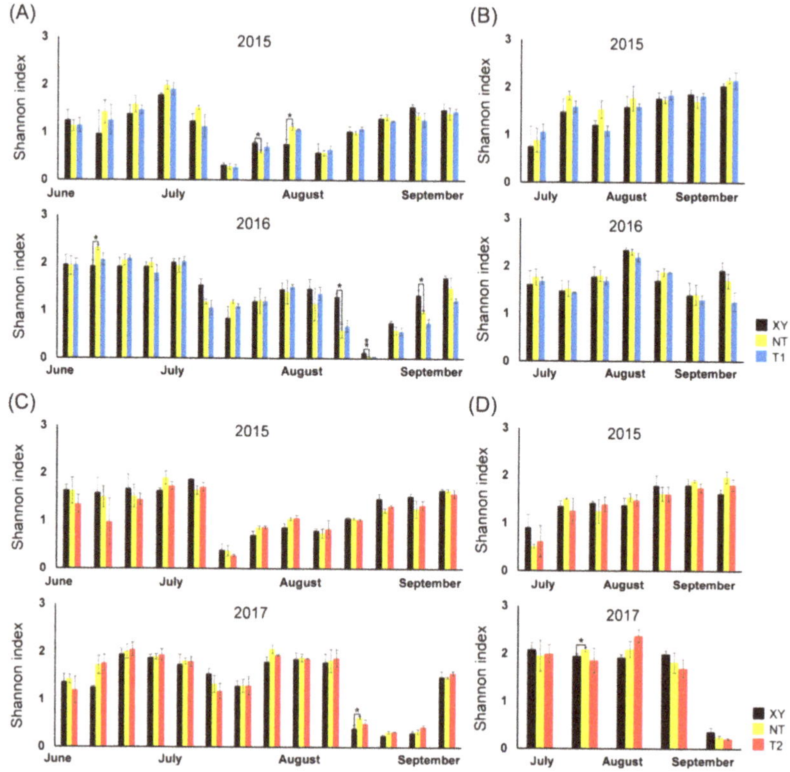

Figure 4. From 2015 to 2017, dynamics of Shannon–Wiener biodiversity index (*H*) of the arthropod community in the plots of cultivated variety XianYu335 (XY), non-transgenic recipient variety DBN318 (NT), transgenic maize DBN9868 (T1) and DBN9936 (T2). (**A**) The *H* of plant-inhabiting arthropods in XY, NT and T1; (**B**) the *H* of ground-dwelling arthropods in XY, NT and T1; (**C**) the *H* of plant-inhabiting arthropods in XY, NT and T2; (**D**) the *H* of ground-dwelling arthropods in XY, NT and T2. Y-axis: Mean ± SE (n = 3) of *H* at each sampling point. We only compared XY with NT and NT with T1 or T2 by using a Student's *t*-test. Asterisks denote significant differences, where $p < 0.05$ (*), $p < 0.01$ (**).

2.3. The Community Similarity Index of Arthropods

The community similarity index (C') appears to be the best approximation of a linear measure of community relationship [29]. Thus, if the same species were in two communities, the C' index would be 1.0; if there were no species in common between the two communities, the C' index would be 0.0. The range from no resemblance to complete identity is appropriately covered by the range from 0.0 to 1.0. The closer the C' index is to 1.00, the more similar the two communities are. The C' of arthropod communities between XY and NT, NT and T1, and NT and T2 from 2015 to 2017 was analyzed (Figure 5). On the plants, 67% of the C' values of "XY vs. NT" were lower than the C' values of "NT vs. T (including T1 and T2)", and the difference between them was significant according to a chi-square test (n = 52, χ^2 = 6.231, p = 0.013). On the ground, 65% of the C' values of "XY vs. NT" were lower than the C' values of "NT vs. T", and the difference between them was not significant according to a chi-square test (n = 26, χ^2 = 2.462, p = 0.117). In total, the community similarity of arthropod between transgenic maize and non-transgenic maize (NT vs. T1; NT vs. T2) was smaller than that between different conventional cultivars (XY vs. NT).

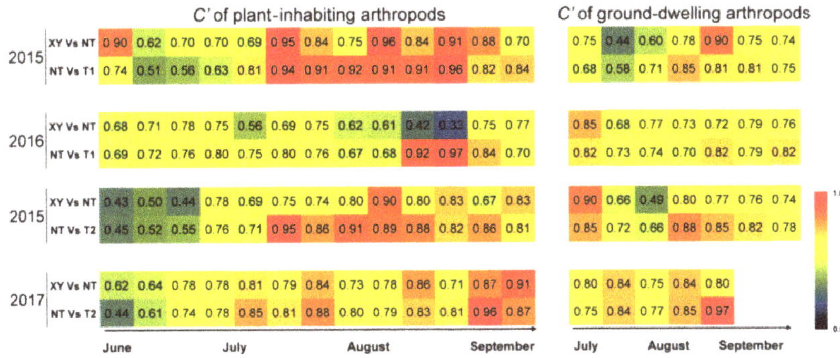

Figure 5. Heat maps of the community similarity index (C') of arthropods in the plots of cultivated variety XianYu335 (XY), non-transgenic recipient variety DBN318 (NT), transgenic maize DBN9868 (T1) and DBN9936 (T2) from 2015 to 2017. Blue indicates a low value (0.0) of the community similarity index and red indicates a high value (1.0).

3. Discussion

There was no evidence to suggest that transgenic crops are toxic to non-target arthropods in the field [10,30–33]. Our study showed that the biodiversity difference of arthropod communities between transgenic maize and non-transgenic maize was smaller than that between different conventional cultivars.

3.1. The Effects of Different Maize Traits on Arthropods

We hypothesized that the differences between arthropod communities between the different maize lines were not caused by transgenes but by differences in agronomic and phenotypic characteristics. For example, there were obvious differences in the characteristics of the husks of the ear between XY and NT. The husks of XY are relatively tight, while the husks of NT are loose (the looseness means aphids can get into the husks to feed, while their natural enemies, ladybeetles, cannot), which means aphids are outside of the control of ladybeetles and outbreaks occur more frequently. Some studies reported that slightly different agronomic and phenotypic characteristics were detected in the transgenic maize hybrids compared to conventional maize hybrids, but none of these phenotypic differences were expected to contribute to biological or ecological changes that would result in increased pest potential or ecological risks [34]. Therefore, the few significant differences in arthropod biodiversity between GM and non-GM maize plots do not indicate that GM crops have an adverse effect on the ecosystem.

3.2. The Effects of GM Maize on Arthropods

If the Bt protein is toxic to non-targets, we would expect those effects in Bt maize. For herbicide-tolerant maize, no effects on non-targets are hypothesized. In this study, T1 (*PAT* and *EPSPS* genes) and T2 (*Cry1Ab* and *EPSPS* genes) did not lead to significant differences in non-target arthropod biodiversity. As a result, Bt effects are unlikely to be present. Most of the plant-inhabiting arthropods (lepidopterans, leaf beetles, ladybeetles, aphids, stink bugs, biting insects, mites and thrips) are able to ingest the insecticide Bt Cry toxins expressed in GM crops [35–39]. Some studies have shown the presence of Cry toxins in some ground-dwelling arthropods (ground beetles, spider mites and Diptera larvae) through direct consumption of the maize or its residues [40–42]. However, the available literature does not provide evidence to support the claim that unexpected interactions between different Bt proteins may lead to adverse effects on non-target species [43]. Our results also demonstrated that arthropod communities are not affected by maize expressing *Cry1Ab*, *EPSPS* and *PAT*. This conclusion is consistent with the results of other studies on the effect of Bt maize on arthropods [6,44–50]. In addition, this study showed that the effect of different maize types on ground-dwelling arthropods was less than that on plant-inhabiting arthropods. This may be because many ground-dwelling arthropods, such as springtails, crickets and ground beetles, depend more on the soil or a weedy environment than on maize plants.

Our results are consistent with other field studies that have reported that the abundance of arthropod species varies greatly between years and months [17,18], which might be a consequence of biotic and abiotic elements in complex agricultural ecosystems, including food uniformity, temperature variation, rainfall distribution and the presence of natural enemies [17,51]. In our study, there was no long-term stable food web in maize-based farmland ecosystems during the growing period. Compared with the significant effects of a complex climate, the effects of GM maize on arthropod communities in the field appear negligible.

3.3. The Effects of GM Maize on the Food Web

Agroecosystems are simple but vulnerable to environmental factors. Multitrophic interactions in communities and food webs are temporary [52]. Food web analyses can provide comprehensive information to understand the effect of GM crops on the structure of ecological communities [12]. Bt maize, expressing *Cry1Ab* insecticidal proteins, kills corn borer and other lepidopteran target pests. In theory, Bt maize can affect the food chain with lepidopterans. In this study, a very low density of lepidopterans was detected in all plots, less than 1% of herbivores. Some studies showed that the abundance of target lepidopteran European corn borer (*Ostrinia nubilalis*) reached a maximum of 0.2% of herbivores [12]. Thus, the effects of GM maize on the food chains and webs with lepidopterans should be assessed.

Aphids do not belong to the leaf-chewing defoliators because of their special feeding behaviors, sucking phloem sap from the leaves and buds. Compared with lepidopterans, Bt maize had little effect on the food chain of aphids and ladybirds. In this study, the differences in aphid populations between maize varieties (XY and NT) were greater than those between GM maize and its counterpart (T1/T2 and NT), which suggested that Bt maize had little effect on the aphid–ladybird food chain. This explains why Bt maize hardly affects arthropod communities in the field. At the same time, there is the question of whether it is necessary to plant Bt maize in areas with low lepidopteran pest densities.

4. Materials and Methods

4.1. Maize Materials

Four maize varieties, including non-transgenic cultivated variety XianYu335 (XY), non-transgenic recipient variety DBN318 (NT), transgenic maize DBN9868 (T1) and DBN9936 (T2), were provided by DNB Biotech Co. Ltd. (Beijing, China). XY is a commonly cultivated variety in Northeast China. NT is the non-transgenic recipient for both T1 and T2. T1, expressing the *PAT* (Phosphinothricin *N*-acetyltransferase) and *EPSPS* (5-enolpyruvylshikimate-3-phosphate synthase) genes, is a glyphosate- and glufosinate-tolerant variety. T2, expressing the *Cry1Ab* and *EPSPS* genes, is an insect-resistant and glyphosate-tolerant variety. Both DBN9868 (T1) and DBN9936 (T2) were declared safe via safety certificates, and our research played a role in that declaration. The biosafety certificate for DBN9936 (T2) was granted by the Ministry of Agriculture, PR China, on 21 January 2020. The expression levels of Cry1Ab, EPSPS and PAT proteins in maize materials, determined by ELISA, are shown in Table S4, and the methods are described in the Supplementary Materials.

4.2. Planting Plot Design

A set of four independent experiments was conducted on the test base in Yitong (Jilin Province, China) (125°26′18.22″ E; 43°14′8.94″ N) from June to September of each year, 2015 to 2017. No crops were planted in the field until the experiment started. XY, NT and T1 were planted in 2015 and 2016; XY, NT and T2 were planted in 2015 and 2017. Each variety was replicated three times on 150 m^2 (10 m $\times$ 15 m), with corridors of 1.0 m between them (the plots design is shown in Figure S3). The experiment was carried out with a randomized blocks design. Maize was sown with a between-row spacing of 60 cm and a within-row spacing of 25 cm. No herbicides or pesticides were applied during the growth period of maize.

4.3. Investigation of Arthropods

The diagonal five-point sampling method was used to determine sampling sites. At each sampling site, 10 maize plants were investigated.

The arthropods on the plants were counted by direct observation. The investigations on arthropods were taken once a week from the 50th day after cultivation to harvest. The families, number and developmental stages of all the arthropods on plants were recorded. Some species that occur in large numbers were counted and collected for later identification in the laboratory. In addition, the food consumption behaviors of arthropods were observed in order to map the food web.

The ground-dwelling arthropods were collected with pitfall traps. Samples were taken once every 14 days from the 50th day after planting to harvest. There were five sample sites within each plot and three traps within each sample site. The traps were made of plastic cups (Ø7 cm $\times$ 8 cm) holding one-third solution (5% detergent + 10% alcohol + H_2O). The traps were embedded in the ground, and the rim of the cup was flush with the ground. Arthropods in the traps were collected the next day and taken to the laboratory for identification and counting.

4.4. Statistical Analyses

Five indices were used to analyze the diversity of the arthropods: the species richness (*S*), the Shannon–Wiener diversity index (*H*), the Pielou evenness index (*J*), the dominance index (C) and the community similarity index (C') [25,26,28,53].

The *S* is the number of species within a defined region.

The *H* of the arthropods was calculated using Equation.

$$H = -\sum_{i=1}^{S} P_i ln P_i$$

The *J* of the arthropods was calculated as per equation.

$$J = \frac{H}{lnS}$$

The C of the arthropods was calculated as per equation.

$$C = \sum_{i=1}^{n} \left(\frac{N_i}{N}\right)^2$$

The C' of the arthropods was calculated as per equation.

$$C' = 2w/(a+b)$$

Pi is the proportion of individuals belonging to a certain species, genus or family to the total number of individuals. *S* is the total number of species in a community. *Ni* is the number of individuals belonging to a certain species, genus or family and *N* is the total number of individuals in the region. *w* is the number of species shared by two samples, and *a* and *b* are the total numbers of species in each of the respective samples. Since it was not possible to identify all species in the field, the number of families was analyzed.

The mean values and SE of each community were calculated and compared using Student's t-test to detect significant differences between XY and NT, NT and T1, and NT and T2. A significance level of $p < 0.05$ was used.

Supplementary Materials: The following supporting information can be downloaded at: https://www.mdpi.com/article/10.3390/plants11172254/s1, Table S1: Sampling dates from 2015 to 2017, Table S2: Orders, family and parts of identified patterns of the arthropod community in maize field, Table S3: From 2015 to 2017, the cumulative number of main arthropod groups in the field of cultivated variety XianYu335 (XY), non-transgenic recipient variety DBN318 (NT), transgenic maize DBN9868 (T1) and DBN9936 (T2), Table S4: The expression levels of Cry1Ab, EPSPS and PAT proteins in maize materials determined by ELISA (μg/g·dwt), Figure S1: From 2015 to 2017, Dynamic of Pielou evenness index (J) of arthropod community in the fields of cultivated variety XianYu335 (XY), non-transgenic recipient variety DBN318 (NT), transgenic maize DBN9868 (T1) and DBN9936 (T2), Figure S2: From 2015 to 2017, Dynamic of the dominance index (C) of arthropod communities in the fields of cultivated variety XianYu335 (XY), non-transgenic recipient variety DBN318 (NT), transgenic maize DBN9868 (T1) and DBN9936 (T2), Figure S3: Plots design in 2015.

Author Contributions: Methodology, B.L. and K.X.; software, Z.R.; investigation, Z.R., M.Y., H.H. and Y.M.; data curation, Z.R.; writing—original draft preparation, Z.R.; writing—review and editing, Y.Z., B.L. and K.X.; visualization, Z.R.; project administration, B.L. and K.X.; funding acquisition, Y.Z., B.L. and K.X. All authors have read and agreed to the published version of the manuscript.

Funding: This research was funded by Key Laboratory of Ecology and Environment in Minority Areas (Minzu University of China), National Ethnic Affairs Commission (No. KLEEMA202102) and National Project on New Transgenic Crops Breeding (No. 2021ZX08013001-002-007).

Data Availability Statement: Data is contained within the Supplementary Material. The data presented in this study are available in [Supplementary Material 2].

Conflicts of Interest: The authors declare no conflict of interest.

References

1. Lucht, J.M. Public Acceptance of Plant Biotechnology and GM Crops. *Viruses* **2015**, *7*, 4254–4281. [CrossRef]
2. Albajes, R.; Lumbierres, B.; Pons, X.; Comas, J. Representative taxa in field trials for environmental risk assessment of genetically modified maize. *Bull. Entomol. Res.* **2013**, *103*, 724–733. [CrossRef]
3. Carpenter, J.E. Impact of GM crops on biodiversity. *GM Crops* **2011**, *2*, 7–23. [CrossRef] [PubMed]
4. Truter, J.; Van Hamburg, H.; Van Den Berg, J. Comparative diversity of arthropods on Bt maize and non-Bt maize in two different cropping systems in South Africa. *Environ. Entomol.* **2014**, *43*, 197–208. [CrossRef] [PubMed]
5. Eckert, J.; Schuphan, I.; Hothorn, L.; Gathmann, A. Arthropods on maize ears for detecting impacts of Bt maize on nontarget organisms. *Environ. Entomol.* **2006**, *35*, 554–560. [CrossRef]
6. Comas, C.; Lumbierres, B.; Pons, X.; Albajes, R. No effects of Bacillus thuringiensis maize on nontarget organisms in the field in southern Europe: A meta-analysis of 26 arthropod taxa. *Transgenic Res.* **2014**, *23*, 135–143. [CrossRef] [PubMed]
7. EFSA. Guidance on the environmental risk assessment of genetically modified plants. *EFSA J.* **2010**, *8*, 1879. [CrossRef]
8. Ahmad, A.; Negri, I.; Oliveira, W.; Brown, C.; Asiimwe, P.; Sammons, B.; Horak, M.; Jiang, C.; Carson, D. Transportable data from non-target arthropod field studies for the environmental risk assessment of genetically modified maize expressing an insecticidal double-stranded RNA. *Transgenic Res.* **2016**, *25*, 1–17. [CrossRef]
9. Al-Deeb, M.A.; Wilde, G.E. Effect of Bt corn expressing the Cry3Bb1 toxin for corn rootworm control on aboveground nontarget arthropods. *Environ. Entomol.* **2003**, *32*, 1164–1170. [CrossRef]
10. Svobodová, Z.; Shu, Y.; Habuštová, O.S.; Romeis, J.; Meissle, M. Stacked Bt maize and arthropod predators: Exposure to insecticidal Cry proteins and potential hazards. *Proc. R. Soc. B-Biol. Sci.* **2017**, *284*, 20170440. [CrossRef]
11. Wolfenbarger, L.L.; Naranjo, S.E.; Lundgren, J.G.; Bitzer, R.J.; Watrud, L.S. Bt crop effects on functional guilds of non-target arthropods: A meta-analysis. *PLoS ONE* **2008**, *3*, e2118. [CrossRef]
12. Szénási, A.; Pálinkás, Z.; Zalai, M.; Schmitz, O.J.; Balog, A. Short-term effects of different genetically modified maize varieties on arthropod food web properties: An experimental field assessment. *Sci. Rep.* **2014**, *4*, 5315. [CrossRef]
13. De la Poza, M.; Pons, X.; Farinós, G.P.; López, C.; Ortego, F.; Eizaguirre, M.; Castañera, P.; Albajes, R. Impact of farm-scale Bt maize on abundance of predatory arthropods in Spain. *Crop Prot.* **2005**, *24*, 677–684. [CrossRef]
14. Wang, Z.Y.; Wu, Y.; He, K.L.; Bai, S.X. Effects of transgenic Bt maize pollen on longevity and fecundity of Trichogramma ostriniae in laboratory conditions. *Bull. Insectol.* **2007**, *60*, 49–53. [CrossRef]
15. Guo, Y.; Feng, Y.; Ge, Y.; Tetreau, G.; Chen, X.; Dong, X.; Shi, W. The cultivation of Bt corn producing Cry1Ac toxins does not adversely affect non-target arthropods. *PLoS ONE* **2014**, *9*, e114228. [CrossRef]

16. Devos, Y.; De Schrijver, A.; De Clercq, P.; Kiss, J.; Romeis, J. Bt-maize event MON 88017 expressing *Cry3Bb1* does not cause harm to non-target organisms. *Transgenic Res.* **2012**, *21*, 1191–1214. [CrossRef]
17. Arias-Martín, M.; García, M.; Castañera, P.; Ortego, F.; Farinós, G.P. Farm-scale evaluation of the impact of Cry1Ab Bt maize on canopy nontarget arthropods: A 3-year study. *Insect Sci.* **2018**, *25*, 87–98. [CrossRef]
18. Xing, Y.; Qin, Z.; Feng, M.; Li, A.; Zhang, L.; Wang, Y.; Dong, X.; Zhang, Y.; Tan, S.; Shi, W. The impact of Bt maize expressing the Cry1Ac protein on non-target arthropods. *Environ. Sci. Pollut. Res. Int.* **2019**, *26*, 5814–5819. [CrossRef]
19. Yin, Y.; Xu, Y.; Cao, K.; Qin, Z.; Zhao, X.; Dong, X.; Shi, W. Impact assessment of Bt maize expressing the Cry1Ab and Cry2Ab protein simultaneously on non-target arthropods. *Environ. Sci. Pollut. Res. Int.* **2020**, *27*, 21552–21559. [CrossRef]
20. Nickson, T.E.; McKee, M.J. Ecological assessment of crops derived through biotechnology. In *Biotechnology and Safety Assessment*; Thomas, J.A., Fuchs, R.L., Eds.; Elsevier: Amsterdam, The Netherlands, 2003; pp. 233–252.
21. Nickson, T.E.; Horak, M.J. Assessing familiarity: The role of plant characterization. In Proceedings of the 9th International Symposium on the Biosafety of Genetically Modified Organisms, Jeju Island, Korea, 24–29 September 2006; Roberts, A., Ed.; International Society for Biosafety Research: Saskatoon, SK, Canada, 2006; pp. 76–80.
22. Nakai, S.; Hoshikawa, K.; Shimono, A.; Ohsawa, R. Transportability of confined field trial data from cultivation to import countries for environmental risk assessment of genetically modified crops. *Transgenic Res.* **2015**, *24*, 929–944. [CrossRef]
23. Horak, M.J.; Rosenbaum, E.W.; Woodrum, C.L.; Martens, A.B.; Mery, R.F.; Cothren, J.T.; Burns, J.A.; Nickson, T.E.; Pester, T.A.; Jiang, C.; et al. Characterization of roundup ready flex cotton, 'MON 88913′, for use in ecological risk assessment: Evaluation of seed germination, vegetative and reproductive growth, and ecological interactions. *Crop Sci.* **2007**, *47*, 268–277. [CrossRef]
24. Raybould, A.; Higgins, L.S.; Horak, M.J.; Layton, R.J.; Storer, N.P.; De La Fuente, J.M.; Herman, R.A. Assessing the ecological risks from the persistence and spread of feral populations of insect-resistant transgenic maize. *Transgenic Res.* **2012**, *21*, 655–664. [CrossRef]
25. Shannon, C.E. A mathematical theory of communication. *Bell Syst. Tech. J.* **1948**, *27*, 379–423. [CrossRef]
26. Pielou, E.C. The measurement of diversity in different types of biological collections. *J. Theor. Biol.* **1966**, *13*, 131–144. [CrossRef]
27. Simpson, E.H. Measurement of Diversity. *Nature* **1949**, *163*, 688.
28. Whittaker, R.H. Dominance and Diversity in Land Plant Communities: Numerical relations of species express the importance of competition in community function and evolution. *Science* **1965**, *147*, 250–260. [CrossRef]
29. Bray, J.R.; Curtis, J.T. An ordination of the upland forest communities of southern Wisconsin. *Ecol. Monogr.* **1957**, *27*, 325–349. [CrossRef]
30. Carstens, K.; Anderson, J.; Bachman, P.; De Schrijver, A.; Dively, G.; Federici, B.; Hamer, M.; Gielkens, M.; Jensen, P.; Lamp, W.; et al. Genetically modified crops and aquatic ecosystems: Considerations for environmental risk assessment and non-target organism testing. *Transgenic Res.* **2012**, *21*, 813–842. [CrossRef]
31. Romeis, J.; Meissle, M.; Bigler, F. Transgenic crops expressing Bacillus thuringiensis toxins and biological control. *Nat. Biotechnol.* **2006**, *24*, 63–71. [CrossRef]
32. Duan, J.J.; Marvier, M.; Huesing, J.; Dively, G.; Huang, Z.Y. A meta-analysis of effects of Bt crops on honey bees (Hymenoptera: Apidae). *PLoS ONE* **2008**, *3*, e1415. [CrossRef]
33. Obrist, L.; Dutton, A.; Albajes, R.; Bigler, F. Exposure of arthropod predators to Cry1Ab toxin in Bt maize fields. *Ecol. Entomol.* **2006**, *31*, 143–154. [CrossRef]
34. Díaz, O.H.; Meza, J.L.A.; Baltazar, B.M.; Bojórquez, G.B.; Espinoza, L.C.; Madrid, J.L.C.; de la Fuente Martínez, J.M.; Pompa, H.A.D.; Escobedo, J.A.; Banda, A.E.; et al. Plant characterization of genetically modified maize hybrids MON-89Ø34-3× MON-88Ø17-3, MON-89Ø34-3× MON-ØØ6Ø3-6, and MON-ØØ6Ø3-6: Alternatives for maize production in Mexico. *Transgenic Res.* **2017**, *26*, 135–151. [CrossRef]
35. Torres, J.B.; Ruberson, J.R. Interactions of *Bacillus thuringiensis* Cry1Ac toxin in genetically engineered cotton with predatory heteropterans. *Transgenic Res.* **2008**, *17*, 345–354. [CrossRef]
36. Meissle, M.; Knecht, S.; Waldburger, M.; Romeis, J. Sensitivity of the cereal leaf beetle *Oulema melanopus* (Coleoptera: Chrysomelidae) to Bt maize-expressed Cry3Bb1 and Cry1Ab. *Arthropod-Plant Interact.* **2012**, *6*, 203–211. [CrossRef]
37. Meissle, M.; Kloos, S.; Romeis, J. Fate of multiple Bt proteins from stacked Bt maize in the predatory lady beetle *Harmonia axyridis* (Pallas) (Coleoptera: Coccinellidae). *Environ. Pollut.* **2021**, *268*, 115421. [CrossRef]
38. Zhang, G.F.; Wan, F.H.; Lövei, G.L.; Liu, W.X.; Guo, J.Y. Transmission of Bt toxin to the predator *Propylaea japonica* (Coleoptera: Coccinellidae) through its aphid prey feeding on transgenic Bt cotton. *Environ. Entomol.* **2006**, *35*, 143–150. [CrossRef]
39. Schünemann, R.; Knaak, N.; Fiuza, L.M. Mode of action and specificity of *Bacillus thuringiensis* toxins in the control of caterpillars and stink bugs in soybean culture. *ISRN Microbiol.* **2014**, *8*, 135675. [CrossRef]
40. García, M.; Ortego, F.; Castañera, P.; Farinós, G.P. Effects of exposure to the toxin Cry1Ab through Bt maize fed-prey on the performance and digestive physiology of the predatory rove beetle Atheta coriaria. *Biol. Control.* **2010**, *55*, 225–233. [CrossRef]
41. Knecht, S.; Nentwig, W. Effect of Bt maize on the reproduction and development of saprophagous Diptera over multiple generations. *Basic Appl. Ecol.* **2010**, *11*, 346–353. [CrossRef]
42. Zwahlen, C.; Hilbeck, A.; Howald, R.; Nentwig, W. Effects of transgenic Bt corn litter on the earthworm Lumbricus terrestris. *Mol. Ecol.* **2003**, *12*, 1077–1086. [CrossRef]
43. Romeis, J.; Meissle, M. Stacked Bt proteins pose no new risks to nontarget arthropods. *Trends Biotechnol.* **2020**, *38*, 234–236. [CrossRef] [PubMed]

44. Leslie, T.W.; Hoheisel, G.A.; Biddinger, D.J.; Rohr, J.R.; Fleischer, S.J. Transgenes sustain epigeal insect biodiversity in diversified vegetable farm systems. *Environ. Entomol.* **2014**, *36*, 234–244. [CrossRef]
45. Rose, R.; Dively, G.P. Effects of insecticide-treated and lepidopteran-active Bt transgenic sweet corn on the abundance and diversity of arthropods. *Environ. Entomol.* **2007**, *36*, 1254–1268. [CrossRef] [PubMed]
46. Farinós, G.P.; de la Poza, M.; Hernández-Crespo, P.; Ortego, F.; Castañera, P. Diversity and seasonal phenology of aboveground arthropods in conventional and transgenic maize crops in Central Spain. *Biol. Control.* **2008**, *44*, 362–371. [CrossRef]
47. Habuštová, O.S.; Svobodová, Z.; Spitzer, L.; Doležal, P.; Hussein, H.; Sehnal, F. Communities of ground-dwelling arthropods in conventional and transgenic maize: Background data for the post-market environmental monitoring. *J. Appl. Entomol.* **2015**, *139*, 31–45. [CrossRef]
48. Twardowski, J.P.; Beres, P.K.; Hurej, M.; Klukowski, Z.; Dabrowski, Z.T.; Sowa, S.; Warzecha, R. The quantitative changes of ground beetles (Col., Carabidae) in Bt and conventional maize crop in southern Poland. *J. Plant Prot. Res.* **2012**, *52*, 404–409. [CrossRef]
49. Balog, A.; Kiss, J.; Szekeres, D.; Szenasi, A.; Marko, V. Rove beetle (Coleoptera: Staphylinidae) communities in transgenic Bt (MON810) and near isogenic maize. *Crop Prot.* **2010**, *29*, 567–571. [CrossRef]
50. Szekeres, D.; Kádár, F.; Kiss, J. Activity density, diversity and seasonal dynamics of ground beetles (Coleoptera: Carabidae) in Bt-(MON810) and in isogenic maize stands. *Entomol. Fennica* **2006**, *17*, 269–275. [CrossRef]
51. Guo, J.; He, K.; Hellmich, R.L.; Bai, S.; Zhang, T.; Liu, Y.; Ahmed, T.; Wang, Z. Field trials to evaluate the effects of transgenic cry1Ie maize on the community characteristics of arthropod natural enemies. *Sci. Rep.* **2016**, *6*, 22102. [CrossRef]
52. Arpaia, S.; Di Leo, G.M.; Fiore, M.C.; Schmidt, J.E.U.; Scardi, M. Composition of arthropod species assemblages in Bt-expressing and near isogenic eggplants in experimental fields. *Environ. Entomol.* **2014**, *36*, 213–227. [CrossRef]
53. Motyka, J.; Dobrzanski, B.; Zawadzki, S. Preliminary studies on meadows in the south-east of the province Lublin. *Ann. Univ. Mariae Curie-Sklodowska* **1950**, *5*, 367–447.

plants MDPI

Article

Impact of Transgenic *Cry1Ab/2Aj* Maize on Abundance of Non-Target Arthropods in the Field

Yan Yang [1,2,†], Yi Chen [3,4,5,†], Jiabao Xue [1,3,4], Yuanyuan Wang [2], Xinyuan Song [6] and Yunhe Li [2,*]

1 Key Laboratory of Genetics and Germplasm Innovation of Tropical Special Forest Trees and Ornamental Plants, Ministry of Education, College of Forestry, Hainan University, Haikou 570228, China
2 State Key Laboratory for Plant Diseases and Insect Pests, Institute of Plant Protection, Chinese Academy of Agricultural Sciences, Beijing 100193, China
3 Institute of Tropical Bioscience and Biotechnology, Chinese Academy of Tropical Agricultural Sciences, Haikou 571025, China
4 Sanya Research Institute of the Chinese Academy of Tropical Agricultural Sciences, Sanya 572022, China
5 Hainan Yazhou Bay Seed Laboratory, Sanya 572025, China
6 Agro-Biotechnology Research Institute, Jilin Academy of Agricultural Sciences, Changchun 130033, China
* Correspondence: liyunhe@caas.cn
† These authors contributed equally to this work.

Abstract: Transgenic *Bacillus thuringiensis* (*Bt*) maize has broad prospects for application in China. Before commercialization, it is necessary to assess possible ecological impacts, including impacts on non-target arthropods (NTAs) in the field. In the present study, transgenic *Bt* maize expressing *cry1Ab/2Aj* and its corresponding non-transformed near isoline were planted under the same environmental and agricultural conditions, and arthropods in the field were collected during the three main growth stages of maize. In a one year trial, the results showed the composition of NTA communities in the transgenic and control maize fields were similar. There were no significant differences for community-level parameters of species richness (*S*), Shannon–Wiener diversity index (H'), evenness index (*J*) and Simpson's dominant concentration (*C*) between the two types of maize fields. Likewise, a Bray–Curtis dissimilarity and distance analysis showed that Cry1Ab/2Aj toxin exposure did not increase community dissimilarities between *Bt* and non-*Bt* maize plots and that the structure of the NTAs community was similar on the two maize varieties. Furthermore, planting of the transgenic *cry1Ab/2Aj* maize did not affect the density or composition of non-target decomposers, herbivores, predators, parasitoids and pollinator guilds. In summary, our results showed that planting of *Bt* maize producing Cry1Ab/Cry2Aj proteins do not adversely affect population dynamics and diversity of NTAs.

Keywords: *Bacillus thuringiensis*; non-target arthropods; biodiversity; community composition; ecological safety assessment

Citation: Yang, Y.; Chen, Y.; Xue, J.; Wang, Y.; Song, X.; Li, Y. Impact of Transgenic *Cry1Ab/2Aj* Maize on Abundance of Non-Target Arthropods in the Field. *Plants* **2022**, *11*, 2520. https://doi.org/10.3390/plants11192520

Academic Editors: Wei Wei and C. Neal Stewart, Jr.

Received: 22 August 2022
Accepted: 22 September 2022
Published: 26 September 2022

1. Introduction

The past 23-year planting history (1996–2018) of genetically engineered (GE) crops have demonstrated the socioeconomic benefits of genetically engineered (GE) crops, including: (1) increased productivity and global food and feed security; (2) support for self-sufficiency in a country's arable land; (3) protect biodiversity, prevent deforestation, and protect biodiversity reserves; (4) mitigate challenges related to climate change; and (5) improve economic, health, and social benefits [1,2]. The global area of GM crops has reached 190.4 million hectares in 2019, including over 29 countries [3]. Among them, GE maize is the third most important crop in terms of GE crop area and has reached 31% of the global maize crop area. With more than 24 years of GE maize commercialization in above ten foreign countries, the planting area reached 60.9 million hectares of GE maize in 2019 [3]. In China several GE maize lines expressing insect-resistant or/and herbicide-resistant traits has obtained biosafety certificates, and they will be approved for agricultural production in the near future [4,5]. China is the second largest maize producer and consumer in the world, with a large cultivated area and distribution throughout the whole country. Maize is not only the staple food crop, but also an important raw material for feed and industry. If GE maize is introduced and planted on a large scale, it may change the ecological environment of the original field and the living environment of pests, thus causing a series of safety problems and non-target effects [4].

The impact of *Bt* crops on non-target arthropods (NTAs) is an important part of an ecological risk assessment. Laboratory studies have been extensively conducted to evaluate the impact of *Bt* crops on NTAs, focusing the functional groups that play important ecological roles in farmland systems, such as natural enemies, pollinators, and economically important insects [6–11]. Systematic analyses of these research data found that the Bt insecticidal proteins produced by GE crops have a very narrow insecticidal spectrum, strong insecticidal specificity and generally have no negative effects on non-target arthropods [11–14]. In addition to carrying out bioassay experiments under controlled conditions in laboratories, it is generally required to further conduct field experiments to evaluate whether the planting of GE crops affects the population structure and density of NTAs [15]. Field experiments were conducted to investigate whether there are differences in NTA species, densities and genotypes when GE crops are grown in the field compared with non-GE crops. Such short-term field investigation conducted before the commercialization of GE crops normally show no effects on arthropod populations [15–18]. However, a number of long-term and large-scale monitoring studies conducted after the commercialization of a GE crop do indicate that planting of *Bt* crops can change the arthropod population structure and dynamics in the field, for example, the increase of certain secondary insect pest populations [19–21]. These effects were widely considered to be caused by the change of farmland practices, such as the reduced application of pesticides associated with reduction of target pest populations. For example, transgenic insect-resistant cotton can effectively control cotton bollworm (*Helicoverpa armigera*); therefore, the use of pesticides has been greatly reduced for controlling this pest, which indirectly results in the break of the non-target pest mirid bug [22].

NTA effects need to be conducted on a case-by-case basis using a weight of evidence approach and considering all relevant information [23]. In recent years, China has developed a number of GE maize lines exhibiting high efficacy in target pest control that have to undergo a strict risk assessment before going to commercial use [24]. A transgenic maize line Shuangkang 12-5 (SK12-5), developed through the *Agrobacterium*-mediated method [25,26] that expresses a *cry1Ab/2Aj* fusion gene and an *EPSPS* gene [27], showed efficient field resistance against lepidopteran pests *Ostrinia furnacalis* (Guenée) and *Helicoverpa armigera* (Hübner) and high herbicide tolerance to glyphosate [26]. The GE maize line has already passed regulatory approval and received a safety certification, and it may enter commercial cultivation soon in China. So far, studies have been conducted with the maize line, mainly focusing on the development of detection methods [28,29], resistance on target pests [26,30] and laboratory assessment of the potential effects on non-target arthropods, including honey bee (*Apis mellifera*) [31], silkworm (*Bombyx mori*) [32], green lacewing (*Chrysoperla sinica*) [33] and effects on microbial diversity [34]. Its potential effects on arthropod populations have rarely been evaluated in the field.

In the present study, the transgenic *cry1Ab/2Aj* maize Shuangkang 12-5 (SK12-5) and its corresponding non-transformed near isoline Lianchuang 303 (LC303) [27] were tested. Number of species, abundance and population characteristics of NTAs collected during three main growth periods of maize were compared between the two types of maize fields, and the species diversity of the arthropod community were further calculated and analyzed. The results will complement knowledge of the impact of transgenic *Bt* maize on the safety of NTAs in the field and will provide a reference for decision-making on commercial application of the SK12-5 maize line in China and maybe in other countries.

2. Results

2.1. Species Composition of NTAs in Different Maize Fields

During the BF (before flowering) stage, 45 species of NTAs belonging to 11 orders and 38 families were detected in non-*Bt* maize fields, and 53 species belonging to 10 orders and 41 families were detected in *Bt* maize fields. There were 39 species of NTAs detected simultaneously in both *Bt* and non-*Bt* maize fields (Table 1), and the Czekanowski Community Similarity Coefficient (C_S) of the species composition of the two communities was 0.73 (Table 2). In both *Bt* and non-*Bt* maize fields, Hemiptera and Coleoptera were the dominant orders, accounting for 51.7 % and 52.8 %, and 13.1 % and 12.6 %, in transgenic and non-transgenic maize, respectively (Figure 1). The compositions of NTAs in *Bt* and non-*Bt* maize fields were not significantly different (Student's *t*-test, all $p > 0.05$).

Table 1. Species composition and the temporal dynamics of non-target arthropods (NTAs) in fields planted with transgenic *Bt* maize and non-*Bt* maize in Jilin, China.

Order	Family	Species	Investigation Date					
			BF [1]		DF [2]		AF [3]	
			Non-*Bt* [a]	*Bt* [b]	Non-*Bt* [a]	*Bt* [b]	Non-*Bt* [a]	*Bt* [b]
Araneae	Agelenidae Araneida Clubionidae Hahniidae Linyphiidae Lycosidae Pisauridae Salticidae Theridiidae Thomisidae	Spider	++	++	++	++	++	++
Coleoptera	Carabidae	/	++	++	+	+	+	++
	Chrysomelidae	/	+	++	++	+	+	+
		Aulacophora femoralis (Motschulsky, 1857)	+	+				+
		Monolepta typographica (Weise, 1915)	++	++	++	++	+	+
		Pachnephorus lewisii (Baly, 1878)	+	+				
	Coccinellidae	/			+		+	
		Harmonia axyridis (Pallas, 1773)	++	++	++	++	+++	+++
		Propylaea japonica (Thunberg, 1781)	++	++	++	++	+	+
		Rodolia rufopilosa (Mulsant, 1850)					+	
		Scymnus hoffmanni (Weise, 1879)		+		+		
	Curculionidae	/				+		+
	Elateridae	/						+
		Aeoloderma agnata (Candeze, 1873)				+		
		Melanotus caudex (Lewis, 1879)				+		
	Melolonthidae	*Holotrichia oblita* (Faldermann, 1835)		+		+		+
	Platypodidae	/				+		
	Rutelidae	*Popillia quadriguttata* (Fabricius, 1787)		+				
	Staphylinidae	*Paederus fuscipes* (Curtis, 1826)	+	+	+	++		+
	Tenebrionidae	*Opatrum subaratum* (Faldermann, 1835)	+	+	+	+		
		Tribolium castaneum (Herbst, 1797)	+	+	+	+	+	+
Dermaptera	Labiduridae	Labidura sp.	++	++	++	++	+	+

Table 1. *Cont.*

Order	Family	Species	Investigation Date					
			BF [1]		DF [2]		AF [3]	
			Non-*Bt* [a]	*Bt* [b]	Non-*Bt* [a]	*Bt* [b]	Non-*Bt* [a]	*Bt* [b]
Dermaptera	Labiduridae	Labidura sp.	++	++	++	++	+	+
Diptera	Asilidae	/		+				
	Cecidomyiidae	*Aphidoletes aphidimyza* (Rondani, 1846)		+				
	Chironomidae	/	+					
	Culicidae	/	++	++	+	+	++	++
	Dolichopodidae	/		+				
	Drosophilidae	*Drosophila macquarti* (Wheeler, 1981)	+	+	+		++	++
	Empododae	/				+		
	Muscidae	/	++	++	++	++	++	++
		Musca domestica (Linnaeus, 1758)	+	+	+	+	++	+
	Sarcophagidae	*Sarcophaga melanura* (Meigen, 1826)	+	+	+			+
	Stratiomyidae	*Hermetia illucens* (Linnaeus, 1758)		+				
	Syrphidae	*Episyrphus balteatus* (De Geer, 1776)	+		+		++	++
		Eupeodes corollae (Fabricius, 1794)						+
		Melanostoma scalare (Fabricius, 1794)						+
		Pipizella varipes (Meigen, 1822)				+		+
		Sphaerophoria menthastri (Linnaeus, 1758)						+
	Tabanidae	/		+			+	++
		Tabanus amaenus (Walker, 1848)				+		
		Tabanus signatipennis (Portschinsky, 1887)		+				
		Tabanus sp.	+					+
	Tephritidae	/	++	++	+	+	+	+
Entomobryomorpha	Isotomidae	*Isotomidae* sp.	++	++	++	++	++	++
Ephemeroptera	Baetidae	/	+	+	+	+		
Hemiptera	Anthocoridae	*Orius strigicollis* (Poppius, 1915)	++	++	+	++	+	+
	Aphididae	/			++	++		++
		Rhopalosiphum maidis (Fitch, 1856)	+++	+++	+++	+++	+++	++
		Rhopalosiphum padi (Linnaeus, 1758)	+++	+++	+++	+++	++	++
	Cicadellidae	*Cicadella viridis* (Linnaeus, 1758)			+	+		
		Macropsis notata (Prohaska, 1923)		+				
		Psammotettix striatus (Linnaeus, 1758)	+	+				
	Cydnidae	*Adrisa magna* (Uhler, 1861)	+	+				
	Delphacidae	/						+
		Laodelphax striatellus (Fallén, 1826)	+	+	+	+	+	+
		Sogatella furcifera (Horváth, 1899)	+	+	+	+	+	+
		Trigonotylus ruficornis (Geoffroy in Fourcroy, 1785)	+	++	+	++	+	+
	Miridae	*Adelphocoris* sp.				+		+
		Apolygus lucorum (Meyer-Dur, 1843)			+			
		Cyrtorhinus lividipennis (Reuter, 1885)	+		+	+		
		Lygus pratensis (Linnaeus, 1758)			+			
	Nabidae	*Nabis stenoferus* (Hsiao, 1964)				+		+

Table 1. *Cont.*

Order	Family	Species	Investigation Date					
			BF [1]		DF [2]		AF [3]	
			Non-*Bt* [a]	*Bt* [b]	Non-*Bt* [a]	*Bt* [b]	Non-*Bt* [a]	*Bt* [b]
Hymenoptera	Formicidae	*Lasius fuliginosus* (Latreille, 1798)	++	+++			++	++
	Apidae	*Apis cerana* (Fabricius, 1793)	+	+	+	+		
	Eumenidae	/					++	++
	Megachilidae	/		+			+	++
	Vespidae	/	+		+			
	Aphelinidae							
	Braconidae							
	Eulophidae		++	++	++	++	+	+
	Ichneumonidae	Parasitic Wasp						
	Pteromalidae							
	Scelionidae							
	Trichogrammatidae							
Neuroptera	Chrysopidae	*Chrysopa pallens* (Rambur, 1838)		+				
		Chrysoperla nipponensis (Okamoto, 1914)	+	+	++	++	++	++
	Hemerobiidae	/			+		+	+
		Micromus timidus (Hagen, 1853)			+		+	+
Orthoptera	Acrididae	/	+		+			
	Coenagrionidae	*Ischnura asiatica* (Brauer, 1865)						+
	Gryllidae	*Gryllidae* sp.			+			
Thysanoptera	Aeolothripidae	*Aeolothrips fasciatus* (Linnaeus, 1758)	+	+		+		
	Phlaeothripidae	*Gynaikothrips uzeli* (Zimmermann, 1900)		+				
		Haplothrips aculeatus (Fabricius, 1803)				+		
	Thripidae	*Anaphothrips obscurus* (Müller, 1776)		+	+	+		

"1" BF—Before Flowering stage; "2" DF—During Flowering stage; "3" AF—After Flowering stage; "a" represents the LC303 fields that planted the non-transgenic near isoline maize of SK12-5; "b" represents the SK12-5 fields that planted transgenic *Cry1Ab/2Aj* maize; "+++" denotes dominant species; "++" denotes common species; "+" denotes rare species.

Table 2. Comparison of community parameters for arthropods between fields planted with *Bt* maize or non-*Bt* maize. Values represent means ± SE, $n = 7$ replicates. The differences in the same maize fields among months were analyzed by a one-way ANOVA followed by the Tukey HSD test (all $p > 0.05$). The differences of the community parameters of arthropods between *Bt* and non-*Bt* maize fields were analyzed by a Student's t-test (an asterisk denotes a significant difference, $p < 0.05$).

Parameter of Community	BF [1]		DF [2]		AF [3]	
	Non-*Bt* [a]	*Bt* [b]	Non-*Bt* [a]	*Bt* [b]	Non-*Bt* [a]	*Bt* [b]
Species richness (S)	8.43 ± 3.14	11.71 ± 2.63	10.71 ± 3.60	11.00 ± 2.72	8.00 ± 1.41	10.86 ± 1.90
Shannon-Wiener diversity index (H')	1.10 ± 0.24	1.50 ± 0.35	1.23 ± 0.26	1.53 ± 0.21	1.15 ± 0.18	1.38 ± 0.17
Evenness index (J)	0.63 ± 0.068	0.63 ± 0.096	0.60 ± 0.11	0.70 ± 0.082	0.58 ± 0.060	0.60 ± 0.050
Simpson's dominant concentration (C)	0.47 ± 0.062	0.45 ± 0.15*	0.45 ± 0.11	0.31 ± 0.064	0.48 ± 0.071	0.41 ± 0.057
Czekanowski Community Similarity Coefficient (C_S)	0.73		0.67		0.7	

"1" BF—Before Flowering stage; "2" DF—During Flowering stage; "3" AF—After Flowering stage; "a" represents the LC303 fields that were planted with the non-transgenic near isoline maize of SK12-5; "b" represents the SK12-5 fields that were planted with transgenic *Cry1Ab/2Aj* maize.

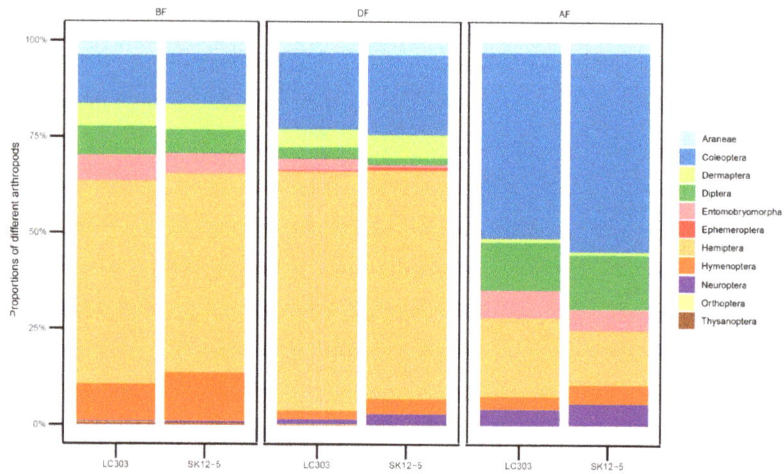

Figure 1. Proportions of all orders of non-target arthropods (NTAs) found in SK12-5 (*Bt*) and LC303 (non-*Bt*) maize fields in three main maize growth stages: BF—Before Flowering stage; DF—During Flowering stage; AF—After Flowering stage.

In the DF (during flowering) stage, the total number of species detected in non-*Bt* maize fields was 47, belonging to 11 orders and 35 families, and in *Bt* maize fields there were 50 species belonging to 10 orders and 39 families. The number of commonly detected species in both *Bt* and non-*Bt* maize fields was 37 (Table 1), and the C_S of the species composition of the two communities was 0.67 (Table 2). In both *Bt* and non-*Bt* maize fields, Hemiptera and Coleoptera were the dominant orders, accounting for 59.5 % and 62.3 %, and 20.7 % and 19.9 %, in transgenic and non-transgenic maize, respectively (Figure 1). NTAs in the order Neuroptera were significantly greater in *Bt* maize fields than in non-*Bt* maize fields (Student's t-test, $F = 36.84$, $p = 0.009$); the composition of other NTAs in *Bt* and non-*Bt* maize fields were not significantly different (Student's *t*-test, all $p > 0.05$). In the AF (after flowering) stage, there were 36 species detected in non-*Bt* maize fields, belonging to 8 orders and 27 families; 51 species were detected in *Bt* maize fields, belonging to 9 orders and 36 families. The number of species detected in both fields were 33 (Table 1), and the C_S of the two communities was 0.70 (Table 2). Further, in both *Bt* and non-*Bt* maize fields, Coleoptera and Hemiptera were the dominant orders, accounting for 51.8 % and 48.3 %, and 14.2 % and 20.4 %, in transgenic and non-transgenic maize, respectively (Figure 1). There were no significant differences in the composition of arthropod species between *Bt* and non-*Bt* maize fields (Student's *t*-test, all $p > 0.05$).

2.2. Community Parameters of NTAs in Different Maize Fields

The NTA community structures in *Bt* and non-*Bt* maize fields were further explored using non-metric multidimensional scaling analysis (NMDS). The distance was estimated using the Bray–Curtis dissimilarity index. Differences in the composition of the NTAs community in the three main maize growth stages were visualized in an NMDS plot (Figure 2). No differences of NTA community structures among *Bt* and non-*Bt* maize fields were observed.

The species richness (S), Shannon–Wiener diversity index (H') and evenness (J) indexes of NTAs in *Bt* maize fields in the BF, DF and AF maize growth stages were all higher than those in non-*Bt* maize fields, while the Simpson's dominant concentration (C) in *Bt* maize fields were all lower than those in non-*Bt* maize fields (Table 2). However, the Simpson's dominant concentration (C) between *Bt* and non-*Bt* maize fields differed significantly in the BF stage (Student's *t*-test, $F = 11.99$, $p = 0.005$); differences in other parameters between *Bt* and non-*Bt* maize fields were not significant (Student's *t*-test, all $p > 0.05$).

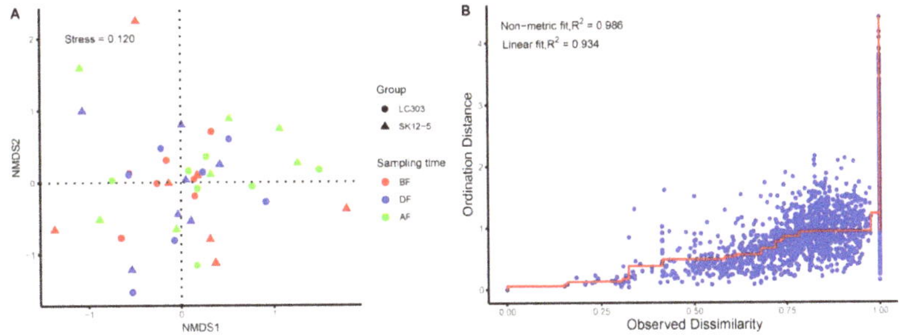

Figure 2. (**A**) Nonmetric Multidimensional Scaling (NMDS based on Bray–Curtis distance) plot of community structures of arthropods for three main growth stages in SK12-5 and LC303 maize fields. Triangles indicate SK12-5 (*Bt*) maize fields, and circles indicate LC303 (non-*Bt*) maize fields. Different colors indicate the three main growth stages: red indicates Before Flowering (BF) stage, blue indicates During Flowering (DF) stage and green indicates After Flowering (AF) stage. (**B**) NMDS Shepard plot.

2.3. Dynamic Comparison of Composition and Community of NTAs at Different Time

In total, the numbers of NTA species in the DF stage were the highest and in the AF stage were the lowest among the three growth stages for both *Bt* and non-*Bt* maize fields; however, there were no significant differences for the three growth stages in both *Bt* and non-*Bt* maize fields (one-way ANOVA followed by Tukey HSD test or Mann–Whitney *U*-test, all $p > 0.05$).

For NTA community parameters (Table 2), the highest species richness (S) was in the DF stage for non-*Bt* maize fields and in the BF stage for *Bt* maize fields; the lowest S values for both *Bt* and non-*Bt* maize fields were in the AF stage. For the Shannon–Wiener diversity index (H'), the highest values for both the *Bt* and non-*Bt* maize fields were in the DF stage, and the lowest for non-*Bt* maize fields in the BF stage and for *Bt* maize fields in the AF stage. The highest evenness index (J) was in the BF stage for non-*Bt* maize fields and in the DF stage for *Bt* maize fields, and the lowest were in the AF stage for both *Bt* and non-*Bt* maize fields. For Simpson's dominant concentration (C), the highest value for non-*Bt* maize fields was in the AF stage and for *Bt* maize fields in the BF stage, and the lowest for both *Bt* and non-*Bt* maize fields in the DF stage. There were no significant differences for the three growth stages among *Bt* and non-*Bt* maize fields (one-way ANOVA followed by Tukey HSD test, all $p > 0.05$).

2.4. Communities Similarity

In the case of multiple surveys and a large number of species, the NMDS model can more accurately reflect the numerical sort of information of the distance matrix. Thus, the similarities of NTAs community composition across maize type and sampling time were visualized using an NMDS based on a Bray–Curtis dissimilarity matrix. A stress function ranged from 0 to 1 was used to assess the goodness of fit between the ordination and the original data of NTAs. The stress values were 0.12, which suggested that the ordination accurately represented the dissimilarity between samples (Figure 2A). A Shepard diagram of non-metric fit illustrated that observed dissimilarities and the ordination distances were highly correlated (non-metric fit was 0.986) (Figure 2B). The samples collected in the NMDS plot were not separated by sampling time and maize type (Figure 2A,B), which was confirmed by a more detailed analysis of similarity (ANOSIM). No significant correlation was detected between arthropod community composition and sampling time (R^2 = 0.023, p = 0.74). No significant correlation was detected between arthropod community composition and maize type (R^2 = 0.0033, p = 0.45).

2.5. Ecosystem Functioning Composition of NTA Communities in Different Maize Fields

Five functional guilds were identified in *Bt* and non-*Bt* maize fields during the study period. The results showed that for the BF and DF stages, the most abundant guilds in *Bt* and non-*Bt* maize fields were herbivores, followed by predators, decomposers, parasitoids and pollinators. For the AF stage, the most abundant guilds in *Bt* and non-*Bt* maize fields were predators, followed by herbivores, decomposers and parasitoids. The number of pollinators increased in the DF stage and was reduced to zero in the AF stage, comprising a rare guild (Figure 3a,e). As decomposers, only *Isotomidae* sp. was observed in the three main maize growth stages (Figure 3b). For herbivores, Aphididae was the most abundant family, although the common species detected were not exactly the same among different stages; *Rhopalosiphum padi* and *R. maidis* were the most abundant aphids in three main growth stages (Figure 3c). For parasitoids, *Trichogramma ostriniae* and Family Ichneumonidae were the abundant groups for different stages (Figure 3d). For predators, the non-*Bt* maize field had eight common groups at the BF stage: *Lasius fuliginosus*, *Labidura* sp. *Propylaea japonica*, *Harmonia axyridis*, *Orius strigicollis*, Linyphiidae, Carabidae and Clubionidae, respectively, and the *Bt* maize field had nine common groups—the extra one was *Paederus fuscipes*. At the DF stage, the number of common groups was higher than at the BF stage, *H. axyridis*, *Labidura* sp. and *P. japonica* were the most abundant groups. At the AF stage, *H. axyridis* was the obvious dominant species (Figure 3f). During the entire study period, the composition of NTA communities was essentially uniform among *Bt* and non-*Bt* maize fields.

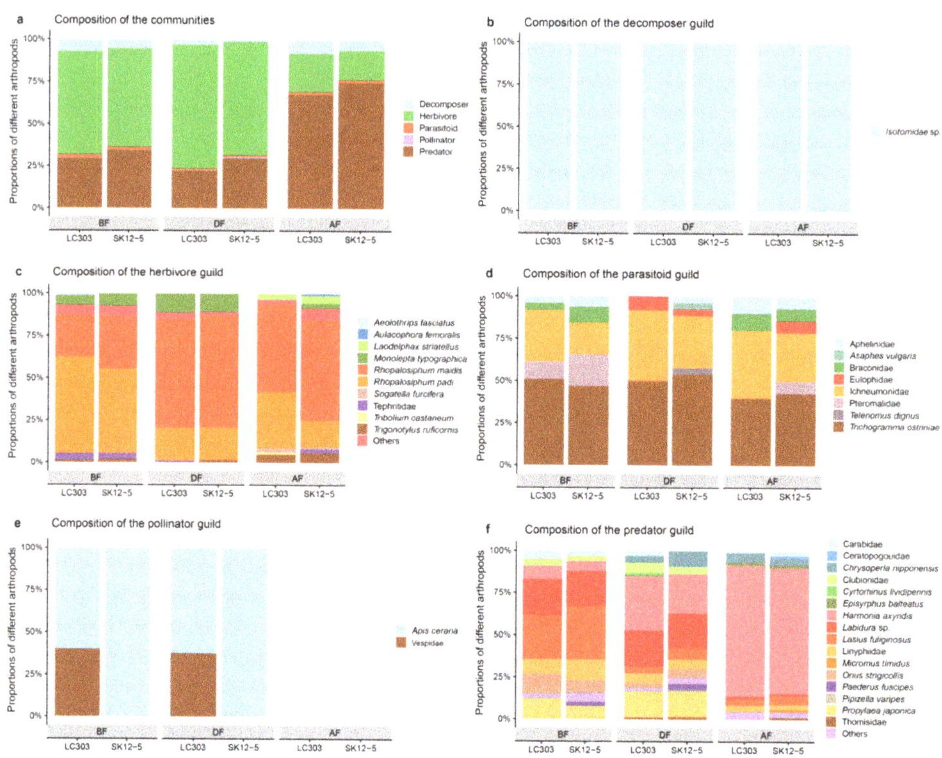

Figure 3. Composition of non-target arthropod (NTA) communities in SK12-5 (*Bt*) and LC303 (non-*Bt*) maize fields for three main maize growth stages: BF—Before Flowering, DF—During Flowering and AF—After Flowering. (**a**) NTA communities; (**b**) decomposers; (**c**) herbivores; (**d**) parasitoids; (**e**) pollinators; and (**f**) predators.

3. Discussion

Transgenic *Bt* crops can bring great advantages for reducing pesticide usage, increasing crop yields and increasing farm income [35,36]. On the other hand, planting of insect-resistant *Bt* crops may pose negative effects on the environment. It is therefore necessary to assess the potential environmental effects of *Bt* crops before being commercialized, including field surveys to evaluate their potential effects on the population dynamics of NTAs [15].

In the current study, SK12-5 transgenic *Cry1Ab/2Aj* maize was selected to assess its potential effects on the composition and community structure of NTAs since it exhibited high efficacy in controlling target pests and it may be commercially planted in China in the near future. The NTA populations were investigated during three main growth stages of maize, that is, the before flowering (BF), during flowering (DF) and after flowering (AF) stages separately. The main reasons for selecting these three periods are that: (1) the highest Cry1Ab/2Aj protein content is expressed in the pollen (data not published) and (2) beneficial insects begin to increase during the maize silking period [37].

During the three maize stages, no non-target lepidopteran species were found in both *Bt* and non-*Bt* fields. For target Lepidopteran species, such as *O. furnacalis*, the total number of the insects captured was only four in *Bt* maize field, and therefore, the lepidopteran insects were not involved for analysis. The results in this study showed no significant differences on the composition, species richness (*S*), Shannon–Wiener diversity index (H'), evenness index (*J*), Simpson's dominant concentration (*C*), community similarity and ecosystem functioning composition of the NTAs community between transgenic insect-resistant maize (SK12-5 transgenic *Cry1Ab/2Aj*) and the non-GE maize. These results are consistent with previous studies showing that *Bt* maize producing Cry1Ab [18,38–44], Cry1Ab and Cry2Ab [45], Cry1A.105 and Cry2Ab [46,47], Cry1Ac [48–50], Cry1A.105 and Cry2Ab and Cry3Bb [51], Cry1Ab and Cry1Ac [52], Cry1Ab and VIP3A [53], Cry1F [54–56], Cry1Ah [57], Cry1Ie [58–60], Cry34Ab1 and Cry35Ab1 [55,61], Cry34Ab1 and Cry35Ab1 and Cry1F [55,61], Cry3Bb [62–65] and VIP 3A [66] insecticidal proteins did not affect populations of NTAs in the field. Similarly, there were no significant effects detected in the majority of such filed experiments with *Bt* cotton and *Bt* rice [37,67–69]. However, there are indeed studies reporting that cultivation of *Bt* crops can alter population structure and dynamics of non-target arthropods in the field [70]. For example, many studies show that growing of *Bt* crops significantly reduces the density of parasitoids that are specific to the target pests of the *Bt* crops, and the reduction is associated with the decrease of target pest populations [11]. In addition, it has also been frequently reported that the long-term growing of *Bt* crops leads to population increases of non-target pest species, for example, the cultivation of *Bt* cotton in China causes a rise of the secondary non-target pest mirid bugs to become the major pest [22]. The same ecological phenomenon was also found for aphids in *Bt* maize fields [71]. The secondary insect pest population increase has been confirmed to be the consequence that they are not susceptible to or have decreased susceptibility to the *Bt* proteins, and such species would have been controlled by the insecticides that were used for controlling target pests before the commercial use of the *Bt* crops [71]. Meanwhile, studies showed that the cultivation of transgenic insect-resistant crops could protect or improve arthropod biodiversity with the reduced use of pesticides in the field [64,72,73]. This information demonstrates that the current data with the transgenic maize SK12-5 showing no negative impact on the number and community structure of NTAs are valuable for decision-making for commercialization of the *Bt* maize line, but it is a short-term field investigation in small-scale field plots that cannot represent the long-term monitoring on large-scale farms, which should be conducted after commercial planting of *Bt* crops.

The current results showed that Hemiptera was the dominant order and aphids were the dominant species at the BF and DF stages, while at the AF stage, Coleoptera was the dominant order and *H. axyridis* were the dominant species. This ecological niche change may be attributed to the tritrophic interaction of aphids and ladybirds. There are microbalances in nature, where species restrict and counter-restrict each other. The number of aphids increasing in the field will subsequently lead to the increase of the ladybug population since aphids are dominant preys of ladybirds [74]. In reverse, the growth of the aphid population will be inhibited by the increased ladybird population, which belongs to the negative feedback regulation in the biological community [75]. Notably, the same negative ecosystem feedback occurred in both *Bt* and non-*Bt* maize fields, illustrating that the cultivation of transgenic *Cry1Ab/2Aj* maize had a similar impact on the ecosystem with the conventional non-transgenic maize. Besides, *Apis cerana* was the dominant pollinator in *Bt* and non-*Bt* maize fields, but Vespidae were not observed in *Bt* field, which may be due to few lepidopteran pests in *Bt* fields, considering lepidopteran pests are the main preys of Vespidae species. Overall, the results showed that composition of NTA species between the *Bt* and non-*Bt* maize fields was similar. By comparing and analyzing the structural dynamics and similarities of the community, the degree of similarity between different communities can be more objectively reflected. C_S, referring to the similarity of species composition between communities, between the *Bt* and non-*Bt* maize fields for the three growth stages were all greater than 0.5, showing that the degree of similarity between communities is high [76]. At the community level, there were no significant differences in the three growth stages in terms of overall analysis of species richness (*S*), Shannon–Wiener diversity index (H'), evenness index (*J*) and Simpson's dominant concentration (*C*) of the NTA community between *Bt* and non-*Bt* maize fields. The temporal trend for each parameter was consistent, and there were no significant differences. The results indicate that transgenic *Cry1Ab/2Aj* maize SK12-5 has no obvious effect on the composition and community parameters of NTAs.

In recent years, NMDS analysis methods have been used for many types of ecological studies and evaluating the impact of transgenic crop cultivation on animal communities in the field [59,60,77]. This study also used this method to analyze the relationship between maize type and NTAs composition and showed that transgenic maize did not have a significant impact on NTAs. The analysis of NTA communities examined at the species and family levels demonstrated that the composition of the dominant, common and rare guilds or species and families was similar in *Bt* and non-*Bt* maize fields. Compared with populations, biological communities have higher structure and more complex diversity. Results of this study show that at the community level, insect-resistant *Cry1Ab/2Aj* maize had no significant ecological impacts on the NTAs community in the field. This finding complements the knowledge of potential effects of insect-resistant GE crops on arthropod populations and provides valuable information for decision-making on commercial application of the *Bt* maize line SK12-5 in China.

4. Material and Methods

4.1. Field Planting and Management Methods

Transgenic *cry1Ab/2Aj* maize Shuangkang 12-5 (SK12-5) and its corresponding non-transformed near-isoline Lianchuang 303 (LC303) seeds were sown simultaneously in experimental fields in the Jilin Academy of Agricultural Sciences, Gongzhuling City, Jilin Province, China (43°19′ N, 124°29′ W). These maize lines were grown in adjacent (50 m^2) plots, with three replicate plots for each maize line. All maize plants were cultivated equally according to the common local agricultural practices in 2018; no chemical pesticides and herbicides were applied throughout the growing season of maize, and other farming operations were the same as the local routine. Thus, the two maize lines were grown under the same environmental and agricultural conditions.

4.2. Sample Collection and Identification

NTAs were sampled for 7 days monthly in three maize growth stages: before flowering stage (abbreviated BF), during flowering stage (abbreviated DF) and after flowering stage (abbreviated AF). Two sampling methods were used to collect NTAs in maize fields—direct sampling and trapping: (1) direct sampling included visual observations to capture taxa occurring on maize plants and sweep nets to capture aerial taxa. In each plot, visual inspections of plants were made row after row to collect all arthropods found every day, and a sweep net was used to capture flying arthropods every day. Sampling was conducted in the morning when arthropods were less active. (2) Trapping sampling used pitfall traps to capture surface- and ground-dwelling taxa. Five pitfall traps (a plastic cup of 11-cm depth half-filled with water and scouring agent) were established in an "X" pattern that covered the whole plot, regularly distributed over the plot length but at least 2 m from the field border. Pitfall traps were set for 24 h, and the trapping agent was changed after gathering the collected arthropods every day.

The collected arthropods were placed separately into centrifuge tubes and immediately frozen at −20 °C in a portable freezer (Alpicool ENX42, Foshan Alpicool Electrical Appliance Co., Ltd., Foshan city, China). All captured arthropods were taken to the laboratory, placed in Petri dishes over dry ice and examined using a Zeiss stereomicroscope (Carl Zeiss Digital Innovation GmbH. Germany) for taxonomic identification. The respective taxonomic levels—species, family, and order—and ecological function were determined based on a database [78]. When samples were too degraded or diagnostic morphological characters were hard to distinguish, identification was performed at the family level.

4.3. Community Parameters and Calculation Formulas

Number of individuals (N), species richness (S), Shannon–Wiener diversity index (H'), Evenness index (J) and Simpson's dominant concentration (C) were used to analyze the structure and temporal dynamics of the NTA communities. S is the number of species within a defined region, which reflects the complexity and heterogeneity of the regional environment. H' and C are the ways to measure the diversity of species in a community; they can reflect changes in the populations of various species. J can be used as a measure of species dominance in a community, a measure of how close different species are to each other in number. The Berger–Parker dominance index (D) was calculated separately based on the different functions in the fields. D is an index reflecting community dominance and is the most sensitive to changes in community diversity. The similarity in species composition of NTA communities was analyzed using the Czekanowski Community Similarity Coefficient (C_S). C_S is used to compare the similarities and differences between biological community structures in different spaces. The formulas for calculation of the above indexes are as follow [79]:

$$H\prime = -\sum_{\mathrm{i}=1}^{\mathrm{S}} \mathrm{P_i} \ln \mathrm{P_i}$$

$$J = \frac{H\prime}{\ln S}$$

$$D = \frac{N_{\mathrm{max}}}{N}$$

$$D = \frac{N_{\mathrm{max}}}{N}$$

$$\mathrm{C} = \sum_{\mathrm{i}=1}^{\mathrm{S}} \left(\frac{\mathrm{n_i}}{N}\right)^2$$

$$C_S = \frac{2A}{a+b}$$

In the formulas: S is the species richness of the community; $\mathrm{P_i}$ is the ratio of the number of individuals of species i in the community to the total number of individuals in the community; $\mathrm{n_i}$ is the number of individuals of species i; N is the total number of individuals in the community; and N_{max} is the number of dominant species. When the species dominance $D \geq 0.1$, it is a dominant species; when $0.01 \leq D < 0.1$, it is a common species; when $D \leq 0.01$, it is a rare species [79]. A is the number of species shared by the two communities *a* and *b*, and *a* and *b* are the numbers of species in the corresponding communities *a* and *b*, respectively.

4.4. Statistical Analysis

All data are presented as mean ± standard error (SE), unless otherwise indicated. A Student's *t*-test was used to compare the composition and community parameters of arthropods in SK12-5 (*Bt*) and LC303 (non-*Bt*) maize fields. Composition and characteristics of NTAs in the same maize fields at different growth stages were analyzed by one-way ANOVA followed by the Tukey HSD test. In addition, the composition of NTAs in the orders Araneae, Dermaptera, Diptera, Entomobryomorpha, Hemiptera, Hymenoptera and Thysanoptera in LC303 (non-*Bt*) maize fields, and the composition of NTAs in the orders of Araneae, Dermaptera, Entomobryomorpha, Hemiptera, Hymenoptera and Neuroptera in SK12-5 (*Bt*) maize fields were analyzed by the Mann–Whitney U-test because of the associated heterogeneity of variance. Differences were considered as significant at $p < 0.05$. These analyses were conducted using the SPSS Version 13.0 statistical software. Community structure was determined with NMDS (non-metric multidimensional scaling) ordinations based on Bray–Curtis dissimilarity. NMDS is a multivariate, nonlinear technique that ranks points such that distance in ordination space represents similarity among sample periods [80] (pp. 91–173). The correspondence of the ordination diagram to the similarity distances is described by a stress value, where 0 is a perfect fit. Furthermore, analysis of similarity (ANOSIM) was used to test if there was a statistical difference among the NTA communities, sample time and maize types [58,77]. This analysis was conducted with the vegan package in R (v.3.2.3; R Development Core Team).

5. Conclusions

This study collected NTAs in *Bt* and non-*Bt* maize fields during three main growing stages around maize flowering periods. By analyzing the compositions of NTAs in these three stages, the ecological niche change from domination by Hemiptera to Coleoptera was observed. Meanwhile, long-term and large-scale planting of *Bt* maize requires attention to the population dynamics of NTAs such as aphids and other pests to avoid pest outbreaks due to less pesticide use. The cultivation of *Bt* maize expressing Cry1Ab/2Aj protein did not show any negative impacts on the densities of non-target decomposers, herbivores, predators, parasitoids and pollinators. The compositions of decomposers, herbivores, predators, parasitoids and pollinators were similar in *Bt* and non-*Bt* maize fields. Taken together, results from our work support the view that planting of *Bt* maize producing Cry1Ab/2Aj toxins does not adversely affect populations of NTAs. Moreover, this study shows that the changes in the abundance and diversity of NTAs in maize fields are driven by time, and the Cry1Ab/2Aj toxin exposure plays a negligible, if any, role in the evolution of these NTA communities. Interactions between maize and NTAs occur over a wide range of timescales from hours to seasons and years and are mostly driven by temperature, insolation or rainfall. Hence, long-term and large-scale studies need to take into account a large variety of environmental parameters, including the effects of insecticide treatments on non-*Bt* crops, and it is also necessary to ensure the long-term efficacy of GE crops with reduced impact on the environment and agricultural ecosystems.

Author Contributions: Y.L. and Y.Y. designed the study, Y.Y. and Y.W. in collecting samples. Y.Y., Y.C. and J.X. analyzed the data. Y.Y. and Y.C. wrote the original draft, X.S. provided the experimental materials. All authors have read and agreed to the published version of the manuscript.

Funding: The project was supported by the Hainan Provincial Natural Science Foundation of China (2019RC137), the National GMO New Variety Breeding Program of China (2016ZX08011-001) and the Hainan Yazhou Bay Seed Laboratory (B22E10303).

Institutional Review Board Statement: Not applicable.

Informed Consent Statement: Not applicable.

Data Availability Statement: Due to privacy the data presented in this study are available on request from the corresponding author.

Conflicts of Interest: The authors declare no conflict of interest.

References

1. Brookes, G.; Barfoot, P. Environmental impacts of genetically modified (GM) Crop use 1996–2016: Impacts on pesticide use and carbon emissions. *GM Crops Food* **2018**, *9*, 109–139. [CrossRef] [PubMed]
2. Brookes, G.; Barfoot, P. Farm income and production impacts of using GM crop technology 1996–2016. *GM Crops Food* **2018**, *9*, 59–89. [CrossRef] [PubMed]
3. ISAAA (International Service for the Acquisition of Agri-biotech Applications). *Global Status of Commercialized Biotech/GM Crops in 2019: Biotech Crops Drive Socio-Economic Development and Sustainable Environment in the New Frontier*; ISAAA Briefs, No. 55; ISAAA: Ithaca, NY, USA, 2019.
4. Zhu, Y.; Jiang, T.; Yang, Y.Z. Research advances in arthropod community in corn fields. *Plant Prot.* **2017**, *43*, 1–5.
5. Ren, Z.T.; Shen, W.J.; Liu, B.; Xue, K. Effects of transgenic maize on biodiversity of arthropod communities in the fields. *Sci. Geol. Sin.* **2017**, *50*, 2315–2325.
6. Duan, J.J.; Marvier, M.; Huesing, J.; Dively, G.; Huang, Y.Z. A meta-analysis of effects of *Bt* crops on honey bees (Hymenoptera: Apidae). *PLoS ONE* **2008**, *3*, e1415. [CrossRef] [PubMed]
7. Li, Y.H.; Zhang, X.J.; Chen, X.P.; Romeis, J.; Yin, X.M.; Peng, Y.F. Consumption of Bt rice pollen containing Cry1C or Cry2A does not pose a risk to *Propylea japonica* (Thunberg) (Coleoptera: Coccinellidae). *Sci. Rep.* **2015**, *5*, 7679. [CrossRef]
8. Vaufleurya, A.D.; Kramarza, P.E.; Bineta, P.; Cortet, J.; Caul, S.; Andersen, M.N.; Plumey, E.; Coeurdassier, M.; Krogh, P.H. Exposure and effects assessments of *Bt* maize on non-target organisms (gastropods, microarthropods, mycorrhizal fungi) in microcosms. *Pedobiologia* **2007**, *51*, 185–194. [CrossRef]
9. Yang, Y.; Liu, Y.; Cao, F.Q.; Chen, X.P.; Cheng, L.S.; Romeis, J.; Li, Y.H.; Peng, Y.F. Consumption of Bt Rice Pollen Containing Cry1C or Cry2A Protein Poses a Low to Negligible Risk to the Silkworm *Bombyx mori* (Lepidoptera: Bombyxidae). *PLoS ONE* **2014**, *9*, e102302. [CrossRef]
10. Yang, Y.; Chen, X.P.; Cheng, L.S.; Cao, F.Q.; Romeis, J.; Li, Y.H.; Peng, Y.F. Toxicological and biochemical analyses demonstrate no toxic effect of Cry1C and Cry2A to *Folsomia candida*. *Sci. Rep.* **2015**, *5*, 15619. [CrossRef]
11. Romeis, J.; Meissle, M.; Bigler, F. Transgenic crops expressing *Bacillus thuringiensis* toxins and biological control. *Nat. Biotechnol.* **2006**, *24*, 63–71. [CrossRef]
12. Hellmich, R.L.; Albajes, R.; Bergvinson, D.; Prasifka, J.R.; Wang, Z.Y.; Weiss, M.J. The present and future role of insect-resistant genetically modified maize in IPM. In *Integration of Insect-Resistant Genetically Modified Crops within IPM Programs. Progress in Biological Control*; Romeis, J., Shelton, A.M., Kennedy, G.G., Eds.; Springer: Dordrecht, The Netherlands; Berlin, Germany, 2008; Volume 5, pp. 119–158. [CrossRef]
13. Li, Y.H.; Romeis, J.; Wu, K.M.; Peng, Y.F. Tier-1 assays for assessing the toxicity of insecticidal proteins produced by genetically engineered plants to non-target arthropods. *Insect Sci.* **2014**, *21*, 125–134. [CrossRef] [PubMed]
14. Li, Y.H.; Hallerman, M.E.; Wu, K.M.; Peng, Y.F. Insect-resistant genetically engineered crops in China: Development, application, and prospects for use. *Annu. Rev. Entomol.* **2020**, *65*, 273–292. [CrossRef] [PubMed]
15. Lu, Z.B.; Dang, C.; Wang, F.; Liu, Z.C.; Chen, J.; Wang, Y.; Yao, H.W.; Fang, Q.; Peng, Y.F.; Gatehouse, A.M.R.; et al. Does long-term Bt rice planting pose risks to spider communities and their capacity to control planthoppers? *Plant Biotechnol. J.* **2020**, *18*, 1851–1853. [CrossRef] [PubMed]
16. Freier, B.; Schorling, M.; Traugott, M.; Juen, A.; Volkmar, C. Results of a 4-year plant survey and pitfall trapping in *Bt* maize and conventional maize fields regarding the occurrence of selected arthropod taxa. *IOBC/WPRS Bull.* **2014**, *27*, 79–84.
17. Jensen, P.D.; Dively, G.P.; Swan, C.M.; Lamp, W.O. Exposure and nontarget effects of transgenic *Bt* corn debris in streams. *Environ. Entomol.* **2010**, *39*, 707–714. [CrossRef]
18. Alcantara, E.P. Postcommercialization monitoring of the long-term impact of *Bt* corn on non-target arthropod communities in commercial farms and adjacent riparian areas in the Philippines. *Environ. Entomol.* **2012**, *41*, 1268–1276. [CrossRef]
19. Lumbierres, B.; Albajes, R.; Pons, X. Transgenic *Bt* maize and *Rhopalosiphum padi* (Hom., Aphididae) performance. *Ecol. Entomol.* **2004**, *29*, 309–317. [CrossRef]
20. Lumbierres, B.; Stary, P.; Pons, X. Effect of *Bt* maize on the plant-aphid-parasitoid tritrophic relationships. *Bio. Control* **2011**, *56*, 133–143. [CrossRef]
21. Pons, X.; Lumbierres, B.; Lopez, C.; Albajes, R. Abundance of non-target pests in transgenic *Bt*-maize: A farm scale study. *Eur. J. Entomol.* **2005**, *102*, 73–79. [CrossRef]
22. Lu, Y.; Wu, K.; Jiang, Y.; Xia, B.; Li, P.; Feng, H.; Guo, Y. Mirid bug outbreaks in multiple crops correlated with wide-scale adoption of *Bt* cotton in China. *Science* **2010**, *328*, 1151–1154. [CrossRef]
23. Romeis, J.; Meissle, M.; Alvarez-Alfageme, F.; Bigler, F.; Bohan, D.A.; Devos, Y.; Malone, L.A.; Pons, X.; Rauschen, S. Potential use of an arthropod database to support the non-target risk assessment and monitoring of transgenic plants. *Transgenic Res.* **2014**, *23*, 995–1013. [CrossRef] [PubMed]

24. Kang, L.; Chen, M. GMO biosafety management, suggestions and biotech public acceptance in China. *Plant Physiol. J.* **2013**, *49*, 637–644. [CrossRef]
25. Zhao, Z.Y.; Gu, W.N.; Cai, T.S.; Tagliani, L.; Hondred, D.; Bond, D.; Schroeder, S.; Rudert, M.; Pierce, D. High throughput genetic transformation mediated by *Agrobacterium tumefaciens* in maize. *Mol. Breed.* **2002**, *8*, 323–333. [CrossRef]
26. Sun, H.W.; Xu, X.H.; Li, F.; Gao, R.; Yang, S.K.; Lu, X.B. Resistances of transgenic maize hybrid "Ruifeng 1-double resistance 12-5" with insect and herbicide resistance against three major lepidopteran pests. *Shandong Agric. Sci.* **2018**, *50*, 109–114.
27. Chang, X.; Lu, Z.B.; Shen, Z.C.; Peng, Y.F.; Ye, G.Y. Bitrophic and tritrophic effects of transgenic *cry1Ab/cry2Aj* maize on the beneficial, nontarget *Harmonia axyridis* (Coleoptera: Coccinellidae). *Environ. Entomol.* **2017**, *46*, 1171–1176. [CrossRef] [PubMed]
28. Chen, Y.; Chen, X.Y.; Peng, C.; Xu, J.F.; Shen, H.; Li, Y.Y.; Wang, X.F. Establishment of a field visual detection method for genetically modified maize "Shuangkang" 12-5 by fluorescence RPA. *Hereditas* **2021**, *43*, 802–812. [CrossRef] [PubMed]
29. Wang, H.Q.; Xiao, F.; Yang, L.; Miao, Q.M.; Zhang, X.D.; Zhang, X.J. Analysis of verification results by PCR methods for genetically modified double-resistant 12-5 event-specific maize. *Biotechnol. Bull.* **2020**, *36*, 48–55.
30. Sun, H.W., Xu, X.H.; Li, F.; Yang, S.K.; Lu, X.B. Insect resistance and tolerance to the target herbicide of the transgenic maize cultivar Shuangkang 12-5-21. *J. Biosaf.* **2021**, *30*, 43–49.
31. Jiang, Y.Y.; Ji, Y.; Lai, Y.M.; Zhu, X.; Chen, X.Y.; Xu, J.F.; Ma, L.J.; Xu, X.L. Potential effect of genetically modified maize DR12-5 on honey bee (*Apis mellifera*). *Acta Agric. Zhejiangensis* **2019**, *31*, 1834–1840.
32. Zhu, X.; Jiang, Y.Y.; Lai, Y.M.; Chen, X.Y.; Wang, X.F.; Xu, X.L.; Xu, J.F. Effects of transgenic maize (*Zea mays*) event DR 12-5 with *Cry1Ab/cry2Aj* and g10evo-epsps genes on silkworm (*Bombyx mori*). *J. Agric. Biotechnol.* **2020**, *28*, 1269–1276.
33. Liu, Y.M.; Li, Y.H.; Chen, X.P.; Song, X.Y.; Shen, P.; Peng, Y.F. No detrimental effect of *Bt* maize pollen containing *Cry1Ab/2Aj* or Cry1Ac on adult green lacewings *Chrysoperla sinica* Tjeder. *J. Integr. Agric.* **2019**, *18*, 893–899. [CrossRef]
34. Sun, H.W.; Zhang, Y.X.; Xu, X.H.; Gao, D.; Li, F.; Yang, S.K.; Lu, X.B. Effects of transgenic maize variety (double resistance 12-5) on enzyme activity and microbial diversity in rhizosphere soil. *Soils* **2019**, *51*, 61–67.
35. Naranjo, S.E. Impacts of *Bt* crops on non-target organisms and insecticide use patterns. *CAB Rev. Perspect. Agric. Vet. Sci. Nutr. Nat. Resour.* **2009**, *4*, 11.
36. Sanvido, O.; Romeis, J.; Bigler, F. An approach for post-market monitoring of potential environmental effects of *Bt*-maize expressing *cry1ab* on natural enemies. *J. Appl. Entomol.* **2010**, *133*, 236–248. [CrossRef]
37. Cattaneo, M.G.; Yafuso, C.; Schmidt, C.; Huang, C.Y.; Rahman, M.; Olson, C.; Ellers-Krik, C.; Orr, B.J.; Marsh, S.E.; Antilla, L. Farm-scale evaluation of the impacts of transgenic cotton on biodiversity, pesticide use, and yield. *Proc. Natl. Acad. Sci. USA* **2006**, *103*, 7571–7576. [CrossRef]
38. Candolfi, M.P.; Brown, K.; Grimm, C.; Reber, B.; Schmidli, H. A faunistic approach to assess potential side-effects of genetically modified *Bt*-corn on non-target arthropods under field conditions. *Biocontrol. Sci. Technol.* **2004**, *14*, 129–170. [CrossRef]
39. Zwahlen, C.; Hilbeck, A.; Nentwig, W. Field decomposition of transgenic Bt maize residue and the impact on non-target soil invertebrates. *Plant Soil* **2007**, *300*, 245–257. [CrossRef]
40. Habuštová, O.; Doležal, P.; Spitzer, L.; Svobodová, Z.; Hussein, H.; Sehnal, F. Impact of Cry1 A b toxin expression on the non-target insects dwelling on maize plants. *J. Appl. Entomol.* **2014**, *138*, 164–172. [CrossRef]
41. Truter, J.; Van, H.H.; Van, D.B.J. Comparative diversity of arthropods on *Bt* maize and non-*Bt* maize in two different cropping systems in South Africa. *Environ. Entomol.* **2014**, *43*, 197–208. [CrossRef]
42. Čerevková, A.; Cagáň, L. Effect of transgenic insect-resistant maize to the community structure of soil nematodes in two field trails. *Helminthologia* **2015**, *52*, 41–49. [CrossRef]
43. Skoková Habuštová, O.; Svobodová, Z.; Spitzer, L.; Doležal, P.; Hussein, H.M.; Sehnal, F. Communities of ground-dwelling arthropods in conventional and transgenic maize: Background data for the post-market environmental monitoring. *J. Appl. Entomol.* **2015**, *139*, 31–45. [CrossRef]
44. Ma, Y.J.; He, H.P.; Shen, W.J.; Liu, B.; Xue, K. Effects of transgenic maize on arthropod diversity. *Biodivers. Sci.* **2019**, *27*, 419. [CrossRef]
45. Yin, Y.; Xu, Y.; Cao, K.; Qin, Z.; Zhao, X.; Dong, X.; Shi, W. Impact assessment of *Bt* maize expressing the *Cry1Ab* and *Cry2Ab* protein simultaneously on non-target arthropods. *Environ. Sci. Pollut. Res.* **2020**, *27*, 21552–21559. [CrossRef] [PubMed]
46. Wang, B.; Chang, L.; Wu, D.; Song, X. Effect of Transgenic Corn Cultivation and Sampling Location on Feeding Habits of Collembola. *J. Agric. Sci. Technol. A* **2014**, *4*, 235–242.
47. Corrales Madrid, J.L.; Martínez Carrillo, J.L.; Osuna Martínez, M.B.; Durán Pompa, H.A.; Alonso Escobedo, J.; Javier Quiñones, F.; Garzón Tiznado, J.A.; Castro Espinoza, L.; Zavala García, F.; Espinoza Banda, A.; et al. Transportability of non-target arthropod field data for the use in environmental risk assessment of genetically modified maize in Northern Mexico. *J. Appl. Entomol.* **2018**, *142*, 525–538. [CrossRef]
48. Guo, Y.; Feng, Y.; Ge, Y.; Tetreau, G.; Chen, X.; Dong, X.; Shi, W. The cultivation of *Bt* corn producing *Cry1Ac* toxins does not adversely affect non-target arthropods. *PLoS ONE* **2014**, *9*, e114228. [CrossRef]
49. Liu, X.Y.; Wang, B.F.; Wang, J.; Feng, S.D.; Song, X.Y. Effects of leaf residue decomposition of *cry1Ac*-transgenic insect-resistant maize on community structure of soil animals. *J. Plant Prot.* **2016**, *43*, 384–390. [CrossRef]
50. Xing, Y.; Qin, Z.; Feng, M.; Li, A.; Zhang, L.; Wang, Y.; Dong, X.; Zhang, Y.; Tan, S.; Shi, W. The impact of *Bt* maize expressing the *Cry1Ac* protein on non-target arthropods. *Environ. Sci. Pollut. Res.* **2019**, *26*, 5814–5819. [CrossRef]

51. Schultheis, E. Faunistic Approach for an Environmental Risk Assessment of a Bt-Maize with Multiple Resistance Genes and Evaluation of *Trigonotylus caelestialium* (KIRKALDY) and *Lumbricus terrestris* LINNAEUS as Model Organisms. Master's Thesis, RWTH Aachen University, Aachen, Germany, 2011.
52. Al-Deeb, M.A.; Wilde, G.E.; Higgins, R.A. No effect of *Bacillus thuringiensis* corn and Bacillus thuringiensis on the predator *Orius insidiosus* (Hemiptera: Anthocoridae). *Environ. Entomol.* **2001**, *30*, 625–629. [CrossRef]
53. Dively, G.P. Impact of transgenic VIP3A× Cry1Ab lepidopteran-resistant field corn on the nontarget arthropod community. *Environ. Entomol.* **2005**, *34*, 1267–1291. [CrossRef]
54. Higgins, L.S.; Babcock, J.; Neese, P.; Layton, R.J.; Moellenbeck, D.J.; Storer, N. Three-year field monitoring of Cry1F, event DAS-Ø15Ø7-1, maize hybrids for nontarget arthropod effects. *Environ. Entomol.* **2009**, *38*, 281–292. [CrossRef] [PubMed]
55. Szénási, A.; Pálinkás, Z.; Zalai, M.; Schmitz, O.J.; Balog, A. Short-term effects of different genetically modified maize varieties on arthropod food web properties: An experimental field assessment. *Sci. Rep.* **2014**, *4*, 1–7. [CrossRef] [PubMed]
56. Resende, D.C.; Mendes, S.M.; Marucci, R.C.; Silva, A.D.C.; Campanha, M.M.; Waquil, J.M. Does *Bt* maize cultivation affect the non-target insect community in the agro ecosystem? *Rev. Bras. Entomol.* **2016**, *60*, 82–93. [CrossRef]
57. Chen, Y.; Ren, M.; Pan, L.; Liu, B.; Guan, X.; Tao, J. Impact of transgenic insect-resistant maize HGK60 with *Cry1Ah* gene on community components and biodiversity of arthropods in the fields. *PLoS ONE* **2022**, *17*, e0269459. [CrossRef] [PubMed]
58. Guo, J.; He, K.; Hellmich, R.L.; Bai, S.; Zhang, T.; Liu, Y.; Ahmed, T.; Wang, Z. Field trials to evaluate the effects of transgenic *cry1Ie* maize on the community characteristics of arthropod natural enemies. *Sci. Rep.* **2016**, *6*, 1–12. [CrossRef]
59. Guo, J.F.; He, K.L.; Bai, S.X.; Zhang, T.T.; Liu, Y.J.; Wang, F.X.; Wang, Z.Y. Effects of transgenic *cry1Ie* maize on non-lepidopteran pest abundance, diversity and community composition. *Transgenic Res.* **2016**, *25*, 761–772. [CrossRef]
60. Fan, C.; Wu, F.; Dong, J.; Wang, B.; Yin, J.; Song, X. No impact of transgenic *cry1Ie* maize on the diversity, abundance and composition of soil fauna in a 2-year field trial. *Sci. Rep.* **2019**, *9*, 1–9. [CrossRef]
61. Balog, A.; Szenasi, A.; Szekeres, D.; Palinkas, Z. Analysis of soil dwelling rove beetles (Coleoptera: Staphylinidae) in cultivated maize fields containing the *Bt* toxins, Cry34/35Ab1 and Cry1F×Cry34/35Ab1. *Biocontrol Sci. Technol.* **2011**, *21*, 293–297. [CrossRef]
62. Al-Deeb, M.A.; Wilde, G.E. Effect of *Bt* corn expressing the Cry3Bb1 toxin for corn rootworm control on aboveground nontarget arthropods. *Environ. Entomol.* **2003**, *32*, 1164–1170. [CrossRef]
63. Al-Deeb, M.A.; Wilde, G.E.; Blair, J.M.; Todd, T.C. Effect of *Bt* corn for corn rootworm control on nontarget soil microarthropods and nematodes. *Environ. Entomol.* **2003**, *32*, 859–865. [CrossRef]
64. Bitzer, R.J.; Rice, M.E.; Pilcher, C.D.; Pilcher, C.L.; Lam, W.K.F. Biodiversity and community structure of epedaphic and euedaphic springtails (Collembola) in transgenic rootworm *Bt* corn. *Environ. Entomol.* **2005**, *34*, 1346–1376. [CrossRef]
65. Ahmad, A.; Negri, I.; Oliveira, W.; Brown, C.; Asiimwe, P.; Sammons, B.; Horak, M.; Jiang, C.; Carson, D. Transportable data from non-target arthropod field studies for the environmental risk assessment of genetically modified maize expressing an insecticidal double-stranded RNA. *Transgenic Res.* **2016**, *25*, 1–17. [CrossRef] [PubMed]
66. Fernandes, O.A.; Faria, M.; Martinelli, S.; Schmidt, F.; Carvalho, V.F.; Moro, G. Short-term assessment of *Bt* maize on non-target arthropods in Brazil. *Sci. Agric.* **2007**, *64*, 249–255. [CrossRef]
67. Li, W.D.; Wu, K.M.; Wang, X.Q.; Guo, Y.Y. Evaluation of impact of pollen grains of Cry1Ac and Cry1A+CpTI transgenic cotton on the growth and development of Chinese tussah silkworm (Antheraea pernyi). *J. Agric. Biotechnol.* **2003**, *11*, 488–493. [CrossRef]
68. Li, F.F.; Ye, G.Y.; Wu, Q.; Peng, Y.F.; Chen, X.X. Arthropod abundance and diversity in *Bt* and non-*Bt* rice fields. *Environ. Entomol.* **2007**, *36*, 646–654. [CrossRef]
69. Liu, Z.C.; Ye, G.Y.; Hu, C.; Datta, S.K. Impact of transgenic indica rice with a fused gene of cry1Ab/cry1Ac on the rice paddy arthropod community. *Acta Entomol. Sin.* **2003**, *46*, 454–465. [CrossRef]
70. Jiang, W.H.; Zhang, D.J.; Liu, J.X.; Li, C.L.; Lu, Z.Y.; Yang, M.S. Comparative Analysis of Arthropod Communities in Transgenic *Bt* and Non-Transgenic Poplar-Cotton Composite Systems. *Sci. Geol. Sin.* **2018**, *54*, 73–79. [CrossRef]
71. Catarino, R.; Ceddia, G.; Areal, F.J.; Park, J. The impact of secondary pests on *Bacillus thuringiensis* (*Bt*) crops. *Plant Biotech. J.* **2015**, *13*, 601–612. [CrossRef]
72. Carpenter, J.E. Impact of GM crops on biodiversity. *GM Crops* **2011**, *2*, 7–23. [CrossRef]
73. Bhatti, M.A.; Duan, J.; Head, G.P.; Jiang, C.J.; McKee, M.J.; Nickson, T.E.; Pilcher, C.L.; Pilcher, C.D. Field evaluation of the impact of corn rootworm (Coleoptera: Chrysomelidae)–protected *Bt* corn on foliage-dwelling arthropods. *Environ. Entomol.* **2005**, *34*, 1336–1345. [CrossRef]
74. Pervez, A.; Omkar. Ecology and biological control application of multicoloured Asian ladybird, *Harmonia axyridis*: A review. *Biocontrol Sci. Technol.* **2006**, *16*, 111–128. [CrossRef]
75. Connell, J.H.; Orias, E. The ecological regulation of species diversity. *Am. Nat.* **1964**, *98*, 399–414. [CrossRef]
76. Liu, Y.F.; Gu, D.X.; Zhang, G.R. The community dynamics of predatory arthropods in both weed habitat and paddy field from a double cropping paddy in Guangdong. *Acta Entomol. Sin.* **2003**, *46*, 591–597. [CrossRef]
77. Yin, J.Q.; Wang, D.M.; Liang, J.G.; Song, X.Y. Negligible Impact of Drought-Resistant Genetically Modified Maize on Arthropod Community Structure Observed in a 2-Year Field Investigation. *Plants* **2022**, *11*, 1092. [CrossRef]
78. Yang, Y. Establishment and Trial Application of a Database of Arthropods in Chinese Main Crops Fields. Doctoral Thesis, Chinese Academy of Agricultural Sciences, Beijing, China, 2018.

79. Zhang, L. Study on Environmental Persistence of Exogenous protein of Transgenic cry1Ab/c Rice and Its Effects on the Hydrobiont. Master's Thesis, Nanjing Agricultural University, Nanjing, China, 2013.
80. Ter Braak, C.J.F. Ordination. In *Data Analysis Community and Landscape Ecology*; Jongman, R.H.G., ter Braak, C.J.F., Van Tongeren, O.F.R., Eds.; Cambridge University Press: London, UK, 1995; pp. 91–173. [CrossRef]

plants MDPI

Article

Negligible Impact of Drought-Resistant Genetically Modified Maize on Arthropod Community Structure Observed in a 2-Year Field Investigation

Jun-Qi Yin [1,†], Da-Ming Wang [1,†], Jin-Gang Liang [2,*] and Xin-Yuan Song [1,*]

1 Agro-Biotechnology Research Institute, Jilin Academy of Agricultural Sciences, Changchun 130033, China; yin_junqi@163.com (J.-Q.Y.); jumina054824@163.com (D.-M.W.)
2 Development Center of Science and Technology, Ministry of Agriculture and Rural Affairs, Beijing 100176, China
* Correspondence: liangjingang@agri.gov.cn (J.-G.L.); songxinyuan1980@163.com (X.-Y.S.)
† These authors contributed equally to this work.

Abstract: Dehydration-responsive element-binding (DREB) transcription factors regulate diverse processes during plant development. Here, a 2-year field study was conducted to assess the potential effects of DREB-genetically modified maize (GM1) on arthropod species and ecological communities. Arthropod abundance, diversity, and community composition in GM1 and its non-transformed counterpart maize variety, Chang 7-2, were compared using whole plant inspection, pitfall trap, and suction sampler methods. Based on Shannon–Wiener diversity, Simpson's diversity, Pielou's indexes, number of species, and total number of individuals, GM1 had a negligible effect on arthropod abundance and diversity. Redundancy analysis indicated that the composition of arthropod community was not associated with maize type in the three investigation methods, while it exhibited significant correlation with year and sampling time in whole plant inspection and suction sample methods, and distinctly correlated with sampling time in the pitfall trap method. Nonmetric multidimensional scaling analysis of variable factors in the three investigation methods showed that sampling time, rather than maize type or year, was closely related to the composition of arthropod community in the field. Our results provide direct evidence to support that DREB-GM maize had negligible effects on arthropods in the Jilin Province under natural conditions.

Keywords: drought-resistant genetically modified maize; arthropods; abundance; diversity; community composition

Citation: Yin, J.-Q.; Wang, D.-M.; Liang, J.-G.; Song, X.-Y. Negligible Impact of Drought-Resistant Genetically Modified Maize on Arthropod Community Structure Observed in a 2-Year Field Investigation. *Plants* **2022**, *11*, 1092. https://doi.org/10.3390/plants11081092

Academic Editors: Wei Wei and C. Neal Stewart Jr.

Received: 14 March 2022
Accepted: 8 April 2022
Published: 18 April 2022

1. Introduction

Worldwide, adverse environmental factors can cause up to 70% loss of crop yields, with drought being the most important abiotic stress factor affecting maize production. Furthermore, global warming has intensified the frequency of occurrence of drought, which, in recent years, has affected 2×10^7 hm^2 annually in China, severely reducing maize yields [1–3].

With the development of genetically modified crops, the safety of transgenic plants and their impact on biodiversity have become increasingly important for research [4]. In China, 10 plant varieties have been approved for agricultural GM safety certificates, including insect-resistant and herbicide-resistant maize. Drought-resistant genetically modified maize has great potential to alleviate conditions of drought, and it is currently being actively researched and developed. Additionally, the arthropod community is important for maize production because the species involved play a crucial role in maintaining the ecological balance in farmlands [5–12]. Drought-resistant genetically modified crops show drought resistance mechanisms that improve their water use efficiency (WUE); however, whether the genetic modifications may have an impact on arthropods is unknown [13], although there are few reports on the impact of drought-resistant crops, especially on arthropods. Thus, China's agricultural genetically modified organisms (GMO) safety management measures require that genetically modified maize must be subjected to a rigorous environmental safety assessment before commercialization, among which the impact assessment on the arthropod community is an important part [14,15]. To date, many laboratories have conducted field trials to study the potential ecological effects of genetically modified crops on non-target organisms, while most of them have focused on insect-resistant or herbicide-resistant crops [16–21]. At present, most studies on genetically modified drought-resistant crops have focused on gene discovery and functional verification, while environmental safety assessments are limited to the evaluation of competitiveness for survival and its effects on available nutrients, enzyme activities, and microbial community diversity in rhizosphere soil [22–29]. However, the effects of drought-resistant genetically modified maize on farmland ecosystems have rarely been reported. Thus, whether drought-resistant maize harms arthropods in the field remains unknown.

The safety assessment of genetically modified maize needs to be conducted for a long-term study. Herein, we used three methods to evaluate the effects of GM1 on arthropods in 2016 and 2017 in a field study that aimed to evaluate whether genetically modified drought-resistant maize, GM1, had a significant impact on the diversity and composition of arthropods. Although many field and laboratory studies have confirmed that the effects of GM crops on arthropods were not significant [30], their influence needs to be analyzed case-by-case as, although some may have no influence on the overall ecological function, they do alter the composition of arthropods and dominant species [4]. This comprehensive study will provide useful data for the commercialization of GMO in the future.

2. Results

2.1. Arthropods in Transgenic and Non-Transgenic Maize Plots

In this study, arthropods that were easy to classify were identified to species level, such as *Rhopalosiphum maidis* and *Monolepta hieroglyphica*. Arthropods which were difficult to classify were identified to family level, such as Tenebrionidae and Formicidae. Through whole plant inspection, 2249 arthropods of 32 species and families in transgenic and 2130 of 33 species and families in non-transgenic maize (CK) were recorded over the 2-year experimental period. *R. maidis* and *M. hieroglyphica* accounted for 34.59% and 24.37%, and 33.24% and 24.98%, in transgenic and non-transgenic maize fields, respectively (Figure 1A). In turn, using the pitfall trap method, we recorded 3585 arthropods of 32 species and families, and 3650 arthropods of 33 species and families, in transgenic and non-transgenic maize, respectively, over 2 years. The occurrences of arthropods in the two types of maize were also similar in terms of taxon and number over the 2 years. Furthermore, in the two types of maize field, Tenebrionidae and Formicidae were the dominant families, accounting for 34.00% and 23.12%, and 41.34% and 18.22%, in transgenic and non-transgenic maize, respectively (Figure 1B). Finally, using the suction sampler method, a smaller number of arthropods was recorded, with a total of 430 arthropods of 16 species and families recorded in transgenic maize and 437 arthropods of 16 species and families in non-transgenic maize, over the 2 years of study. *R. maidis*, *M. hieroglyphica*, and *Harmonia axyridis* were the dominant taxa, accounting for 24.88%, 19.07%, and 12.33%, and 24.26%, 21.05%, and 13.04% in transgenic and non-transgenic maize fields, respectively (Figure 1C). Further, we analyzed the difference of proportion of *R. maidis*, which all occurred largely in the three methods, and the results show that there was no significant difference in the proportion of *R. maidis* ($p > 0.05$).

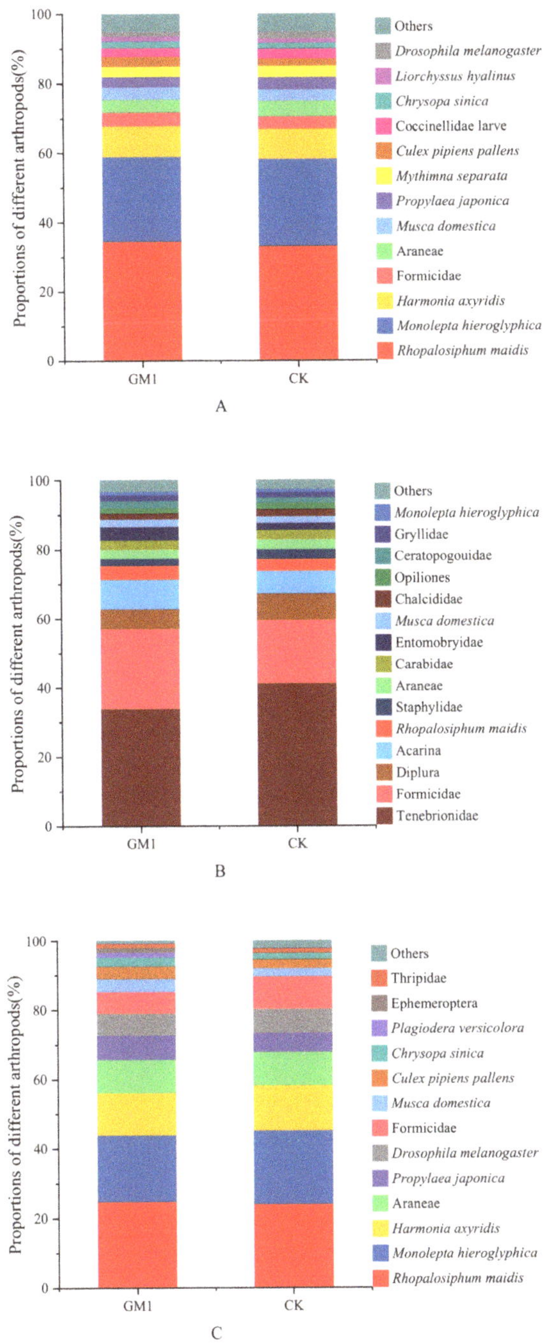

Figure 1. Proportions of arthropods found in GM1 and CK maize plots in 2016 and 2017 by (**A**) whole plant inspection, (**B**) pitfall trap, and (**C**) suction sampler. The taxa with a proportion greater than 1% are shown individually, while taxa with a proportion smaller than 1% are shown together (others).

2.2. Impacts of Maize Type, Year, and Sampling Time on Abundance and Diversity of Arthropods

We sampled arthropods using the whole plant inspection, pitfall trap, and suction sampler methods. The effects of maize type, year, sampling time, and their interactions on arthropod abundance and diversity indices were evaluated using Shannon–Wiener index (H'), Simpson's diversity index (D), number of species (S), Pielou's evenness index (J), and the total number of individuals (N). Unequally spaced repeated-measures ANOVA results showed that maize type had no significant effect on H', D, J, S, or N of arthropods, while sampling time did have a significant impact on all these parameters; however, the interactions between maize type and year, maize type and sampling time, and between maize type and year and sampling time did not influence any of them, regardless of sampling method. Furthermore, ANOVA showed that year had a significant impact on H', D, and J, when investigated using whole plant inspection or pitfall trap, but only year influenced D and J significantly when the suction sampler method was used. The above results indicate that arthropods were affected by sampling time but not by maize type. Meanwhile, two critical arthropod groups with different feeding habits (*Rhopalosiphum maidis* feed on maize, Formicidae do not eat maize directly) were selected to analyze the difference of their abundance in order to verify the impact of maize types on biodiversity of critical groups, and the results show that the differences were not significant ($p > 0.05$). Specifically, genetically modified maize, GM1, did not influence arthropod abundance or diversity in the field (Table 1).

2.3. Effects of Maize Type, Year, and Sampling Time on the Community Composition of Arthropods

2.3.1. Time-Dependent Effects of GM1 on Arthropod Community

The effects of maize type, year, and sampling time on the composition of arthropod community were examined using redundancy analysis (RDA). With the whole-plant inspection method, maize type, year, and sampling time explained 36% of the total variation in the composition of the arthropod community. RDA also showed that both sampling time ($p = 0.0020$, $F = 60.90$, and 999 Monte Carlo permutations) and year ($p = 0.0120$, $F = 3.01$, and 999 Monte Carlo permutations) were significantly correlated with the composition of the arthropod community, whereas maize type ($p = 0.9440$, $F = 0.44$, and 999 Monte Carlo permutations) was not correlated with it (Table 2, Figure 2).

Table 1. Multi-factor effects of year (2016 and 2017), maize type (GM1 and CK), and sampling time on diversity and abundance of arthropods.

Investigation Method	Factor	Shannon–Weiner's Index (*H'*)		Simpson's Diversity Index (*D*)		Pielou's Evenness Index (*J*)		Number of Species (*S*)		Total Abundance (*N*)		Abundance of Critical Groups			
												Rhopalosiphum maidis		Formicidae	
		F	*p*	*F*	*p*	*F*	*p*	*F*	*p*	*F*	*p*	*F*	*p*	*F*	*p*
Whole plant inspection method	Year	18.823	**0.002 ****	61.658	**0.000 *****	80.475	**0.000 *****	3.964	0.082	0.212	0.657	0	1	2.462	0.155
	Maize type	0.064	0.807	0.038	0.85	0.823	0.391	0.367	0.561	1.156	0.314	2.709	0.138	0.427	0.532
	Sampling time	19.56	**0.000 *****	38.71	**0.000 *****	84.41	**0.000 *****	21.41	**0.000 *****	244.91	**0.000 *****	306.456	**0.000 *****	10.178	**0.000 *****
	Maize type with Year	0.202	0.665	0.035	0.857	0.274	0.615	0.187	0.677	0.001	0.979	0.142	0.717	1.094	0.326
	Year with Sampling time	15.32	**0.000 *****	38.31	**0.000 *****	54.47	**0.000 *****	2.93	**0.041 ***	20.33	**0.000 *****	53.152	**0.000 *****	3.427	**0.032 ***
	Maize type with Sampling time	0.30	0.863	0.10	0.964	0.32	0.868	0.70	0.588	1.09	0.38	1.881	0.179	0.98	0.42
	Maize type with Year and Sampling time	0.21	0.921	0.20	0.905	0.38	0.824	0.35	0.828	2.32	0.07	2.158	0.141	0.344	0.8
	Mean ± SD (GM1)	2.312 ± 0.171		0.733 ± 0.028		0.756 ± 0.019		8.866 ± 1.063		37.484 ± 3.085		12.967 ± 1.026		1.467 ± 0.503	
	Mean ± SD (CK)	2.288 ± 0.170		0.730 ± 0.033		0.765 ± 0.022		8.517 ± 0.917		35.499 ± 2.451		11.800 ± 1.293		1.300 ± 0.626	
Pitfall trap method	Year	23.087	**0.001 *****	14.533	**0.005 ****	12.843	**0.007 ****	5.911	**0.041 ***	3.275	0.108	0.181	0.682	1.852	0.211
	Maize type	2.756	0.135	2.831	0.131	0.024	0.882	4.034	0.079	0.078	0.787	0.647	0.445	4.605	0.064
	Sampling time	60.849	**0.000 *****	85.218	**0.000 *****	163.107	**0.000 *****	23.25	**0.000 *****	213.789	**0.000 *****	46.295	**0.000 *****	31.194	**0.000 *****
	Maize type with Year	1.695	0.229	1.798	0.217	0.114	0.745	3.229	0.11	1.719	0.226	4.942	0.057	4.274	0.073
	Year with Sampling time	0.6	0.7	0.884	0.436	1.05	0.373	0.352	0.878	7.037	**0.000 *****	8.438	**0.006 ****	1.332	0.292
	Maize type with Sampling time	1.376	0.254	3.29	0.061	1.51	0.251	1.02	0.420	1.738	0.148	0.659	0.506	1.52	0.25
	Maize type with Year and Sampling time	1.339	0.268	2.37	0.123	1.69	0.216	1.48	0.219	0.902	0.489	1.971	0.181	0.743	0.483
	Mean ± SD (GM1)	2.831 ± 0.148		0.799 ± 0.027		0.769 ± 0.023		13.389 ± 0.977		99.584 ± 8.970		4.000 ± 0.920		23.028 ± 5.218	
	Mean ± SD (CK)	2.733 ± 0.101		0.780 ± 0.020		0.767 ± 0.021		12.333 ± 0.802		101.390 ± 8.549		3.528 ± 0.721		18.472 ± 2.981	
Suction sampler method	Year	0.578	0.469	5.416	**0.048 ***	12.626	**0.007 ****	1.84	0.212	5.788	**0.043 ***	1.903	0.205	8.696	**0.018 ***
	Maize type	0.05	0.829	2.476	0.154	2.288	0.169	0.115	0.743	0.027	0.874	0.004	0.954	4.261	0.073
	Sampling time	9.825	**0.000 *****	22.19	**0.000 *****	41.108	**0.000 *****	14.973	**0.000 *****	104.592	**0.000 *****	49.17	**0.000 *****	17.171	**0.000 *****
	Maize type with Year	0.078	0.787	0.319	0.587	2.917	0.126	0.013	0.913	0.197	0.669	0.291	0.604	4.261	0.073
	Year with Sampling time	1.122	0.364	0.451	0.653	2.568	0.057	0.50	0.739	3.107	**0.029 ***	5.148	**0.003 ****	7.805	**0.000 *****
	Maize type with Sampling time	0.816	0.524	0.73	0.501	1.31	0.286	0.42	0.794	1.536	0.215	1.36	0.27	2.341	0.076
	Maize type with Year and Sampling time	0.412	0.799	0.62	0.555	1.71	0.173	0.68	0.614	0.483	0.748	0.556	0.696	0.39	0.814
	Mean ± SD (GM1)	2.277 ± 0.227		0.860 ± 0.020		0.903 ± 0.017		6.300 ± 0.937		14.333 ± 2.014		3.567 ± 0.699		0.900 ± 0.307	
	Mean ± SD (CK)	2.244 ± 0.212		0.831 ± 0.040		0.888 ± 0.024		6.100 ± 0.806		14.567 ± 1.665		3.533 ± 0.911		1.367 ± 0.482	

The values in bold font are statistically significant (* $p < 0.05$; ** $p < 0.01$; *** $p < 0.001$).

Table 2. Monte Carlo permutation test for variable factors in the whole plant inspection method.

Variable Factor	Proportion of Variance Explained (%)	p	F
Sampling time	34	**0.0020** **	60.90
Year	2	**0.0120** *	3.01
Maize type	0	0.9440	0.44
Total	36		

The values in bold font are statistically significant (* $p < 0.05$; ** $p < 0.01$).

Using the pitfall trap method, RDA revealed that maize type, year, and sampling time, together explained 37% of the variation in the composition of arthropod community. Moreover, RDA showed that the composition of arthropod community was significantly correlated with sampling time ($p = 0.0020$, $F = 38.50$, and 999 Monte Carlo permutations), whereas it was not influenced by maize type ($p = 0.7660$, $F = 0.62$, and 999 Monte Carlo permutations) or year ($p = 0.1180$, $F = 1.49$, and 999 Monte Carlo permutations) (Table 3 and Figure 3).

Using the suction sampler method, RDA indicated that maize type, year, and sampling time together explained 33% of the variation of composition of arthropod community. Furthermore, RDA showed that both sampling time ($p = 0.0020$, $F = 23.71$, and 999 Monte Carlo permutations) and year ($p = 0.0480$, $F = 2.10$, and 999 Monte Carlo permutations) significantly correlated with arthropod community composition but not with maize type ($p = 0.8880$, $F = 0.50$, and 999 Monte Carlo permutations) (Table 4, Figure 4).

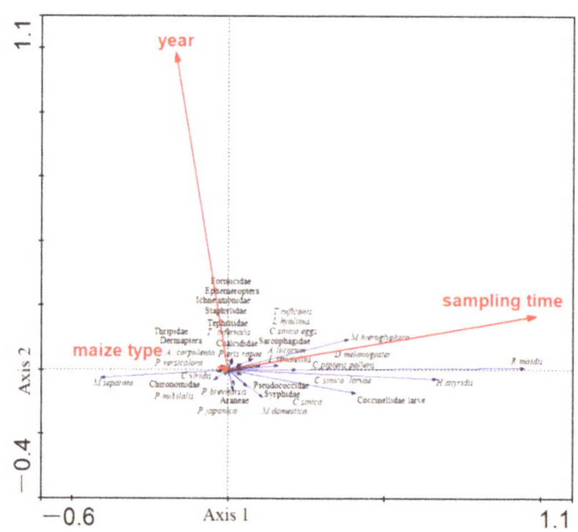

Figure 2. RDA for the composition of arthropod community in the whole plant inspection method, and the relationship of the composition of arthropod community and environmental factors. The red arrows represent variable factors, and the blue arrows represent arthropod species collected in the fields. The length of the blue arrows indicates the magnitude of the effect of environmental factors on arthropods. The cosine of the angle between the red and the blue arrows represents the correlation between variable factors and arthropods.

Table 3. Monte Carlo permutation test for the variable factors in the pitfall trap method.

Variable Factor	Proportion of Variance Explained (%)	p	F
Sampling time	35	**0.0020** **	38.50
Year	1	0.1180	1.49
Maize type	1	0.7660	0.62
Total	37		

The values in bold font are statistically significant (** $p < 0.01$).

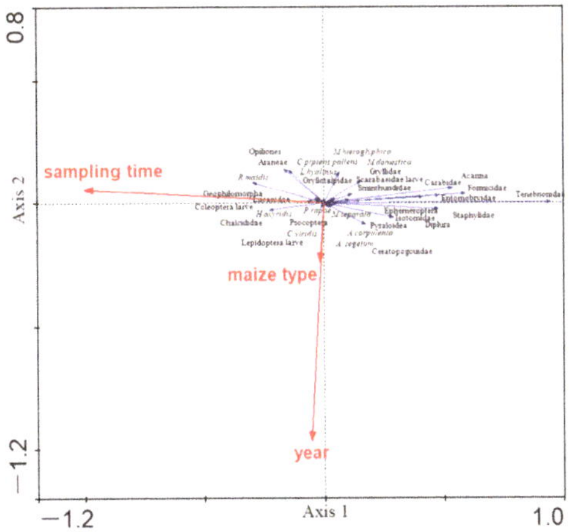

Figure 3. RDA for the composition of arthropod community in the pitfall trap method, and the relationship of the composition of arthropod community and environmental factors. The red arrows represent variable factors, and the blue arrows represent arthropod species collected in the fields. The length of the blue arrows indicates the magnitude of the effect of environmental factors on arthropods. The cosine of the angle between the red and the blue arrows represents the correlation between variable factors and arthropods.

Table 4. Monte Carlo permutation test of the suction sampler method.

Variable Factor	Proportion of Variance Explained (%)	p	F
Sampling time	29	**0.0020** **	23.71
Year	3	**0.0480** *	2.10
Maize type	1	0.8880	0.50
Total	33		

The values in bold font are statistically significant (* $p < 0.05$; ** $p < 0.01$).

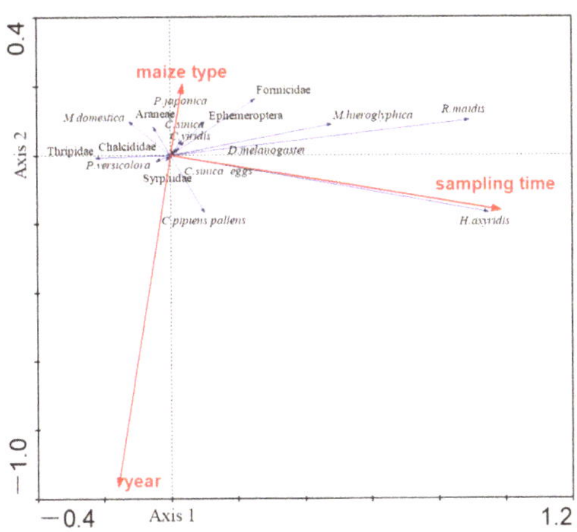

Figure 4. RDA for the composition of arthropod community in the suction sampler method, and the relationship of the composition of arthropod community and environmental factors. The red arrows represent variable factors, and the blue arrows represent arthropod species collected in the fields. The length of the blue arrows indicates the magnitude of the effect of environmental factors on arthropods. The cosine of the angle between the red and the blue arrows represents the correlation between variable factors and arthropods.

2.3.2. Similarity of Arthropod Communities in Transgenic and Non-Transgenic Maize Fields

The arthropod community structures in transgenic and non-transgenic maize fields were further explored using non-metric multidimensional scaling analysis (nMDS). The distance between the two sampling points was estimated using the Bray–Curtis dissimilarity index. Differences in the composition of arthropod community among all animal samples were visualized in the nMDS plot. The samples collected in the nMDS plot were separated by sampling time but not by maize type or year (Figures 5–7), which was confirmed by a more detailed analysis of similarity (ANOSIM). A significant correlation was detected between arthropod community composition and sampling time, but arthropods were not affected by maize type or year (Table 5).

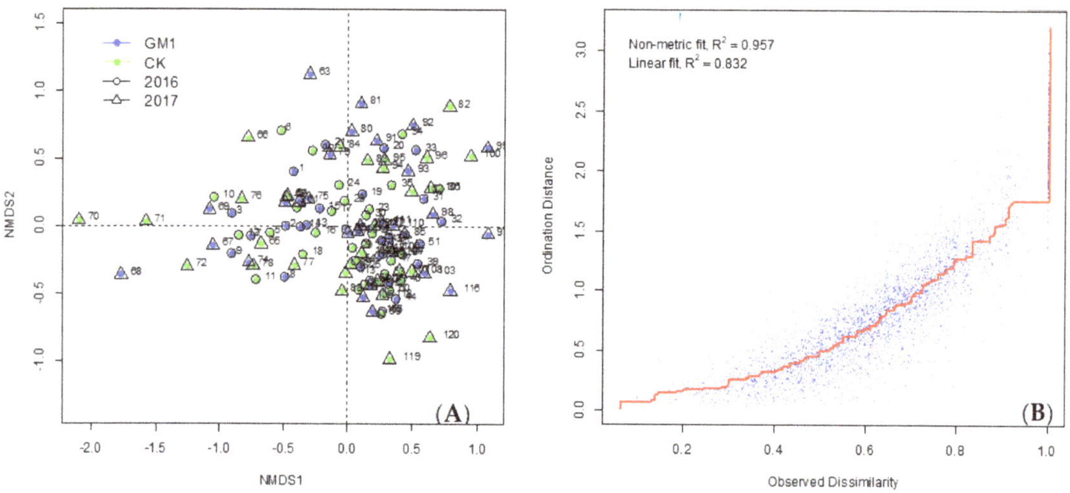

Figure 5. nMDS plot of community structures of arthropods from GM1 and CK in each sampling time in 2016 and 2017 using the whole plant inspection method. (**A**): Blue indicates transgenic maize, and green indicates non-transgenic maize. The circles with the numbers 1 to 60 indicate the sampling points in 2016. Numbers 1–3, 7–9, 13–15, 19–21, 25–27, 31–33, 37–39, 43–45, 49–51, 55–57: transgenic maize from 1st to 10th stage respectively; while 4–6, 10–12, 16–18, 22–24, 28–30, 34–36, 40–42, 46–48, 52–54, 58–60: non-transgenic maize from 1st to 10th stage respectively. The triangles with the numbers 61 to 120 indicate the sampling points in 2017. Numbers 61–63, 67–69, 73–75, 79–81, 85–87, 91–93, 97–99, 103–105, 109–111, 115–117: transgenic maize from 1st to 10th stage respectively; while 64–66, 70–72, 76–78, 82–84, 88–90, 94–96, 100–102, 106–108, 112–114, 118–120: non-transgenic maize from 1st to 10th stage respectively. (**B**): Shepard stress plot.

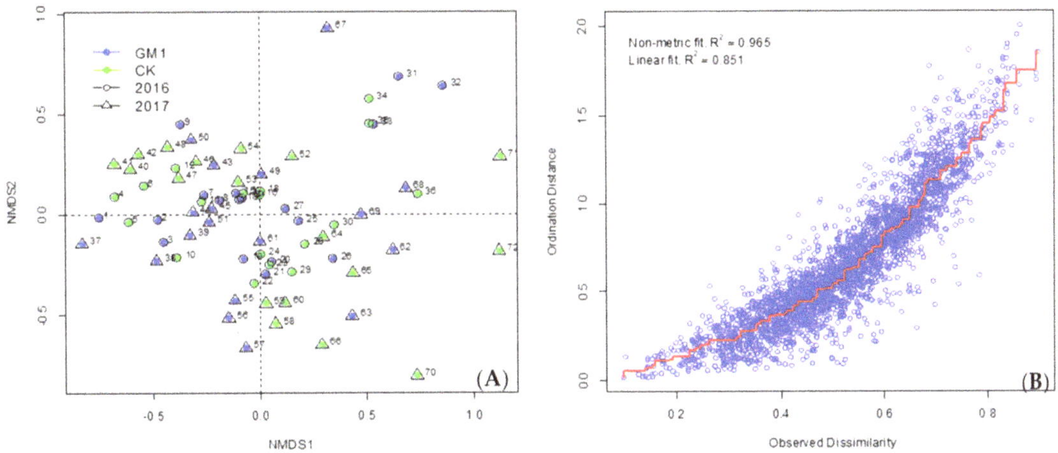

Figure 6. nMDS plot of community structures of arthropods from GM1 and CK in each sampling time in 2016 and 2017 using the pitfall trap method. (**A**): Blue indicates transgenic maize, and green indicates non-transgenic maize. The circles with the numbers 1 to 36 indicate the sampling points in 2016. Numbers 1–3, 7–9, 13–15, 19–21, 25–27, 31–33: transgenic maize from 1st to 6th stage respectively; while 4–6, 10–12, 16–18, 22–24, 28–30, 34–36: non-transgenic maize from 1st to 6th stage respectively. The triangles with the numbers 37 to 72 indicate the sampling points in 2017. Numbers 37–39, 43–45, 49–51, 55–57, 61–63, 67–69: transgenic maize from 1st to 6th stage respectively, while 40–42, 46–48, 52–54, 58–60, 64–66, 70–72: non-transgenic maize from 1st to 6th stage respectively. (**B**): Shepard stress plot.

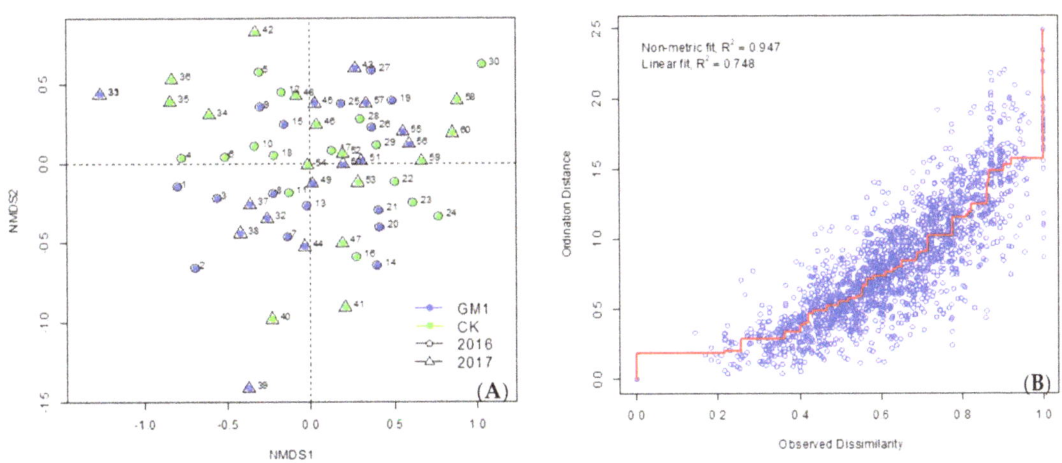

Figure 7. nMDS plot of community structures of arthropods from GM1 and CK in each sampling time in 2016 and 2017 by the suction sampler method. (**A**): Blue indicates transgenic maize, and green indicates non-transgenic maize. The circles with the numbers 1 to 30 indicate the sampling points in 2016. Numbers 1–3, 7–9, 13–15, 19–21, 25–27: transgenic maize from 1st to 5th stage respectively; while 4–6, 10–12, 16–18, 22–24, 28–30: non-transgenic maize from 1st to 5th stage respectively. The triangles with the numbers 31 to 60 indicate the sampling points in 2017. Numbers 31–33, 37–39, 43–45, 49–51, 55–57: transgenic maize from 1st to 5th stage respectively; while 34–36, 40–42, 46–48, 52–54, 58–60: non-transgenic maize from 1st to 5th stage respectively). (**B**): Shepard stress plot.

Table 5. Effects of maize type (GM1 and CK), year, and sampling time on the community structure of arthropods (nMDS structure) in 2016 and 2017.

Correlation with nMDS Structure	Whole Plant Inspection Method		Pitfall Trap Method		Suction Sampler Method	
	R^2	p	R^2	p	R^2	p
Sampling time	0.80	**0.001** ***	0.67	**0.001** ***	0.49	**0.001** ***
Year	0.00	0.07	0.00	0.26	0.00	0.112
Maize type	0.00	0.994	0.00	0.897	0.00	0.986

The values in bold font are statistically significant (*** $p < 0.001$).

3. Discussion

Biodiversity of arthropod communities is an important evaluation index for the safety of genetically modified crops. We present the analysis of the arthropod communities affected by the cultivation of the genetically modified drought-resistant maize cultivar, GM1. We provide basic safety data for the risk assessment associated to the commercialization of drought-resistant maize, GM1.

Although GM crops have significantly improved crop quality and resistance, they have also sparked intense debate regarding their safety [31]. With the increase in commercial applications and the expansion of the planting area of GM crops, their biosafety to arthropods has become a major concern. To date, studies on genetically modified insect- and herbicide-resistant maize have shown that maize type has no significant effect on arthropod diversity [32–38]. Moreover, it has been demonstrated that maize type and year do not have significant impacts on biodiversity [39–42]. Genetically modified wheat, TB4, has no significant effects on the content of available nutrients, enzyme activities, or diversity of the microbial community in the soil rhizosphere [25]. Similarly, drought-tolerant wheat (*Triticum aestivum* L.) and genetically modified dehydration-responsive element-binding 3 (DREB3) drought-resistant soybeans have low potential weediness [26,27]. A significant effect on the number of bacteria, actinomycetes, fungi, *Trichoderma*, or *Azotobacter* in the rhizosphere of transgenic drought resistant soybean was not observed [28]. In our 2-year field trial conducted under natural conditions, we found that genetically modified drought-resistant maize, GM1, had no significant effect on arthropod biodiversity. This result is consistent with previous studies showing that genetically modified maize did not affect arthropod biodiversity in the field.

In this study, the effects of genetically modified maize showed that sampling time and year were more related to the diversity and richness of arthropods than maize type. The results of Guo and Fan support this view [17,18,40]. Furthermore, RDA showed that sampling time and year were the leading causes of variation in the arthropod community composition, which may be explained as follows: (1) sampling times differ with weather conditions, (2) sampling years differ in climatic conditions, and (3) arthropods are influenced by climate conditions in a complex manner. The multivariate permutation test for analyzing the effects of maize type, year, and sampling time showed that sampling time had a close relationship with Bray–Curtis distances but not maize type or year. Arthropod diversity and composition were also affected by environmental factors, such as climate, humidity, temperature, and light [41], which may explain why sampling time was the main reason for variation.

Analysis of five parameters, including Shannon–Wiener, Pielou's, Simpson's indexes, number of species, and the total number of individuals, showed that GM1 had a negligible effect on arthropod abundance and diversity. Furthermore, RDA and nMDS of the variable factors in the three investigation methods indicated that the arthropod community composition was not associated with maize type. Moreover, the three arthropod investigation methods used in this study complemented each other to provide a more comprehensive collection of arthropods [43–45]. However, using these three methods simultaneously is an intensive and complex operation; hence, we are trying to simplify the sampling method without compromising a comprehensive sample collection.

This field study evaluated the environmental safety of drought-resistant genetically modified maize, GM1, in Jilin in 2016 and 2017, when annual precipitation was normal (drought). Results revealed that the maize type had no impact on the diversity, abundance, or composition of the community of arthropods. Therefore, we concluded that planting drought-resistant genetically modified maize GM1 in Jilin Province will not adversely affect the environment.

4. Materials and Methods

4.1. Experimental Materials

Genetically modified DREB maize (GM1) and its non-transformed counterpart (CK) were used in this study.

4.2. Experimental Field

Safety assessment trials were conducted in 2016 and 2017 at the Jilin Academy of Agricultural Sciences in Gongzhuling, Jilin Province, China (43°30′ N, 124°49′ E).

Jilin Province is a vital spring maize production region in China; however, Jilin is a rainfed agricultural area, in which various degrees of drought occur annually [46]. In the 2 years of study, precipitation was normal, which was 772.5 and 546.9 mm in 2016 and 2017, respectively, and it mostly occurred from May to September (84.3% in 2016 and 94.5% in 2017). Meteorological data were obtained from the Gongzhuling Meteorological Observatory in Jilin.

4.3. Experimental Design

Maize seeds were planted on 5 May in 2016 and 2017. Three replicate plots for each cultivar (GM1 and Chang 7-2 (CK)) were established in a completely randomized design [17,18,39,40,47]. According to the National Standards of the People's Republic of China, Ministry of Agriculture Announcement No. 2122-10.4-2014, the following experimental set up was established: Each plot was 10 m wide and 15 m long and contained 25 rows with 60 cm spacing. There were 40 plants in each row, 25 cm apart. No pesticides were applied during the growth period [48]. This study was approved by the Ministry of Agriculture of the People's Republic of China and by the Genetically Modified Organism Safety Team at the Jilin Academy of Agricultural Sciences, China. The field study did not involve endangered or protected species.

4.4. Sample Collection Methods

The Ministry of Agriculture Announcement No. 2122-10.4-2014 requires the use of whole plant inspection, pitfall trap, and suction samplers to evaluate the impact of biodiversity [48]. The whole plant inspection method was more flexible and simpler, the suction sampler was very effective in catching small and flying arthropods, and the pitfall trap method could continuously collect ground-dwelling arthropods in the daytime and at night [43–45]. The use of these methods provides a more comprehensive collection of arthropods.

4.4.1. Whole Plant Inspection Method

Arthropods were investigated every 7 d (depending on the weather) from day 10 after the final plant thinning to maturity. A five-spot sampling method was used to investigate the arthropods in each plot: one was located at the center of the plot, and the other four were located at the middle of the lines that connected the center of the plot and the plot corners. At each sampling spot, we randomly and gently turned over five maize plants one by one and quickly counted the visible arthropods on the plant surface and on the ground within 1 m^2 around the plant. During the investigation, the number of active insects and spiders were counted.

4.4.2. Pitfall Trap Method

Arthropods were investigated every 10 d (depending on the weather) from day 10 after final thinning to maturity. A five-spot sampling method was used. Three pitfall traps with a spacing of 0.5 m were established at each sampling spot in every plot. For each trap, a plastic cup (15 cm in diameter × 10 cm in depth) was buried in the soil, with the upper rim of the cup level with the ground. Each cup was then placed in 5% detergent diluent (no more than 1/3 of the cup volume). Traps were exposed to the field for approximately 24 h, after which they were removed, and the trapped arthropods were collected and stored in 75% ethanol for species and number identification.

4.4.3. Suction Sampler Method

The first investigation was conducted 15 days after final thinning of maize, and then it was repeated at the mid-whorl and late whorl stages, at the peak of the silking stage, and the late grain-filling stage. Arthropods were investigated five times in total using this method. A five-spot sampling method was used again in this case. At each sampling site, five maize plants were randomly selected, and the arthropods on the whole maize plant and on the ground within 1 m^2 of the plant were extracted with a suction sampler (John W. Hock Co. Gainesville, FL, USA).

The unknown species sampled by the above methods were collected and stored in a 5 mL plastic tube containing 75% alcohol and brought to the laboratory for identification under an insect anatomical lens (Motic, Xiamen City, China).

4.5. Statistical Analysis

As a single method cannot comprehensively evaluate the arthropod community, it is necessary to use a combination of methods to obtain a more comprehensive understanding. Therefore, diversity indices (Shannon–Weiner and Simpson's indexes, and for evenness, Pielou's index), abundance indices, redundancy analysis (RDA), and nonmetric multidimensional scaling (nMDS) were used in this study.

Arthropod diversity and abundance were analyzed using the Data Processing System (DPS, version 2006 package, China). The Shannon–Wiener index (H'), Simpson's diversity index (D), and Pielou's index (J) were calculated as follows:

$$H' = -\sum_{i=1}^{S} P_i \ln(P_i) \tag{1}$$

where P_i is the proportion of individuals belonging to the ith taxon in a single plot.

$$D = 1 - \sum_{i=1}^{S} \frac{N_i(N_i - 1)}{N(N-1)} \tag{2}$$

where N_i is the individual number in the ith taxon in one plot, and N is the total number of animals in one plot.

$$J = H'/\ln S \tag{3}$$

where, S is the genus number of the collected arthropods in one plot.

In this study, maize type, year, and sampling time were considered as the factors presumably influencing arthropod abundance and diversity. They were all used as fixed factors when analyzing for repeated-measures ANOVA (SPSS 23.0).

According to Guo's research in 2016 [17,18], this study used RDA, a canonical analysis for studying environmental safety and analyze the relationships among arthropods and the relationships between arthropods and maize type, year, or sampling time, to identify the influencing factors of arthropod community. RDA can be performed using CANOCO 4.5 [41,49]. In addition, nonmetric multidimensional scaling (nMDS) was used to indicate the effects of each influencing factor on arthropod community, which is an indirect ordination method that can reveal the similarity of arthropod samples through metric multidimensional scale analysis using the Bray–Curtis distance between sampling points [50,51]. nMDS was conducted using the Vegan package in R (v.4.0.4; R Development Core Team).

5. Conclusions

Maize variety did not influence arthropod abundance, diversity, or community composition over 2 consecutive years of study in the field. Further, RDA indicated that maize variety did not significantly affect arthropod community composition. These results strongly suggest that the cultivation of DREB-GM maize does not affect the arthropod community. However, considering the limitations of time and area of influence inherent to this study, long-term field experimentation in a larger area is necessary to guarantee the environmental and ecological safety of genetically modified crop plants.

Author Contributions: Conceptualization: J.-Q.Y., J.-G.L. and X.-Y.S.; supervision: J.-Q.Y., J.-G.L. and X.-Y.S.; investigation: J.-Q.Y. and D.-M.W.; data analysis: J.-Q.Y. and D.-M.W.; writing—original draft: J.-Q.Y.; writing—review and editing: J.-G.L., J.-Q.Y. and X.-Y.S. All authors have read and agreed to the published version of the manuscript.

Funding: This research was funded by the National Nature Science Fund of China (32071546).

Institutional Review Board Statement: Not applicable.

Informed Consent Statement: Not applicable.

Data Availability Statement: The data presented in this study are available on request from the corresponding author. The data are not publicly available due to privacy.

Conflicts of Interest: The authors declare that they have no conflict of interest.

References

1. Li, B.; Wang, Z.C.; Sun, Z.G.; Chen, Y.; Yang, F. Resources and sustainable resource exploitation of salinized land in China. *Agric. Res. Arid. Areas* **2005**, *23*, 154–158.
2. Ji, J.H.; Li, Y.Y.; Liu, S.Q.; Tong, Y.X. Effect of Plastic Film Mulch and Drip Irrigation under Plastic Film Mulch on Growth and Development of Maize and Water use Efficiency. *Water Sav. Irrig.* **2015**, *3*, 22-24+27.
3. Yu, G.R.; Zhang, W.; Du, W.P.; Song, J.; Chen, Q.; Xu, L.Y. Transgenic Drought-resistant Gene IrrE Maize Material Obtaining and Resistance Identification. *Southwest China J. Agric. Sci.* **2016**, *29*, 1782–1786. [CrossRef]
4. Liu, B.; Zeng, Q.; Yan, F.M.; Xu, H.G.; Xu, C.R. Effects of transgenic plants on soil microorganisms. *Plant Soil* **2005**, *271*, 1–13. [CrossRef]
5. Truter, J.; Hamburg, H.V.; Berg, J.V.D. Comparative diversity of arthropods on Bt maize and non-Bt maize in two different cropping systems in South Africa. *Environ. Entomol.* **2014**, *43*, 197–208. [CrossRef] [PubMed]
6. Yu, H.L.; Li, Y.H.; Li, X.J.; Wu, K.M. Arthropod Abundance and Diversity in Transgenic Bt Soybean. *Environ. Entomol.* **2014**, *43*, 1124–1134. [CrossRef]
7. Bal, H.K.; Dhawan, A.K. Impact of intercropping on arthropod diversity in Bt and non-Bt cotton. *J. Insect Sci.* **2010**, *23*, 136–140.
8. Bal, H.K.; Dhawan, A.K. Impact of transgenic cotton on diversity of non-target arthropod communities in cotton agro-ecosystem. *J. Insect Sci.* **2009**, *22*, 130–138.
9. Romeis, J.; Hellmich, R.L.; Candolfi, M.P.; Carstens, K.; Schrijver, A.D.; Gatehouse, A.M.R.; Herman, R.A.; Huesing, J.E.; Mclean, M.A.; Raybould, A. Recommendations for the design of laboratory studies on non-target arthropods for risk assessment of genetically engineered plants. *Transgenic Res.* **2011**, *20*, 1–22. [CrossRef]
10. Romeis, J.; Raybould, A.; Bigler, F.; Candolfi, M.P.; Hellmich, R.L.; Huesing, J.E.; Shelton, A.M. Deriving criteria to select arthropod species for laboratory tests to assess the ecological risks from cultivating arthropod-resistant genetically engineered crops. *Chemosphere* **2013**, *90*, 901–909. [CrossRef]
11. Romeis, J.; Meissle, M.; Alvarez-Alfageme, F.; Bigler, F.; Bohan, D.A.; Devos, Y.; Malone, L.A.; Pons, X.; Rauschen, S. Potential use of an arthropod database to support the non-target risk assessment and monitoring of transgenic plants. *Transgenic Res.* **2014**, *23*, 995–1013. [CrossRef] [PubMed]
12. Szénási, A.; Pálinkás, Z.; Zalai, M.; Schmitz, O.J.; Adalbert, B. Short-term effects of different genetically modified maize varieties on arthropod food web properties: An experimental field assessment. *Sci. Rep.* **2014**, *4*, 5315. [CrossRef] [PubMed]
13. Hu, H.H.; Xiong, L.Z. Genetic Engineering and Breeding of Drought-Resistant Crops. *Annu. Rev. Plant Biol.* **2014**, *65*, 715–741. [CrossRef] [PubMed]
14. Kang, Q.; Vahl, C.I. Statistical Analysis in the Safety Evaluation of Genetically-Modified Crops: Equivalence Tests. *Crop Sci.* **2014**, *54*, 2183–2200. [CrossRef]

15. Zhang, M.D.; Sun, L.; Xiong, Q.F. The Safety Evaluation of the Genetically Modified Crops. *Hubei Agric. Sci.* **2015**, *6*, 98–104. [CrossRef]
16. Ren, Z.T.; Shen, W.J.; Liu, B.; Xue, K. Effects of Transgenic Maize on Biodiversity of Arthropod Communities in the Fields. *Sci. Agric. Sin.* **2017**, *50*, 2315–2325. [CrossRef]
17. Guo, J.F.; He, K.L.; Bai, S.X.; Zhang, T.T.; Liu, Y.J.; Wang, F.X.; Wang, Z.Y. Effects of transgenic cry1Ie maize on non-lepidopteran pest abundance, diversity and community composition. *Transgenic Res.* **2016**, *25*, 761–772. [CrossRef]
18. Guo, J.F.; He, K.L.; Hellmich, R.L.; Bai, S.X.; Zhang, T.T.; Liu, Y.J.; Ahmed, T.; Wang, Z.Y. Field trials to evaluate the effects of transgenic cry1Ie maize on the community characteristics of arthropod natural enemies. *Sci. Rep.* **2016**, *6*, 22102. [CrossRef]
19. Guo, Y.Y.; Feng, Y.J.; Ge, Y.; Tetreau, G.; Chen, X.W.; Dong, X.H.; Shi, W.P. The Cultivation of Bt Corn Producing Cry1Ac Toxins Does Not Adversely Affect Non-Target Arthropods. *PLoS ONE* **2014**, *9*, e114228. [CrossRef]
20. Habustova, O.; Dolezal, P.; Spitzer, L.; Svobodova, Z.; Hussein, H.; Sehnal, F. Impact of Cry1Ab toxin expression on the non-target insects dwelling on maize plants. *J. Appl. Entomol.* **2014**, *138*, 164–172. [CrossRef]
21. Rauschen, S.; Schultheis, E.; Hunfeld, H.; Schaarschmidt, F.; Schuphan, I.; Eber, S. Diabrotica-resistant, Bt-maize, DKc5143 event MON88017 has no impact on the field densities of the leafhopper, Zyginidia scutellaris. *Environ. Biosaf. Res.* **2010**, *9*, 87–99. [CrossRef] [PubMed]
22. Xu, Z.S.; Chen, M.; Li, L.C.; Ma, Y.Z. Functions of the ERF transcription factor family in plants. *Botany* **2008**, *86*, 969–977. [CrossRef]
23. Xu, Z.S.; Chen, M.; Li, L.C.; Ma, Y.Z. Functions and application of the AP2/ERF transcription factor family in crop improvement. *J. Integr. Plant Biol.* **2011**, *53*, 570–585. [CrossRef] [PubMed]
24. Xu, Z.S.; Ni, Z.Y.; Liu, L.; Nie, L.N.; Li, L.C.; Chen, M.; Ma, Y.Z. Characterization of the TaAIDFa gene encoding a CRT/DRE-binding factor responsive to drought, high-salt, and cold stress in wheat. *Mol. Genet. Genom.* **2008**, *280*, 497–508. [CrossRef]
25. Jiang, Q.Y.; Li, X.H.; Hu, Z.; Ma, Y.Z.; Zhang, H.; Xu, Z.S. Safety Assessment of Weediness of Transgenic Drought-Tolerant Wheat (*Triticum aestivum* L.). *Sci. Agric. Sin.* **2015**, *48*, 2096–2107.
26. Yu, M.Z.; Dai, R.Y.; Wu, J.R.; Xu, J.H.; Du, J.; Chen, S.L.; Shi, J.R. Analysis of available nutrient, enzyme activities and microorganism community diversity in rhizospheric soil of TaDREB4 transgenic wheat with drought resistance. *Jiangsu J. Agric. Sci.* **2013**, *29*, 938–945.
27. Wang, J.Y.; Ding, W.; Muhammad, S.K.; Cheng, Z.; Dai, H.Y. Survival competition between transgenic drought-resistant soybean and weeds with different carbon metabolism pathways under drought stress. *Jiangsu Agric. Sci.* **2018**, *46*, 73–76. [CrossRef]
28. Cao, Y.; Ding, W.; Li, X.H.; Ma, Y.Z.; Wang, Z.H.; Li, W.B. Effect of transgenic drought resistant soybean on soil microbial community and beneficial microorganism. *J. Northeast Agric. Univ.* **2011**, *42*, 17–20. [CrossRef]
29. Xing, S.; Yao, Y.; Lei, X.; Hu, X.; Yang, L. Advances of Drought-resistant Genes and Drought-resistant Transgenic of Main Crops of Gramineae Plants. *Chin. Agric. Sci. Bull.* **2014**, *42*, 251–258.
30. Shirai, Y. Influence of transgenic insecticidal crops on non-target arthropods: A review. *Jpn. J. Ecol.* **2004**, *54*, 47–65. [CrossRef]
31. Cui, K.; Shoemaker, S.P. Public perception of genetically modified (GM) food: A Nationwide Chinese consumer study. *Sci. Food* **2018**, *2*, 10. [CrossRef] [PubMed]
32. Li, Y.H.; Zhang, X.J.; Chen, X.P.; Romeis, J.; Yin, X.M.; Peng, Y.F. Consumption of Bt rice pollen containing Cry1C or Cry2A does not pose a risk to Propylea japonica (Thunberg) (Coleoptera: Coccinellidae). *Sci. Rep.* **2015**, *5*, 7679. [CrossRef] [PubMed]
33. Wang, Y.Y.; Li, Y.H.; Romeis, J.; Chen, X.P.; Zhang, J.; Chen, H.Y.; Peng, Y.F. Consumption of Bt rice pollen expressing Cry2Aa does not cause adverse effects on adult Chrysoperla sinica Tjeder (Neuroptera: Chrysopidae). *Biol. Control* **2012**, *61*, 246–251. [CrossRef]
34. Yang, Y.; Liu, Y.; Cao, F.Q.; Chen, X.P.; Chen, L.S.; Romeis, J.; Li, Y.H.; Peng, Y.F. Consumption of Bt Rice Pollen Containing Cry1C or Cry2A Protein Poses a Low to Negligible Risk to the Silkworm Bombyx mori (Lepidoptera: Bombyxidae). *PLoS ONE* **2014**, *9*, e102302. [CrossRef]
35. Ahmad, A.; Wilde, G.E.; Zhu, K.Y. Detectability of Coleopteran-specific Cry3Bb1 Protein in Soil and Its Effect on Nontarget Surface and Below-Ground Arthropods. *Environ. Entomol.* **2005**, *34*, 385–394. [CrossRef]
36. Li, X.G.; Liu, B.; Desneux, N. A 2-Year Field Study Shows Little Evidence That the Long-Term Planting of Transgenic Insect-Resistant Cotton Affects the Community Structure of Soil Nematodes. *PLoS ONE* **2013**, *8*, e61670. [CrossRef]
37. Romeis, J.; Bartsch, D.; Bigler, F.; Candolfi, M.P.; Gielkens, M.M.C.; Hartley, S.E.; Hellmich, R.L.; Huesing, J.E.; Jepson, P.C.; Layton, R.; et al. Assessment of risk of insect-resistant transgenic crops to nontarget arthropods. *Nat. Biotechnol.* **2008**, *26*, 203–208. [CrossRef]
38. Li, Y.H.; Romeis, J.; Wu, K.M.; Peng, Y.F. Tier-1 assays for assessing the toxicity of insecticidal proteins produced by genetically engineered plants to non-target arthropods. *Insect Sci.* **2014**, *21*, 125–134. [CrossRef]
39. Wang, B.F.; Wu, F.C.; Yin, J.Q.; Jiang, Z.L.; Song, X.Y.; Reddy, G.V.P. Use of taxonomic and trait-based approaches to evaluate the effect of bt maize expressing cry1ie protein on non-target collembola: A case study in northeast china. *Insects* **2021**, *12*, 88. [CrossRef]
40. Fan, C.M.; Wu, F.C.; Dong, J.Y.; Wang, B.F.; Yin, J.Q.; Song, X.Y. No impact of transgenic cry1Ie maize on the diversity, abundance and composition of soil fauna in a 2-year field trial. *Sci. Rep.* **2019**, *9*, 10333. [CrossRef]
41. Song, X.Y.; Chang, L.; Reddy, G.V.P.; Zhang, L.; Fan, C.M.; Wang, B.F. Use of taxonomic and trait-based approaches to evaluate the effects of transgenic cry1ac corn on the community characteristics of soil collembola. *Environ. Entomol.* **2018**, *48*, 263–269. [CrossRef] [PubMed]

42. Carpenter, J.E. Impact of GM crops on biodiversity. Communities of ground-dwelling arthropods in conventional and transgenic maize: Background data for the post-market environmental monitoring. *J. Appl. Entomol.* **2015**, *139*, 31–45.
43. Bogya, S.; Markó, V. Effect of pest management systems on ground-dwelling spider assemblages in an apple orchard in Hungary. *Agric. Ecosyst. Environ.* **1999**, *73*, 7–18. [CrossRef]
44. Liu, J.; Chen, J. Effects of transgenetic Bt cotton on ground-dwelling spider assemblages by pitfall traps. *J. Plant Prot.* **2015**, *42*, 59–65. [CrossRef]
45. Wang, Y.; Chen, J.; Xiao, D.H.; Ma, F.G.; Hua, H.X. Assessing the efficacy of different sampling methods for arthropods in rice field. *J. Environ. Entomol.* **2016**, *38*, 1090–1098. [CrossRef]
46. Mu, J.; Qiu, M.J.; Gu, Y.; Ren, J.Q.; Liu, Y. Applicability of five drought indices for agricultural drought evaluation in Jilin Province, China. *Chin. J. Appl. Ecol.* **2018**, *29*, 2624–2632. [CrossRef]
47. Liang, J.G.; Meng, F.; Sun, S.; Wu, C.X.; Wu, H.Y.; Zhang, M.R.; Zhang, H.F.; Zheng, X.B.; Song, X.Y.; Zhang, Z.G. Community structure of arbuscular mycorrhizal fungi in rhizospheric soil of a transgenic high-methionine soybean and a near isogenic variety. *PLoS ONE* **2015**, *10*, e0145001. [CrossRef]
48. Ministry of Agriculture of the PRC. NY/T 720.3-2003 Environmental impact testing of genetically modified maize-Part 3:testing the effects on biodiversity. Standards Press of China: Beijing, China, 2003.
49. Braak, C.J.F.T.; Smilauer, P. CANOCO reference manual andCanoDraw for Windows user's guide: Software for canonical communityordination (version 4.5). Microcomputer Power: Ithaca, NY, USA, 2002.
50. Jackson, M.M.; Turner, M.G.; Pearson, S.M.; Ives, A.R. Seeing the forest and the trees: Multilevel models reveal both species and community patterns. *Ecosphere* **2012**, *3*, 1–16. [CrossRef]
51. Clarke, K.R. Non-parametric multivariate analyses of changes in community structure. *Aust. J. Ecol.* **2010**, *18*, 117–143. [CrossRef]

plants

MDPI

Review

Environmental Behaviors of *Bacillus thuringiensis* (*Bt*) Insecticidal Proteins and Their Effects on Microbial Ecology

Yujie Li [1,2,3,†], Cui Wang [2,3,†], Lei Ge [2,3], Cong Hu [2,3], Guogan Wu [2,3], Yu Sun [2,3], Lili Song [2,3], Xiao Wu [2,3], Aihu Pan [2,3], Qinqing Xu [4], Jialiang Shi [5], Jingang Liang [6,*] and Peng Li [2,3,7,*]

1 College of Food Sciences and Technology, Shanghai Ocean University, Shanghai 201306, China; lyj1187571462@163.com
2 Biotechnology Research Institute, Shanghai Academy of Agricultural Sciences, Shanghai 201106, China; cuiwang87@126.com (C.W.); gl13084233857@163.com (L.G.); hucong819@163.com (C.H.); wgghappy@126.com (G.W.); yvessuen@hotmail.com (Y.S.); songlili@saas.sh.cn (L.S.); gwuxiao@126.com (X.W.); aihup@163.com (A.P.)
3 Shanghai Key Laboratory of Agricultural Genetics and Breeding, Shanghai 201106, China
4 Shandong County Agricultural Technology Extension Center, Jinan 250003, China; xuqinqing1982@163.com
5 Dezhou Academy of Agricultural Sciences, Dezhou 253000, China; shijialiang2004@163.com
6 Development Center of Science and Technology, Ministry of Agriculture and Rural Affairs, Beijing 100176, China
7 Shanghai Co-Elite Agricultural Sci-Tech (Group) Co., Ltd., Shanghai 201106, China
* Correspondence: liangjingang@agri.gov.cn (J.L.); pengli13@fudan.edu.cn (P.L.)
† These authors contributed equally to this work.

Citation: Li, Y.; Wang, C.; Ge, L.; Hu, C.; Wu, G.; Sun, Y.; Song, L.; Wu, X.; Pan, A.; Xu, Q.; et al. Environmental Behaviors of *Bacillus thuringiensis* (*Bt*) Insecticidal Proteins and Their Effects on Microbial Ecology. *Plants* **2022**, *11*, 1212. https://doi.org/10.3390/plants11091212

Academic Editors: Wei Wei and C. Neal Stewart Jr.

Received: 24 March 2022
Accepted: 28 April 2022
Published: 29 April 2022

Abstract: *Bt* proteins are crystal proteins produced by *Bacillus thuringiensis* (*Bt*) in the early stage of spore formation that exhibit highly specific insecticidal activities. The application of *Bt* proteins primarily includes *Bt* transgenic plants and *Bt* biopesticides. Transgenic crops with insect resistance (via *Bt*)/herbicide tolerance comprise the largest global area of agricultural planting. After artificial modification, *Bt* insecticidal proteins expressed from *Bt* can be released into soils through root exudates, pollen, and plant residues. In addition, the construction of *Bt* recombinant engineered strains through genetic engineering has become a major focus of *Bt* biopesticides, and the expressed *Bt* proteins will also remain in soil environments. *Bt* proteins expressed and released by *Bt* transgenic plants and *Bt* recombinant strains are structurally and functionally quite different from *Bt* prototoxins naturally expressed by *B. thuringiensis* in soils. The former can thus be regarded as an environmentally exogenous substance with insecticidal toxicity that may have potential ecological risks. Consequently, biosafety evaluations must be conducted before field tests and production of *Bt* plants or recombinant strains. This review summarizes the adsorption, retention, and degradation behavior of *Bt* insecticidal proteins in soils, in addition to their impacts on soil physical and chemical properties along with soil microbial diversity. The review provides a scientific framework for evaluating the environmental biosafety of *Bt* transgenic plants, *Bt* transgenic microorganisms, and their expression products. In addition, prospective research targets, research methods, and evaluation methods are highlighted based on current research of *Bt* proteins.

Keywords: *Bt* insecticide protein; environmental behaviors; *Bt* crops; *Bt* biopesticides; soil microorganisms

1. Introduction

Bt protein is a δ-endotoxin insecticidal crystal protein derived from the *Bt* protoxin produced by *Bacillus thuringiensis* (*Bt*) in the early stage of endospore formation. The crystal protein exhibits highly specific insecticidal activity due to protease hydrolysis [1]. Indeed, *Bt* exhibits poisoning effects on Lepidoptera, Diptera, Coleoptera, Hymenoptera, Hemiptera, Orthoptera, Mallophaga, and other insects, in addition to snails, nematodes, and protozoa [2–4]. Importantly, *Bt* proteins are selective for insects, but they are harmless to humans, vertebrates, and plants. Consequently, *Bt* accounts for about 90% of the production and use of biopesticides and has become the most widely used microbially produced pesticide globally, with widespread use in agriculture, forestry, fruit trees, vegetables, and environmental sanitation pest control [5,6]. However, *Bt* insecticides feature disadvantages, including narrow insecticidal spectra, poor stability, short residual effect periods, and the generation of insect resistance, which all limit their wider application.

The artificial design and optimization of *Bt* insecticidal proteins have led to the construction of high-efficiency and broad-spectrum recombinant insecticidal proteins, with recombinant engineered *Bt* strains being an effective means to solve the above problems [7]. For example, Wang et al. introduced the Cry3Aa7 gene encoding a coleopteran-specific insecticide into the wild *B. thuringiensis* strain G03 using electroporation, with the insecticide exhibiting high toxicity to lepidopteran pests [8]. The resultant recombinant engineered strain (G033A) was the first instance of a transgenic *Bt* engineered strain that was approved and registered as a pesticide in China [9]. In addition, *Bt* insecticidal proteins are also widely used in *Bt* transgenic crops that use genetic engineering technology to introduce exogenous beneficial genes into crop genomes. These introduced genes can be stably inherited, resulting in crops that have better agronomic traits and economic value [10,11].

The global large-scale commercial cultivation of genetically modified crops beginning in 1996 has led to the expansion of planted areas with each passing year, reaching 190.4 million hectares in 2019 (a ~112-fold increase compared to 1996). Among these, insect resistance (*Bt* gene)/herbicide tolerance composite traits within genetically modified crops comprise the largest global area of planted crops. In particular, transgenic insect-resistant cotton is a genetically modified *Bt* crop that has been approved for commercial cultivation in China [12]. *Bt* is the most widely used insect resistance gene and exhibits the greatest potential and application prospects for plant genetic engineering research. The *Bt* gene has been successfully introduced into tobacco [13], corn [14], and cotton [15] plants, among others, leading to numerous transgenic plant varieties or germplasm resources with good insect resistance. Rui et al. observed that the content of the *Bt* toxin secreted by rhizospheres can reach 200–300 ppb [16], while Valldor et al. reported that the Cry1Ab protein contents released by *Bt* transgenic maize Mon810 through root systems into soils reached 165 g/hectare [17]. Residual accumulated *Bt* proteins in the environment exceed the consumption by insect larvae and degradation by environmental factors, leading to potential impacts on the abundances, community structures, and functions of natural soil microbial communities (Figure 1A) [18,19]. Crecchio and Stotzky observed that the *Bt* proteins secreted by *Bt* maize roots can intimately associate with clay minerals and humic acids, among other soil substrates, leading to retained insecticidal activity in soils. These complex forms are more difficult to biodegrade compared with free state Bt, also suggesting the potential for sustained ecological risks [20].

In addition to the development of *Bt* recombinant engineered strains with higher insecticidal activity and wider insecticidal spectrum to delay insect resistance evolution, researchers have also commonly artificially modified the codons or promoter sequences of *Bt* to promote the expression of *Bt* in transgenic plants [21]. Consequently, the *Bt* proteins expressed by *Bt* transgenic plants are quite different from the *Bt* prototoxins expressed by the original *Bt* bacteria in terms of protein structure and function. Thus, *Bt* proteins that are expressed by *Bt* recombinant engineered strains and *Bt* transgenic plants represent exogenous environmental substances with insecticidal toxicity [22,23]. In addition, because *Bt* proteins can be complex with clays and humic acids when entering soils, they are not easily degraded by microorganisms and retain their biological activity for extended periods of time (Figure 1A). Consequently, the environmental behavior of exogenous *Bt* proteins in soil ecosystems and their impact on soil biodiversity should be systematically evaluated [24]. To improve the safety of genetically modified organisms and their products, China has implemented a safety evaluation system for transgenic plants (including *Bt* transgenic plants) and transgenic microorganisms (including those used in plant crops) [25]. Only transgenic products that have been evaluated for safety can now be planted in fields and commercially produced at a large scale. This paper summarizes the structure and mechanism of *Bt* insecticidal proteins, in addition to the retention, adsorption, and degradation of *Bt* insecticidal proteins in soils and their effects on soil physical and chemical properties along with soil microbial diversity. Moreover, we highlight prospects for potential *Bt* research targets in the future, enabling a scientific framework for the environmental and ecological security evaluation of *Bt* transgenic plants, *Bt* transgenic microorganisms, and their expression products.

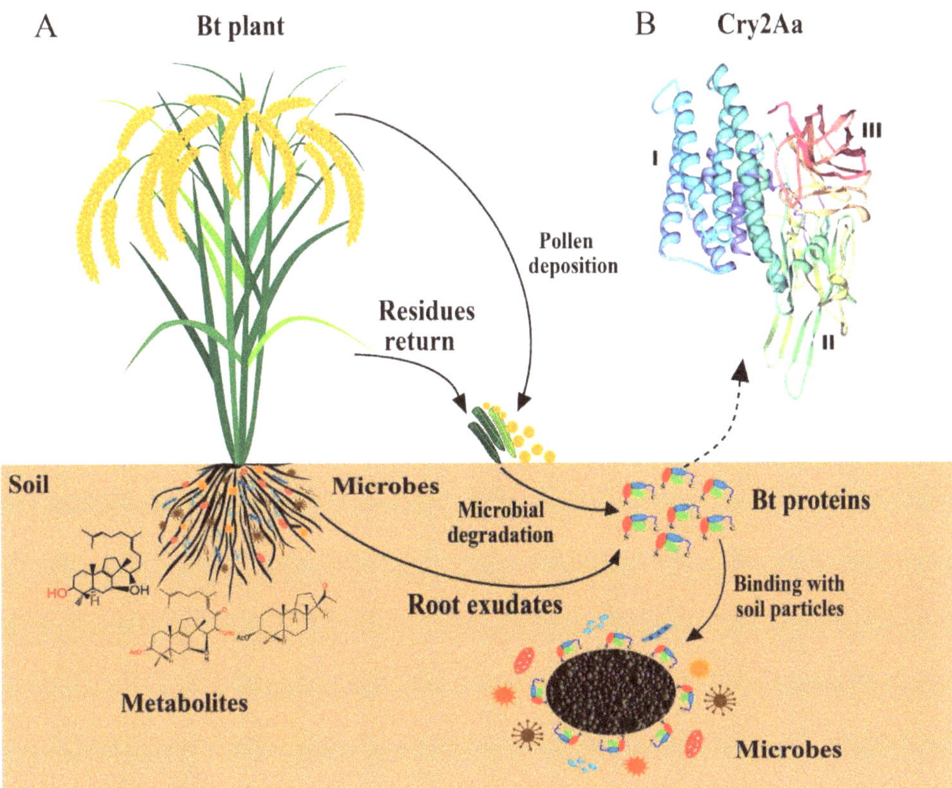

Figure 1. Environmental behaviors of *Bt* protein (**A**) and its three-dimensional structures (**B**). I, II, and III: domains I, II, and III.

2. Expression and Mechanism of *Bt* Insecticidal Proteins

2.1. Formation and Structure of Bt Insecticidal Proteins

Crystals are formed at the same time as endospore formation by the *Bt* bacterium. The main components of these crystals are proteins with insecticidal activity, known as ICPs or δ-endotoxins. *Bt* proteins exhibit various forms, including bipyramidal (Cry1), cuboidal (Cry2), flat rectangle (Cry3A), irregular (Cry3B), spherical (Cry4A and Cry4B), bar-shape (Cry11A), amorphous, and mosaic morphologies (https://www.bpprc.org/, 27 April 2022) [26,27]. The shape, structure, and size of the *Bt* proteins are closely related to their insecticidal virulence and specificity [28,29]. Using transmission electron microscopy, Liu and Guo observed that the *Bt* strain SFZZ03 conducts cell division in addition to the formation and development of sporulation and parasporal crystals at different fermentation times. Most of the crystals produced by strain SFZZ-03 were blunt diamonds, although square and long rhombic forms were also produced, in addition to a few hexagonal, triangular, oval, and irregular polygons. Regardless of the crystal forms, mosaic shapes were widespread and may be related to the regulation of synthetic crystals produced by the *Bt* SFZZ-03 strain [30]. Li et al. investigated the relationship between different crystal morphologies and their insecticidal activities. Their results revealed that crystal shapes among different strains and crystal compositions were not identical, with different forms of crystals exhibiting different virulence characteristics against insects. For example, insecticidal crystal proteins that are highly virulent to Lepidoptera larvae are generally in the form of bipyramidal crystals [31]. In addition, *Bt* can produce vegetative insecticidal proteins (Vips) that show no homology with ICPs. Because of their novel spectrum of activity and lack of cross-resistance with ICPs, Vips have been considered second-generation insecticidal proteins. As the primary group of vip genes, vip3 genes encode proteins of about 88.5 kDa. Vip3 toxins also play an important role in killing a broad spectrum of Lepidoptera larvae [3]. Meanwhile, Vip3A transgenic crops have been developed, e.g., corn MIR162 and cotton COT102, which should be valuable for resistance management [32,33].

Bt crystal proteins contain three different domains: Domain I, Domain II, and Domain III; (Figure 1B) [34]. Domain I (the pore-forming domain) is located at the N-terminal of the Cry active protein and consists of a bundle of seven antiparallel α-helices. The domain is cleaved by proteolytic enzymes during toxin activation and is associated with toxins entering cell membranes and pore formation [35]. Domain II (the central domain) comprises three antiparallel β-folds and is a receptor recognition and binding site that determines the specificity of insecticidal protein targets [36]. Domain III (the galactose-binding domain) comprises two antiparallel β-folds and is related to receptor binding and pore formation [35]. Schnepf et al. identified that the active core region of the prototoxin activated by midgut protease (Domains I, II, and III) contained five conserved amino acid sequences, with three conserved regions outside of the active core [27]. In addition, the homology analysis of the amino acid sequences of different *Bt* proteins revealed that the amino acid sequence homology of Cry1Aa and Cry3Aa was about 36%, while the three-dimensional structures of Cry1Aa and Cry3Aa exhibit some similarities. In contrast, the homology of amino acid sequences in Cyt2Aa, Cry1Aa, and Cry3Aa is less than 20%. Cyt2Aa has a peculiar structural form, with its protein three-dimensional structure comprising a domain consisting of two outer α-helices wrapped around a mixed β-fold [37]. Cao et al. discovered that the Cry78Aa protein exhibits high insecticidal activity against both *Laodelphax striatellus Fallén* (Hemiptera: Delphacidae) and *Nilaparvata lugens Stål* (Hemiptera: Delphacidae). The Cry78Aa crystal structure comprises two independent domains, including a β-trefoil domain at the N-terminal similar to the S-type Ricin B lectin, while a β-pore forming domain of the aerolysin family is present at the C-terminal. The insecticidal activity of Cry78Aa depends on the synergistic action of these two domains [38].

Given the large area of *Bt* transgenic crop application and the widespread use of *Bt* insecticides, many insect populations have developed resistance to *Bt* insecticidal crystal proteins, albeit to varying degrees [39,40]. Consequently, identifying new *Bt* insecticidal proteins with high insecticidal activity and wide insecticidal spectra are important research foci globally. Further, the establishment of a new method for effectively gene mining and identifying *Bt* insecticidal proteins is critical. For example, a method combining PCR restriction fragment length polymorphism (PCR-RFLP) and single-oligonucleotide nested-PCR has been used for such purposes. Using this method, a novel *Cry* gene (*Cry30Fa1*) was isolated and identified that encodes 687 amino acid residues and exhibits a molecular weight of 77.1 kDa, while carrying significant insecticidal effects against *Plutella xylostella L.* (Lepidoptera: Plutellidae) and *Aedes aegypti L.* (Diptera: Culicidae) [41]. Liu et al. re-designed the previous tool to provide a novel, high-throughput, and local software BtToxin_Digger, which can be directly used to handle large-scale genomic and metagenomic data to predict all kinds of putative toxin genes. It is more suitable for large-scale toxin gene mining, and at the same time, it can easily implement the high-throughput analysis [42]. The system was validated by using 21 *Bt* strains from the laboratory. Among them, three potentially represent novel *cry* gene types (primary ranks) and five of them became *cry* holotypes [43].

2.2. Mechanism of Action for Bt *Insecticidal Proteins*

Mechanistic investigations of *Bt* insecticidal proteins have primarily focused on lepidopteran insects [36] and evaluated three different models, namely, the perforated classic model, the signaling pathway model, and the continuous binding model [44]. Bravo et al. proposed the classical model to explain the mechanism of *Bt* insecticidal crystal protein using the Cry1Ab protein as the focus. In this model, *Bt* prototoxin is ingested by insects and is then solubilized in the alkaline intestinal environment, followed by hydrolysis via insect midgut proteases, leading to the production of low molecular proteins (~60 kDa) with insecticidal activity [45]. These low molecular weight *Bt* proteins can bind to the specific receptor Aminopeptidase N (APN), Alkaline phosphatase (ALP), and Cadherin (CAD), forming polymeric structures. The polymer then penetrates the lipid membrane, causing membrane perforation and disrupting the osmotic pressure balance within the insect cells, leading to cellular dehydration and death [35,45]. In the signaling pathway model, the *Bt* insecticidal protein activates an Mg^{2+}-dependent signal transduction system and protein kinase A (PKA) through specific binding to CAD. The induction of the adenylate cyclase (AC)/protein kinase A (PKA) signaling pathway then initiates a series of cytological events that include membrane blebbing, the appearance of nuclear ghosts, and cell swelling that are followed by cell lysis, and ultimately cell death [46]. In the continuous binding model, *Bt* prototoxin becomes a low molecular protein after activation by midgut protease. The protein then binds to CAD analogs and undergoes conformational changes, including the hydrolyzed oligomer structure of the domain, followed by binding to secondary receptors like Aminopeptidase that can promote the insertion of the toxin into cell membranes, ultimately leading to cell apoptosis and insect death [5,47]. The perforated classic model has been most invoked in studies of *Bt* proteins. Nevertheless, many questions regarding the model require resolution, including how do insect cell membranes form perforated channels and how do monomers and polymers exert toxic effects and alter the strength of toxicity?

The signaling pathway model has not gained widespread acceptance due to its lack of scientific demonstration. Further, it is unclear how the anterior and posterior processes in the continuous binding model are associated with each other, while the specific insecticidal mechanism also remains unclear [48]. Notably, numerous proteases are present in the midgut cavities of insects. The compositions of different proteases will directly affect the degradation and activation activities of *Bt* proteins, thereby playing a critical role in the specificity of insecticides. Moreover, the active site of the Cry protein is the midgut epithelial cells of insects. The possibility of different Cry insecticidal proteins binding the same receptor protein in the same insect requires further investigation [49].

Wang et al. developed the perforated classic model and discovered a novel "double channel" mechanism for *Bt* insecticidal protein activity against *Hübner* (Lepidoptera: Noctuidae), while also first identifying a pair of highly structurally similar and functionally overlapping *Bt* receptors, ABCC2 and ABCC3. After binding and interacting with *Bt* insecticidal proteins, the receptors form permeability pores on the membrane of the midgut, leading to damage and shedding of midgut cells that then causes larvae to stop feeding and die [50]. Sun et al. studied the contribution of two paralogous ATP-binding cassette (ABC) transporters and two APN to *Bt* Cry1Ac toxicity in the diamondback moth, *P. xylostella*, using CRISPR/Cas9 to generate a series of homozygous polygenic knockout strains. The results showed that a double-gene knockout strain, in which the two paralogous ABC transporters ABCC2 and ABCC3 were deleted, exhibited 4482-fold resistance to the Cry1A toxin, and a double-gene knockout strain in which APN1 and APN3a were deleted exhibited 1425-fold resistance to the Cry1Ac toxin. Furthermore, genetic crosses of the two double-gene knockouts yielded a hybrid strain in which all four receptor genes were deleted, and this resulted in a >34,000-fold resistance, indicating that while both types of receptors need to be present for the toxin to be fully effective, there is a level of functional redundancy between them [51]. Guo et al. elaborated the mechanism of *Bt* Cry toxin resistance in *P. xylostella*, and they confirmed that the MAPK-mediated differential expression of APN and other midgut genes did indeed lead to Cry1Ac resistance in *P. xylostella* [52]. In further study, they carried out a genome-wide characterization of all of the MAPK orthologs in *P. xylostella* to define their phylogenetic relationships and confirm their evolutionary conserved modules. The results from quantitative phosphoproteomic analyses, combined with functional validation studies using specific inhibitors and dsRNAs, lead them to establish a MAPK "road map", where p38 and ERK MAPK signaling pathways, in large part, mount a resistance response against *Bt* toxins through regulating the differential expression of multiple Cry toxin receptors and their non-receptor paralogs in the *P. xylostella* midgut [53]. Batool et al. and Zhao et al. analyzed the regulatory mechanism of *Bt* proteins through the C-type lectin CTLGA9 and ATP-binding protein of *A. eegypti*. These proteins may compete with *Bt* proteins to bind ALP1 and APN receptors, thereby inhibiting their insecticidal toxicity. These results could provide new insights into an in-depth understanding of the mechanism of Bt [54,55]. Gao et al. used eukaryotic and prokaryotic expression systems to express the cadherin gene (HaCad and PxCad) and other extracellular fragments (HaCad-TBR and PxCad-TBR) of *H. armigera* and *P. xylostella* in Sf9 cells and *Escherichia coli*, respectively. The authors also analyzed their binding characteristics with different *Bt* toxins, cytotoxicity mediation, and synergistic effects with toxins. The results suggested that PxCad may be the functional receptor of the Cry1Ac toxin, although the binding affinity for Cry1Ac toxin and cytotoxicity mediation are significantly weaker than for HaCad. Both cadherins can bind to some of the toxins Cry2A, Cry1B, Cry1C, and Cry1F, but they are not functional receptors for the toxins. Notably, although PxCad is not a functional receptor for the Cry1F toxin, the PxCad-TBR fragment exhibits a synergistic effect with both Cry1Ac and Cry1F toxin against *P. xylostella* [56].

Xie et al. designed four novel genetically engineered antibodies (GEAbs), wherein GEAb-dV_L comprises two light chains incorporating overlapping binding sites with Cry1A and Cry1B proteins. The GEAb also exhibited a high binding affinity with Brush border membrane vesicles (BBMVs) in *P. xylostella* midguts. GEAb-dV_L is structurally different from the Cry toxin in that it does not have the α-helix perforation structure and exhibits weak binding activity against CAD and other Cry toxin receptors. Nevertheless, GEAb-dV_L can still stably bind *P. xylostella* BBMVs, resulting in midgut cell damage and insect death. Interestingly, GEAb-dV_L exhibits a different insecticidal mechanism compared to the Cry toxin perforation pattern, providing an important contrast for the development of next-generation biological control pesticide products and key insect-resistance genes [57].

3. Environmental Fate of *Bt* Insecticidal Proteins in Soils

3.1. *Adsorption, Retention, and Degradation of* Bt *Insecticidal Proteins in Soils*

After *Bt* proteins are released into soils, they closely bind cohesive soil and humus particles by adsorption, where they can remain for a long time (Figure 1A). Consequently, the potential environmental safety of *Bt* proteins is a concern worthy of widespread attention [58]. Tapp and Stotzky first proposed that active particles in soils exhibit adsorption effects on *Bt* insecticidal proteins, wherein different soil types have different adsorption effects on *Bt* proteins. Cohesive particles in the soil can rapidly adsorb >70% *Bt* protein within 30 min. During adsorption, the concentration of *Bt* protein first increases, followed by a later decrease, ultimately achieving equilibrium within 5–6 h [59]. Stotzky and Crecchio observed that the adsorption of soil surface-active particles can effectively inhibit degradation by soil microorganisms. Soil clays, humic acids, and montmorillonite–humic acid–polymeric aluminum hydroxide complexes closely bind *Bt* proteins, preventing *Bt* protein from being used as a carbon source by microorganisms. Indeed, microbial growth in media containing *Bt* protein was significantly reduced by the above complexes compared with growth in the presence of free *Bt* proteins [60].

Conde and Patino observed that the interactions between *Bt* proteins and the surfaces of soil particles are maintained by physical and chemical processes, including cation exchange, electrostatic interactions, hydrophobic forces, hydrogen bonding, and van der Waals forces [61]. To better understand the mechanism of *Bt* protein adsorption by different soil types, Helassa et al. investigated the adsorption and degradation of purified Cry1Aa proteins in sandy and clay soils. Decreased Cry1Aa protein concentrations were not significantly correlated with soil microbial degradation, but rather to physicochemical interactions that occurred on soil surfaces, wherein hydrophobic interactions may play an important role in determining the interactions of Cry1Aa proteins with surfaces [62]. Wang also analyzed the adsorption of the *Bt* protein using five types of soils as substrates, revealing that the adsorption capacity of the active soil particles was negatively correlated with soil pH. These conclusions are consistent with those of Tapp, wherein the adsorption capacity of montmorillonite is inversely proportional to the acidity and alkalinity of soils over a certain pH range, with higher soil pH leading to lesser adsorption of *Bt* proteins [22]. Further, Yao et al. identified a linear relationship between soil organic matter content and the soil adsorption capacity for *Bt* protein, wherein soil cohesive particles adsorbed *Bt* proteins, and the higher organic matter content in soil particles led to greater *Bt* protein adsorption [63]. The studies of Zhou et al. and She et al. both observed that the adsorption capacity for *Bt* proteins differed among various soil particles, with the adsorption capacity for *Bt* protein by goethite, kaolin, and silicon dioxide sequentially decreasing. Further, the insecticidal activity and anti-UV degradation ability of *Bt* were enhanced after protein adsorption by attapulgite [64,65]. Zhou et al. investigated the effects of *Bt* protein application on Pb(II) adsorption in Ultisols and Vertisols, observing that Pb(II) adsorption decreased in both soil types at *Bt* toxin concentrations from 0 to 10 mg/L. This dynamic may be related to competition for adsorption sites and the formation of Pb toxin complexes. The adsorption capacity of Pb(II) in Vertisols was higher than in Ultisols, while the influential trend of *Bt* toxins was the opposite of the maximum adsorption capacity of Pb(II) in the two soils. In addition, the adsorption capacity of Pb(II) decreased when *Bt* protein was applied in the Ultisols and Vertisols, thereby increasing the environmental risk of Pb(II) [66]. Consequently, these results suggest that close attention should be paid to the potential risk of *Bt* proteins in soil environments and functioning [58].

The degradation of *Bt* proteins has been widely studied in soils, and its degradation rate in soils is affected by the type and concentration of *Bt* protein, soil types, soil microbial community composition, and soil pH [67]. Wang et al. analyzed the soil degradation of the *Bt* protein released by four different *Bt* transgenic maize (34B24, NK58-D1, R×601RR/YG, and Nongda 61). At 25 °C, the degradation of *Bt* protein occurred over three stages, namely, the early release stage, the rapidly declining stage, and the low, stable stage for all four *Bt* maize straw types. A comparison of the degradation in sterile and non-sterile soils indicated that biotic decomposition was one of the main factors contributing to degradation [68]. Stotzky also determined the insecticidal activity of residual proteins in soils that were subjected to the application of exogenous Cry1Ab protein and planted with *Bt* transgenic maize, revealing the insecticidal activity of Cry1Ab can persist for 120–180 d. Further, *Bt* protein activity could still be detected after 1–2 years within *Bt* transgenic maize straw [69]. In contrast, Zhang et al. observed that *Bt* proteins rapidly decompose in soils and do not accumulate. The authors measured Cry1Ac protein content after *Bt* transgenic cotton straw was applied, revealing a lack of Cry1Ac protein activity in soils after 180 d, and that Cry1Ac proteins did not continuously accumulate in the field [70]. Feng et al. used a shift-log model to describe the degradation of Cry1Ab proteins under different environmental conditions, observing results consistent with those of Wang et al. [68]. They also evaluated the effects of soil temperature, water content, and pH on Cry1Ab protein degradation. Among these, soil temperature significantly affected the degradation of Cry1Ab, with the degradation rate accelerating at higher temperatures. In contrast, soil moisture content and pH had no significant effects on the degradation of Cry1Ab [71]. Deng et al. also found that temperature conditions can alter the soil microbial activity, thus further affecting the persistence of Cry toxins. When the ambient temperature is low and even reaches freezing point at night in winter, it will inhibit the microbial activity and delay the degradation process [72]. To verify the hypothesis that soil type might influence the degradation rate of *Bt* protein in soils, Zhou investigated the effects of soils from four different regions on the adsorption of *Bt* proteins in laboratory soil culture experiments. Soils from Zhengzhou and Gongzhuling exhibited similar degradation trends for soil *Bt* proteins, while soils from Beijing and Jinan exhibited similar later degradation of soil *Bt* proteins [73]. Thus, the adsorption and degradation of soil *Bt* proteins are closely related to soil types.

3.2. *Transformation Fate of* Bt *Insecticidal Proteins in Soils*

The proportion of free *Bt* proteins extracted from paddy soils using the commonly employed phosphate-buffered saline with Tween 20 (PBST) method was relatively low in one study, although the proportions of *Bt* proteins bound to soil particles were high, difficult to extract, and could not be determined by enzyme-linked immunosorbent assays (ELISAs) [74]. ELISA tests cannot currently distinguish the *Bt* proteins from exogenous *Bt* proteins released by transgenic plants compared to the *Bt* prototoxins produced by soil *Bt* bacteria. Consequently, the environmental behavior and biological effects of exogenous *Bt* proteins cannot be accurately evaluated at present. Consequently, rather than measuring the low content and free states of the *Bt* protein in soils to assess its accumulation and retention in soils, environmental safety assessments of exogenous *Bt* insecticidal proteins should focus on the transformation direction and ecological effects of *Bt* proteins that are expressed in *Bt* transgenic crops and *Bt* recombinant engineered strains in soil environments [17,74]. Valldor et al. investigated the fate of *Bt* insecticidal proteins in sandy soils and clays using Cry1Ab labeled with ^{14}C radioisotopes. The $^{14}CO_2$ production rate due to ^{14}C-Cry1Ab mineralization decreased with soil cultivation time in the two soil types, with the ^{14}C flux accounting for 15–40% of the total ^{14}C flux. At 29 and 37 days of soil culture, 16–23% of the ^{14}C flux entered into microbial carbon, suggesting that microorganisms can use Cry1Ab as a substrate for their growth. The proportion of ^{14}C in Cry1Ab remaining in soil due to adsorption by soil particles was the highest (approximately 40–80%). Due to the high degradation efficiency of ^{14}C-Cry1Ab in sandy soils, the *Bt* protein content was very low in clay based on ELISA detection [17]. In another study using ^{14}C-labelled Cry protein, similar rates of mineralization were found for a Cry1Ac protein in loamy sand with a comparably low pH [75]. Thus, it is speculated that clay particle adsorption is the main factor limiting the microbial degradation of *Bt* proteins. *Bt* protein can be absorbed onto soil particles and humic acid to form bound *Bt* protein, leading to the retaining of insecticidal activity. Bound *Bt* protein is more difficult to biodegrade compared to free *Bt* protein and carries more potential ecological risks [20,76]. The desorption rate of *Bt* protein by soil particles and humic acid is affected by soil properties, agricultural practices, and other factors [77,78]. Consequently, these factors should be given more attention to understand the behavior and fate of exogenous *Bt* proteins in agricultural ecosystems, rather than using ELISA tests to estimate the retention time of water-soluble *Bt* proteins in soils.

4. Effects of *Bt* Insecticidal Proteins on Soil Microbial Ecology

4.1. *Effects of* Bt *Insecticidal Proteins on Soil Physicochemical Properties*

Soil is an important site for the exchange of matter and energy in terrestrial ecosystems. The life activities of soil organisms largely depend on the physicochemical properties of soils. Consequently, an important consideration is whether exogenous *Bt* protein in soils affects the physical and chemical properties of soils that could then affect nutrient transformation processes. Such considerations could represent important indicators for evaluating the safety of field-released *Bt* transgenic plants and *Bt* transgenic microorganisms [79]. Liu applied two insecticidal crystal proteins (Cry1Ab and Cry1Ac) to three soil types, including paddy soils originating from fluvo-aquic soil, red soils, and yellow-brown soils. The application of *Bt* proteins increased soil pH in the short term compared to control soils, while no significant differences were identified between the control and experimental soils after long-term cultivation. Further, *Bt* protein application decreased the degradation rate of organic matter in the paddy soils that originated from fluvo-aquic soils. Moreover, the addition of *Bt* protein increased the nitrate–nitrogen content of the red soils, wherein the amount of total nitrogen in the red soils was higher than in the control after cultivation over a long period of time. Lastly, the ratio of C/N in the yellow-brown soils was lower in the experimental soils compared to the control [79]. Chen et al. analyzed the changes in untargeted metabolomics for the soil metabolite profiling of transgenic and non-transgenic maize. As predicted, the metabolomic profile greatly differed between transgenic and non-transgenic maize cultivars at all stages, and the difference was more prominent in the middle stage. These results suggest that genetic modification with the *cry1Ah* gene-altered maize can alter soil metabolism [80]. Our group also analyzed the dynamic changes of the physicochemical properties associated with soil cultivation by applying different concentrations of *Bt* protein to paddy soils at different times. No significant differences were observed in the physicochemical indices when comparing soils with different concentrations of *Bt* protein and control soils without *Bt* protein in the early stages of soil cultivation (i.e., at 1, 5, and 10 d). In contrast, NH_4^+-N and NO_2^--N were significantly higher at the later stages of soil cultivation (50 and 100 d) compared to control soils without *Bt* protein application, although no significant difference was observed for NO_3^--N in soils when *Bt* was applied at 500 ng/g. The above results suggest that the evaluation of environmental effects of *Bt* proteins should consider the concentration of *Bt* protein and its persistence time in soils (unpublished data).

Few studies have evaluated the effects of *Bt* proteins on soil physicochemical properties, and those that have primarily focused on the effects of root exudates of *Bt* transgenic crops and from straw returned to soils. Chen et al. analyzed changes in the available nutrient content in the rhizosphere soils of four transgenic *Bt* and non-*Bt* cotton species at different growth stages. The nitrate–nitrogen content of transgenic *Bt* cotton was significantly different from conventional cotton at many life stages. Further, the content of ammonium–nitrogen was significantly different from conventional cotton in the squaring, flowering, and boll stages. In addition, the content of available phosphorus differed, with the Lumian No. 28 variety exhibiting significantly higher levels than conventional cotton in the squaring stage, and the Lumian No. 36 variety exhibiting significantly higher levels than conventional cotton at the seeding stage. Thus, the planting and growth period of transgenic *Bt* cotton were the primary factors affecting *Bt* protein residues and the content of available nutrients in rhizosphere soils [81]. To thoroughly investigate the effects of transgenic *Bt* crop straw returned to soils on soil physicochemical properties, Yu investigated the effects of rice straw that was returned to fields on the soil physicochemical properties, denitrifying enzymes, denitrifying rates, and microbial diversity. The experiments were conducted under flooded conditions and used two transgenic *Bt* rice and two non-transgenic rice straw materials. Compared with non-transgenic *Bt* rice, the transgenic *Bt* rice straw returned to the field promoted soil denitrification, but no significant differences were observed in total nitrogen, organic matter, and other nutrient content in the paddy soils [82]. Zhang et al. also analyzed four types of *Bt* cotton with different levels of insect resistance, in addition to one non-transgenic conventional cotton variety (Simian no. 3). After one or two years of planting, all of the cotton straws were mechanically crushed and returned to the field in situ, followed by the measurement of changes in soil nutrient contents. The levels of organic matter, available phosphorus, available nitrogen, available potassium, and total nitrogen remarkably increased in soils within both the first and second years of cotton planting, as did soil pH. However, significant differences were not observed in the variation of all nutrient contents between *Bt* transgenic treatments and non-transgenic treatments [83].

4.2. *Effects of* Bt *Insecticidal Proteins on Soil Microbial Community Diversity*

Soil microbial communities are the most active soil biota and participate in soil organic matter decomposition, humus synthesis, and nutrient transformation, and promote soil development and formation. Further, they are important biological indicators for evaluating soil health [84–86]. Exogenous *Bt* protein enters soils and rapidly binds clays and humic acids to form bound *Bt* protein, leading to significant changes in its chemical composition and structure that may then adversely affect non-target organisms [24]. Rui et al. observed that the application of the *Bt* protein did not significantly alter the number of phosphate-dissolving and potassium-dissolving bacteria in soils based on plate counting measurements. Further, the abundances of nitrogen-fixing bacteria significantly decreased when *Bt* toxin concentrations were 500 ng/g or higher. The highest concentration of *Bt* toxin in the rhizosphere of *Bt* cotton NuCOTN99^{B} was 300 ng/g. Thus, differences in the numbers of functional soil bacteria between *Bt* cotton and non-*Bt* cotton treatments did not result from *Bt* toxin toxicity [16]. Li et al. added one of the most commonly used *Bt* proteins in *Bt* crops, Cry1Ac, to soils and evaluated the changes in soil bacterial, fungal, and archaeal diversities and community structures using terminal restriction fragment length polymorphism (T-RFLP) and real-time PCR (qPCR). The application of Cry1Ac did not significantly change the community diversity of microorganisms. In contrast, other studies have reported that the application of exogenous *Bt* protein can significantly alter the structure of soil microbial communities [87]. For example, Liu applied two insecticidal crystal proteins (Cry1Ab and Cry1Ac) to three soil types using paddy soils originating from fluvo-aquic soil, red soils, and yellow-brown soils, and then analyzed the changes in soil microbial genetic diversity using denaturing gradient gel electrophoresis (DGGE). Cry1Ac application increased diversity based on the Shannon index in paddy soils that originated from fluvo-aquic soils. Further, the application of both Cry1Ab and Cry1Ac increased microbial diversity in the yellow-brown soils [79]. Guan et al. performed a two-year field experiment that featured the cultivation of three *Bt* transgenic oilseed rape (*Brassica napus* L.) plant lines to evaluate their effects on rhizosphere microorganisms. The Shannon's diversity, richness, and evenness of the soil microbial community increased significantly in transgenic GT1 and GT5 lines during the flowering stage [88]. Frouz et al. conducted a three-year field and laboratory study on the effect of *Bt* corn on the soil microbial community and decomposition rates of corn post-harvest residues, and found that *Bt* corn may have a deleterious effect on decomposers in the laboratory; however, this effect was minor and restricted to the initial stages of decomposition, and was undetectable in long-term field experiments [89]. Hong et al. investigated the effect of planting insect-resistant and herbicide-tolerant transgenic maize on rhizospheric microbial communities, and the results showed that at the genus level, planting transgenic maize significantly decreased the relative abundance of rhizospheric *Candidatus Nitrososphaera* at the jointing and mature stages [90]. These studies suggest that *Bt* proteins can impact microbial community structures and microbial diversity present within soils. The inconsistent results of the above studies may be due to the varying experiments using different *Bt* protein types, soil types, soil incubation times, and the analytical methods used in the experiments.

Previously, our group evaluated the impact of transgenic *Bt* crop cultivation on microbial community diversity. Specifically, microbial community structures were analyzed in rhizosphere soils using DGGE at the seedling, tiller, booting, heading, and mature stages of *Bt* transgenic rice in comparison with non-transgenic control rice. *Bt* rice cultivation had little effect on the dominant rhizosphere bacterial, fungal, and actinobacterial communities [91]. In addition, we evaluated the responses of fungal population sizes and community compositions via 18S rRNA gene sequencing when growing the transgenic insect-resistant cotton SGK321 variety expressing *Cry1Ac/CpTI* proteins. The population sizes of rhizosphere fungi associated with SGK321 taproots at the seedling stage were significantly higher compared to within lateral roots, suggesting that root microhabitats were influential towards fungal communities. Further, no significant differences were observed for fungal community sizes when comparing the rhizospheres of SGK321 and SY321 from the same root zones, suggesting that fungal abundances were not affected by *Bt* protein production in the root tissues of transgenic *Bt* cotton [92]. Plate-count enumeration and high-throughput sequencing were used by Li et al. and Qi et al. to evaluate the abundances of soil microorganisms associated with the long-term cultivation of transgenic insect-resistant cotton expressing Cry1Ac. A lack of significant variation was observed for the number of bacteria, fungi, *Azotobacter*, denitrifying bacteria, ammonia-oxidizing bacteria, and diversity indices of microbial communities when comparing transgenic cotton expressing Cry1Ac and non-transgenic cotton [93,94]. Xu et al. employed stable isotope probing and high-throughput sequencing to identify the active microorganisms involved in *Bt*-containing straw decomposition. Their results suggested that *Bt* rice exerts a significant, but transient, impact on soil microorganisms during microbial straw decomposition [95]. Zuo et al. also observed that transgenic poplars did not affect the physical and chemical properties of soils or soil microbial community structures, although community structures were affected by location and season [96]. Zhang et al. found that the Shannon–Wiener index of soil microbial communities in transgenic *Bt* cotton fields was significantly higher compared to those of non-transgenic cotton fields 30 days after cotton straw was returned to the fields [70]. Wu et al. used the high-throughput sequencing of bacterial 16S rRNA genes to evaluate the effects on the bacterial community diversity in soils of *Bt* rice grown for three consecutive years. No significant changes in bacterial community structures were observed for rhizosphere soils between *Bt* rice and non-*Bt* rice soils, while planting time had a greater effect on rhizosphere soil bacterial community composition than did cultivar type [97]. Wang et al. explicitly investigated the effect of transgenic rice on its active rhizosphere microflora by combining high-throughput sequencing with the SIP technique. The high-throughput sequencing of the bacterial 16S rRNA gene showed that transgenic rice did not significantly change the soil bacterial community structure compared with its parental variety. The soil bacterial community structure of transgenic and parental-labeled microbes was not significantly different, but was significantly different from those of the non-parental varieties, indicating that planting transgenic *Bt* rice had a limited impact on the soil microbiome [98]. In a two-year-long trial, Fan et al. found that insect-resistant transgenic maize carrying the *cry1Ie* gene did not affect the soil fauna [99]. Zhang et al. employed PCR-DGGE to monitor the rhizosphere soil microbial communities after three years cultivation of NC 33B in northern China. The results showed that the population sizes and community structures of eubacteria, fungi, and actinomycetes in rhizosphere soil were markedly affected by natural variations in the environment related to cotton growth stages. However, there was no significant difference in the eubacterial, fungal, or actinomycete population size and community structures in rhizosphere soil between NC 33B and its non-transgenic parent [100]. Even where *Bt* transgenic rice had been planted for 8 years, soil enzymatic activities and microbial biomass were also observed with no consistently significant changes [101]. Chen et al. examined and compared the effects of a non-transgenic maize cultivar and an insect-resistant transgenic maize cultivar genetically engineered with *cry1Ah* gene from *Bt* on the rhizosphere bacterial community using 16S rDNA amplicon sequencing and soil metabolome profile, using UPLC/MS analysis at six different growth stages. They found no significant differences in bacterial community composition or diversity at any growth stage between the two cultivars. These studies suggest that *Bt* crops have no significant effects on the soil microbial community [80].

4.3. *Effects of* Bt *Insecticidal Proteins on Functional Diversity of Soil Microorganisms*

The functional diversity of soil microorganisms refers to the functions that soil microbial communities can perform and the results from these functions, including, for example, nutrient decomposition, nutrient transformation, and promoting or inhibiting plant growth. These functions carry important significance for soil ecological functions and natural elemental cycles [102,103]. Liu observed that the application of *Bt* proteins inhibited the activity of nitrate reductases in paddy soils originating from fluvo-aquic and yellow-brown soils. Further, application of *Bt* proteins inhibited the activity of nitrite reductases in three soil types, including paddy soils originating from fluvo-aquic soils, red soils, and yellow-brown soils [79]. Yaqoob et al. found a reduced rate of phosphate solubilization and auxin biosynthesis during the maturity stages of both years when compared with that of the early stages, although these had no harmful effect on the biochemical or molecular characteristics of isolated soil bacteria from *Bt* cotton rhizosphere [104]. The effects of transgenic *Bt* crops on the functional diversity of soil microorganisms have also been extensively reported. Wu et al. determined that returning transgenic *Bt* rice straw to fields altered the soil dehydrogenase activity [105]. Liu et al. analyzed the rhizosphere soils of transgenic *Bt* rice grown for two years and observed seasonal changes corresponding to significant differences in phosphatase activity, dehydrogenase activity, respiration, methanogenesis, and fungal community composition in rhizosphere soils, although these metrics were not significantly affected by transgenic *Bt* rice cultivation [106]. Our group observed that the activities of dehydrogenases, invertases, phenol oxidases, acid phosphatases, ureases, and proteases were not significantly different between the soils of *Bt* and non-*Bt* rice. In addition, a Biolog system was used to evaluate the effects of *Bt* rice on the functional diversity of microbial communities. Although differences were observed in carbon substrate utilization between *Bt* and non-*Bt* rice at the seedling, tillering, and heading stages, these differences were transient [91]. Zhang et al. observed that *Bt* transgenic cotton significantly improved the utilization of carbon sources such as amino acids, amines, and carbohydrates by soil microbial communities in comparison to non-transgenic cotton 30 days after returning the straw to the fields [70]. Liang et al. observed a lack of differences in the functional diversity of microbial communities and the utilization of carbon sources in rhizosphere soils between transgenic maize IE09S034 and non-transgenic maize soils after two consecutive years of field experiments [107]. Further, Luan collected rhizosphere soils at the seedling, flowering, and maturity stages of transgenic *Bt* maize over two consecutive years, observing that the planted transgenic insect-resistant maize did not significantly affect the functional diversity of the rhizosphere soil microbial communities [108]. Xin observed that different maize growth stages and experimental years affect the functional diversity of microbial communities, and this may be related to the year of transgenic crop planting, in addition to the growth and development stages [109]. Li et al. explored how the responses of soil enzymatic activity varied across *Bt* crops or in different growth periods. The results showed that the activities of dehydrogenase, β-glucosidase, urease, nitrate reductase, alkaline phosphatase, and arylsulfatase significantly increased under *Bt* crop cultivation with residues incorporation. Further, the response ratios of soil enzymes varied with *Bt* crop types and growth periods [110]. Escalas et al. reported that numerous factors influence the functional diversity of soil microbial communities, including soil properties, climate types, individual microbial metabolisms, and community sizes [103]. Therefore, long-term systematic monitoring studies are necessary for evaluating the effects of *Bt* insecticidal proteins on the functional diversity of soil microorganisms at large scales.

5. Perspectives

Investigating the environmental behavior and ecological effects of exogenous *Bt* insecticidal proteins in soils is extremely important for understanding changes in soil structures and physicochemical properties, in addition to changes in microbial communities and the responses of functional diversity during the degradation and transformation of *Bt* insecticidal proteins. Understanding these dynamics will provide a scientific framework and theoretical reference for evaluating the environmental safety of transgenic *Bt* plants and *Bt*-recombinant engineered strains. Future studies should focus on conducting research in the two critical areas described below.

5.1. Evaluating Environmental Behaviors of Bt *Insecticidal Proteins Using Stable Isotope Tracing*

Current ELISA methods for measuring *Bt* protein content cannot distinguish between exogenous recombinant *Bt* proteins and *Bt* protoxins produced by soil *Bt* bacteria, and thus cannot accurately evaluate the environmental behaviors and the biological effects of exogenous *Bt* proteins [17]. Valldor et al. obtained ^{14}C-labeled Cry1Ab from batch fermentations with recombinant *E. coli* using ^{14}C-labeled glycerol as the cultivation carbon source [111]. Although radioisotopes feature high sensitivity and low production costs, they are hazardous to operators, thereby limiting their research applications [112]. Stable isotope-labeled peptides and proteins are used to investigate subjects using physical, chemical, and biological methods by evaluating their traces, retention sites, and abundances in study systems using mass spectrometry and emission spectroscopy. Consequently, the method is commonly used in the detection and tracking of biological samples [113]. In particular, stable isotope mass spectrometry can be used to trace and quantitatively analyze the transformation, distribution, and dynamics of *Bt* protein carbon and nitrogen components among different forms of carbon and nitrogen, while effectively circumventing the influence of original *Bt* proteins in soils. Many stable isotope-labeled small-molecule compounds are commercially produced, although the development of singly labeled *Bt* proteins using ^{13}C and ^{15}N stable isotopes has not been reported. Our group recently used M9 medium with glucose as the sole carbon source and ammonium chloride as the sole nitrogen source to produce $^{13}C/^{15}N$ singly labeled *Bt* proteins with high purity and insecticidal activity from natural biosynthesis pathways. The resultant description, entitled "A method of preparing Cry protein with stable isotope ^{13}C labeling", has been patented (202010227154. X) and provides experimental materials and technical support for evaluating the environmental safety and ecological effects of *Bt* proteins [50]. Nevertheless, it is unknown whether the structure and stability of the above stable isotope-labeled *Bt* proteins differ from natural *Bt* proteins, thus requiring further analysis. In addition, soil physicochemical properties are important factors that affect the adsorption and degradation of *Bt* insecticidal proteins. In subsequent experiments, it will be necessary to analyze the environmental behavioral differences of *Bt* proteins in different types of soil.

5.2. Analysis of Microbial Ecological Effects of Bt *Insecticidal Proteins Using Microbiome Techniques*

The selection of a suitable analytical method to comprehensively and accurately analyze microbial communities and functional diversity response mechanisms is critical for evaluating the degradation and transformation of *Bt* proteins in soil. Many microorganisms in natural environments cannot be cultivated under conventional experimental conditions, although the isolation and cultivation of pure microbial cultures remain the basis for studies of microbial physiology, functioning, and genetics. High-throughput sequencing technology based on microbiomics can illuminate the community structures of soil microorganisms at a broad level, while also resolving changes in the relative abundances of specific microorganisms involved in certain functions. Combined with changes in key soil elements, these techniques can reveal important molecular regulatory mechanisms of microbial physiological and ecological processes in soils [114]. Consequently, current studies of microorganisms in environments have advanced from relying on pure cultures of microorganisms to studying the metabolisms of microbial communities by combining pure cultures and high-throughput sequencing [115]. Our group has performed the macrogenomic analysis of the microbial community functions involved in the degradation of Cry1Ab/Ac proteins, revealing that *Thermobifida, Streptomyces, Achromobacter, Noviherbaspirillum,* and *Pseumdomonas* were significantly involved in these processes. We speculate that these microorganisms are involved in the degradation and carbonaceous transformation of Cry1Ab/Ac. Further, among these taxa, *Streptomyces, Achromobacter,* and *Pseumdomonas* can all produce protein hydrolytic enzymes that can catalyze the hydrolysis of peptides or proteins into amino acids. A strain of *Streptomyces griseus* was subsequently isolated to purity, and a single cultivation experiment was performed using Cry1Ab/Ac proteins as substrates, with initial results suggesting that *Streptomyces griseus* can catabolize and utilize Cry1Ab/Ac proteins (unpublished data). In addition, soil metabolites can be considered phenotypes or signatures of soil microbial community changes, because alterations experienced at the organism and enzyme level will manifest as modified metabolite profiles [116]. Resolving soil metabolomes will contribute to a better understanding of the soil microbial ecological effects of exogenous *Bt* proteins. Consequently, the combined analysis of microbiomics and metabolomics is an important direction for further research to comprehensively identify the correlations between soil microorganisms, metabolites, and exogenous *Bt* proteins from multiple perspectives and levels.

Author Contributions: J.L. and P.L. designed and supervised the review, Y.L. and C.W. performed the review and wrote the manuscript, L.G., C.H., G.W., Y.S., L.S., X.W., A.P., Q.X. and J.S. read and approved the manuscript. All authors have read and agreed to the published version of the manuscript.

Funding: This work was financially supported by the National Natural Science Foundation of China (Grant No. 32071657), the Natural Science Foundation of Shanghai (Grant No. 19ZR1434400), "Science and Technology Innovation Action Plan"—Technical Standards of Shanghai (20DZ2202900), Talent Project of SAAS (202108), SAAS Excellent Research Team No. 2022 (B-016) and KFKT2022-02.

Data Availability Statement: Not applicable.

Conflicts of Interest: The authors declare no conflict of interest.

References

1. Hannay, C.L. Crystalline inclusions in aerobic spore-forming bacteria. *Nature* **1953**, *172*, 1004. [CrossRef] [PubMed]
2. Faust, R.M.; Abe, K.; Held, G.A.; Iizuka, T.; Bulla, L.A.; Meyers, C.L. Evidence for plasmid-associated crystal toxin production in *Bacillus thuringiensis* subsp. israelensis. *Plasmid* **1983**, *9*, 98–103. [CrossRef]
3. Estruch, J.J.; Warren, G.W.; Mullins, M.A.; Nye, G.J.; Craig, J.A.; Koziel, M.G. Vip3A, a novel *Bacillus thuringiensis* vegetative insecticidal protein with a wide spectrum of activities against lepidopteran insects. *Proc. Natl. Acad. Sci. USA* **1996**, *93*, 5389–5394. [CrossRef] [PubMed]
4. Barloy, F.; Lecadet, M.M.; Delécluse, A. Cloning and sequencing of three new putative toxin genes from *Clostridium bifermentans* CH18. *Gene* **1998**, *211*, 293–299. [CrossRef]
5. Shan, Y.M. Discovery and Activity Analysis of Novel Insecticidal Genes from *Bacillus Thuringiensis*. Ph.D. Dissertation, Northeast Agricultural University, Harbin, China, 2019. (In Chinese with English Abstract)
6. Deng, L.X. Biopesticide *Bacillus thuringiensis* insecticide and its synergist. *Chem. Teach.* **2004**, *3*, 29–31. (In Chinese with English Abstract)
7. Ding, X.Z.; Luo, Z.H.; Xia, L.Q.; Gao, B.D.; Sun, Y.J.; Fu, Z.J.; Liu, F.; Hu, S.B.; Mo, X.T.; Zhang, Y.M. Cloning and expression of the *cry1Ac-tchiB* fusion gene from *Bacillus Thuringinesis* and Tobacco and its insecticidal synergistic effect. *Acta Microbiol. Sin.* **2007**, *47*, 1002–1008. (In Chinese with English Abstract)
8. Wang, G.; Zhang, J.; Song, F.; Wu, J.; Feng, S.L.; Huang, D.F. Engineered *Bacillus thuringiensis* GO33A with broad insecticidal activity against lepidopteran and coleopteran pests. *Appl. Microbiol. Biot.* **2006**, *72*, 924–930. [CrossRef]
9. Liu, H.M.; Xu, G.J. Microbial pesticide *Bacillus thuringiensis* G033A. *Pestic. Sci. Admin.* **2018**, *39*, 59–60. (In Chinese with English Abstract)
10. Timmons, A.M.; Charters, Y.M.; Crawford, J.W.; Burn, D.; Scott, S.E.; Dubbels, S.J.; Wilson, N.J.; Robertson, A.; O'Brien, E.T.; Squire, G.R.; et al. Risks from transgenic crops. *Nature* **1996**, *380*, 487. [CrossRef]
11. Wang, T.; Yang, Y.; Li, J.P.; Du, K. Research progress in the effects of genetically modified crops on soil microbial community. *Biotechnol. Bull.* **2021**, *37*, 255–265. (In Chinese with English Abstract)
12. James, C. *Global Status of Commercialized Biotech/GM Crops: ISAAA Brief No. 55*; ISAAA: Ithaca, NY, USA, 2019.
13. Sun, L.Y.; Li, J.; Yuan, H.W.; Yao, J.H.; Li, W.; Chen, X.Y. Expression analysis of *Bt-BADH-GA20ox-rolB* multigenes in transgenic tobacco plants. *J. Beijing Fores. Univ.* **2011**, *33*, 86–90. (In Chinese with English Abstract)
14. Sun, S.S.; Wang, C.S.; Zhu, T.T.; Zhao, H.L.; Jiao, Y.; Qu, S.W.; Wang, P. Construction of bivalent plant expressed vector of *NPR1* and *Cry1Ab13-1* genes and is transformation in maize. *Mol. Plant Breed.* **2021**, 1–14. (In Chinese with English Abstract)
15. Dong, H.; Li, W. Variability of endotoxin expression in *Bt* transgenic cotton. *J. Agron. Crop. Sci.* **2007**, *193*, 21–29. [CrossRef]
16. Rui, Y.K.; Yi, G.X.; Zhao, J.; Wang, B.M.; Li, Z.H.; Zhai, Z.X.; He, Z.P.; Li, Q.X. Changes of Bt toxin in the rhizosphere of transgenic *Bt* cotton and its influence on soil functional bacteria. *World J. Microbiol. Biot.* **2005**, *21*, 1279–1284. [CrossRef]
17. Valldor, P.; Graff, R.; Martens, R.; Tebbe, C.C. Fate of the insecticidal Cry1Ab protein of GM crops in two agricultural soils as revealed by ^{14}C-tracer studies. *Appl. Microbiol. Biot.* **2015**, *99*, 7333–7341. [CrossRef]
18. Zhou, X.Y.; Liu, N.; Zhao, M.; Li, H.; Zhou, L.; Tang, Z.W.; Cao, F.; Li, W. Advances in effects of insecticidal crystal proteins released from transgenic *Bt* crops on soil ecology. *Hereditas* **2011**, *33*, 443–448. (In Chinese with English Abstract) [CrossRef]
19. Souza, C.S.F.; Silveira, L.C.P.; Souza, B.H.S.; Nascimento, P.T.; Damasceno, N.C.R.; Mendes, S.M. Efficiency of biological control for fall armyworm resistant to the protein Cry1F. *Braz. J. Biol.* **2020**, *81*, 154–163. [CrossRef]
20. Crecchio, C.; Stotzky, G. Insecticidal activity and biodegradation of the toxin from *Bacillus thuringiensis* subsp. Kustuki bound to humic acids from soil. *Soil Biol. Biochem.* **1998**, *30*, 463–470. [CrossRef]
21. Li, X.Y.; Li, S.Y.; Lang, Z.H.; Zhang, J.; Zhu, L.; Huang, D.F. Chloroplast-targeted expression of the codon-optimized truncated *cry1Ah* gene in transgenic tobacco confers a high level of protection against insects. *Plant. Cell Rep.* **2013**, *32*, 1299–1308. [CrossRef]
22. Wang, H.Y. Study on Environmental Behavior and Bioeffects of *bt* Transgenic Rice and Expressed Products of Its Exogenous gene. Ph.D. Dissertation, Zhejiang University, Hangzhou, China, 2007. (In Chinese with English Abstract)

23. Song, Y.Y.; Liu, J.W.; Li, L.K.; Liu, M.Q.; Chen, X.Y.; Chen, F.J. Evaluating the effects of transgenic *Bt* rice cultivation on soil stability. *Environ. Sci. Pollut. Res.* **2020**, *27*, 17412–17419. [CrossRef]
24. Stotzky, G.; Saxena, D. Insecticidal toxin from *Bacillus thuringiensis* is released from roots of transgenic *Bt* corn in vitro and in situ. *FEMS Microbiol. Ecol.* **2000**, *33*, 35–39.
25. Ministry of Agriculture and Rural Affairs of the People's Republic of China (No. 2 of 2022). Decision of the Ministry of Agriculture and Rural Affairs on Amending the Regulations on the Administration of Safety Evaluation of Agricultural Genetically Modified Organisms. Available online: http://www.moa.gov.cn/govpublic/CYZCFGS/202201/t20220124_6387560.htm (accessed on 27 April 2022).
26. Ren, G.X.; Feng, X.C.; Feng, W.X. The Shape and Antigenic Characteristics of the Parasporal Crystal of *Bacillus thuringiensis*. *Acta Microbiol. Sin.* **1983**, *23*, 57–62. (In Chinese with English Abstract)
27. Schnepf, E.; Crickmore, N.; Van, R.J.; Lereclus, D.; Baum, J.; Feitelson, J.; Zeigler, D.R.; Dean, D.H. *Bacillus thuringiensis* and its pesticidal Crystal proteins. *Microbiol. Mol. Biol. Rev.* **1998**, *62*, 775–806. [CrossRef]
28. Huang, S.F.; Zhu, Y.J.; Liu, B.; Ruan, C.Q.; Lin, J. Study on morphological characteristics of parasporal crystals of *Bacillus thuringiensis* strain LSZ9408. *WuYi Sci. J.* **2006**, *22*, 37–40. (In Chinese with English Abstract)
29. Shan, Y.; Shu, C.; He, K.; Cheng, X.; Zhang, J. Characterization of a novel insecticidal protein Cry9Cbl from *Bacillus thuringiensis*. *J. Agric. Food Chem.* **2019**, *67*, 3781–3788. [CrossRef]
30. Liu, C.; Guo, Z.H. Observation on development process of *Bacillus thuringiensis* SFZZ-03. *J. Chin. Elect. Microbol. Soci.* **2016**, *35*, 436–439. (In Chinese with English Abstract)
31. Li, M.Y.; Gao, J.H.; Song, C.M.; Li, T.F.; Wang, G. Study on parasporal Crystal of *Bacillus thuringiensis*. *J. Yunnan Nation. Univ. (Nat. Sci.)* **2006**, *15*, 324–327. (In Chinese with English Abstract)
32. Burkness, E.C.; Dively, G.; Patton, T.; Morey, A.C.; Hutchison, W.D. Novel Vip3A *Bacillus thuringiensis* (Bt) maize approaches high-doseefficacy against Helicoverpazea (Lepidoptera: Noctuidae) under field conditions: Implications for resistance management. *GM Crops* **2010**, *1*, 337–343. [CrossRef]
33. Adamczyk, J.J.; Mahaffey, J.S. Efficacy of Vip3A and Cry1Ab transgenic traits in cotton against various lepidopteran pests. *Fla. Entomol.* **2008**, *91*, 570–575.
34. Palma, L.; Muñoz, D.; Berry, C.; Murillo, J.; Caballero, P. *Bacillus thuringiensis* toxins: An overview of their biocidal activity. *Toxins* **2014**, *6*, 3296–3325. [CrossRef]
35. Xu, C.; Wang, B.C.; Yu, Z.N.; Sun, M. Structural insights into *Bacillus thuringiensis* Cry, Cyt and parasporin toxins. *Toxins* **2014**, *6*, 2732–7270. [CrossRef] [PubMed]
36. Bravo, A.; Gill, S.S.; Soberón, M. Mode of action of *Bacillus thuringiensis* Cry and Cyt toxins and their potential for insect control. *Toxicon* **2007**, *49*, 423–435. [CrossRef] [PubMed]
37. Bai, Y.Y.; Jiang, M.X.; Cheng, J.A.; Jiang, Y.H. Advances in safety studies of soil Bt toxin proteins released from transgenic *Bt* crops. *Chin. J. Appl. Ecol.* **2003**, *14*, 2062–2066. (In Chinese with English Abstract)
38. Cao, B.; Shu, C.; Geng, L.; Song, F.; Zhang, J. Cry78Ba1, One novel crystal protein from *Bacillus thuringiensis* with high activity against rice planthopper. *J. Agric. Food Chem.* **2020**, *68*, 2539–2546. [CrossRef] [PubMed]
39. McGaughey, W.H. Insect resistance to the biological insecticide *Bacillus thuringiensis*. *Science* **1985**, *229*, 193–195. [CrossRef]
40. Jin, L.; Wang, J.; Guan, F.; Zhang, J.P.; Yu, S.; Liu, S.Y.; Xue, Y.Y.; Li, L.L.; Wu, S.W.; Wang, X.L.; et al. Dominant point mutation in a tetraspanin gene associated with field-evolved resistance of cotton bollworm to transgenic *Bt* cotton. *Proc. Natl. Acad. Sci. USA* **2018**, *115*, 11760–11765. [CrossRef]
41. Tan, F.R.; Zheng, A.P.; Zhu, J.; Wang, L.X.; Li, S.C.; Deng, Q.M.; Wang, S.Q.; Li, P.; Tang, X.M. Rapid cloning, identification, and application of one novel crystal protein gene *cry30Fa1* from *Bacillus thuringiensis*. *FEMS Microbiol. Lett.* **2010**, *302*, 46–51. [CrossRef]
42. Liu, H.L.; Zheng, J.S.; Bo, D.X.; Yu, Y.; Ye, W.X.; Peng, D.H.; Sun, M. BtToxin_Digger: A comprehensive and high-throughput pipeline for mining toxin protein genes from *Bacillus thuringiensis*. *Bioinformatics* **2022**, *38*, 250–251. [CrossRef]
43. Ye, W.X.; Zhu, L.; Liu, Y.Y.; Crickmore, N.; Peng, D.H.; Ruan, L.F.; Sun, M. Mining new crystal protein genes from *Bacillus thuringiensis* on the basis of mixed plasmid-enriched genome sequencing and a computational pipeline. *Appl. Environ. Microbol.* **2012**, *78*, 14. [CrossRef]
44. Vachon, V.; Laprade, R.; Schwartz, J.L. Current models of the mode of action of *Bacillus thuringiensis* insecticidal crystal proteins: A critical review. *J. Invertebr. Pathol.* **2012**, *111*, 1–12. [CrossRef]
45. Bravo, A.; Gómez, I.; Conde, J.; Muñoz-Garay, C.; Sánchez, J.; Miranda, R.; Zhuang, M.; Gill, S.S.; Soberón, M. Oligomerization triggers binding of a *Bacillus thuringiensis* Cry1Ab pore-forming toxin to aminopeptidase N receptor leading to insertion into membrane microdomains. *BBA-Biomembr.* **2004**, *1667*, 38–46. [CrossRef] [PubMed]
46. Zhang, X.B.; Candas, M.; Griko, N.; Taussig, R.; Bulla, A. A mechanism of cell death involving an adenylyl cyclase/PKA signaling pathway is induced by the Cry1Ab toxin of *Bacillus thuringiensis*. *Proc. Natl. Acad. Sci. USA* **2006**, *103*, 9897–9902. [CrossRef] [PubMed]
47. Pardo-López, L.; Soberón, M.; Bravo, A. *Bacillus thuringiensis* insecticidal three-domain Cry toxins: Mode of action, insect resistance and consequences for crop protection. *FEMS Microbiol. Rev.* **2013**, *37*, 3–22. [CrossRef] [PubMed]
48. Xu, C.C. Structural and Functional Research on *Bacillus thuringiensis* Cry51Aa1 Insecticidal Crystal Protein. Ph.D. Dissertation, Huazhong Agricultural University, Wuhan, China, 2015. (In Chinese with English Abstract)

49. Tan, F.R.; Zhu, J.; Tang, J.; Tang, X.M.; Wang, S.Q.; Zheng, A.P.; Li, P. Cloning and characterization of two novel crystal protein genes, *cry54Aa1* and *cry30Fa1*, from *Bacillus thuringiensis* strain BtMC28. *Curr. Microbiol.* **2009**, *58*, 654–659. [CrossRef] [PubMed]
50. Wang, J.; Ma, H.; Zhao, S.; Huang, J.; Yang, Y.; Tabashnik, B.E.; Wu, Y. Functional redundancy of two ABC transporter proteins in mediating toxicity of *Bacillus thuringiensis* to cotton bollworm. *PLOS Pathog.* **2020**, *16*, e1008427. [CrossRef]
51. Sun, D.; Zhu, L.H.; Guo, L.; Wang, S.L.; Wu, Q.J.; Crickmore, N.; Zhou, X.G.; Bravo, A.; Soberón, M.; Guo, Z.J.; et al. A versatile contribution of both aminopeptidases N and ABC transporters to Bt Cry1Ac toxicity in the diamondback moth. *BMC Biol.* **2022**, *20*, 33. [CrossRef]
52. Guo, Z.J.; Kang, S.; Sun, D.; Gong, L.J.; Zhou, J.L.; Qin, J.Y.; Guo, L.; Zhu, L.H.; Bai, Y.; Ye, F.; et al. MAPK-dependent hormonal signaling plasticity contributes to overcoming *Bacillus thuringiensis* toxin action in an insect host. *Nat. Commun.* **2020**, *11*, 3003. [CrossRef]
53. Guo, Z.J.; Kang, S.; Wu, Q.J.; Wang, S.L.; Crickmore, N.; Zhou, X.G.; Bravo, A.; Soberón, M.; Zhang, Y.J. The regulation landscape of MAPK signaling cascade for thwarting *Bacillus thuringiensis* infection in an insect host. *PLoS Pathog.* **2021**, *17*, e1009917. [CrossRef]
54. Batool, K.; Alam, I.; Jin, L.; Xu, J.; Wu, C.; Wang, J.; Huang, E.; Guan, X.; Yu, X.Q.; Zhang, L. CTLGA9 Interacts with ALP1 and APN receptors to modulate Cry11Aa toxicity in *Aedes aegypti*. *J. Agric. Food Chem.* **2019**, *67*, 8896–8904. [CrossRef]
55. Zhao, G.H.; Liu, J.N.; Hu, X.H.; Batool, K.; Zhang, L.L. Cloning, expression, and activity of ATP binding protein in Bt toxicity modulation against Aedes aegypti. *Parasites Vectors* **2019**, *12*, 319. [CrossRef]
56. Gao, M.J.; Dong, S.; Hu, X.D.; Zhang, X.; Liu, Y.; Zhong, J.F.; Lu, L.; Wang, Y.; Chen, L.; Liu, X.J. Roles of midgut cadherin from two moths in different Bt action mechanisms: Correlation among toxin binding, cellular toxicity, and synergism. *J. Agric. Food Chem.* **2019**, *67*, 13237–13246. [CrossRef] [PubMed]
57. Xie, Y.J.; Xu, C.X.; Gao, M.J.; Zhang, X.; Lu, L.; Hu, X.D.; Chen, W.; Jurat-Fuentes, J.L.; Zhu, Q.; Liu, Y.; et al. Docking-based generation of antibodies mimicking Cry1A/1B protein binding sites as potential insecticidal agents against diamondback moth (*Plutella xylostella*). *Pest. Manag. Sci.* **2021**, *77*, 4593–4606. [CrossRef] [PubMed]
58. Liu, J.; Liang, Y.S.; Hu, T.; Zeng, H.; Gao, R.; Wang, L.; Xiao, Y.H. Environmental fate of Bt proteins in soil: Transport, adsorption/desorption, and degradation. *Ecotoxicol. Environ. Saf.* **2021**, *226*, 112805. [CrossRef] [PubMed]
59. Tapp, H.; Stotzky, G. Persistence of the insecticidal toxin from *Bacillus thuringiensis* subsp. kurstaki in soil. *Soil Biol. Biochem.* **1998**, *30*, 471–476. [CrossRef]
60. Stotzky, G.; Crecchio, C. Biodegradation and insecticidal activity of the toxin from *Bacillus thuringiensis* subsp. kurstaki bound on complexes of montmorillonite humic acids-Al hydroxy polymers. *Soil Biol. Biochem.* **2001**, *33*, 573–581.
61. Conde, J.M.; Patino, J.M.R. The effect of enzymatic treatment of a sunflower protein isolate on the rate of adsorption at the air-water interface. *J. Food Eng.* **2007**, *78*, 1001–1009. [CrossRef]
62. Helassa, N.; M'Charek, A.; Quiquampoix, H.; Noinville, S.; Déjardin, P.; Frutos, R.; Staunton, S. Effects of physicochemical interactions and microbial activity on the persistence of Cry1Aa Bt (*Bacillus thuringiensis*) toxin in soil. *Soil Biol. Biochem.* **2011**, *43*, 1089–1097. [CrossRef]
63. Yao, Y.L.; Cui, H.R.; Lu, M.Z.; Jin, J.Q. Adsorption/desorption of Cry1Ab Crystal protein in different soils. *Acta Pedol. Sin.* **2007**, *2*, 312–317. (In Chinese with English Abstract)
64. Zhou, X.Y.; Gao, J.B.; Cai, P.; Huang, Q.Y. Adsorption and Desorption of Bt toxin on three kinds of minerals. *Chin. J. Appl. Ecol.* **2008**, *19*, 1144–1148. (In Chinese with English Abstract)
65. She, C.M.; Zhou, X.Y.; Zheng, C.Y.; Liu, H.F. Adsorption characteristics and bioactivity of Bt (*Bacillus thuringiensis*) insecticidal protein on *Attapulgite*. *Acta Mine Ralog. Sin.* **2017**, *37*, 67–74. (In Chinese with English Abstract)
66. Zhou, X.; Li, H.; Liu, D.; Hao, J.; Liu, H.; Lu, X. Effects of toxin from *Bacillus thuringiensis* (Bt) on sorption of Pb (II) in red and black soils: Equilibrium and kinetics aspects. *J. Hazard. Mater.* **2018**, *360*, 172–181. [CrossRef]
67. Wang, J.W.; Luo, S.M.; Feng, Y.J.; Cindy, N. Environmental fate and ecological effects of Bt toxin from transgenic *Bt* crops in soil. *Acta Ecol. Sin.* **2003**, *4*, 797–804. (In Chinese with English Abstract)
68. Wang, J.W.; Feng, Y.J.; Luo, S.M. Studies on spatial-temporal dynamics of insecticidal protein expression of *Bt* corn and its degradation in soil. *Sci. Agric. Sin.* **2003**, *11*, 1279–1286. (In Chinese with English Abstract)
69. Stotzky, G. Persistence and biological activity in soil of the insecticidal proteins from *Bacillus thuringiensis*, especially from transgenic plants. *Plant. Soil.* **2005**, *266*, 77–89. [CrossRef]
70. Zhang, M.J.; Feng, M.C.; Xiao, L.J.; Song, X.Y.; Ding, G.W.; Yang, W.D. Persistence of Cry1Ac Protein from Transgenic *Bt* Cotton Cultivation and Residue Returning in Fields and Its Effect on Functional Diversity of Soil Microbial Communities. *Pedosphere* **2019**, *29*, 114–122. [CrossRef]
71. Feng, Y.; Lin, L.; Fan, H.; Liu, Y.; Tan, F.; Shu, Y.; Wang, J. Effects of temperature, water content, and pH on degradation of Cry1Ab protein released from *Bt* corn straw in soil. *Soil Biol. Biochem.* **2011**, *43*, 1600–1606. [CrossRef]
72. Deng, J.; Wang, Y.; Yang, F.; Liu, Y.; Liu, B. Persistence of insecticidal Cry toxins in *Bt* rice residues under field conditions estimated by biological and immunological assays. *Sci. Total Environ.* **2019**, *679*, 45–51. [CrossRef]
73. Zhou, L. Effects of Different Soil Conditions on the Degradation of Bt Protein. Mater's Thesis, Harbin Normal University, Harbin, China, 2018. (In Chinese with English Abstract)
74. Li, P.; Ye, S.F.; Liu, H.; Pan, A.H.; Ming, F.; Tang, X.M. Cultivation of drought-tolerant and insect-resistant rice affects soil bacterial, but not fungal, abundances and community structures. *Front. Microbiol.* **2018**, *9*, 1390. [CrossRef]

75. Accinelli, C.; Koskinen, W.C.; Becker, J.M.; Sadowsky, M.J. Mineralization of the *Bacillus thuringiensis* Cry1Ac endotoxin in soil. *J. Agric. Food Chem.* **2008**, *56*, 1025–1028. [CrossRef]
76. Saxena, D.; Flores, S.; Stotzky, G. Insecticidal toxin in root exudates from *Bt* corn. *Nature* **1999**, *402*, 6761. [CrossRef]
77. Madliger, M.; Sander, M.; Schwarzenbach, R.P. Adsorption of transgenic insecticidal Cry1Ab protein to SiO_2. 2. Patch-controlled electrostatic attraction. *Environ. Sci. Technol.* **2010**, *44*, 8870–8876. [CrossRef] [PubMed]
78. Madliger, M.; Gasser, C.A.; Schwarzenbach, R.P.; Sander, M. Adsorption of transgenic insecticidal Cry1Ab protein to silica particles. Effects on transport and bioactivity. *Environ. Sci. Technol.* **2011**, *45*, 4377–4384. [CrossRef] [PubMed]
79. Liu, J.P. Effect of Bt Protein on Denitrification and Microbial Diversity in Paddy Soil. Mater's Thesis, Huazhong Agricultural University, Wuhan, China, 2013 (In Chinese with English Abstract)
80. Chen, Y.J.; Pan, L.B.; Ren, M.Y.; Li, J.S.; Guan, X.; Tao, J. Comparison of genetically modified insect-resistant maize and non-transgenic maize revealed changes in soil metabolomes but not in rhizosphere bacterial community. *GM Crops Food* **2022**, *13*, 1–14. [CrossRef] [PubMed]
81. Chen, Y.J.; Ren, M.Y.; Guan, X.; Zhu, W.J.; Shen, Q. Influence of insect-resistant transgenic *Bt* cotton on Bt protein residue and available nutrient contents in rhizosphere soil. *Agric. Res. Arid. Areas* **2019**, *37*, 123–128. (In Chinese with English Abstract)
82. Yu, X. Effects of Transgenic *Bt* Gene Rice Denitrification and Microbial Diversity Straw on in Paddy Soil. Mater's Thesis, Huazhong Agricultural University, Wuhan, China, 2012. (In Chinese with English Abstract)
83. Zhang, F.L.; Yin, Q.; Mao, J.F.; Guo, L.A.; Lei, S.R.; Liu, W.J.; Luo, P.; Wang, D.; Song, J.; Yang, X.F. Effects of *Bacillus thuringiensis* transgenic cotton straw returning to field on soil fertility. *Chin. J. Eco-Agric.* **2020**, *28*, 734–744.
84. Teng, Y.; Huang, C.Y. Ecological effect of heavy metals on microbes and research advances on the mechanisms of bioremediation. *Soil Environ. Sci.* **2002**, *11*, 85–89. (In Chinese with English Abstract)
85. Cao, R.; Wu, F.Z.; Yang, W.Q.; Xu, Z.F.; Tan, B.; Wang, B.; Li, J.; Chang, C.H. Effects of altitudes on soil microbial biomass and enzyme activity in alpine-gorge regions. *Chin. J. Appl. Ecol.* **2016**, *27*, 1257–1264. (In Chinese with English Abstract)
86. Li, P.; Wu, G.G.; Li, Y.J.; Hu, C.; Ge, L.; Zheng, X.Q.; Zhang, J.Q.; Chen, J.; Zhang, H.L.; Bai, N.L.; et al. Long-term rice-crayfish-turtle co-culture maintains high crop yields by improving soil health and increasing soil microbial community stability. *Geoderma* **2022**, *413*, 115745. [CrossRef]
87. Li, Z.L.; Bu, N.S.; Chen, X.P.; Cui, J.; Xiao, M.Q.; Song, Z.P.; Nie, M.; Fang, C.M. Soil incubation studies with Cry1Ac protein indicate no adverse effect of *Bt* crops on soil microbial communities. *Ecotoxicol. Environ. Saf.* **2018**, *152*, 33–41.
88. Guan, Z.J.; Wei, W.; Stewart, C.N.J.; Tang, Z.X. Effects of transgenic oilseed rape harboring the *Cry1Ac* gene on microbial communities in the rhizosphere soil. *Eur. J. Soil Biol.* **2021**, *103*, 103277. [CrossRef]
89. Jan, F.; Dana, E.; Monika, H.; Frantisek, K. The Effect of *Bt*-corn on Soil Invertebrates, Soil Microbial Community and Decomposition Rates of Corn Post-Harvest Residues Under Field and Laboratory Conditions. *J. Sustain. Agr.* **2008**, *32*, 645–655.
90. Hong, X.; Han, C.; Kong, F.; Zhou, F.W.; Wu, S.S.; Zhong, W.H.; Liu, B. Effects of Insect-resistant and Herbicide-tolerant Transgenic Maize on Rhizospheric Bacterial and Fungal Communities. *J. Ecolo. Rural. Environ.* **2020**, *36*, 358–366. (In Chinese with English Abstract)
91. Wei, M.; Tan, F.R.; Hong, Z.; Cheng, K.; Xiao, W.; Ling, X.; Wang, J.; Kai, Z.; Tang, X.M. Impact of *Bt*-transgenic rice (SHK601) on soil ecosystems in the rhizosphere during crop development. *Plant. Soil Environ.* **2012**, *58*, 217–223. [CrossRef]
92. Li, P.; Xue, Y.; Shi, J.L.; Pan, A.H.; Tang, X.M.; Ming, F. The response of dominant and rare taxa for fungal diversity within different root environments to the cultivation of *Bt* and conventional cotton varieties. *Microbiome* **2018**, *6*, 184. [CrossRef]
93. Li, X.G.; Liu, B.; Cui, J.J.; Liu, D.F.; Ding, S.; Ben, G.; Luo, J.Y.; Fang, Z.X.; Cao, W.; Han, Z.M. No evidence of persistent effects of continuously planted transgenic insect-resistant cotton on soil microorganisms. *Plant Soil* **2011**, *339*, 247–257. [CrossRef]
94. Qi, X.M.; Liu, B.; Song, Q.X.; Zou, B.J.; Bu, Y.; Wu, H.P.; Ding, L.; Zhou, G.H. Assessing Fungal Population in Soil Planted with Cry1Ac and CPTI Transgenic Cotton and Its Conventional Parental Line Using 18S and ITS rDNA Sequences over Four Seasons. *Front. Plant Sci.* **2016**, *7*, 1023. [CrossRef]
95. Xu, Q.C.; Dai, R.B.; Ruan, Y.; Ren, S.C.; Liu, M.Q.; Guo, S.W.; Ling, N.; Shen, Q.R. Probing active microbes involved in *Bt*-containing rice straw decomposition. *Appl. Microbiol. Biot.* **2018**, *102*, 10273–10284. [CrossRef]
96. Zuo, L.H.; Yang, R.L.; Zhen, Z.X.; Liu, J.X.; Huang, L.S.; Yang, M.S. A 5-year field study showed no apparent effect of the *Bt* transgenic poplar on the arthropod community and soil bacterial diversity. *Sci. Rep.* **2018**, *8*, 1956. [CrossRef]
97. Wu, N.; Shi, W.C.; Liu, W.W.; Gao, Z.; Han, L.Z.; Wang, X.F. Differential impact of *Bt*-transgenic rice plantings on bacterial community in three niches over consecutive years. *Ecotoxicol. Environ. Saf.* **2021**, *223*, 112569. [CrossRef]
98. Wang, J.; Chapman, S.J.; Ye, Q.; Yao, H. Limited effect of planting transgenic rice on the soil microbiome studied by continuous $^{13}CO_2$ labeling combined with high-throughput sequencing. *Appl. Microbiol. Biotechnol.* **2019**, *103*, 4217–4227. [CrossRef]
99. Fan, C.; Wu, F.; Dong, J.; Wang, B.; Yin, J.; Song, X. No impact of transgenic *cry1Ie* maize on the diversity, abundance and composition of soil fauna in a 2-year field trial. *Sci. Rep.* **2019**, *9*, 10333. [CrossRef] [PubMed]
100. Zhang, Y.J.; Xie, M.; Wu, G.; Peng, D.L.; Yu, W.B. A 3-year field investigation of impacts of Monsanto's transgenic *Bt*-cotton NC 33B on rhizosphere microbial communities in northern China. *Appl. Soil Ecol.* **2015**, *89*, 18–24. [CrossRef]
101. Li, Z.L.; Bu, N.S.; Cui, J.; Chen, X.P.; Xiao, M.Q.; Wang, F.; Song, Z.P.; Fang, C.M. Effects of long-term cultivation of transgenic *Bt* rice (Kefeng-6) on soil microbial functioning and C cycling. *Sci. Rep.* **2017**, *7*, 4647.
102. Lin, X.G.; Hu, J.L. Scientific connotation and ecological service function of soil microbial diversity. *Acta Pedol. Sin.* **2008**, *45*, 9. (In Chinese with English Abstract)

103. Escalas, A.; Hale, L.; Voordeckers, J.W.; Yang, Y.F.; Firestone, M.K.; Alvarez-Cohen, L.; Zhou, J.Z. Microbial functional diversity: From concepts to applications. *Nat. Ecol. Evol.* **2019**, *9*, 12000–12016. [CrossRef]
104. Yaqoob, A.; Ahmad, A.; Shahid, I.B.; Salisu, S.A.; Mukhtar, A.; Abdul, Q.R. Effects of Cry toxins on non-target soil bacteria during a 2-year follow up study. *Span. J. Agric. Res.* **2019**, *17*, e0303. [CrossRef]
105. Wu, W.X.; Ye, Q.F.; Min, H.; Duan, X.J.; Jin, W.M. *Bt* transgenic rice straw affects the culturable microbiota and dehydrogenase activity in flooded soil. *Soil Biol. Biochem.* **2004**, *36*, 289–295.
106. Liu, W.; Lu, H.H.; Wu, W.X.; Wei, Q.K.; Chen, Y.X.; Thies, J.E. Transgenic *Bt* rice does not affect enzyme activities and microbial composition in the rhizosphere during crop development. *Soil Biol. Biochem.* **2008**, *40*, 475–486. [CrossRef]
107. Liang, J.G.; Xin, L.T.; Luan, Y.; Song, X.Y.; Zhang, Z.G. Effect of Cry1Ie *Bt* Maize on Carbon Source Metabolism of Rhizosphere Microorganisms. *J. Agr. Sci. Tech.* **2019**, *21*, 104–110. (In Chinese with English Abstract)
108. Luan, Y. Effects of Transgenic BT-Maize with Cry1Ab-Ma Gene on the Microbial Community Structure and Functional Diversity of Rhizosphere Soil. Mater's Thesis, Nanjing Agricultural University, Nanjing, China, 2019. (In Chinese with English Abstract)
109. Xin, L.T. Effect of Transgenic Bt-Maize with Cry1Ie Gene on the Structure and Functional Diversity of Rhizosphere Microbial Communities. Mater's Thesis, Nanjing Agricultural University, Nanjing, China, 2017. (In Chinese with English Abstract)
110. Li, Z.L.; Cui, J.; Mi, Z.R.; Tian, D.S.; Wang, J.S.; Ma, Z.L.; Wang, B.X.; Chen, H.Y.H.; Niu, S.L. Responses of soil enzymatic activities to transgenic *Bacillus thuringiensis* (Bt) crops—A global meta-analysis. *Sci. Total Environ.* **2018**, *651*, 1830–1838. [CrossRef]
111. Valldor, P.; Graff, R.; Dockhorn, S.; Martens, R.; Tebbe, C.C. Production of the ^{14}C-labeled insecticidal protein Cry1Ab for soil metabolic studies using a recombinant *Escherichia coli* in small-scale batch fermentations. *Appl. Microbiol. Biot.* **2012**, *96*, 221–229. [CrossRef] [PubMed]
112. Wang, Z.P.; Chen, Q.S. Recently photosynthesized carbon allocation and turnover: A minor review of the literature. *Acta Phytoecol. Sin.* **2005**, *29*, 845–850. (In Chinese with English Abstract)
113. Ouyang, Y.; Sun, X.H.; Yan, J.; Chu, J.F.; Yan, C.Y. Progress in relative quantitative analysis of biological molecules with stable isotope labeling. *Chin. Sci. Bull.* **2013**, *58*, 2762–2778. (In Chinese with English Abstract) [CrossRef]
114. Zheng, Y.; Jia, Z.J. Next generation sequencing and stable isotope probing of active microorganisms responsible for aerobic methane oxidation in red paddy soils. *Acta Microbiol. Sin.* **2013**, *53*, 173–184. (In Chinese with English Abstract)
115. Quan, Z.X. Seeking of "missed" microorganisms. *Microbiol. China* **2013**, *40*, 34–43. (In Chinese with English Abstract)
116. Zhang, H.; Huang, M.; Zhang, W.; Gardea-Torresdey, J.L.; White, J.C.; Ji, R.; Zhao, L. Silver nanoparticles alter soil microbial community compositions and metabolite profiles in unplanted and cucumber-planted soils. *Environ. Sci. Technol.* **2020**, *54*, 3334–3342. [CrossRef]

plants MDPI

Review

Assessing Impacts of Transgenic Plants on Soil Using Functional Indicators: Twenty Years of Research and Perspectives

Vadim Lebedev [1,*], Tatyana Lebedeva [2], Elena Tikhonova [3] and Konstantin Shestibratov [1,3,4]

1 Branch of the Shemyakin-Ovchinnikov Institute of Bioorganic Chemistry of the Russian Academy of Sciences, Prospekt Nauki 6, 142290 Pushchino, Russia
2 Institute of Physicochemical and Biological Problems in Soil Science of the Russian Academy of Sciences, Instituskaya Str. 2, 142290 Pushchino, Russia
3 Department of Landscape Architecture and Soil Science, Voronezh State University of Forestry and Technologies Named after G.F. Morozov, 8 Timiryazeva Str., 394087 Voronezh, Russia
4 Pushchino State Institute of Natural Sciences, Prospekt Nauki 3, Pushchino, 142290 Moscow, Russia
* Correspondence: vglebedev@mail.ru

Abstract: Assessment of the effects of transgenic plants on microbiota and soil fertility is an important part of the overall assessment of their biosafety. However, the environmental risk assessment of genetically modified plants has long been focused on the aboveground effects. In this review, we discuss the results of two decades of research on the impact of transgenic plants on the physicochemical properties of soil, its enzyme activities and microbial biomass. These indicators allow us to assess both the short-term effects and long-term effects of cultivating transgenic plants. Most studies have shown that the effect of transgenic plants on the soil is temporary and inconsistent. Moreover, many other factors, such as the site location, weather conditions, varietal differences and management system, have a greater impact on soil quality than the transgenic status of the plants. In addition to the effects of transgenic crop cultivation, the review also considers the effects of transgenic plant residues on soil processes, and discusses the future prospects for studying the impact of genetically modified plants on soil ecosystems.

Keywords: biosafety; transgenic plants; risk assessment; soil enzyme activity; microbial biomass; soil fertility

Citation: Lebedev, V.; Lebedeva, T.; Tikhonova, E.; Shestibratov, K. Assessing Impacts of Transgenic Plants on Soil Using Functional Indicators: Twenty Years of Research and Perspectives. *Plants* **2022**, *11*, 2439. https://doi.org/10.3390/plants11182439

Academic Editors: Wei Wei and C. Neal Stewart, Jr.

Received: 20 July 2022
Accepted: 15 September 2022
Published: 19 September 2022

1. Introduction

Socioeconomic benefits of genetically modified plants (GMP) have favored their wide cultivation in the world so that most cotton (79%) and soybean (74%), as well as a significant part of maize (31%) and canola (27%), are biotech crops [1]. The emergence of transgenic plants has raised a debate about their biosafety, both for human health and the environment. Soon after the commercial release of the first genetically altered crop in 1996, Snow and Moran-Palma [2] divided environmental risks into four groups: (1) transgene flow to wild relatives, (2) the evolution of resistant pests, (3) the effects on non-target organisms and (4) the effects on soil biota and fertility. Therefore, assessment of the risks to soil due to the cultivation of GMPs should be an important part of an overall safety assessment of transgenic plants. However, the environmental risk assessment of transgenic plants has long been focused mainly on the aboveground effects. In the early 2000s, several reviews summarized the research on transgene flow to wild and/or weedy relatives [3], the evolution of resistant pests [4] or the effects on non-target organisms [5]. At the same time, studies on the effects of transgenic plants on soil microorganisms were just beginning. Currently, this aspect is very important because microorganisms are the dominant underground soil organisms: they account for more than 80% of the total biomass (without roots) and largely determine the functioning of terrestrial ecosystems [6]. The disregard for the effects of transgenic plants on the underground components is mainly explained by the difficulties inherent in the study of soil microbiota [7]. The abundance of various biotic and abiotic factors has so far prevented researchers from reaching a consensus on whether GMPs can affect soil microorganisms [8]. The available studies provide contradictory results. Most studies have assessed the effects as insignificant, although some studies have noted significant effects, although these are transitory. On the other hand, there are known publications on the negative effects of transgenic plants on the physical and chemical properties of soil, its microbial biomass, enzyme activities and microbial biodiversity [9–11]. Our review summarizes two decades of research on the effects of GMPs on the physicochemical properties of soil, its enzyme activities and microbial biomass. We also note the involvement of external factors in these interactions and outline the prospects for further research on these issues.

2. Potential Risks of Transgenic Plants for Soil Ecosystems

Soil microorganisms are essential components of soil biological activity and are involved in important biochemical processes such as the decomposition of organic matter, humus formation and the transformation and cycling of nutrients [12]. Rhizosphere, the root–soil interface, is the key site where microorganisms integrate soil and plants [8]. Plants release up to 20% of the photosynthetically fixed carbon into soil with root exudates, which are the main source of C for microorganisms [13]. These exudates contain primary and secondary metabolites and, via solubilization and mineral desorption, provide microorganisms with nutrients, thus playing a key role in establishing plant–microorganism interactions [14]. Therefore, any intended or unintended alteration in the composition or quantity of root exudates can affect the soil microbiota (Figure 1). In addition to the well-known components (sugars, organic acids, amino acids, etc.), the root exudates of transgenic plants can also contain new substances, including toxins [15]. Microorganisms also use plant residues—shoots and roots—as a source of C and nutrients. There is no clear understanding of what is the cause of changes in nutrient cycling in soil under GM crops—differences in the composition and quantity of root exudates or plant residues (proteins, carbohydrates, lignins, etc.) [16].

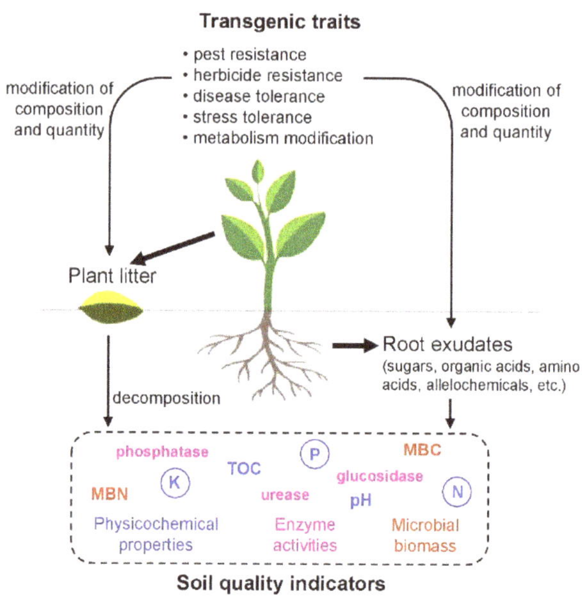

Figure 1. Potential impact of transgenic plants on soil quality indicators.

Different authors differently define the potential effects of GMPs on soil [15,16], but generally these effects can be divided into three groups: (1) direct effects via new GMP-produced substances, e.g., Bacillus thuringiensis (Bt) toxins; (2) indirect effects due to intended or unintended changes in plant metabolism that alter the quality and quantity of root exudates or the composition and quantity of biomass, mostly underground; (3) changes in management systems associated with the introduction of GM crops, e.g., the use of other herbicides. Any of them can boost or inhibit the growth of certain groups of microorganisms and thus may ultimately affect the processes of carbon and nutrient cycling on the ecosystem level [17]. At the very beginning of research in this area, however, the question was asked as to whether a transgenic plant with as little as one or two genes that made it differ from the original plant could have significant effects on soil microorganisms [18]. It looks more likely that fluctuations in temperature or precipitation, or changes in plant management would have a much greater impact on such a heterogeneous system as soil than an altered genetic trait could [16]. An accurate assessment of changes in soil ecosystem requires selecting the proper indicators.

3. Indicators of Soil Quality and Fertility

In the extremely complex "plant–soil–microbes" system, each component should be evaluated with its specific dedicated methods. Most often, a transgenic plant is already well characterized by the beginning of experiments, and the only task is to ascertain some parameters under specific growing conditions. Soil quality research traditionally focused on chemical and physical properties [19]. The number of such parameters is quite limited (pH, organic matter, macro- and some micronutrients) and they can be determined by simple analytical methods. A wide variety of methods assess soil microorganisms, their diversity, abundance and function. Microbial diversity is studied using molecular tools, particularly amplicon sequencing, terminal restriction fragment length polymorphism (T-RFLP) and denaturing gradient gel electrophoresis (DGGE), while the phospholipid fatty acids (PLFA) analysis provides information about the overall structure of a microbial community [20]. The total microbial biomass is assessed by quantifying the biophilic elements in microbial cells, most often C (microbial biomass C, MBC), less often N (microbial biomass N, MBN) and even less often P (microbial biomass P, MBP). The functionality is evaluated by measuring the activities of soil enzymes, which are very diverse.

Great attention is given to the effects of GMPs on biodiversity. The review by Guan et al. [8] provides a detailed discussion of the effects of transgenic plants on the soil microbial diversity as assessed using PLFA, DGGE, T-RFLP and other methods. Understanding the relationship between microbial composition and functionality is necessary to predict changes in ecosystem functioning in response to various environmental disturbances [21], including the impact of transgenic plants. However, relationships between microbial diversity and soil functions are still debated [22]. The concept that biodiversity promotes the functioning of ecosystems has long been adopted for animals and plants. If directly extended to microorganisms, however, the concept faces a number of serious issues [23]. These issues originate from fundamental differences between macro- and microorganisms: the latter are characterized by small sizes but an immense richness of microbes, faster metabolism and physiological versatility, and rapid and colonial growth [24,25]. There is no direct evidence that the microbial diversity of soil is related to soil ecosystem functioning. The existing methods and techniques are either not effective enough to obtain relevant evidence, or the available data may be insufficient for valid conclusions [8]. A recent study was the first to compare five levels of soil microbial diversities of taxonomy and function responding to biodiversity loss based on global soil metagenomes across diverse biomes [26]. It showed that the relative abundance of microbial functioning can remain stable despite a sharp reduction in taxonomic species that leads to biotic homogenization but functional stability. This stability suggests a decoupling of taxonomy and function. The cause of this stability is that microbial communities have high taxonomic variability but a stable functional structure [27]. Assuming that changes in functionality are more likely to be the consequence of diversity disturbance than vice versa, we focused on the analysis of GMP effects on the physicochemical properties of soil, its enzyme activities and microbial biomass.

3.1. Soil Physicochemical Properties

The composition of a soil microbial community is the result of the soil's physical and chemical properties, which develop at different time scales, over a long period of soil formation, as well as more recent, in response to local weather conditions and management systems [28]. For instance, pH is one of the most powerful factors that affects the composition of a soil microbial community [29]. Another example is the relationship between the electrical conductivity (EC) of soil and its microbial biomass [30]. Thus, changes in soil's physicochemical properties can directly or indirectly affect the activities of soil enzymes and its microbial biomass [31]. Plant roots and soil microbes can, in turn, alter the physical and chemical properties of soil in the rhizosphere [32]. For example, organic acids from root exudates not only alter pH but also play an important part in the availability of phosphorus. Changes in the composition and quantity of root exudates in plants with new genetic traits may have a direct effect on transformations of soil P and/or an indirect effect on the availability of P via shifts in the microbial community and the activities of microorganisms inhabiting the rhizosphere [16].

3.2. Soil Enzyme Activity

The enzymatic activity of soil plays a crucial role in the formation and decomposition of soil organic matter, as well as in nutrient cycling [33]. In many studies, soil enzymes are used as indicators of microbial activity and soil fertility. Their activity is considered an early and sensitive indicator of natural or anthropogenic disturbances [34]. The list of microbial genotypic function traits important to biogeochemistry, ecology and environmental sciences includes enzymes such as chitinase xylanase (carbon degradation in carbon cycling), urease (N mineralization in N cycling), etc. [23].

The most important enzymes associated with changes in soil quality are hydrolases and oxidoreductases. The best studied ones include the intracellular enzyme dehydrogenase (DHA), an oxidoreductase, and a number of extracellular hydrolases (β-glucosidase (BGL), phosphatase (PHO), urease, arylsulfatase (ARS) and others), which are directly involved in transformations of organic compounds and the release of C and nutrients, such as N, P and S [35]. The hydrolytic degradation of complex soil components is an important step in several biogeochemical cycles [36]. The thus obtained carbon and nutrients are then assimilated by microbial cells and used in several metabolic pathways controlled by intracellular enzymes, including DHA. Even before studies with transgenic plants, soil quality was often assessed based on enzyme activities, such as DHA (general biochemical parameter), PHO, BGL and urease (P, C and N cycles) [37]. The same enzymes, as well as protease and arylsulfatase, are most often used to assess the effects of transgenic plants on soil microorganisms. The DHA activity is an important indicator of oxidative metabolism in soils and a sensitive marker of soil microbial activity since this intracellular enzyme is associated with viable cells [38]. β-glucosidase catalyzes cellobiose hydrolysis to glucose and dominates among other enzymes involved in the degradation of carbohydrates in soils [39]. Urease plays an essential role in the effective use of urea in soil by hydrolyzing it to NH_3 and CO_2; changes in urease activity are an indirect indicator of changes in the pool of potentially available N in soil [40]. Protease is another essential enzyme in the N cycle in soil; it breaks proteins down to amino acids and is often considered to be the rate-limiting step of N mineralization [41]. Phosphatases catalyze the cycling and transformation of P in soil ecosystems, and they are a good indicator of organic P mineralization and soil activity [42]. Most often, studies assess the activities of acid PHO, which is mainly produced by plant roots, and alkaline PHO, which originates from microorganisms and fauna [43]. Arylsulfatase is an important soil enzyme catalyzing the hydrolysis of sulfate esters [44].

Less often assessed are hydrolases, such as cellulase and invertase, which hydrolyze cellulose and sucrose to monosaccharides, respectively [45,46], and oxidoreductases, such as polyphenol oxidase and catalase, which degrade recalcitrant aromatic compounds and hydrogen peroxide, respectively [47,48].

3.3. Microbial Biomass

Soil microbial biomass is a critical component of most terrestrial ecosystems because it regulates nutrient cycling and acts as a highly labile source of nutrients available to plants [49]. For instance, MBC is more sensitive to changes in the status of organic matter than the total organic C is [50]. MBC, MBN and MBP are the active components of C, N and P in soil, respectively, and, as such, participate in the cycling of these elements in the ecosystems [51]. MBC is the driving force of the decomposition of soil organic matter, while MBN is critical to regulating the N flow into soil. MBP governs the mineralization and fixation of soil P, reflects the capacity and intensity of its cycling and is an important source of available soil P [11].

The microbiological parameters of soil (enzyme activities and biomass) are considered more sensitive to changes in management and environmental conditions than chemical or physical properties are [7,52]. For example, TOC is relatively insensitive to environmental changes and reflects the cumulative result of changes in affecting factors over a relatively long period [53]. Thus, biochemical indicators show the early response of soil to exposure, while physicochemical indicators reflect longer-term trends. Together, they complement each other and provide information about the intensity and direction of changes in a soil ecosystem.

4. Effects of Transgenic Plants' Cultivation

According to recent data, the most cultivated transgenic plants in the world in 2019 were those with stacked traits with insect resistance and herbicide tolerance, and herbicide-tolerant and insect-resistant crops, which occupied 45%, 43% and 12% of the global biotech crop area, respectively [1]. All other GM crops—virus-resistant or salt-tolerant plants, canola with modified oils, low gossypol cotton, etc.—accounted for less than 0.5% of the global biotech crop area. Despite the dominance of herbicide-resistant crops, most studies on the GMP effects on soil were conducted with insect-resistant plants because they produce toxic Bt proteins (Table 1). Plants with other traits are studied less frequently, although theoretically they may also cause unintended changes unrelated to the new gene product but able to affect soil processes.

Table 1. Risk assessment of insect-resistant transgenic plants on soil quality.

Species	Gene	Growth Conditions	Indicators	Additional Factors	References
maize	*Cry3Bb*	field (2 years)	MBC [1]	growth stage insecticide	[54]
maize	*Cry3Bb*	field (3 years)	MBC	growth stage insecticide	[17]
maize	*Cry1Ab, Cry3Bb1*	field (4 years)	N, P(2), S, DHA [2]	variety	[55]
rice	*Cry1Ab*	field (2 years)	P, DHA	growth stage insecticide	[56]
maize	*Cry1Ab*	field	N, DHA	growth stage	[57]
cotton	*Cry1Ac*	net house	NH4, NO3, N, P [3] DHA	growth stage	[58]
cotton	*Cry1Ac*	net house	organic C N, P(2) MBC, MBN, MBP	growth stage	[59]
maize	*Cry1Ab*	field (7 years)	C, N, texture		[60]
cotton	*Cry1Ac*	field (3 years)	N(2), P, DHA		[61]

Table 1. *Cont.*

Species	Gene	Growth Conditions	Indicators	Additional Factors	References
cotton	*Cry1Ac* *Cry1Ac + CpTI*	greenhouse (4 years)	C, N(3), P(2), S, DHA, CAT MBC		[9]
rice	*Cry1Ac*	open air (pots)	C(2), N(2), P, DHA	growth stage	[62]
cotton	*Cry1Ac*	pots	org. matter, N, P, K C, N(2), P, DHA	growth stage	[63]
maize	*Cry1Ab*	climate chamber	DHA MBC, MBN	variety	[64]
maize	*Cry1Ab*	climate chamber	DHA MBC, MBN	soil type	[65]
maize	*Cry1Ab*	field (5 years)	C MBC	insecticide crop rotation	[66]
cotton	*cry1Ac*	field	N, DHA MBC		[40]
cotton	*CrylAc + CpTI*	field (2 years)	N, P, DHA	growth stage	[67]
sugarcane	*Cry1Ac*	field	C, N(2), P	growth stage	[68]
cotton	*Cry1Ac*	field	pH, EC, org. C, NO3, NH4, P, K MBC	fertilization	[32]
cotton	*Cry1Ac*	field (6 years)	C, N(3), P(2), S, DHA MBC		[10]
rice	*Cry1Ac*	field (8 years)	org. C, N, P, C/N P, N, DHA, CAT MBC, MBN	growth stage	[53]
cotton	*Cry1Ac*	field (2 years)	N, P(2), CAT MBC	growth stage salinity	[12]
maize	*Cry1Ab*	field	pH, org. C, NO3, NH4, P C, N, P	irrigation	[69]
poplar	*Cry1Ac, Cry3A*	field (5 years)	pH, org. matter, N, P, K	growth stage	[70]
cotton	*Bt*	field	N, P, K P, DHA MBC	crop rotation	[71]
poplar	*Cry1Ah1*	field (4 years)	pH, N, P MBC, MBN, MBP		[11]

[1] Microbial biomass (MBC, MBN, MBP). [2] Enzyme activity: element cycle (enzyme number). [3] Physicochemical soil properties.

4.1. Insect-Resistant Transgenic Plants

Pot studies of Bt cotton (*Cry1Ac*) in India found no differences in NH4 [58] and TOC [59] in soil. However, Bt cotton showed significant variations in available P, with the availability levels both lower (in mid-vegetation) and higher (at the end of vegetation) versus a control [58]. In another pot experiment conducted in China, Bt cotton with the same gene (*Cry1Ac*) had no effect on the content of organic matter, total N, available N or K throughout the growing period [63]. The study also noted a significant decrease in available P in Bt cotton, although only during a flowering period.

According to a later short-term field study, Bt cotton did not show any significant adverse effects on the physicochemical properties of soil as compared with non-transgenic plants. Indian studies did not reveal any significant effect of Bt cotton on the content of available N, P and K [71]. Studies in Pakistan showed that Bt cotton had no effect on pH and EC, while its effects on other parameters depended on NPK fertilizers [32]. TOC did not differ among treatments without fertilizers, but was significantly higher in two clones in treatments with fertilizers. Phosphorus, conversely, did not differ among treatments with fertilizers, but was significantly higher in all four clones in the presence of fertilizers.

The content of available K in various clones, regardless of fertilizers, could be either higher or lower than in the control [32].

The long-term field tests of Bt plants had ambiguous results. Field cultivation of Bt maize for 7 years did not significantly alter the total C or total N, or the soil texture [60]. An 8-year cultivation of Bt rice did not bring any consistent changes in soil properties: the control samples contained significantly less TOC and TN mid-season and significantly less available P at the end of the season [53]. Long-term field tests of Bt poplar showed no effect on the content of N [11,70]. The effect on P, however, differed: while the Bt poplar showed no effect in one study [70], all five Bt poplar lines significantly reduced the content of available P in another [11]. The authors suggest that transgenic trees had a negative impact on the activity of phosphate solubilizing microorganisms and thus affected the transformation of soil P. On the whole, cultivation of Bt plants did not change the physical and chemical properties of soil; however, many researchers noted their influence on the content of available P.

One- to two-year field tests revealed no effect on soil enzyme activities in crops such as maize [57], sugarcane [68] or cotton [12]. In general, after a 4-year cultivation in the field, there were no consistent significant differences in the activities of N-, P- and S-cycle enzymes and DHA in maize with *Cry1Ab* or *Cry3Bb1* genes and a non-transgenic control [55]. Random significant differences were not stable and did not persist. A number of studies reported the effect of Bt plants on individual enzymes. For example, acid and alkaline PHOs were significantly increased in Bt cotton pots by the end of cultivation [59]. A 3-year cultivation of Bt cotton in the field had no effect on the activities of extracellular enzymes, whereas the DHA activity increased significantly [61]. According to another report, three field-grown Bt cotton clones significantly increased the urease and DHA activities [40]. The authors believe that the increase in the DHA activity could be due to a higher microbial activity stimulated by the increased root density in Bt cotton compared with the control. On the other hand, field-grown Bt maize had no effect on urease but significantly reduced the activities of BGL and acidic PHOs [69]. This indicates that some bacterial species could have been inhibited and did not participate in the metabolic activity of soil.

Unlike many studies, Chen et al. [9] showed an inhibitory effect of cotton with pest resistance genes, when cultivated in pots in the greenhouse, on enzymes of the N, P and S cycles, as well as DHA and catalases, with the BGL activity being the only one unchanged. The authors explain the discrepancy in findings by the absence or too low levels of Cry proteins in the soils of previous researchers. They attribute the reduction in enzyme activities in the soil of transgenic cotton to decreased enzyme synthesis by microorganisms or to a competition between the Cry1Ac proteins and the CpTI enzymes for adsorption sites in soil. A subsequent field assessment showed a significant increase in the DHA activity in soil growing Bt cotton and its significant inhibition after residue incorporation in soil [10]. Thus, similarly to its effects in the greenhouse, Bt cotton inhibited the growth and activities of soil microorganisms. The observed stimulation of extracellular enzymes could have been associated with the adsorption of Cry1Ac proteins on soil particles, the release of a certain amount of enzymes and the increase in their activities. The increased enzyme activities can accelerate the C, N, P and S cycles in soil and should therefore be considered as a potential unintended risk of transgenic Bt cotton associated with adding its residues into soil [10].

Bt plants had ambiguous effects on soil microbial biomass. Devare et al. [17,54] reported the absence of any significant effect of Bt maize with the *Cry3Bb* gene on MBC after two or three years of field tests in the USA. Five-year tests of Bt maize with the *Cry1Ab* gene also showed no effect on MBC [66], the same as 2-year tests of Bt cotton [12]. Nor were there any differences in MBC and MBN after 8 years of growing Bt rice in the field [53]. On the other hand, there were reports of microbial biomass stimulation in soil from Bt cotton. Significantly higher values of MBC, MBN and MBP were found in soil from Bt cotton grown in pots under net house conditions [59]. Field tests also confirmed the stimulating effect of three Bt cotton lines on MBC in a layer of 0 to 15 cm, but only for one line in a layer of

15 to 30 cm [40]. A significant increase in MBC in the field in one of four Bt cotton lines, both with and without fertilizer, suggests that it was peculiar to this specific transgenic genotype [32].

In contrast to those results, three and four years of greenhouse pot cultivation of Bt and Bt+CpTI cotton resulted in a significant reduction in MBC, which indicates the inhibition of microbial activity in the soil of transgenic plants [9]. Further field studies confirmed the significant inhibition of MBC by Bt cotton plants [10]. Four-year field tests of Bt poplar in China showed a significant effect on soil microbial biomass: in soil samples from all five clones, MBC was significantly higher, while MBN and MBP were significantly lower, compared with the control [11]. These changes modify the ability of soil microorganisms to metabolize C, N and P and thus can ultimately affect the plant growth. The lower MBN and MBP in the soil from Bt poplars indicates that soil microorganisms are stressed by nutrient deficiencies. Moreover, there were also changes in the structure of the soil microbial community. In a control, the MBC/MBN ratio was about 4.6, which indicates the dominant role of bacteria; in Bt poplars, it was about 9.2, showing the predominance of fungi [11].

4.2. Herbicide-Resistant Transgenic Plants

The cultivation of herbicide-resistant plants (Table 2) does not imply an a priori effect on soil microflora, and a greenhouse-grown oilseed rape with the *pat* gene did not affect ARS or MBN, although there were significant changes in the activities of invertase, phosphatase and urease [72]. The latter were probably caused by changes in the composition and/or concentration of root exudate due to unintended alterations in the transformation process. The change in exudation is also evidenced by a high invertase activity in the rhizospheres of senescent transgenic plants, which is indicative of increased sucrose concentrations in the root zone [72].

Although herbicide-resistant plants occupy the largest part of GM crop areas, there have not been many studies with them. This fact, however, is offset by the large scale of those studies. Field tests in Canada and Brazil lasted for 3 to 9 years, were carried out on several sites with different soil and climatic conditions, and included various management systems and crop rotations (Table 2). In those studies, GMPs resistant to glyphosate or imidazolinone showed no significant effect on MBC, MBN [73–76] or soil physicochemical properties (Ca, Mg, K, organic matter, N, P, cation exchange capacity, Mn, Fe, Cu, Zn, soil density and granulometry), except for pH [76] or enzymatic activity [73,74,77].

Table 2. Risk assessment of herbicide-resistant transgenic plants on soil quality.

Species	Gene	Growth Conditions	Indicators	Additional Factors	References
oilseed rape	*pat*	greenhouse	MBN C, N, P, S	growth stage	[72]
wheat canola maize	*epsps* *epsps*	field (4 years) field (5 years)	MBC DHA MBC C	location crop rotation herbicide crop rotation	[73] [74]
soybean	*ahas*	field (3 years)	MBC, MBN	location	[75]
soybean	*epsps*	field (3 years)	MBC, MBN C, P	location herbicide	[77]
soybean	*epsps*	field (8–9 years)	pH, org. matter, N, P, microelem., texture MBC, MBN	location	[76]

4.3. Disease-Tolerant Transgenic Plants

One of the first disease-tolerant transgenic plants authorized for commercial use back in 1996 was virus-tolerant papaya expressing the coat protein gene of the Papaya ringspot virus (PRSV) (Table 3). Studies on pot-grown papaya did not reveal any effect on pH, organic matter, P, K, Ca or Mg, but the line showed a significant increase in EC, and a significant decrease in the content of N and S, 2.2 and 1.1 times, respectively [78]. Since most parameters remained unchanged, the observed changes were attributed to the introduction of litter and root exudates into the soil. Another study with virus-resistant papaya, however, obtained different results. It also found no differences in pH and C, but nor did it find any differences in N [78]. The possible causes might have been due to differences in cultivation conditions or plant age (9 months and 9 years), as well as different genotypes used in these two experiments.

Table 3. Risk assessment of disease-tolerant transgenic plants on soil quality.

Species	Gene	Growth Conditions	Indicators	Additional Factors	References
papaya	*PRSV CP*	field	pH, org. C, N		[79]
papaya	*PRSV RP*	open air (pots)	pH, EC, org. matter, N, P, K, microel. C(3), N(2), P(3), S, DHA, CAT		[78]
spruce	*ech42*	greenh. (5 years)	fungal biomass		[80]
tobacco	*McChit1*	chamber house	pH N(2), CAT	growth stage	[81]
wheat	*WYMV-Nib8*	field (2 years)	C, N, DHA	growth stage location	[82]
melon	*AFP + CHI*	greenhouse	pH, EC, org. matter, P, K C(2), N(2), P(2), S, DHA, CAT		[83]
oilseed rape	*NiC*	greenhouse	C (2), N, P, S	variety	[84]
rice	*OsCK1*	field	pH, EC, org. matter, N, P, K, microel.	growth stage	[85]
potato	*Rpi-vnt1.1*	field (2 years)	pH, org. C, N, C/N	variety fungicide location	[28]

Pot cultivation of tobacco with a chitinase gene revealed significant pH fluctuations compared with a control, possibly due to root exudates [81]. Recent field studies did not find any effect on soil physicochemical properties in rice with the *OsCK1* (cholinkinase) gene resistant to rice pyriculariosis and bacterial blight [85], as well as in potato with R-genes resistant to late blight [28].

Transgenic papaya also had a notable effect on soil enzyme activities [78]. A significant increase in activity was observed for alkaline PHO, ARS and invertase, while the protease, polyphenol oxidase and urease activities showed a significant reduction. The activities of acidic PHOs, cellulase, catalase and DHA did not change. The most sensitive enzyme was arylsulfatase, which is involved in the S mineralization in soil; their level grew 5.4 times in the soil of virus-resistant papaya [78]. This was due to the improved growth of the transgenic line; the plant tissues produced more sulfur-containing papain and chymopapain, which required enhanced immobilization of N and S. This, in turn, led to significant differences in the content of N and S, as well as in the ARS activity. Changes in the activities of other enzymes could have been caused by the impact of transgenic papaya on the activity of microorganisms.

The effects of other disease-tolerant transgenic plants on soil biochemistry were insignificant. Field tests of wheat resistant to wheat yellow mosaic virus were conducted for 2 years in two regions of China; cultivation of the GM wheat did not alter the activities of urease, DHA or sucrase [82]. Chitinase-expressing tobacco grown in pots in the chamber house had no effect on protease but showed some effect on catalase, although at the rosette stage only, and on urease, at the stubble stage only [81]. Greenhouse-cultivated oilseed rape plants with a chitinase gene showed no difference from their parent line in C, N, P and S cycle enzymes [84]. Five-year-old white spruce plants (Picea glauca (Moench) Voss), an important commercial species whose wood is used in construction and pulp wood production, were evaluated in the greenhouse [80]. A transformation with the *ech42* gene encoding endochitinase increased the enzyme activity in the spruce roots and root exudates 6 and 2–10 times, respectively, compared to the control; however, the biomass of soil fungi did not change. As has been repeatedly noted in a number of studies, the results of greenhouse assessments of the effects of transgenic trees on soil may differ from those obtained in the field, and therefore long-term field tests are necessary for a conclusive safety assessment of such plants.

4.4. Stress-Tolerant Transgenic Plants

Although transgenic plants with tolerance to abiotic stresses are still not common in commercial plantations, their effects on soil have been studied for quite a long time (Table 4). In greenhouse conditions, the roots of transgenic alfalfa plants overexpressing the gene of nodule-enhanced malate dehydrogenase (neMDH)—which confers tolerance to aluminum—produced 7.1 times more organic acids than a control [86]. Subsequent field tests showed a significant increase in the content of P, K, Mn, Cu and Zn in soil from transgenic plants, and a significant decrease in Mg; only the content of Ca and Fe did not change [87]. These observations demonstrate that organic acids produced by plant roots significantly affect the microbial diversity of the rhizosphere and increase the availability of macro- and microelements.

Table 4. Risk assessment of stress-tolerant transgenic plants on soil quality.

Species	Gene	Growth Conditions	Indicators	Additional Factors	References
alfalfa	*MDH*	field	pH, P, K, micro		[87]
potato	*DREB1A*	greenhouse	C, N, P, S	salinity	[88]
tobacco	*MCM6*	greenhouse	P, DHA	salinity	[89]
rice	*PDH45*	greenhouse	pH, EC, org. C, N, P, K, microel., texture	salinity	[90]
			N(2), P, DHA	soil type	
maize	*BADH*	greenhouse	pH, EC, org. C, N	growth stage	[91]
			C, N, CAT	soil type	
cotton	*CBF1*	field (3 years)	pH, EC, org. matter, N, P, K		[92]
maize	*BADH*	field (3 years)	C, N, DHA	growth stage	[93]

Greenhouse studies did not show any effect of salt-tolerant transgenic plants on the physicochemical properties of soil. These properties (pH, EC, organic C, macro- (N, P, K) and microelements (S, Ca, Mg, Na)), as well as soil texture and density, were not altered by the cultivation of rice with a pea DNA helicase 45 (*PDH45*) gene [89]. No definite trends in soil properties (pH, EC, organic C and N) were observed at any growth stage of two lines of maize with a betaine aldehyde dehydrogenase (*BADH*) gene, either in neutral or in saline–alkaline soil [90]. A 3-year field cultivation of abiotic stress-resistant cotton containing Arabidopsis transcription factor CBF1 did not affect the soil pH, EC, organic matter, P or K [91]. A significant reduction was found only in N, and only in the second year out of three.

Multiple greenhouse studies did not show any effect of stress-tolerant transgenic plants on soil enzyme activities. Two potato lines with a *DREB1A* gene did not alter the activities of urease and β-glucosidase, although some differences were noted in the activities of ARS and alkaline PHOs [87]. Those differences, however, were not confirmed in the second test. Since alkaline PHO is produced only by soil microorganisms [43], changes in its activity are due solely to changes in the microbial activity. These changes could have been caused by fluctuations in temperature and other environmental factors, which could have affected the soil microflora both directly and via changes in plant physiology, e.g., changes in the transgene expression levels [87]. The enzymatic activity of soil was also not changed by the cultivation of various salt-tolerant transgenic species, such as tobacco [88], rice [89] and maize [90].

4.5. Transgenic Plants with Modified Metabolic Pathways

An important area in plant genetic engineering is the modification of a crop's qualitative composition or the content of a certain component. The modifications are most often aimed to improve raw materials for the industry. Generally, a variety of metabolic pathways can be modified in such plants (Table 5), and the resultant new substances or quantitative changes in the existing ones can affect soil microorganisms. For instance, modification of bioenergy crops is used to improve the conversion of lignocellulose biomass into biofuels by manipulating genes of lignin biosynthesis because it is this complex phenolic polymer that prevents access to fermentable polysaccharides [94,95]. Plants with a modified lignin content/composition can affect the soil by altering the uptake of nutrients and/or the composition of plant residues and root exudate [96,97]. Hybrid poplar trees (P. tremula × P. alba) with antisense *CAD* and *COMT* genes for inhibition of lignin biosynthesis were grown in the field for 4 years [98]. The CAD line had significantly lower lignin content, while the COMT lines had a significantly reduced S/G lignin monomer ratio. Yet, the trees showed no effect on total C and N, or MBC in soil. The authors attributed this to a spatial variability of soil properties in the field. Greenhouse cultivation of three tobacco lines with suppressed *CAD* and *COMT* genes of lignin biosynthesis (separately or jointly) did not alter the content of C, soluble carbohydrates or the C/N ratio in the roots, but all the lines had significantly higher MBN [99]. As shown by the measured activities of soil C-cycle enzymes, the cellulase activity in the soil of transgenic plants did not differ from a non-transgenic control, but the roots of the CAD-suppressed line contained significantly more N, and the invertase and xylanase activities were significantly higher in the soil of this line.

Table 5. Risk assessment of metabolic engineered transgenic plants on soil quality.

Species	Gene	Growth Conditions	Indicators	Additional Factors	References
poplar	*CAD, COMT* (AS) [1]	field (4 years)	C, N MBC	location	[98]
potato	*GBSS* (RNAi)	field (3 years)	C(2), PER	growth stage variety location	[100]
potato	*cphA*	field (3 years)	C(4), N(2), P	growth stage	[101]
rice	*pacrtB + pacrtE + pacrtY + pacrtI*	field	C, N(2), CAT	growth stage	[102]
soybean	*AtD-CGS*	field	C, N		[103]
tobacco	*CAD, COMT, CCR*	greenhouse	MBN C(3)		[99]
potato	*GBSS* (RNAi) [2]	field	org. matter	growth stage location	[104]
rice	*AhSTS1*	field	pH, EC, org. matter, N, P, K, microel.	growth stage	[105]
switchgrass	*COMT* (RNAi)	field (5 years)	pH, org. matter, P, K, microel.		[106]
switchgrass	*COMT* (RNAi) *MYB*	field (5–6 years)	organic C		[107]

[1] AS = antisense. [2] RNAi = RNA interference.

The possible effects of modified bioenergy crops on rhizospheric processes, especially those related to C accumulation in soil, were assessed in a field study on two transgenic millet lines (*Panicum virgatum* L.) with downregulation of caffeic acid 3-O-methyltransferase (COMT) [106]. In the first two years, the plants did not affect the pH and concentrations of 19 soil elements. The roots of 5-year-old plants did not differ in lignin content from the control, but, due to a lower content of syringyl (S) monomers, the S/G ratio of the two lines decreased by 40.1% and 42.7% versus the control. However, this did not affect the total SOC content in the upper (0–15 cm) and deeper soil layers (15–30 cm) after 5 years of cultivation [102]. Since changes in SOC occur very slowly, the subsequent study focused on the active fractions of SOC, namely, dissolved organic carbon (DOC) and permanganate oxidizable carbon (POXC), which are more sensitive to changes [107]. In addition to two lines with the suppressed *COMT* gene, the study also used three millet lines overexpressing the PvMYB4 (MYB4) enzyme, the transcription repressor of many lignin biosynthesis genes. After 5–6 years of cultivation, there were no differences in total SOC, DOC and POXC between the COMT and MYB4 lines and the control. Since the aboveground biomass was removed at the end of each growing season, SOC was only dependent on C coming from the roots [107]. The absence of differences in SOC was probably due to the absence of changes in the lignin content in millet roots with suppressed COMT [106]. It should be noted that the aboveground biomass cannot always be removed in production systems; therefore, future studies should include assessments of SOC dynamics at different rates of aboveground biomass removal [107].

Field tests showed that transgenic plants with the modified biosynthesis of less important compounds had an insignificant effect on soil properties. Soybean expressing an Arabidopsis cystathionine-c-synthase gene, which increases the methionine content, had no effect on the total C and N in soil, nor did it differ from a non-transgenic control in the content of amino acids in root exudates [103]. Rice modified to produce resveratrol, a non-flavonoid polyphenol, which is not normally synthesized in cereals due to the absence of resveratrol synthase, had no significant effect on the soil pH, EC, available P, Ca, K, Mg, and Na cations, total N and organic matter [105]. There was also no effect of such plants on the activity of microorganisms. Potato modified to produce amylase-free starch, which may be important for a range of industrial applications, had no effect on laccase, cellulose and Mn-peroxidase [100], or on the microbial catabolic diversity [104]. Potato was also modified to synthesize cyanophycin, a protein polymer used to obtain polyaspartate, which is a biodegradable substitute for the synthetic polymer polyacrylate [101]. Potato tubers left in soil for three winter seasons were shown to have no effect on six enzymes representing the main pathways of the C, N and P cycles in soil.

5. Factors Influencing the Interaction of Transgenic Plants with Soil

Most studies have not shown any significant effect of transgenic plants on soil. Some authors reported a number of significant changes but, for the most part, these were inconsistent and transient. The lack of result consistency is probably due to a large variety of interactions among plant parameters (genotype, inserted gene, stage of development) and multiple external factors. These factors include the location, main soil type and climatic conditions, including weather changes during the season and from year to year (temperatures and precipitation), which often explain most observed changes, as well as management systems, which include the use of fertilizers and pesticides, as well as various crop rotation systems [8,15]. All these factors can mask the effects of genetic modification, and it has been repeatedly shown that their influence is greater than that of the transgenic status.

To assess the dynamics of changes in soil, one needs several samples taken during the season, either at different stages of plant development or at certain intervals (which is a less common practice). Rare reports of significant changes in pH, normally a very stable soil parameter, were associated with season. Such findings were reported for pot-cultivated tobacco with a chitinase gene [81] and for field-grown Bt poplars [70]. Supposedly, these changes were caused not by the plants' transgenic nature but rather by environmental factors. A significant decrease in the content of N in the soil of Bt cotton in the middle of the growing season [58] or that of available P during flowering [63] suggests higher nutrient uptake at certain development phases compared to controls. Changes in TOC, total N and available P depending on the growth stage were also reported for Bt rice [53] and salt-tolerant maize [91]. They could have been caused by, e.g., increased growth.

A recent study by Chen et al. [108] did not investigate a limited number of elements but rather performed a comprehensive profiling of the soil metabolomes of Bt maize and a non-transgenic variety. Soil metabolomics profiling generated a total of 1730 compounds that differed at each of the six growth stages. The degree of changes in metabolites increased up to the fourth stage (heading), and then decreased. This was probably due to changes in plant physiology in transition from vegetative to reproductive growth, which altered the root metabolism. Despite obvious differences in the soil metabolic profiles of the two maize varieties, changes in the rhizosphere bacterial community were associated with the development stages rather than with the genetic modification of plants [108].

A number of authors have reported that within-season changes in soil enzyme activities in Bt plants have little to do with their genetic transformation, but are more likely caused by such factors as differences in the soil water content, temperature fluctuations, use of fertilizers, etc. [55,62,68]. The small magnitude of the effect of transgenic status on soil microbial communities, compared to that of growth stage, was also reported for herbicide-resistant plants [8]. Since plants can alter their exudate composition depending on their development stage, this may be another explanation for changes in soil microbial activity during the season [109].

Long-term tests demonstrated significant effects not only of growth stage but also of year, as was observed for Bt maize [17,66] and Bt cotton [9]. For example, a significant reduction in MBC was observed in maize plants in a year with poorly distributed rainfall [54], while the highest MBC value was in the hottest year in a five-year study [74]. Moreover, not only were the growing season conditions important, but also those of the winter. The activities of enzymes involved in the cycles of C, N and P significantly differed among three winter periods when cyanophycin-producing GM potato tubers were left in soil [105].

Classical breeding varieties are often obtained by crossing various genotypes within, or sometimes between, species, which leads to a large genetic diversity. It is obvious that such varieties can have a significant impact on soil, incomparable with that of a variety obtained by the insertion of one transgene. In most cases, the effect of transgenic modification was found to be insignificant compared to the variability among varieties obtained by traditional breeding. For instance, the effect of Bt maize was within the variation range of the effects of ten conventional varieties [64]. In the study by Khan et al. [84], two oilseed rape varieties significantly differed in the activities of BGL (2.3 times) and ARS (1.5 times), while transgenic plants did not differ from their parents belonging to different varieties.

Fluctuations in the soil content of organic C and macroelements due to the cultivation of transgenic plants could be dependent on the availability of nutrients and water. Field tests with Bt plants showed a significant increase in P under dry land conditions [69] or a significant reduction in NO_3 and P without the use of fertilizers [32]. Possibly, the differences between transgenic and non-transgenic plants appeared only under stress (drought or nutrient deficiencies) and disappeared under favorable conditions. Small changes in pH at the end of a greenhouse experiment with a transgenic melon [83] were associated with changes in the soil buffer capacity due to added fertilizers. Organic amendments added to the soil were reported to have changed the microbial activity and the bacteria-to-fungi ratio [110]. According to [111], the physicochemical properties of soil, its microbial biomass and activity were sensitive to the introduction of N and P, and the response largely depended on the rate of their application.

Along with the decomposition of organic compounds and nutrient cycling, soil microorganisms are also responsible for the decomposition of pesticides [75]. Therefore, when evaluating GMPs with such traits as resistance to pests, herbicides or diseases, it is also important to take into account the effects of new technologies involving the use of new pesticides or reducing the doses and number of treatments with the old ones. Studies with Bt crops showed no differences among transgenic, control and insecticide-treated control plants in MBC [54], PHO [56], or MBC and BGL [66]. Other studies demonstrated significant differences in MBC [17] or DHA [57], but those differences were not stable during the season or from year to year. The absence of differences between the effects of imazapyr and conventional herbicides on MBC and MBN in the field was confirmed in various soil and climatic conditions covering the main biomes of Brazil [75]. A fungicide treatment of potato plants with late blight resistance also had no effect on the physicochemical properties of soil in 2-year field trials in Ireland and the Netherlands [28].

Large-scale studies, which included the assessment of a number of factors, such as site location, management system and growth stage, demonstrated their greater impact on soil compared with that of an inserted transgene. The effects of glyphosate-resistant plants on soil microorganisms were minor and inconsistent compared to the effects of growing site location and crop rotation [73]. The effect of glyphosate-resistant soybean on soil microbial biomass was insignificant compared to those of site, growing season and soybean cultivar [76]. The potato plant growth stage and field location affected the soil enzyme activities more than the starch modification of tubers [100]. Compared to harvest year, plant growth stage and cotton cultivar, the pest resistance trait had a negligible effect on the activities of DHA, urease and phosphatase in soil [67]. The year and stage of growth significantly affected the urease, DHA and sucrase activities in the field trials of virus-resistant wheat [82].

In hybrid poplar (*Populus alba* × *P. tremula*) expressing a prokaryotic *tzs* gene, the level of cellular cytokinin (trans-zeatin) increased 20-fold, which caused an increase in the aboveground biomass and a number of changes in the plant growth, development and biochemical composition [112]. Field tests of three poplar lines were carried out in three locations in South Korea that differed significantly in pH, NPK content and soil texture [113]. A significant difference in microbial biomass was found in one of the three locations and in one clone. The change was possibly caused by the genetic transformation (changes in the composition of exudates), but it was temporary and associated with location and genotype.

6. Effects of Transgenic Plant Residues

Decomposition of plant residues is a key function in element cycling, and any change in their composition can affect the functions of soil [15]. There were reports of unintended effects such as a 33% to 97% increase in the lignin content in maize [114], changes in the C and N content [66,104], and increases in the underground [65] and aboveground [53] biomass. Most studies of litter decomposition have assessed the decomposition rate by measuring weight loss and C emission, and only a few assessed the effects on microorganisms. In the study by Wu et al. [115], PHO was insensitive to the decomposition of Bt rice straw, which contained significantly more N, P, K and unchanged C. At the same time, the activity of DHA sharply increased in the initial phase of the experiment (the first two weeks out of 12). The higher content of macronutrients in the Bt straw possibly contributed to the significant growth of the microbial population, but the effect was temporary. The effect instability was also noted in [116], where they studied the decomposition of salt-resistant maize straw in neutral or in saline–alkali soil for 7 months. The transgenic straw did not differ from the control in cellulose and lignin, but its C/N ratio was lower. The early stage of its decomposition was faster in the saline–alkali soil and this significantly increased MBC and MBN; by the end of the experiment, however, there were no differences in microbial biomass between the transgenic straw and control. Potato tubers with modified starch composition did not differ in the content of lignin, cellulose and non-cellulose polysaccharides [117]. When they were left to decompose in litterbags in the growing soil, there were significant differences from the control in the activities of laccase, Mn-peroxidase and cellulose. It is not clear, however, if the enzymes in the residue sphere were produced there or were leaching out of the bags.

Of particular interest is understanding how soil processes are affected by litter from transgenic trees that grow in one place for a long time. Stems and roots with a modified lignin composition were shown to initially decompose faster due to a lower protection of labile plant components from enzymatic attack [94,118]. Trunk segments of 4-year-old poplars with suppression of lignin biosynthesis genes *CAD* and *COMT* were left to decompose in soil from three different locations [119]. After 552 days of the experiment, the adhering layer of soil (detritisphere) was analyzed for MBN. While the original wood samples did not differ in C and N, there was a significant difference in soil MBN by the end of the experiment. However, the effects of the genetic transformation on MBN were not consistent across different soils. Fungi are known to have a more important part to play than bacteria in litter decomposition [120]. A study by Vauramo et al. [121] determined the fungal biomass in decomposing leaves from birch with an antifungal chitinase gene.

The C/N ratio in the transgenic leaves did not differ from the control, and the 11-month experiment failed to find any effect on the fungal biomass.

In hot climate countries, plant residues can be used as mulch to preserve moisture in the soil, control weeds and improve nutrient availability [122]. The effects of Bt cotton mulch on soil properties (pH, EC, macro- and microelements), weed growth and productivity of winter crops—wheat, Egyptian clover and canola—were studied in Pakistan [123]. Toxins released by Bt mulch reduced weed density, but negatively affected the winter crop productivity and did not significantly affect the properties of soil. This should be taken into account when selecting crops for rotation.

7. Future Prospects

To date, the effects on soil processes have been studied for GMPs with various traits (Tables 1–5). A large number of greenhouse and field studies were carried out to identify possible deviations from non-transgenic plants, which also assessed a number of factors related to both the plant (growth stage, genotype features) and growing conditions (site location, changes in management systems). In most reports, the detected changes were within the limits of statistical error, and in the case of statistical significance, they were most often temporary and were not reproduced at the next sampling in the season or the next year. No specific effect of a particular type of transgenic plant (e.g., pest or herbicide-resistant) has been identified, and the lack of generally accepted experimental design and evaluation criteria makes such comparison difficult. For example, some researchers noted the effect of Bt plants on the content of available P, but this could also be caused by increased growth, and biomass was not always measured. In general, no unequivocally negative or positive effect of any transgenic genotype on the physicochemical or microbial properties of the soil has been shown. However, some groups of transgenic plants or their possible effects have not been given due attention, and research in these areas can be expanded.

The development of transgenic plants with increased productivity by improving the nutrient use efficiency or photosynthesis is a popular area in plant genetic engineering. A wide range of genes and approaches have been applied to improve the plant use efficiency of N [124,125], P [126] and other nutrients [127,128], as well as to optimize photosynthesis [129,130]. Such plants would of course need more nutrients for increased biomass production, yet their impact on soil processes has barely been studied and further research is needed. Our group studied the effects of transgenic birch and aspen plants with the pine glutamine synthetase *GS1* gene on the physicochemical properties, enzyme activity and microbial biomass of soil. Four-year pot experiments showed that transgenic plants differed in growth rate and C and N content, but differences in enzyme activity and microbial biomass were temporary and inconsistent. However, by the end of the experiments, we observed a decrease in soil K, possibly due to its enhanced uptake to neutralize secondary NH4 reassimilated via glutamine synthetase (unpublished data).

In addition to common substances such as carbohydrates, proteins, organic acids and amino acids, root exudates of some plants may contain allelochemicals involved in rhizospheric interactions between plants and other organisms [131]. These substances can also affect soil microorganisms. The neem tree (*Azadirachta indica* Juss) from tropical Asia contains a number of allelochemicals, the most toxic of which is the alkaloid azadirachtin, which has insecticidal activity [132]. When added into soil, azadirachtin granules did not affect DHA but significantly altered the PHO activity: increased at the recommended dose (1×) and inhibited at five times (5×) the recommended dose [133]. For environmentally safe weed control, it was proposed to use allelopathic rice capable of inhibiting the growth of neighboring plants [134]. This is achieved owing to root exudates containing allelochemicals, the most important of which is flavone O-glycoside. In different growth phases, the effect of allelopathic rice on soil enzymes (DHA, polyphenol oxidase, urease and invertase) was positive or neutral, although the flavone O-glycoside concentration in soil was always about five times that in control. The potential mechanism of this effect remains unclear [134].

The influence of genetic transformation on changes in the composition and/or content of allelochemicals is still poorly understood. This issue takes on particular importance with regard to artificial tree plantations, where a prolonged cultivation and the dominance of one species can lead to the accumulation of allelochemicals to toxic levels [135]. Allelopathic tree species include, among others, some important fruit (walnut) and forest (eucalypts, some coniferous species) species [136]. Eucalypt species are currently the most important in the plantation forestry. They are widely used for genetic transformation, and a fast growing eucalypt was approved for commercial use in Brazil in 2015 [137]. Most eucalypt species are known to have an allelopathic influence in nature [138]. To date, allelopathy tests for environment biosafety have been conducted on salt-tolerant eucalypts overexpressing various genes [138–140] and on eucalypt with reduced lignin content [141]. The tests assessed lettuce seed germination on agar with added dried leaves (sandwich method) or in soil used in growing transgenic eucalypts. The abundance of microorganisms and their function were not studied, so further research should be conducted.

Long-term cultivation of GMPs, particularly the woody ones, in one place can affect not only the very labile microbiological parameters (enzyme activity, microbial biomass), but also such lengthy processes as the decomposition of plant residues, mineralization of organic matter and accumulation of soil carbon. Unlike the case with annual crops, both root exudates and residues of trees can accumulate in soil, and even minor changes can build up over time. However, long-term field tests are quite expensive and time-consuming and they still cannot take account of all possible soil and climatic conditions in which the plants can be grown, as well as various applicable management systems. Meanwhile, long-term effects on soil can be assessed using a mathematical modeling of various scenarios. For example, modeling of the 30- and 60-year cultivation of transgenic aspen plantations with a modified wood composition under Northern Eurasia conditions showed 5–7% changes in soil C and N pools, which do not exceed the effects of a standard silvicultural management [142].

Unfortunately, there have been virtually no studies that have simultaneously assessed the effects of transgenic plants on the quantitative (microbial biomass), functional (soil enzymes) and qualitative (diversity) characteristics of microbial communities. Conducting such studies would help towards a better understanding of the relationships between the diversity and functions of microorganisms in soil used to cultivate transgenic plants.

Author Contributions: Conceptualization, V.L.; writing—original draft preparation, V.L. and T.L.; writing—review and editing, V.L. and K.S.; funding acquisition, E.T. and K.S. All authors have read and agreed to the published version of the manuscript.

Funding: This research was carried out within the state program of the Ministry of Science and High Education of the Russian Federation (theme "Plant molecular biology and biotechnology: their cultivation, pathogen and stress protection (BIBCH)" (No 0101-2019-0037)).

Institutional Review Board Statement: Not applicable.

Informed Consent Statement: Not applicable.

Data Availability Statement: Not applicable.

Conflicts of Interest: The authors declare no conflict of interest.

References

1. ISAAA. *Global Status of Commercialized Biotech/GM Crops in 2019: Biotech Crops Drive Socio-Economic Development and Sustainable Environment in the New Frontier*; ISAAA Brief No. 55; ISAAA: Ithaca, NY, USA, 2019.
2. Snow, A.A.; Palma, P.M. Commercialization of transgenic plants: Potential ecological risks. *BioScience* **1997**, *47*, 86–96. [CrossRef]
3. Messeguer, J. Gene flow assessment in transgenic plants. *Plant Cell Tissue Organ Cult.* **2003**, *73*, 201–212. [CrossRef]
4. Sisterson, M.S.; Carrière, Y.; Dennehy, T.J.; Tabashnik, B.E. Evolution of resistance to transgenic crops: Interactions between insect movement and field distribution. *J. Econ. Entomol.* **2006**, *98*, 1751–1762. [CrossRef] [PubMed]
5. Romeis, J.; Meissle, M.; Bigler, F. Transgenic crops expressing Bacillus thuringiensis toxins and biological control. *Nat. Biotechnol.* **2006**, *24*, 63–71. [CrossRef]

6. Bruinsma, M.; Kowalchuk, G.A.; van Veen, J.A. Effects of genetically modified plants on microbial communities and processes in soil. *Biol. Fertil. Soils* **2003**, *37*, 329–337. [CrossRef]
7. Singh, A.K.; Dubey, S.K. 8—Transgenic Plants and Soil Microbes. In *Current Developments in Biotechnology and Bioengineering*; Dubey, S.K., Pandey, A., Sangwan, R.S., Eds.; Elsevier: Amsterdam, The Netherlands, 2017; pp. 163–185. [CrossRef]
8. Guan, Z.; Lu, S.; Huo, Y.; Guan, Z.-P.; Liu, B.; Wei, W. Do genetically modified plants affect adversely on soil microbial communities? *Agric Ecosyst. Environ.* **2016**, *235*, 289–305. [CrossRef]
9. Chen, Z.; Chen, L.; Wu, Z. Relationships among persistence of Bacillus thuringiensis and Cowpea trypsin inhibitor proteins, microbial properties and enzymatic activities in rhizosphere soil after repeated cultivation with transgenic cotton. *Appl. Soil Ecol.* **2012**, *53*, 23–30. [CrossRef]
10. Chen, Z.; Wei, K.; Chen, L.; Wu, Z.; Luo, J.; Cui, J. Effects of the consecutive cultivation and periodic residue incorporation of Bacillus thuringiensis (Bt) cotton on soil microbe-mediated enzymatic properties. *Agric Ecosyst. Environ.* **2017**, *239*, 154–160. [CrossRef]
11. Wei, H.W.; Movahedi, A.; Liu, G.; Kiani-Pouya, A.; Rasouli, F.; Yu, C.; Chen, Y.; Zhong, F.; Zhang, J. Effects of field-grown transgenic Cry1Ah1 poplar on the rhizosphere microbiome. *Res. Sq.* **2022**, preprint. [CrossRef]
12. Luo, J.-Y.; Zhang, S.; Zhu, X.-Z.; Lu, L.-M.; Wang, C.-Y.; Li, C.-H.; Cui, J.-J.; Zhou, Z.-G. Effects of soil salinity on rhizosphere soil microbes in transgenic Bt cotton fields. *J. Integr. Agric.* **2017**, *16*, 1624–1633. [CrossRef]
13. Haichar, F.Z.; Marol, C.; Berge, O.; Rangel-Castro, J.I.; Prosser, J.I.; Balesdent, J.; Achouak, W. Plant host habitat and root exudates shape soil bacterial community structure. *ISME J.* **2008**, *2*, 1221–1230. [CrossRef] [PubMed]
14. Guyonnet, J.P.; Cantarel, A.A.M.; Simon, L.; Haichar, F.Z. Root exudation rate as functional trait involved in plant nutrient-use strategy classification. *Ecol. Evol.* **2018**, *8*, 8573–8581. [CrossRef] [PubMed]
15. Hannula, S.E.; de Boer, W.; van Veen, J.A. Do genetic modifications in crops affect soil fungi? *A review. Biol. Fertil. Soils* **2014**, *50*, 433–446. [CrossRef]
16. Mandal, A.; Sarkar, B.; Owens, G.; Thakur, J.K.; Manna, M.C.; Niazi, N.K.; Jayaraman, S.; Patra, A.K. Impact of genetically modified crops on rhizosphere microorganisms and processes: A review focusing on Bt cotton. *Appl. Soil Ecol.* **2020**, *148*, 103492. [CrossRef]
17. Devare, M.; Londono-R, L.M.; Thies, J.E. Neither transgenic Bt maize (MON863) nor tefluthrin insecticide adversely affect soil microbial activity or biomass: A 3-year field analysis. *Soil Biol. Biochem.* **2007**, *39*, 2038–2047. [CrossRef]
18. Dunfield, K.E.; Germida, J.J. Impact of genetically modified crops on soil- and plant-associated microbial communities. *J. Environ. Qual.* **2004**, *33*, 806–815. [CrossRef]
19. Larson, W.E.; Pierce, F.J. Conservation and enhancement of soil quality. In *Evaluation for Sustainable Land Management in the Developing World*; Vol. 2: Technical Papers; International Board for Research and Management: Bangkok, Thailand, 1991; IBSRAM Proceedings No. 12; pp. 175–203.
20. Beule, L.; Vaupel, A.; Moran-Rodas, V.E. Abundance, diversity, and function of soil microorganisms in temperate alley-cropping agroforestry systems: A review. *Microorganisms* **2022**, *10*, 616. [CrossRef]
21. McGill, B.J.; Enquist, B.J.; Weiher, E.; Westoby, M. Rebuilding community ecology from functional traits. *Trends Ecol. Evol.* **2006**, *21*, 178–185. [CrossRef]
22. Singh, B.K.; Quince, C.; Macdonald, C.A.; Khachane, A.; Thomas, N.; Al-Soud, W.A.; Sørensen, S.J.; He, Z.; White, D.; Sinclair, A.; et al. Loss of microbial diversity in soils is coincident with reductions in some specialized functions. *Environ. Microbiol.* **2014**, *16*, 2408–2420. [CrossRef]
23. Escalas, A.; Hale, L.; Voordeckers, J.W.; Yang, Y.; Firestone, M.K.; Alvarez-Cohen, L.; Zhou, J. Microbial functional diversity: From concepts to applications. *Ecol. Evol.* **2019**, *9*, 12000–12016. [CrossRef]
24. Peter, H.; Beier, S.; Bertilsson, S.; Lindstrom, E.S.; Langenheder, S.; Tranvik, L.J. Function-specific response to depletion of microbial diversity. *ISME J.* **2011**, *5*, 351–361. [CrossRef] [PubMed]
25. Plante, C.J. Defining disturbance for microbial ecology. *Microb. Ecol.* **2017**, *74*, 259–263. [CrossRef] [PubMed]
26. Chen, H.; Ma, K.; Huang, Y.; Yao, Z.; Chu, C. Stable soil microbial functional structure responding to biodiversity loss based on metagenomic evidences. *Front. Microbiol.* **2021**, *12*, 716764. [CrossRef]
27. Louca, S.; Jacques, S.M.S.; Pires, A.P.F.; Leal, J.S.; Srivastava, D.S.; Parfrey, L.W.; Farjalla, V.F.; Doebeli, M. High taxonomic variability despite stable functional structure across microbial communities. *Nat. Ecol. Evol.* **2016**, *1*, 15. [CrossRef]
28. Krause, S.M.B.; Näther, A.; Cortes, V.O.; Mullins, E.; Kessel, G.J.T.; Lotz, L.A.P.; Tebbe, C.C. No tangible effects of field-grown cisgenic potatoes on soil microbial communities. *Front. Bioeng. Biotechnol.* **2020**, *8*, 603145. [CrossRef]
29. Lauber, C.L.; Hamady, M.; Knight, R.; Fierer, N. Pyrosequencing-based assessment of soil pH as a predictor of soil bacterial community structure at the continental scale. *Appl. Environ. Microbiol.* **2009**, *75*, 5111–5120. [CrossRef]
30. Lee, Y.H.; Ahn, B.K.; Sonn, Y.K. Effects of electrical conductivity on the soil microbial community in a controlled horticultural land for strawberry cultivation. *Korean J. Soil Sci. Fert.* **2011**, *44*, 830–835. [CrossRef]
31. Zhang, C.; Liu, G.; Xue, S.; Song, Z. Rhizosphere soil microbial activity under different vegetation types on the Loess Plateau, China. *Geoderma* **2011**, *161*, 115–125. [CrossRef]
32. Ahamd, M.; Abbasi, W.M.; Jamil, M.; Iqbal, M.; Hussain, A.; Akhtar, M.F.; Nazli, F. Comparison of rhizosphere properties as affected by different Bt- and non-Bt-cotton (Gossypium hirsutum L.) genotypes and fertilization. *Environ. Monit. Assess.* **2017**, *189*, 278. [CrossRef]

33. Song, Y.; Song, C.; Yang, G.; Miao, Y.; Wang, J.; Guo, Y. Changes in labile organic carbon fractions and soil enzyme activities after marshland reclamation and restoration in the Sanjiang Plain in Northeast China. *Environ. Manag.* **2012**, *50*, 418–426. [CrossRef]
34. Hinojosa, M.B.; Carreira, J.A.; García-Ruiz, R.; Dick, R.P. Soil moisture pre-treatment effects on enzyme activities as indicators of heavy metal-contaminated and reclaimed soils. *Soil Biol. Biochem.* **2004**, *36*, 1559–1568. [CrossRef]
35. Lino, I.A.N.; Santos, V.M.; Escobar, I.E.C.; Silva, D.K.A.; Maia, L.C. Soil enzymatic activity in Eucalyptus grandis plantations of different ages. *Land Degrad. Dev.* **2016**, *27*, 77–82. [CrossRef]
36. Brzezińska, M.; Lipiec, J.; Frąc, M.; Oszust, K.; Szarlip, P.; Turski, M. Quantitative interactions between total and specific enzyme activities and C and N contents in earthworm-affected pear orchard soil. *Land Degrad. Devel.* **2018**, *29*, 3379–3389. [CrossRef]
37. Gil-Sotres, F.; Trasar-Cepeda, C.; Leiros, M.C.; Seoane, S. Different approaches to evaluating soil quality using biochemical properties. *Soil Biol. Biochem.* **2005**, *37*, 877–887. [CrossRef]
38. Garcia, C.; Hernandez, T.; Costa, F. Potential use of dehydrogenase activity as an index of microbial activity in degraded soils. *Commun. Soil Sci. Plant Anal.* **1997**, *28*, 123–134. [CrossRef]
39. Eivazi, F.; Tabatabai, M.A. Glucosidases and galactosidases in soils. *Soil Biol. Biochem.* **1988**, *20*, 601–606. [CrossRef]
40. Velmourougane, K.; Sahu, A. Impact of transgenic cottons expressing cry1Ac on soil biological attributes. *Plant Soil Environ.* **2013**, *3*, 108–114. [CrossRef]
41. Jan, M.T.; Roberts, P.; Tonheim, S.K.; Jones, D.L. Protein breakdown represents a major bottleneck in nitrogen cycling in grassland soils. *Soil Biol. Biochem.* **2009**, *41*, 2272–2282. [CrossRef]
42. Kramer, S.; Green, D.M. Acid and alkaline phosphatase dynamics and their relationship to soil microclimate in a semiarid woodland. *Soil Biol. Biochem.* **2000**, *32*, 179–188. [CrossRef]
43. Tarafdar, J.C.; Claassen, N. Organic phosphorus compounds as a phosphorus source for higher plants through the activity of phosphatase produced by plant roots and microorganisms. *Biol. Fertil. Soils* **1988**, *5*, 308–312. [CrossRef]
44. Kertesz, M.A.; Mirleau, P. The role of soil microbes in plant sulphur nutrition. *J. Exp. Bot.* **2004**, *55*, 1939–1945. [CrossRef] [PubMed]
45. Frankenberger, W.T., Jr.; Johanson, J.B. Method of measuring invertase activity in soils. *Plant Soil* **1983**, *74*, 301–311. [CrossRef]
46. Gander, L.K.; Hendricks, C.W.; Doyle, J.D. Interferences, limitations and an improvement in the extraction and assessment of cellulase activity in soil. *Soil Biol. Biochem.* **1994**, *26*, 65–73. [CrossRef]
47. Floch, C.; Alarcon-Gutiérrez, E.; Criquet, S. ABTS assay of phenol oxidase activity in soil. *J. Microbiol. Methods* **2007**, *71*, 319–324. [CrossRef] [PubMed]
48. Trasar-Cepeda, C.; Camina, F.; Leirós, C.; Gil-Sotres, F. An improved method to measure catalase activity in soils. *Soil Biol. Biochem.* **1999**, *31*, 483–485. [CrossRef]
49. Singh, J.S.; Raghubanshi, A.S.; Singh, R.S.; Srivastava, S.C. Microbial biomass acts as a source of plant nutrients in dry tropical forest and savanna. *Nature* **1989**, *338*, 499–500. [CrossRef]
50. Sparling, G.P. Ratio of microbial biomass carbon to soil organic carbon as a sensitive indicator of changes in soil organic matter. *Aust. J. Soil Res.* **1992**, *30*, 195–207. [CrossRef]
51. Chen, H.; Zhao, X.; Chen, X.; Lin, Q.; Li, G. Seasonal changes of soil microbial C, N, P and associated nutrient dynamics in a semiarid grassland of north China. *Appl. Soil Ecol.* **2018**, *128*, 89–97. [CrossRef]
52. Geisseler, D.; Horwath, W.R. Short-term dynamics of soil carbon, microbial biomass, and soil enzyme activities as compared to longer-term effects of tillage in irrigated row crops. *Biol. Fertil. Soils* **2009**, *46*, 65–72. [CrossRef]
53. Lee, Z.L.; Bu, N.S.; Cui, J.; Chen, X.P.; Xiao, M.Q.; Wang, F.; Song, Z.P.; Fang, C.M. Effects of long-term cultivation of transgenic Bt rice (Kefeng-6) on soil microbial functioning and C cycling. *Sci. Rep.* **2017**, *7*, 4647. [CrossRef]
54. Devare, M.H.; Jones, C.M.; Thies, J.E. Effect of Cry3Bb transgenic corn and tefluthrin on the soil microbial community: Biomass, activity, and diversity. *J. Environ. Qual.* **2004**, *33*, 837–843. [CrossRef] [PubMed]
55. Icoz, I.; Saxena, D.; Andow, D.A.; Zwahlen, C.; Stotzky, G. Microbial populations and enzyme activities in soil in situ under transgenic corn expressing cry proteins from Bacillus thuringiensis. *J. Environ. Qual.* **2008**, *37*, 647–662. [CrossRef] [PubMed]
56. Liu, W.; Lu, H.H.; Wu, W.X.; Wei, Q.K.; Chen, Y.X.; Thies, J.E. Transgenic Bt rice does not affect enzyme activities and microbial composition in the rhizosphere during crop development. *Soil Biol. Biochem.* **2008**, *40*, 475–486. [CrossRef]
57. Oliveira, A.P.; Pampulha, M.E.; Bennett, J.P. A two-year field study with transgenic Bacillus thuringiensis maize: Effects on soil microorganisms. *Sci. Total Environ.* **2008**, *405*, 351–357. [CrossRef]
58. Sarkar, B.; Patra, A.K.; Purakayastha, T.J. Transgenic Bt-cotton affects enzyme activity and nutrient availability in a sub-tropical inceptisol. *J. Agron. Crop Sci.* **2008**, *194*, 289–296. [CrossRef]
59. Sarkar, B.; Patra, A.K.; Purakayastha, T.; Megharaj, M. Assessment of biological and biochemical indicators in soil under transgenic Bt and non-Bt cotton crop in a sub-tropical environment. *Environ. Monit. Assess.* **2009**, *156*, 595–604. [CrossRef] [PubMed]
60. Kravchenko, A.N.; Hao, X.; Robertson, G.P. Seven years of continuously planted Bt corn did not affect mineralizable and total soil C and total N in surface soil. *Plant Soil* **2009**, *318*, 269–274. [CrossRef]
61. Mina, U.; Chaudhary, A.; Kamra, A. Effect of Bt cotton on enzymes activity and microorganisms in rhizosphere. *J. Agric. Sci.* **2011**, *3*, 96–104. [CrossRef]
62. Wei, M.; Tan, F.; Zhu, H.; Cheng, K.; Wu, X.; Wang, J.; Zhao, K.; Tang, X. Impact of Bt-transgenic rice (SHK601) on soil ecosystems in the rhizosphere during crop development. *Plant Soil Environ.* **2012**, *58*, 217–223. [CrossRef]

63. Yang, W.; Zhang, M.; Ding, G. Effect of transgenic Bt cotton on bioactivities and nutrients in rhizosphere soil. *Commun. Soil Sci. Plant Anal.* **2012**, *43*, 689–700. [CrossRef]
64. Fließbach, A.; Messmer, M.; Nietlispach, B.; Infante, V.; Mäder, P. Effects of conventionally bred and Bacillus thuringiensis (Bt) maize varieties on soil microbial biomass and activity. *Biol. Fertil. Soils* **2012**, *48*, 315–324. [CrossRef]
65. Fließbach, A.; Nietlispach, B.; Messmer, M.; Rodriguez-Romero, A.-S.; Maeder, P. Microbial response of soils with organic and conventional management history to the cultivation of Bacillus thuringiensis (Bt)-maize under climate chamber conditions. *Biol. Fertil. Soils* **2013**, *49*, 829–837. [CrossRef]
66. Lupwayi, N.Z.; Blackshaw, R.E. Soil microbial properties in Bt (Bacillus thuringiensis) corn cropping systems. *Appl. Soil Ecol.* **2013**, *63*, 127–133. [CrossRef]
67. Zhang, Y.N.; Xie, M.; Li, C.Y.; Wu, G.; Peng, D.L. Impacts of the transgenic CrylAc and CpTI insect-resistant cotton SGK321 on selected soil enzyme activities in the rhizosphere. *Plant Soil Environ.* **2014**, *60*, 401–406. [CrossRef]
68. Zhou, D.; Xu, L.; Gao, S.; Guo, J.; Luo, J.; You, Q.; Que, Y. Cry1Ac transgenic sugarcane does not affect the diversity of microbial communities and has no significant effect on enzyme activities in rhizosphere soil within one crop season. *Front. Plant Sci.* **2016**, *7*, 265. [CrossRef]
69. van Wyk, D.A.B.; Adeleke, R.; Rhode, O.H.J.; Bezuidenhout, C.C.; Mienie, C. Ecological guild and enzyme activities of rhizosphere soil microbial communities associated with Bt-maize cultivation under field conditions in North West Province of South Africa. *J. Basic Microbiol.* **2017**, *57*, 781–792. [CrossRef]
70. Zuo, L.H.; Yang, R.L.; Zhen, Z.X.; Liu, J.X.; Huang, L.S.; Yang, M.S. A 5-year field study showed no apparent effect of the Bt transgenic 741 poplar on the arthropod community and soil bacterial diversity. *Sci. Rep.* **2018**, *8*, 1956. [CrossRef]
71. Mandal, A.; Thakur, J.K.; Sahu, A.; Manna, M.C.; Rao, A.S.; Sarkar, B.; Patra, A.K. Effects of Bt-cotton on biological properties of Vertisols in central India. *Arch. Agron. Soil Sci.* **2019**, *65*, 670–685. [CrossRef]
72. Sessitsch, A.; Gyamfi, S.; Tscherko, D.; Gerzabek, M.H.; Kandeler, E. Activity of microorganisms in the rhizosphere of herbicide treated and untreater transgenic glufosinate-tolerant and wildtype oilseed rape grown in containment. *Plant Soil* **2005**, *266*, 105–116. [CrossRef]
73. Lupwayi, N.Z.; Hanson, K.G.; Harker, K.N.; Clayton, G.W.; Blackshaw, R.E.; Donovan, J.T.; Johnson, E.N.; Gan, Y.; Irvine, R.B.; Monreal, M.A. Soil microbial biomass, functional diversity and enzyme activity in glyphosate-resistant wheat–canola rotations under low-disturbance direct seeding and conventional tillage. *Soil Biol. Biochem.* **2007**, *39*, 1418–1427. [CrossRef]
74. Lupwayi, N.Z.; Blackshaw, R.E. Soil microbiology in glyphosate-resistant corn cropping systems. *Agron. J.* **2012**, *104*, 1041–1048. [CrossRef]
75. Souza, R.A.; Babujia, L.C.; Silva, A.P.; Guimarães, M.F.; Arias, C.A.; Hungria, M. Impact of the ahas transgene and of herbicides associated with the soybean crop on soil microbial community. *Transgenic Res.* **2013**, *22*, 877–892. [CrossRef] [PubMed]
76. Babujia, L.C.; Silva, A.P.; Nakatani, A.S.; Cantao, M.E.; Vasconcelos, A.T.R.; Visentainer, J.V.; Hungria, M. Impact of long-term cropping of glyphosate-resistant transgenic soybean [Glycine max (L.) Merr.] on soil microbiome. *Transgenic Res.* **2016**, *25*, 425–440. [CrossRef] [PubMed]
77. Nakatani, A.S.; Fernandes, M.F.; Souza, R.A.; Silva, A.P.; Reis-Junior, F.B.; Mendes, I.C.; Hungria, M. Effects of the glyphosate-resistance gene and of herbicides applied to the soybean crop on soil microbial biomass and enzymes. *Field Crops Res.* **2014**, *162*, 20–29. [CrossRef]
78. Wei, X.D.; Zou, H.L.; Chu, L.M.; Liao, B.; Ye, C.M.; Lan, C.Y. Field released transgenic papaya affects microbial communities and enzyme activities in soil. *Plant Soil* **2006**, *285*, 347–358. [CrossRef]
79. Hsieh, Y.-T.; Pan, T.-M. Influence of planting papaya ringspot virus resistant transgenic papaya on soil microbial biodiversity. *J. Agric. Food Chem.* **2006**, *54*, 130–137. [CrossRef]
80. Stefani, F.O.P.; Tanguay, P.; Pelletier, G.; Piche, Y.; Hamelin, R.C. Impact of endochitinase-transformed white spruce on soil fungal biomass and ectendomycorrhizal symbiosis. *Appl. Environ. Microbiol.* **2010**, *76*, 2607–2614. [CrossRef]
81. Wang, B.; Shen, H.; Yang, X.; Guo, T.; Zhang, B.; Yan, W. Effects of chitinase-transgenic (McChit1) tobacco on the rhizospheric microflora and enzyme activities of the purple soil. *Plant Soil Environ.* **2013**, *59*, 241–246. [CrossRef]
82. Wu, J.; Yu, M.; Xu, J.; Du, J.; Ji, F.; Dong, F.; Li, X.; Shi, J. Impact of transgenic wheat with wheat yellow mosaic virus resistance on microbial community diversity and enzyme activity in rhizosphere soil. *PLoS ONE* **2014**, *9*, e98394. [CrossRef]
83. Bezirganoglu, I.; Uysal, P. Impact of transgenic AFPCHI (Cucumis melo L. Silver Light) fungal resistance melon on soil microbial communities and enzyme activities. *J. Plant Biotechnol.* **2017**, *44*, 156–163. [CrossRef]
84. Khan, M.S.; Sadat, S.U.; Jan, A.; Munir, I. Impact of transgenic Brassica napus harboring the antifungal synthetic chitinase (NiC) gene on rhizosphere microbial diversity and enzyme activities. *Front. Plant Sci.* **2017**, *8*, 1307. [CrossRef] [PubMed]
85. Sohn, S.-I.; Oh, Y.-J.; Ahn, J.-H.; Kang, H.-J.; Cho, W.-S.; Cho, Y.; Lee, B.K. Effects of disease resistant genetically modified rice on soil microbial community structure according to growth stage. *Korean J. Environ. Agric.* **2019**, *38*, 185–196. [CrossRef]
86. Tesfaye, M.; Temple, S.J.; Allan, D.L.; Vance, C.P.; Samac, D.A. Over-expression of malate dehydrogenase in transgenic alfalfa enhances organic acid synthesis and confers tolerance to aluminum. *Plant Physiol.* **2001**, *127*, 1836–1844. [CrossRef] [PubMed]
87. Tesfaye, M.; Dufault, N.S.; Dornbusch, M.R.; Deborah, L.; Allan, D.L.; Vance, C.P.; Samac, D.A. Influence of enhanced malate dehydrogenase expression by alfalfa on diversity of rhizobacteria and soil nutrient availability. *Soil Biol. Biochem.* **2003**, *35*, 1103–1113. [CrossRef]

88. Mimura, M.; Lelmen, K.E.; Shimazaki, T.; Kikuchi, A.; Watanabe, K.N. Impact of environmental stress-tolerant transgenic potato on genotypic diversity of microbial communities and soil enzyme activities under stress conditions. *Microbes Environ.* **2008**, *23*, 221–228. [CrossRef] [PubMed]
89. Chaudhry, V.; Dang, H.Q.; Tran, N.Q.; Mishra, A.; Chauhan, P.S.; Gill, S.S.; Nautiyal, C.S.; Tuteja, N. Impact of salinity-tolerant MCM6 transgenic tobacco on soil enzymatic activities and the functional diversity of rhizosphere microbial communities. *Res. Microbiol.* **2012**, *163*, 511–517. [CrossRef] [PubMed]
90. Sahoo, R.K.; Tuteja, N. Effect of salinity tolerant PDH45 transgenic rice on physicochemical properties, enzymatic activities and microbial communities of rhizosphere soils. *Plant Signal. Behav.* **2013**, *8*, e24950. [CrossRef]
91. Bai, X.; Zeng, X.; Huang, S.; Liang, J.; Dong, L.; Wei, Y.; Li, Y.; Qu, J.; Wang, Z. Marginal impact of cropping BADH transgenic maize BZ-136 on chemical property, enzyme activity, and bacterial community diversity of rhizosphere soil. *Plant Soil* **2019**, *436*, 527–541. [CrossRef]
92. Tian, W.-H.; Yi, X.-L.; Liu, S.-S.; Zhou, C.; Wang, A.-Y. Effect of transgenic cotton continuous cropping on soil bacterial community. *Ann. Microbiol.* **2020**, *70*, 61. [CrossRef]
93. Zeng, X.; Pei, T.; Song, Y.; Guo, P.; Zhang, H.; Li, X.; Li, H.; Di, H.; Wang, Z.A. Three-year plant study of salt-tolerant transgenic maize showed no effects on soil enzyme activity and nematode community. *Life* **2022**, *12*, 412. [CrossRef]
94. Li, Q.; Song, J.; Peng, S.; Wang, J.P.; Qu, G.Z.; Sederoff, R.R.; Chiang, V.L. Plant biotechnology for lignocellulosic biofuel production. *Plant Biotechnol. J.* **2014**, *12*, 1174–1192. [CrossRef] [PubMed]
95. Lebedev, V.G.; Shestibratov, K.A. Genetic engineering of lignin biosynthesis in trees: Compromise between wood properties and plant viability. *Russ. J. Plant Physiol.* **2021**, *68*, 596–612. [CrossRef]
96. Motavalli, P.P.; Kremer, R.J.; Fang, M.; Means, N.E. Impact of genetically modified crops and their management on soil microbially mediated plant nutrient transformations. *J. Environ. Qual.* **2004**, *33*, 816–824. [CrossRef] [PubMed]
97. Kolseth, A.-K.; D'Hertefeldt, T.; Emmerich, M.; Forabosco, F.; Marklund, S.; Cheeke, T.E.; Hallin, S.; Weih, M. Influence of genetically modified organisms on agro-ecosystem processes. *Agric. Ecosyst. Environ.* **2015**, *214*, 96–106. [CrossRef]
98. Pilate, G.; Guiney, E.; Holt, K.; Petit-Conil, M.; Lapierre, C.; Leplé, J.-C.; Pollet, B.; Mila, I.; Webster., E.A.; Marstorp, H.G.; et al. Field and pulping performances of transgenic trees with altered lignification. *Nat. Biotechnol.* **2002**, *20*, 607–612. [CrossRef]
99. Tilston, E.L.; Halpin, C.; Hopkins, D.W. Simultaneous down-regulation of enzymes in the phenylpropanoid pathway of plants has aggregated effects on rhizosphere microbial communities. *Biol. Fertil. Soils* **2014**, *50*, 455–463. [CrossRef]
100. Hannula, S.E.; de Boer, W.; van Veen, J. A 3-Year study reveals that plant growth stage, season and field site affect soil fungal communities while cultivar and GM-trait have minor effects. *PLoS ONE* **2012**, *7*, e33819. [CrossRef]
101. Lahl, K.; Unger, C.; Emmerling, C.; Broer, I.; Thiele-Bruhn, S. Response of soil microorganisms and enzyme activities on the decomposition of transgenic cyanophycin-producing potatoes during overwintering in soil. *Eur. J. Soil Biol.* **2012**, *53*, 1–10. [CrossRef]
102. Li, P.; Dong, J.; Yang, S.; Bai, L.; Wang, J.; Wu, G.; Wu, X.; Yao, Q.; Tang, X. Impact of β-carotene transgenic rice with four synthetic genes on rhizosphere enzyme activities and bacterial communities at different growth stages. *Eur. J. Soil Biol.* **2014**, *65*, 40–46. [CrossRef]
103. Liang, J.; Sun, S.; Ji, J.; Wu, H.; Meng, F.; Zhang, M.; Zheng, X.; Wu, C.; Zhang, Z. Comparison of the rhizosphere bacterial communities of zigongdongdou soybean and a high-methionine transgenic line of this cultivar. *PLoS ONE* **2014**, *9*, e103343. [CrossRef]
104. Brolsma, K.M.; Vonk, J.A.; Hoffland, E.; Mulder, C.; de Goede, R.G.M. Effects of GM potato Modena on soil microbial activity and litter decomposition fall within the range of effects found for two conventional cultivars. *Biol. Fertil. Soils* **2015**, *51*, 913–922. [CrossRef]
105. Sohn, S.-I.; Oh, Y.-J.; Kim, B.-Y.; Kweon, S.-J.; Cho, H.-S.; Ryu, T.-H. Effect of genetically modified rice producing resveratrol on the soil microbial communities. *J. Korean Soc. Appl. Biol. Chem.* **2015**, *58*, 795–805. [CrossRef]
106. Debruyn, J.M.; Bevard, D.A.; Essington, M.E.; Mcknight, J.Y.; Schaeffer, S.M.; Baxter, H.L.; Mazarei, M.; Mann, D.G.J.; Dixon, R.A.; Chen, F.; et al. Field-grown transgenic switchgrass (Panicum virgatum L.) with altered lignin does not affect soil chemistry, microbiology, and carbon storage potential. *GCB Bioenergy* **2017**, *9*, 1100–1109. [CrossRef]
107. Xu, S.; Ottinger, S.L.; Schaeffer, S.M.; DeBruyn, J.M.; Stewart, C.N., Jr.; Mazarei, M.; Jagadamma, S. Effects of field-grown transgenic switchgrass carbon inputs on soil organic carbon cycling. *PeerJ.* **2019**, *16*, e7887. [CrossRef]
108. Chen, Y.; Pan, L.; Ren, M.; Li, J.; Guan, X.; Tao, J. Comparison of genetically modified insect-resistant maize and non-transgenic maize revealed changes in soil metabolomes but not in rhizosphere bacterial community. *GM Crops Food* **2022**, *13*, 1–14. [CrossRef]
109. Lakshmanan, V. Root microbiome assemblage is modulated by plant host factors. In *Plant Microbe Interactions*; Bais, H., Sherrier, J., Eds.; Academic Press Ltd-Elsevier Science Ltd.: London, UK, 2015; pp. 57–79.
110. Ng, E.L.; Patti, A.F.; Rose, M.T.; Schefe, C.R.; Wilkinson, K.; Cavagnaro, T.R. Functional stoichiometry of soil microbial communities after amendment with stabilized organic matter. *Soil Biol. Biochem.* **2014**, *7*, 170–178. [CrossRef]
111. Zi, H.; Hu, L.; Wang, C. Differentiate responses of soil microbial community and enzyme activities to nitrogen and phosphorus addition rates in an alpine meadow. *Front. Plant Sci.* **2022**, *13*, 829381. [CrossRef]
112. Choi, Y.I.; Noh, E.W.; Choi, K.S. Low level expression of prokaryotic tzs gene enhances growth performance of transgenic poplars. *Trees* **2009**, *23*, 741–750. [CrossRef]

113. Nam, K.J.; Kim, D.Y.; Nam, K.-H.; Pack, I.S.; Park, J.H.; Jeong, S.-C.; Choi, Y.I.; Noh, E.W.; Kim, C.-G. Effects of transgenic poplars expressing increased levels of cellular cytokinin on rhizosphere microbial communities. *Eur. J. Soil Biol.* **2014**, *63*, 78–82. [CrossRef]
114. Saxena, D.; Stotzky, G. Bt corn has a higher lignin content than non-Bt corn. *Am. J. Bot.* **2001**, *88*, 1704–1706. [CrossRef]
115. Wu, W.-X.; Ye, Q.-F.; Min, H.; Duan, X.-J.; Jin, W.-M. Bt-transgenic rice straw affects the culturable microbiota and dehydrogenase and phosphatase activities in a flooded paddy soil. *Soil Biol. Biochem.* **2004**, *36*, 289–295. [CrossRef]
116. Huang, S.; Zeng, X.; Wei, Y.; Bai, X.; Jin, Z.; Zhang, M.; Wang, Z.; Wang, H.; Qu, J.; Di, H. Decomposition of betaine aldehyde dehydrogenase transgenic maize straw and its effects on soil microbial biomass and microbiota diversity. *Appl. Soil Ecol.* **2020**, *153*, 103582. [CrossRef]
117. Hannula, S.E.; de Boer, W.; Baldrian, P.; van Veen, J.A. Effect of genetic modification of potato starch on decomposition of leaves and tubers and on fungal decomposer communities. *Soil Biol. Biochem.* **2013**, *58*, 88–98. [CrossRef]
118. Hopkins, D.W.; Webster, E.A.; Chudek, J.A.; Halpin, C. Decomposition of stems from tobacco plants with genetic modifications to lignin biosynthesis. *Soil Biol. Biochem.* **2001**, *33*, 1455–1462. [CrossRef]
119. Tilston, E.L.; Halpin, C.; Hopkins, D.W. Genetic modifications to lignin biosynthesis in field-grown poplar trees have inconsistent effects on the rate of woody trunk decomposition. *Soil Biol. Biochem.* **2004**, *36*, 1903–1906. [CrossRef]
120. Deacon, L.J.; Pryce-Miller, E.J.; Frankland, J.C.; Bainbridge, B.W.; Moore, P.D.; Robinson, C.H. Diversity and function of decomposer fungi from a grassland soil. *Soil Biol. Biochem.* **2006**, *38*, 7–20. [CrossRef]
121. Vauramo, S.; Pasonen, H.-L.; Pappinen, A.; Setala, H. Decomposition of leaf litter from chitinase transgenic silver birch (Betula pendula) and effects on decomposer populations in a field trial. *Appl. Soil Ecol.* **2006**, *32*, 338–349. [CrossRef]
122. Qin, W.; Hu, C.; Oenema, O. Soil mulching significantly enhances yields and water and nitrogen use efficiencies of maize and wheat: A meta-analysis. *Sci. Rep.* **2015**, *5*, 16210. [CrossRef]
123. Marral, M.W.R.; Khan, M.B.; Ahmad, F.; Farooq, S.; Hussain, M. The influence of transgenic (Bt) and nontransgenic (non-Bt) cotton mulches on weed dynamics, soil properties and productivity of different winter crops. *PLoS ONE* **2020**, *15*, e0238716. [CrossRef]
124. Dellero, Y. Manipulating amino acid metabolism to improve crop nitrogen use efficiency for a sustainable agriculture. *Front. Plant Sci.* **2020**, *11*, 602548. [CrossRef]
125. Lebedev, V.G.; Popova, A.A.; Shestibratov, K.A. Genetic engineering and genome editing for improving nitrogen use efficiency in plants. *Cells* **2021**, *10*, 3303. [CrossRef] [PubMed]
126. Baker, A.; Ceasar, S.A.; Palmer, A.J.; Paterson, J.B.; Qi, W.; Muench, S.P.; Baldwin, S.A. Replace, reuse, recycle: Improving the sustainable use of phosphorus by plants. *J. Exp. Bot.* **2015**, *66*, 3523–3540. [CrossRef] [PubMed]
127. Teng, W.; He, X.; Tong, Y.-P. Transgenic approaches for improving use efficiency of nitrogen, phosphorus and potassium in crops. *J. Integr. Agric.* **2017**, *16*, 2657–2673. [CrossRef]
128. Iqrar, S.; Ashrafi, K.; Kiran, U.; Fatima, S.; Kamaluddin; Abdin, M.Z. Chapter seven—Transgenic plants with improved nutrient use efficiency. In *Transgenic Technology Based Value Addition in Plant Biotechnology*; Kiran, U., Abdin, M.Z., Kamaluddin, Eds.; Academic Press: Cambridge, MA, USA, 2020; pp. 157–185. [CrossRef]
129. Singer, S.D.; Soolanayakanahally, R.Y.; Foroud, N.A.; Kroeber, R. Biotechnological strategies for improved photosynthesis in a future of elevated atmospheric CO_2. *Planta* **2020**, *251*, 24. [CrossRef]
130. Araus, J.L.; Sanchez-Bragado, R.; Vicente, R. Improving crop yield and resilience through optimization of photosynthesis: Panacea or pipe dream? *J. Exp. Bot.* **2021**, *72*, 3936–3955. [CrossRef]
131. Weir, T.L.; Park, S.W.; Vivanco, J.M. Biochemical and physiological mechanisms mediated by allelochemicals. *Curr. Opin. Plant Biol.* **2004**, *7*, 472–479. [CrossRef]
132. Schmutterer, H.; Singh, R.P. List of insect pests susceptible to neem products. In *The Neem Tree: Azadirachta Indica A. Juss and Other Meliacious Plants—Sources of Unique Natural Products for Integrated Pest Management, Medicine, Industry and Other Purposes*, 2nd ed.; Schmutterer, H., Ed.; Neem Foundation: Mumbai, India, 2002; pp. 411–456.
133. Gopal, M.; Gupta, A.V.; Arunachalam, V.; Magu, S.P. Impact of azadirachtin, an insecticidal allelochemical from neem on soil microflora, enzyme and respiratory activities. *Bioresour. Technol.* **2007**, *98*, 3154–3158. [CrossRef]
134. Gu, Y.; Wang, P.; Kong, C.H. Urease, invertase, dehydrogenase and polyphenoloxidase activities in paddy soil influenced by allelopathic rice variety. *Eur. J. Soil Biol.* **2009**, *45*, 436–441. [CrossRef]
135. Reigosa, M.S.; Gonzalesy, L.; Souto, X.C.; Pastoriza, J.E. Allelopathy in forest ecosystem. In *Allelopathy in Ecological Agriculture and Forestry*; Narwal, S.S., Hoagland, R.E., Dilday, R.H., Reigosa, M.J., Eds.; Kluwer Academic Pubishersl: Dordrecht, The Netherlands, 2000; pp. 183–193.
136. Lebedev, V.G.; Krutovsky, K.V.; Shestibratov, K.A. Fell Upas sits, the hydra-tree of death, or the Phytotoxicity of trees. *Molecules* **2019**, *24*, 1636. [CrossRef]
137. ISAAA. ISAAA's GM Approval Database. 2022. Available online: http://www.isaaa.org/gmapprovaldatabase/ (accessed on 4 July 2022).
138. Kikuchi, A.; Yu, X.; Shimazaki, T.; Kawaoka, A.; Ebinuma, H.; Watanabe, K.N. Allelopathy assessments for the environmental biosafety of the salt-tolerant transgenic Eucalyptus camaldulensis, genotypes codA12-5B, codA12-5C, and codA20C. *J. Wood Sci.* **2009**, *55*, 149–153. [CrossRef]

139. Yu, X.; Kikuchi, A.; Shimazaki, T.; Yamada, A.; Ozeki, Y.; Matsunaga, E.; Ebinuma, H.; Watanabe, K.N. Assessment of the salt tolerance and environmental biosafety of Eucalyptus camaldulensis harboring a mangrin transgene. *J. Plant Res.* **2013**, *126*, 141–150. [CrossRef] [PubMed]
140. Tran, N.-H.T.; Oguchi, T.; Matsunaga, E.; Kawaoka, A.; Watanabe, K.N.; Kikuchi, A. Evaluation of potential impacts on biodiversity of the salt-tolerant transgenic Eucalyptus camaldulensis harboring an RNA chaperonic RNA-Binding-Protein gene derived from common ice plan. *Transgenic Res.* **2021**, *30*, 23–34. [CrossRef] [PubMed]
141. Shimazaki, T.; Kikuchi, A.; Matsunaga, E.; Nanto, K.; Shimada, T.; Watanabe, K.N. Establishment of a homogenized method for environmental biosafety assessments of transgenic plants. *Plant Biotechnol.* **2009**, *26*, 143–148. [CrossRef]
142. Lebedev, V.; Larionova, A.; Bykhovets, S.; Shanin, V.; Komarov, A.; Shestibratov, K. Model assessment of transgenic trees impact on nitrogen and carbon cycles in forest plantations. In Proceedings of the IUFRO Tree Biotechnology Conference 2015 "Forests: The importance to the planet and society", Florence, Italy, 8–12 June 2015; pp. 186–188. [CrossRef]

plants

MDPI

Article

Effects of Insect-Resistant Maize 2A-7 Expressing mCry1Ab and mCry2Ab on the Soil Ecosystem

Shuke Yang [1], Xin Liu [1,2], Xiaohui Xu [1,*], Hongwei Sun [1], Fan Li [1], Chaofeng Hao [1] and Xingbo Lu [1,2,*]

1 Shandong Key Laboratory of Plant Virology, Institute of Plant Protection, Shandong Academy of Agricultural Sciences, Jinan 250100, China
2 College of Life Sciences, Shandong Normal University, Jinan 250014, China
* Correspondence: xuxiaohui1023@163.com (X.X.); luxb99@sina.com (X.L.)

Abstract: Transgenic maize 2A-7 expressing mCry1Ab and mCry2Ab has excellent resistance to lepidopteran pests. Previous studies have investigated the effects of several *Bacillus thuringiensis* (Bt) proteins on the soil. However, the effects of artificially modified Bt proteins on soil ecosystems are still unclear. To evaluate the effects of transgenic maize 2A-7 on soil, the physicochemical properties, enzyme activities and functional diversities of the microbial communities in rhizosphere soils from 2A-7 and its near-isogenic non-transgenic control Dongdan 6531 were analyzed at different developmental stages under field conditions. The alteration of six physicochemical properties (pH, total nitrogen, total phosphorus, organic matter, available phosphorus and alkali-hydrolyzed nitrogen) and six functional enzymes (catalase, alkaline phosphatase, sucrase, acid phosphatase, urease and alkaline protease) activities in the rhizosphere soils between the two maize cultivars were drastically correlated with plant growth stage, but not affected by the artificially modified Bt transgenes. An analysis of time-course Biolog data revealed that the functional diversity of microbial communities in the rhizosphere soil of 2A-7 and its control were similar at each developmental stage. The results suggest that transgenic maize 2A-7 has no significant impact on the soil ecosystem and provide valuable information on scientific safety assessments of 2A-7 and its commercial applications.

Keywords: transgenic insect-resistant maize 2A-7; soil ecosystem; physical and chemical properties; enzyme activity; microbial functional diversity

Citation: Yang, S.; Liu, X.; Xu, X.; Sun, H.; Li, F.; Hao, C.; Lu, X. Effects of Insect-Resistant Maize 2A-7 Expressing mCry1Ab and mCry2Ab on the Soil Ecosystem. *Plants* **2022**, *11*, 2218. https://doi.org/10.3390/plants11172218

Academic Editors: Wei Wei and C. Neal Stewart, Jr.

Received: 18 July 2022
Accepted: 20 August 2022
Published: 26 August 2022

1. Introduction

The global area of genetically modified (GM) crops has increased year by year since its first commercial release in the United States in 1996 [1,2]. In 2019, over 16 million farmers in 29 countries grew GM crops commercially, and the global area of GM crops reached 190.4 million hectares, over 112 times more than that of 1996 [3]. China is not only the seventh largest producer of GM crops, but also the second largest consumer of GM crops [3]. In China, approximately 7–20% maize yield loss is caused by insect pests every year. The application of insect-resistant GM crops can effectively reduce the threat from various insects, including members of Lepidoptera, Diptera, Coleoptera and Hymenoptera. Importantly, the use of insect-resistant plants has greatly reduced the application of pesticides, which has brought significant environmental, social, and economic benefits [4].

The acquisition of insect-resistant trait in GM crops is due to the expression of a class of insecticidal crystal proteins derived from the Gram-positive bacterium *Bacillus Thuringiensis* (Bt) [5–7]. These proteins are broken down into small active toxin fragments in the target organism that can bind to the organism's intestinal epithelial cells and cause perforations, thereby affecting the osmotic balance, eventually resulting in the death of the target organism [8–10]. The large-scale planting of GM crops worldwide has an impact on the ecological environment of the soil over the past two decades [11,12]. In general, Bt crops can accumulate Bt proteins in various tissues and organs, such as the leaf, stem, root and seed. Bt proteins can be released into soils in two ways: directly through plant rhizosphere exudates and through plant residues after GM crop-derived straw is returned to the soil [13,14]. Irrespective of how Bt proteins enter the soil, they can directly or indirectly cause changes in soil physical and chemical properties, affect soil enzyme activities and change the levels of soil microbial functional diversity and bacterial dominant flora [15,16]. Varying levels of Bt proteins have been detected in the rhizospheres of Bt cotton cultivars NuCOTN99B and SGK321 during their whole growth stages, with the highest concentrations reaching 200 ng/g and 300 ng/g, respectively [17]. In the rhizosphere soil of insect-resistant maize Mon810, the Cry1Ab protein content can reach 165 g/hectare [18]. In addition, with the application of GM crop-derived straw during cultivation, Bt proteins can enter the soil from the straw residue, which may have impacts on the soil ecological environments. The Bt proteins in root exudates of GM crops have been associated with clay minerals and humic acids of soil substrates, leading to retained insecticidal activity in the soil [19]. Additionally, the Cry1Ab protein in transgenic maize Mon810 accumulates in the rhizosphere soil at different plant developmental stages, and after the crop-derived straw is returned to the field, the degradation of the Bt proteins has two patterns: most Bt proteins degrade rapidly during the early stage, but a few Bt proteins degrade stably during the later stage [20].

Insect-resistant transgenic maize 2A-7, which expresses artificially modified codons of two Bt proteins Cry1Ab and Cry2Ab, was developed to control lepidopteran pests. Insect-resistant assays performed in both the field and laboratory revealed that 2A-7 is highly resistant to lepidopteran pests, such as the corn borer, armyworm and bollworm, and it also shows good resistance to *Spodoptera frugiperda* [21]. A recent study showed that the application of 2A-7 had no adverse effect on the non-target arthropod communities [22]. Therefore, 2A-7 is a candidate with good application prospects. The modifications of the two Bt proteins in 2A-7 may produce distinct characteristics, including expression patterns, protein stability levels and degradation patterns. Although numerous studies have investigated the effects of GM crops expressing the Cry1Ab or Cry2Ab protein on the soil [23–25], none have reported the effects of GM crops containing modified Cry1Ab and Cry2Ab proteins on the soil ecosystem.

By comparing the physical and chemical properties, enzyme activities and microbial diversity levels between the rhizosphere soil of the GM maize and that of its corresponding non-transgenic control in the field under natural conditions at different growth stages, the impact of the GM maize on the soil ecological environment can be assessed [26–28]. Thus, in this paper, the effects of transgenic maize 2A-7 on the soil environment were systematically evaluated. Our results provide scientific data related to the evaluation of the environmental safety of 2A-7 and establish a basis for government decision making of 2A-7 commercialization.

2. Results

2.1. Effects of Transgenic Insect-Resistant Maize 2A-7 on the Physical and Chemical Properties of Rhizosphere Soil

In this study, the pH, total nitrogen, total phosphorus, organic matter, available phosphorus and alkaline-hydrolyzed nitrogen of the rhizosphere soils of transgenic maize 2A-7 and its non-transgenic control Dongdan 6531 were measured at five different developmental stages: the seedling, jointing, tasseling, silking and full ripe stages. The tested indices in the rhizosphere soils of the two maize inbred lines were similar at the same developmental stages (Table 1). Specifically, during the growth and development of both maize lines, the pH values of the rhizosphere soils showed gradual upward trends, from acidic (pH between 6.7 and 6.8) to alkaline (pH between 7.7 and 7.8). The total nitrogen contents were distributed between 0.12% and 0.15%, showing sharp increases during the seedling stages, followed by slightly decreases during the other stages. The total phosphorus contents remained stable among the five stages, with distributions between 0.08% and 0.12%. The organic matter contents ranged from 17 g/kg to 27 g/kg during the five different developmental stages, but there was no significant difference between the two maize lines at each of the developmental stages. The available phosphorus contents in the rhizosphere soils of the two maize inbred lines gradually decreased as the development process advanced, ranging from 57.27 mg/kg to 98.63 mg/kg. The alkali-hydrolyzed nitrogen contents were distributed between 75.47 mg/kg and 91.27 mg/kg, and they remained stable during each period. Based on the above results, the maize developmental period, but not the transgenic event, appeared to be a vital factor affecting the physical and chemical properties of the rhizosphere soil.

Table 1. Changes in the physical and chemical properties of the rhizosphere soils of transgenic maize 2A-7 and non-transgenic maize at different developmental stages.

Developmental Stage	Variety	pH	Total Nitrogen (%)	Total Phosphorus (%)	Organic Matter (g/Kg)	Available Phosphorus (mg/Kg)	Alkali-Hydrolyzed Nitrogen (mg/Kg)
Seedling stage	2A-7	6.70 ± 0.10 [a]	0.15 ± 0.01 [a]	0.10 ± 0.01 [a]	22.3 ± 2.70 [a]	96.10 ± 4.62 [a]	75.47 ± 5.67 [a]
	CK	6.80 ± 0.00 [a]	0.14 ± 0.01 [a]	0.08 ± 0.03 [a]	21.63 ± 1.34 [a]	98.63 ± 10.73 [a]	94.60 ± 16.32 [a]
Jointing stage	2A-7	7.17 ± 0.06 [a]	0.12 ± 0.01 [a]	0.10 ± 0.02 [a]	21.00 ± 4.66 [a]	90.83 ± 3.50 [a]	89.13 ± 11.48 [a]
	CK	7.27 ± 0.06 [a]	0.13 ± 0.01 [a]	0.10 ± 0.01 [a]	23.23 ± 0.55 [a]	88.70 ± 0.35 [a]	82.20 ± 9.00 [a]
Tasseling stage	2A-7	7.40 ± 0.10 [a]	0.14 ± 0.01 [a]	0.10 ± 0.01 [a]	21.60 ± 0.61 [a]	80.00 ± 8.15 [a]	85.47 ± 8.59 [a]
	CK	7.50 ± 0.00 [a]	0.13 ± 0.01 [a]	0.11 ± 0.00 [a]	24.93 ± 0.74 [a]	80.80 ± 2.05 [a]	79.93 ± 11.22 [a]
Silking stage	2A-7	7.47 ± 0.06 [a]	0.14 ± 0.02 [a]	0.08 ± 0.03 [a]	26.60 ± 6.32 [a]	73.47 ± 14.72 [a]	91.27 ± 17.07 [a]
	CK	7.60 ± 0.00 [a]	0.12 ± 0.02 [a]	0.11 ± 0.00 [a]	17.57 ± 0.32 [a]	57.27 ± 1.80 [a]	88.30 ± 16.63 [a]
Full ripe stage	2A-7	7.73 ± 0.06 [a]	0.13 ± 0.00 [a]	0.12 ± 0.00 [a]	20.60 ± 3.21 [a]	59.33 ± 2.40 [a]	85.10 ± 1.39 [a]
	CK	7.80 ± 0.00 [a]	0.12 ± 0.01 [a]	0.12 ± 0.01 [a]	17.17 ± 3.27 [a]	59.17 ± 4.28 [a]	85.63 ± 5.59 [a]

The data have been shown as mean ± SE. Letters following the data indicate significant differences between the rhizosphere soil samples of transgenic maize 2A-7 and its non-transgenic control during different developmental stages (Student's t-test was employed, $p < 0.05$). 2A-7 represents the transgenic maize 2A-7, and CK represents the non-transgenic control Dongdan 6531.

2.2. Effects of Transgenic Insect-Resistant Maize 2A-7 on the Enzyme Activity Levels in Rhizosphere Soil

The activities of six functional soil enzymes in the rhizosphere soil of transgenic maize 2A-7 and its non-transgenic control Dongdan 6531 at different developmental stages were analyzed. As shown in Figure 1, the activity levels of each enzyme were not significantly different in the rhizosphere soils of the two maize inbred lines during the same growth stage. During the five developmental periods, the catalase activities in the 2A-7 and the control maize rhizosphere soils were relatively stable (Figure 1A). The alkaline phosphatase activity showed a gradual upward trend during the first three stages and then decreased in the latter two stages (Figure 1B). The sucrase and acid phosphatase activities in the rhizosphere soils of the two maize inbred lines had similar trends throughout the maize growth period, showing gradual decreases from the seedling to silking stages, followed by slight increases (Figure 1C,D). The urease activities in the rhizosphere soils of the two maize inbred lines peaked during the silking stage, and the activity levels were similar for the rest of the developmental stages (Figure 1E). The alkaline protease activities in the rhizosphere soils of the two maize inbred lines were not significantly different at each developmental stage (Figure 1F). Thus, the transgenic maize 2A-7 did not appear to have a significant impact on the soil enzymes, whereas the growth stages appeared to have been highly influential.

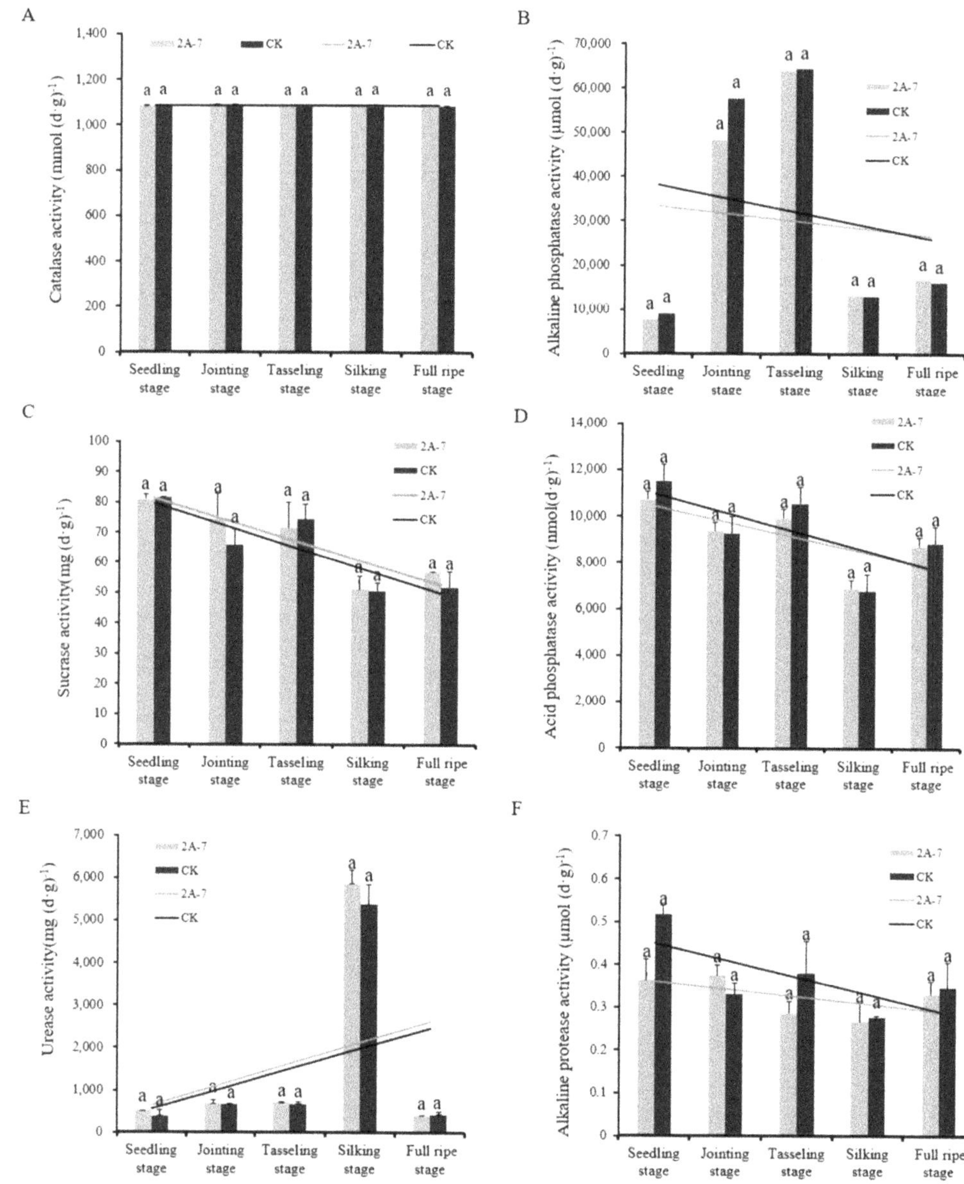

Figure 1. Effects of transgenic maize 2A-7 on the enzyme activities in the rhizosphere soil at different developmental stages. (**A**) Catalase activity; (**B**) alkaline phosphatase activity; (**C**) sucrase activity; (**D**) acid protease activity; (**E**) urease activity; (**F**) alkaline protease activity. 2A-7 represents the transgenic maize 2A-7, and CK represents the non-transgenic control Dongdan 6531. Letters above the columns indicate significant differences in rhizosphere soil treatments between the two maize lines at the same developmental stage ($p < 0.05$). The linear trend lines were graphed for the two maize cultivars at the five developmental stages.

2.3. Metabolic Functional Diversity of Microorganisms in the Rhizosphere Soils of Transgenic Maize 2A-7 and Its Control

2.3.1. Metabolic Activity Changes in Microorganisms Inhabiting the Rhizosphere Soils of Transgenic Maize 2A-7 and ITS Control

Changes in values of the average well color development (AWCD) may reflect changes in the metabolic activity of a microbial community. More species and a greater abundance of soil microflora lead to more types and greater amounts of carbon sources that can be utilized in the plate cells. Generally speaking, the AWCD value is positively correlated with the growth of the cultured microbes. As shown in Figure 2, the changes in the AWCD values during different developmental stages showed a similar pattern, which was a linear increase during the early incubation period followed by a stationary stage during the latter periods. In each growth period, the AWCD values of the soil microorganisms of the two maize inbred lines increased slowly in the first 24 h, followed by rapid growth during the next 72 h, indicating that the microbes had high metabolic activities. After 96 h of culture, the increase rates of the AWCD values slowed and then plateaued. The AWCD curves of the rhizosphere soil microorganisms of the two maize inbred lines showed good coincidence, except during the silking stage. The results indicated that no significant differences exist in the microbial metabolic activities in the rhizosphere soils of the two maize inbred lines during the seedling, jointing, tasseling and full ripe stages.

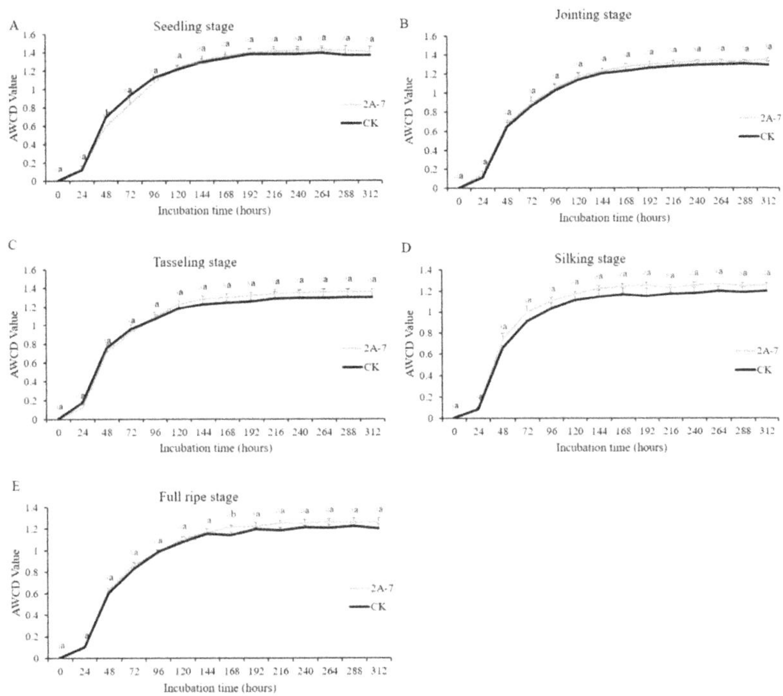

Figure 2. Changes in the microbial average well color development (AWCD) values of the rhizosphere soils of transgenic maize 2A-7 and its non-transgenic control during different developmental periods. (**A**) Changes of AWCD values at the seedling stage; (**B**) Changes of AWCD values at the jointing stage; (**C**) Changes of AWCD values at the tasseling stage; (**D**) Changes of AWCD values at the silking stage; (**E**) Changes of AWCD values at the full ripe stage. 2A-7 represents the transgenic maize 2A-7, and CK represents the non-transgenic control Dongdan 6531. Letters above the curves indicate significant differences in rhizosphere soil treatments between the two maize lines at the same developmental stage ($p < 0.05$).

2.3.2. Changes in the Rhizosphere Soil Microorganisms' Utilization of Different Carbon Sources between the Two Maize Lines

After 96h incubation, the microfloral utilization rate of the carbon sources slowed and then remained relatively steady (Figure 1). Thus, the carbon source utilization at this time point may reflect the differences in the utilization of individual carbon sources by microbes. From data collected after a 96 h incubation, the AWCD value of each type of carbon source was calculated. As shown in Figure 3, there were no significant differences in the utilization levels of the four types of carbon sources between the rhizosphere soil microorganisms of the two maize inbred lines at each developmental stage (Figure 3). The soil microbes from both types of maize utilized carbon source carbohydrates and their derivatives at a relatively stable state at the different developmental stages (Figure 3A), whereas the utilization of the other three types of carbon sources showed trends of first increasing and then decreasing, with the highest utilization rate at the jointing stage (Figure 3B–D).

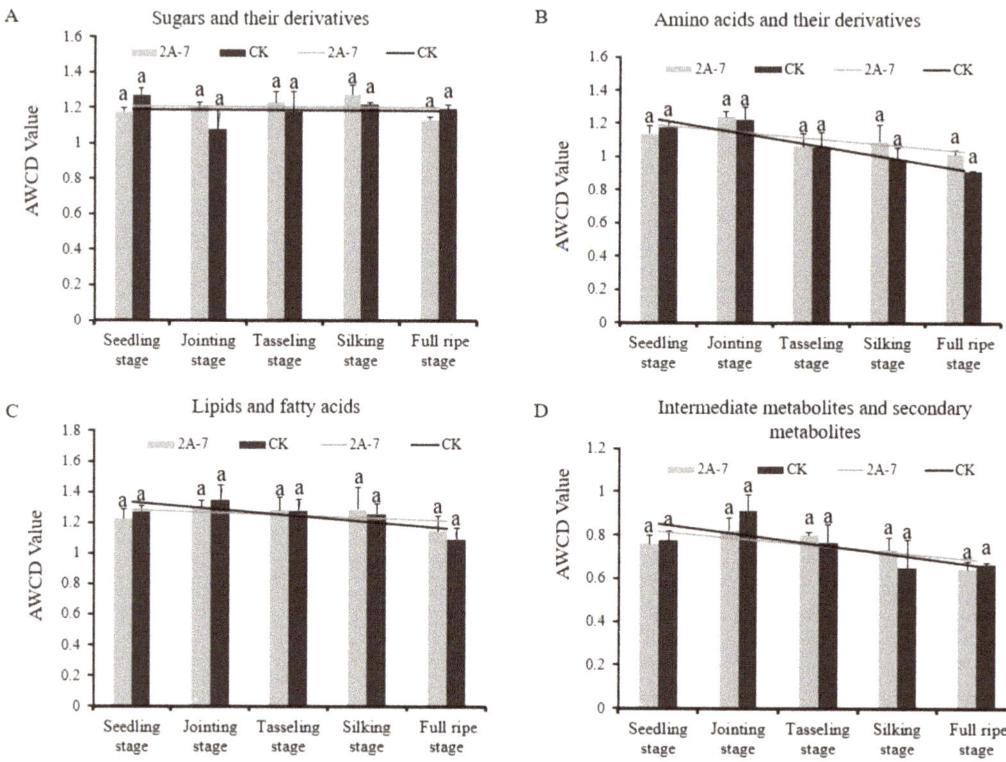

Figure 3. The utilization levels of four types of carbon sources by rhizosphere soil microorganisms of transgenic maize 2A-7 and its non-transgenic control during different developmental periods. (**A**) Sugars and their derivatives; (**B**) amino acid and their derivatives; (**C**) lipids and fatty acids; (**D**) intermediate metabolites and secondary metabolites. 2A-7 represents the transgenic maize 2A-7, and CK represents the non-transgenic control Dongdan 6531. Letters above columns indicate significant differences in rhizosphere soil treatments between the two maize lines at the same time node ($p < 0.05$). The linear trend lines were graphed for the two maize cultivars at the five developmental stages.

2.3.3. Diversity Index Analysis of the Rhizosphere Soil Microbial Communities of Transgenic Maize 2A-7 and Its Control

The Shannon index (H′), Simpson index (D) and McIntosh index (U) indices of the rhizosphere soil microorganisms from the two maize lines during each developmental period were calculated using the 96 h soil microorganism carbon source utilization data. As shown in Table 2, at all the tested periods, the three diversity indices showed no significant differences between the microbial communities of the two maize lines. Therefore, we inferred that the transgenic maize 2A-7 had no significant impact on the species' richness, dominance or homogeneity of the rhizosphere soil microorganisms.

Table 2. Comparison of metabolic functional diversity indices of microbial communities in rhizosphere soils of transgenic maize 2A-7 and its non-transgenic control during different developmental periods.

Period	Variety	Shannon Index (H′)	Simpson Index (D)	McIntosh Index (U)
Seedling stage	2A-7	3.34 ± 0.00 [a]	0.96 ± 0.00 [a]	6.40 ± 0.20 [a]
	CK	3.35 ± 0.01 [a]	0.96 ± 0.00 [a]	6.65 ± 0.21 [a]
Jointing stage	2A-7	3.34 ± 0.01 [a]	0.96 ± 0.00 [a]	6.33 ± 0.22 [a]
	CK	3.32 ± 0.02 [a]	0.96 ± 0.00 [a]	6.20 ± 0.52 [a]
Tasseling stage	2A-7	3.34 ± 0.03 [a]	0.96 ± 0.00 [a]	6.57 ± 0.21 [a]
	CK	3.34 ± 0.04 [a]	0.96 ± 0.00 [a]	6.35 ± 0.34 [a]
Silking stage	2A-7	3.34 ± 0.00 [a]	0.96 ± 0.00 [a]	6.71 ± 0.10 [a]
	CK	3.31 ± 0.01 [a]	0.96 ± 0.00 [a]	6.25 ± 0.27 [a]
Full ripe stage	2A-7	3.33 ± 0.01 [a]	0.96 ± 0.00 [a]	5.91 ± 0.14 [a]
	CK	3.33 ± 0.01 [a]	0.96 ± 0.00 [a]	5.92 ± 0.17 [a]

The data have been shown as mean ± SE. Letters following the data indicate significant differences between the rhizosphere soil samples of transgenic maize 2A-7 and its non-transgenic control during different developmental stages (Student's *t*-test was employed, $p < 0.05$). 2A-7 represents the transgenic maize 2A-7, and CK represents the non-transgenic control Dongdan 6531.

3. Discussion

3.1. Effects of Transgenic Maize 2A-7 on Physical and Chemical Properties of Rhizosphere Soil

The physical and chemical properties of soil are important factors affecting soil quality and are used as important indicators for evaluating the level of soil fertility [29,30]. However, with the expansion of GM crop planting areas, more people have become worried about whether these crops pose a threat to the soil ecological environment. Exogenous proteins secreted from their roots or straw residue may change the physical and chemical properties of the soil, thereby affecting crop growth [31]. Changes in total nitrogen, total phosphorus and pH values in soil are not provoked by the planting of transgenic Bt maize, but they do occur with the return of GM crop-derived straw to the field [32]. In addition, the diversity of communities of Arbuscular Mycorrhizal Fungi in the soil is not changed by Bt maize [33]. Consistently, the planting of Bt cotton does not change the contents of total nitrogen, available phosphorus and available potassium in the soil, nor does it affect soil characteristics [34,35]. Here, our data provide evidence that the physical and chemical properties of the rhizosphere soil can be affected by the stages of maize development, but not affected by the transgenic event of Bt (Table 1). These results suggest that the planting of transgenic maize 2A-7 has no significant effects on the physical and chemical properties of the rhizosphere soil.

3.2. Effects of Transgenic Maize 2A-7 on Soil Enzyme Activities in Rhizosphere

Soil enzymes mainly originate from soil microorganisms, plant root exudates and the decomposition of animal and plant residues. Enzymes, such as phosphatase, urease, protease, invertase, cellulose and amylase, are key components in various biochemical reactions in soil, including the biotransformation of nitrogen, phosphorus, organic matter and carbon in soil, and their activities can be used to evaluate indicators of soil biotransformation direction and strength [36–38]. Enzyme activity levels are affected by environment, climate and planting conditions, such as temperature and humidity [39]. In general, soil enzyme activity is positively correlated with soil fertility. Transgenic maize and wheat do not affect the activities of urease, dehydrogenase and sucrase in rhizosphere soil at different developmental stages [40–42]. In this study, soil enzyme levels of one oxidoreductase and five hydrolases in the rhizosphere soils of transgenic maize 2A-7 and its control were determined. The activities of all six major soil functional enzymes did not significantly differ between the rhizosphere soils of the two maize lines during each developmental period (Figure 1). This result indicated that the changes in soil enzyme activities may be related to the maize developmental period.

3.3. Effects of Transgenic Maize 2A-7 on the Functional Diversity of Rhizosphere Soil Microorganisms

Microbes are involved in metabolic activities, such as the use of carbon and nitrogen sources, and microbial functional diversity is one of the important characteristics of biological communities in the soil ecological environment [43]. At present, GM crops have not been found to have caused significant impacts on soil microbial diversity [44,45]. The change trend of the AWCD value between phosphorus-efficient transgenic rice and its control is consistent, and the change in the microbial diversity index has only been related to fertilizer application and growth stages [46]. The rhizosphere microbial diversity index and a principal component analysis of transgenic maize Shuangkang 12-5 showed that the diversity level of its rhizosphere soil microorganisms is similar to that of its control maize [47]. The effects of transgenic insect-resistant maize expressing CryIIe on the functional diversity levels of microbial communities in rhizosphere soils are not significantly different to those of its control maize Zong31, but the environmental conditions and plant growth stage appear to have stronger effects than the cultivar [48]. In this study, the changes in the AWCD values of the transgenic maize 2A-7 and its control Dongdan 6531 were similar. In the early stage of incubation (0–24 h), at all the tested periods, the AWCD values of the two maize inbred lines were low, close to 0, indicating that the metabolic activities of microbial communities in the rhizosphere soil of the transgenic maize 2A-7 and its control were low, with little of the carbon source being used [49–51]. After a 24 h incubation, the increase in the AWCD rate accelerated, indicating that the microbial community metabolic capacity increased, and the soil microorganisms gradually used a large amount of the carbon source. At 96 h, the AWCD curve's slope for each treatment was the largest, indicating that the metabolic activity of the soil microbial community reached the highest level. From 144 h to 312 h, the increase in the AWCD rate gradually slowed and remained at a steady level, indicating that, due to factors such as competition and the limitation of a single carbon source, the increase in the number of microorganisms gradually decreased (Figure 2). Using the data collected after a 96 h incubation, the carbon source utilization capacity and three microbial diversity indices were analyzed. The utilization levels of the four types of carbon sources by the two maize lines were similar (Figure 3). The utilization of lipids and their derivatives, amino acids and their derivatives, intermediate metabolites and secondary metabolites was the highest at the jointing stage and then decreased (Figure 3). This result indicated that the rhizophere soil microbe communities of both the transgenic maize 2A-7 and its control require greater carbon sources during the vegetative stage compared with the reproductive stage. In addition, the utilization levels of carbohydrates and their derivatives remained stable throughout the whole growth period (Figure 3), implying the carbohydrates and their derivatives, as carbon sources, are necessary for the microbe communities. Three microbial diversity indices, H′, D and U, were not significantly different between transgenic 2A-7 and the non-transgenic control (Table 2). This result indicates that the species' richness, dominance and homogeneity of the rhizosphere soil microbe communities were not affected by the transgenic maize 2A-7.

4. Materials and Methods

4.1. Plant Materials

Insect-resistant maize 2A-7 with Dongdan 6531 background and its control Dongdan 6531 were provided by China Agricultural University. The transgenic materials were first obtained by using the homozygous transgenic line 2A-7 with B73 background as the male parent, and 83B28 as the recurrent female parent. After successive backcrossing for 5 generations and then self-crossing for 2 generations, the homozygous transgenic line 2A-7 with 83B28 background was obtained. Thereafter, the hybrids containing 2A-7 event were obtained by crossing the 83B28 (2A-7) transgenic lines and PH6WC. The control maize lines Dongdan 6531 were the F1 hybrid of 83B28 and PH6WC.

4.2. Test Design

The transgenic maize 2A-7 and its non-transgenic control Dongdan 6531 were planted using a random block design, with three replicates. Each plot covered an area of 150 m^2, with a row spacing of 60 cm and a plant spacing of 25 cm. There was a 1.0 m wide isolation belt between treatments, and the fields were under routine management. A five-point sampling method was used in this study, the whole root system of three plants at each point were dug out to collect rhizosphere soil samples in each plot. The rhizosphere soil samples were collected using the shake method [52] at five different developmental stages: the seedling (V3-V4 stage, the third leaf exposes 3 cm from the heart), jointing (V7-V8 stage, the total length of the stem nodes is 2–3 cm), tasseling (Vn stage, the tip of the plant tassel exposes 3–5 cm of the parietal leaves), silking (R1 stage, the filaments of the female ears protrude about 2 cm from the bracts) and full ripe (R6 stage, dry and hard kernels) stages. After filtering through 40-mesh sieves to remove impurities, the pure soils from one plot were mixed together as a biological replicate. The soil samples were put into Ziplock bags, transferred to the laboratory with ice bags, and then each sample was divided into two parts: one ($\geq$120 g dry soils per sample) was air-dried indoors for the soil physicochemical properties determination, and the other part ($\geq$15 g soil per sample) was activated immediately at room temperature for 24 h for a microbial functional diversity test.

4.3. Determination of Physical and Chemical Properties of the Rhizosphere Soil

The soil pH value was measured using the potentiometric method [53]. The contents of soil total nitrogen, total phosphorus, organic matter, available phosphorus and alkali-hydrolyzed nitrogen were measured using the methods described in a previous study [54].

4.4. Determination of Enzyme Activities in the Rhizosphere Soil

The soil enzyme activity levels were tested in accordance with the producer manual instructions. The soil catalase activities were determined with soil catalase (S-CAT) activity detection kits (BC0105, Solerbio, Beijing, China). The soil urease activities were determined with soil urease (S-UE) activity detection kits (BC0125, Solerbio, Beijing, China). The soil sucrase activities (S-SC) were determined with soil sucrase (S-SC) activity detection kits (BC0245, Solerbio, Beijing, China). The soil alkaline phosphatase activities (S-AKP/ALP) were determined with soil alkaline phosphatase (S-AKP/ALP) activity detection kits (BC0285, Solerbio, Beijing, China). The soil acid phosphate activities (S-ACP) were determined with soil acid phosphatase (S-ACP) activity detection kits (BC0140, Solerbio, Beijing, China). The soil alkaline protease activities were determined with soil alkaline protease (S-ALPT) activity detection kits (BC0885, Solerbio, Beijing, China).

4.5. Determination of Microbial Functional Diversity in the Rhizosphere Soil

A Biolog Eco plate was used for the microbial functional diversity determination, and the Eco plate inoculum was prepared using the centrifugal decarbonization method [55]. The AWCD value of each well was calculated to reflect the overall metabolic activity of soil microorganisms to 31 carbon sources using the method described in a previous study [56]. The diversity index analysis of the soil microbial community was calculated after a 96 h incubation. Three diversity indices, Shannon index (H′), Simpson index (D) and McIntosh index (U), were used to evaluate the species richness, dominance and homogeneity, respectively. The calculation formulae were as follows:

$$H = -\sum Pi(\ln Pi) \tag{1}$$

$$D = 1 - \sum (Pi)^2 \tag{2}$$

$$U = (\sum (Ni)^2)^{\frac{1}{2}} \tag{3}$$

4.6. Data Processing

Data processing was performed using Microsoft Excel 2019 for data organization and graphing, and a GraphPad Prism (version 6.0) was used for the Student's *t*-test analyses.

5. Conclusions

This study provided comprehensive information on the effects of transgenic maize 2A-7 on the soil ecosystem. All the data implied that the changes in the physicochemical properties, enzyme activities and microbial diversities are directly related to the growth period but not to the presence of exogenous genes. Therefore, to evaluate whether a GM crop has impacts on its rhizosphere soil ecosystem, it is necessary to comprehensively consider multiple factors, such as its genetic background, planting environment and the presence of transferred or modified genes. Overall, the results suggest that transgenic maize 2A-7 expressing mCry1Ab and mCry2Ab proteins do not have significant impacts on soil physicochemical properties, enzyme activities and microbial functional diversity.

Author Contributions: Conceptualization, X.X. and X.L (Xingbo Lu).; methodology, S.Y., X.L. (Xin Liu) and C.H.; investigation, S.Y., H.S. and F.L.; data curation, X.L. (Xin Liu) and H.S.; writing—original draft preparation, X.L. (Xin Liu); writing—review and editing, X.X and X.L. (Xingbo Lu); funding acquisition, X.X. All authors have read and agreed to the published version of the manuscript.

Funding: This research was funded by the Agricultural Scientific and Technological Innovation project of Shandong Academy of Agricultural Sciences (No. CXGC2022C02); Genetically Modified Organisms Breeding Major Projects of China (No. 2019ZX08013004-002-005 and No. 2019ZX08013009-002-005).

Data Availability Statement: The data presented in this study are available in the graphs and tables provided in the manuscript.

Acknowledgments: We thank Liwen Bianji (Edanz) (www.liwenbianji.cn) and Hao Chen (North Carolina State University) for editing the language of a draft of this manuscript.

Conflicts of Interest: The authors declare no conflict of interest.

References

1. Anderson, J.A.; Ellsworth, P.C.; Faria, J.C.; Head, G.P.; Owen, M.D.K.; Pilcher, C.D.; Shelton, A.M.; Meissle, M. Genetically engineered crops: Importance of diversified integrated pest management for agricultural sustainability. *Front. Bioeng. Biotechnol.* **2019**, *7*, 24. [CrossRef]
2. Ramankutty, N.; Mehrabi, Z.; Waha, K.; Jarvis, L.; Kremen, C.; Herrero, M.; Rieseberg, L.H. Trends in global agricultural land use: Implications for environmental health and food security. *Annu. Rev. Plant Biol.* **2018**, *69*, 789–815. [CrossRef] [PubMed]
3. ISAAA. *Global Status of Commercialized Biotech/GM Crops in 2019 (ISAAA Brief No. 55)*; ISAAA: Ithaca, NY, USA, 2019.
4. Brookes, G.; Barfoot, P. Farm income and production impacts of using GM crop technology 1996–2016. *GM Crops Food* **2018**, *9*, 59–89. [CrossRef] [PubMed]
5. Campos, S.O.; Santana, I.V.; Silva, C.; Santos-Amaya, O.F.; Guedes, R.N.C.; Pereira, E.J.G. *Bt*-induced hormesis in *Bt*-resistant insects: Theoretical possibility or factual concern? *Ecotoxicol. Environ. Saf.* **2019**, *183*, 109577. [CrossRef] [PubMed]
6. Liu, T.; Chen, X.; Qi, L.; Chen, F.; Liu, M.; Whalen, J.K. Root and detritus of transgenic *Bt* crop did not change nematode abundance and community composition but enhanced trophic connections. *Sci. Total Environ.* **2018**, *644*, 822–829. [CrossRef]
7. Cheng, M.M.; Shu, Y.H.; Wang, J.W. Effect of *Bt* rice straw returning in soil on the growth and reproduction of *Eisenia fetida*. *Ying Yong Sheng Tai Xue Bao* **2016**, *27*, 3667–3674.

8. Zhang, W.; Cao, Z.; Wang, M.; Chen, X.; Wang, B. Absorption, translocation, and effects of *Bt Cry1Ac* peptides from transgenic cotton to the intercrops and soil functional bacteria. *Sci. Rep.* **2020**, *10*, 17294. [CrossRef]
9. Song, X.; Chang, L.; Reddy, G.V.P.; Zhang, L.; Fan, C.; Wang, B. Use of taxonomic and trait-based approaches to evaluate the effects of transgenic *Cry1Ac* corn on the community characteristics of soil collembola. *Environ. Entomol.* **2019**, *48*, 263–269. [CrossRef]
10. Fan, C.; Wu, F.; Dong, J.; Wang, B.; Yin, J.; Song, X. No impact of transgenic *cry1Ie* maize on the diversity, abundance and composition of soil fauna in a 2-year field trial. *Sci. Rep.* **2019**, *9*, 10333. [CrossRef]
11. Tabashnik, B.E.; Carrière, Y. Surge in insect resistance to transgenic crops and prospects for sustainability. *Nat. Biotechnol.* **2017**, *35*, 926–935. [CrossRef]
12. Li, J.; Shu, Y.; Wang, F.; Wang, J. Effects of Cry1Ab-expressing Bt rice straw return on juvenile and adult *Eisenia fetida*. *Ecotoxicol. Environ. Saf.* **2019**, *169*, 881–893. [CrossRef] [PubMed]
13. Xiao, Y.; Wu, K. Recent progress on the interaction between insects and *Bacillus thuringiensis* crops. *Philos. Trans. R Soc. Lond. B Biol. Sci.* **2019**, *374*, 20180316. [CrossRef] [PubMed]
14. Shu, Y.; Zhang, Y.; Zeng, H.; Zhang, Y.; Wang, J. Effects of Cry1Ab Bt maize straw return on bacterial community of earthworm *Eisenia fetida*. *Chemosphere* **2017**, *173*, 1–13. [CrossRef] [PubMed]
15. Liu, Y.; Li, J.; Luo, Z.; Wang, H.; Liu, F. The fate of fusion Cry1Ab/1Ac proteins from Bt-transgenic rice in soil and water. *Ecotoxicol. Environ. Saf.* **2016**, *124*, 455–459. [CrossRef]
16. Van Wyk, D.A.B.; Adeleke, R.; Rhode, O.H.J.; Bezuidenhout, C.C.; Mienie, C. Ecological guild and enzyme activities of rhizosphere soil microbial communities associated with Bt-maize cultivation under field conditions in North West Province of South Africa. *J. Basic Microbiol.* **2017**, *57*, 781–792. [CrossRef] [PubMed]
17. Rui, Y.K.; Yi, G.X.; Zhao, J.; Wang, B.M.; Li, Z.H.; Zhai, Z.X.; He, Z.P.; Li, Q.X. Changes of Bt toxin in the rhizosphere of transgenic Bt cotton and its influence on soil functional bacteria. *World J. Microbiol. Biotechnol.* **2005**, *21*, 1279–1284. [CrossRef]
18. Valldor, P.; Miethling-Graff, R.; Martens, R.; Tebbe, C.C. Fate of the insecticidal Cry1Ab protein of GM crops in two agricultural soils as revealed by ^{14}C-tracer studies. *Appl. Microbiol. Biotechnol.* **2015**, *99*, 7333–7341. [CrossRef]
19. Crecchio, C.; Stotzky, G. Insecticidal activity and biodegradation of the toxin from *Bacillus thuringiensis* subsp. Kustuki bound to humic acids from soil. *Soil Biol. Biochem.* **1998**, *30*, 463–470.
20. Li, F.; Wang, M.; Sun, H.W.; Yang, S.K.; Lu, X.B. Dynamics of Cry1Ab protein content in the rhizosphere soil and straw debris of transgenic Bt corn. *Chin. J. Appl. Ecol.* **2013**, *24*, 1907–1913.
21. Liu, W.; Zhao, H.; Miao, C.; Jin, W. Integrated proteomics and metabolomics analysis of transgenic and gene-stacked maize line seeds. *GM Crops Food* **2021**, *12*, 361–375. [CrossRef]
22. Yin, Y.; Xu, Y.; Cao, K.; Qin, Z.; Zhao, X.; Dong, X.; Shi, W. Impact assessment of *Bt* maize expressing the Cry1Ab and Cry2Ab protein simultaneously on non-target arthropods. *Environ. Sci. Pollut. Res. Int.* **2020**, *27*, 21552–21559. [CrossRef]
23. Mulder, C.; Wouterse, M.; Rutgers, M.; Posthuma, L. Transgenic maize containing the Cry1Ab protein ephemerally enhances soil microbial communities. *Ambio* **2007**, *36*, 359–361. [CrossRef]
24. Barriuso, J.; Valverde, J.R.; Mellado, R.P. Effect of Cry1Ab protein on rhizobacterial communities of *Bt*-maize over a four-year cultivation period. *PLoS ONE* **2012**, *7*, e35481. [CrossRef] [PubMed]
25. Baumgarte, S.; Tebbe, C.C. Field studies on the environmental fate of the *Cry1Ab Bt*-toxin produced by transgenic maize (MON810) and its effect on bacterial communities in the maize rhizosphere. *Mol. Ecol.* **2005**, *14*, 2539–2551. [CrossRef] [PubMed]
26. Li, Y.; Wang, C.; Ge, L.; Hu, C.; Wu, G.; Sun, Y.; Song, L.; Wu, X.; Pan, A.; Xu, Q.; et al. Environmental behaviors of *Bacillus thuringiensis* (*Bt*) insecticidal proteins and their effects on microbial ecology. *Plants* **2022**, *11*, 1212. [CrossRef] [PubMed]
27. Chen, Y.; Pan, L.; Ren, M.; Li, J.; Guan, X.; Tao, J. Comparison of genetically modified insect-resistant maize and non-transgenic maize revealed changes in soil metabolomes but not in rhizosphere bacterial community. *GM Crops Food* **2022**, *13*, 1–14. [CrossRef]
28. Liu, N.; Zhu, P.; Peng, C.; Kang, L.; Gao, H.; Clarke, N.J.; Clarke, J.L. Effect on soil chemistry of genetically modified (GM) vs. non-GM maize. *GM Crops* **2010**, *1*, 157–161. [CrossRef] [PubMed]
29. Sun, B.Y.; Li, Z.B.; Xiao, J.B.; Zhang, L.T.; Ma, B.; Li, J.M.; Cheng, D.B. Research progress on the effects of freeze-thaw on soil physical and chemical properties and wind and water erosion. *Ying Yong Sheng Tai Xue Bao* **2019**, *30*, 337–347. [PubMed]
30. Lei, L.; Guo, Q.S.; Wang, C.L.; Li, X.; An, J. Effect of compound planting on soil physical and chemical properties and soil enzyme activities of *Salvia miltiorrhiza*. *Zhongguo Zhong Yao Za Zhi* **2018**, *43*, 2480–2488.
31. Chen, Q.; Yang, B.; Liu, X.; Chen, F.; Ge, F. Long-term cultivation of Bt rice expressing the Cry1Ab/1Ac gene reduced phytoparasitic nematode abundance but did not affect other nematode parameters in paddy fields. *Sci. Total Environ.* **2017**, *607–608*, 463–474. [CrossRef]
32. Griffiths, B.S.; Heckmann, L.H.; Caul, S.; Thompson, J.; Scrimgeour, C.; Krogh, P.H. Varietal effects of eight paired lines of transgenic Bt maize and near-isogenic non-Bt maize on soil microbial and nematode community structure. *Plant Biotechnol. J.* **2007**, *5*, 60–68. [CrossRef] [PubMed]
33. Zeng, H.; Zhong, W.; Tan, F.; Shu, Y.; Feng, Y.; Wang, J. The influence of Bt maize cultivation on communities of arbuscular mycorrhizal fungi revealed by MiSeq sequencing. *Front. Microbiol.* **2018**, *9*, 3275. [CrossRef] [PubMed]
34. Chang, X.; Ning, D.; Mao, L.; Wang, B.; Fang, Q.; Yao, H.; Wang, F.; Ye, G. Metabolic analysis reveals *Cry1C* gene transformation does not affect the sensitivity of rice to rice dwarf virus. *Metabolites* **2021**, *11*, 209. [CrossRef] [PubMed]

35. Ahamd, M.; Abbasi, W.M.; Jamil, M.; Iqbal, M.; Hussain, A.; Akhtar, M.F.; Nazli, F. Comparison of rhizosphere properties as affected by different Bt- and non-Bt-cotton (*Gossypium hirsutum* L.) genotypes and fertilization. *Environ. Monit. Assess.* **2017**, *189*, 278. [CrossRef]
36. Du, J.; Hou, F.; Zhou, Q. Response of soil enzyme activity and soil bacterial community to PCB dissipation across different soils. *Chemosphere* **2021**, *283*, 131229. [CrossRef] [PubMed]
37. Li, Q.; Sun, Y.N.; Lin, L.; Li, Y.K.; Du, Y.G.; Guo, X.W.; Yang, Y.S.; Cao, G.M. Changes of soil enzyme activities and nutrients across different succession stages of grazing alpine Kobresia grassland. *Ying Yong Sheng Tai Xue Bao* **2019**, *30*, 2267–2274. [PubMed]
38. Feng, H.L.; Xu, C.S.; He, H.H.; Zeng, Q.; Chen, N.; Li, X.L.; Ren, T.B.; Ji, X.M.; Liu, G.S. Effect of biochar on soil enzyme activity & the bacterial community and its mechanism. *Huan Jing Ke Xue* **2021**, *42*, 422–432.
39. Liang, J.G.; Xin, L.T.; Luan, Y.; Song, X.Y.; Zhang, Z.G. Effect of Cry1Ie Bt maize on carbon source metabolism of rhizosphere microorganisms. *J. Agric. Sci. Technol. Iran* **2019**, *21*, 104–110.
40. Zeng, X.; Pei, T.; Song, Y.; Guo, P.; Zhang, H.; Li, X.; Li, H.; Di, H.; Wang, Z. A three-year plant study of salt-tolerant transgenic maize showed no effects on soil enzyme activity and nematode community. *Life* **2022**, *12*, 412. [CrossRef]
41. Wu, J.; Yu, M.; Xu, J.; Du, J.; Ji, F.; Dong, F.; Li, X.; Shi, J. Impact of transgenic wheat with *wheat yellow mosaic virus* resistance on microbial community diversity and enzyme activity in rhizosphere soil. *PLoS ONE* **2014**, *9*, e98394.
42. Trivedi, P.; Leach, J.E.; Tringe, S.G.; Sa, T.; Singh, B.K. Plant-microbiome interactions: From community assembly to plant health. *Nat. Rev. Microbiol.* **2020**, *18*, 607–621. [CrossRef] [PubMed]
43. Wu, N.; Shi, W.; Liu, W.; Gao, Z.; Han, L.; Wang, X. Differential impact of Bt-transgenic rice plantings on bacterial community in three niches over consecutive years. *Ecotoxicol. Environ. Saf.* **2021**, *223*, 112569. [CrossRef] [PubMed]
44. Li, P.; Li, Y.; Shi, J.; Yu, Z.; Pan, A.; Tang, X.; Ming, F. Impact of transgenic Cry1Ac + CpTI cotton on diversity and dynamics of rhizosphere bacterial community of different root environments. *Sci. Total Environ.* **2018**, *637–638*, 233–243. [CrossRef] [PubMed]
45. Xu, Q.; Dai, R.; Ruan, Y.; Rensing, C.; Liu, M.; Guo, S.; Ling, N.; Shen, Q. Probing active microbes involved in Bt-containing rice straw decomposition. *Appl. Microbiol. Biotechnol.* **2018**, *102*, 10273–10284. [CrossRef] [PubMed]
46. Chen, J.Y.; Liu, R.H.; Wang, L.L.; Li, J.; Li, G.; Yang, D.L.; Zhao, J.N. Effects of continuous planting of phosphorus-efficient transgenic rice on soil microbial functional diversity. *J. Tianjin Agric. Univ.* **2019**, *26*, 1–7.
47. Sun, H.W.; Zhang, Y.X.; Xu, X.H.; Gao, R.; Li, F.; Yang, S.K.; Lu, X.B. Effects of transgenic maize variety (double resistance 12-5) on enzyme activity and microbial diversity in rhizosphere soil. *Soils* **2019**, *51*, 61–67.
48. Jiang, Z.; Zhou, L.; Wang, B.; Wang, D.; Wu, F.; Yin, J.; Song, X. Toxicological and biochemical analyses demonstrate no toxic effect of Bt maize on the *Folsomia candida*. *PLoS ONE* **2020**, *15*, e0232747. [CrossRef]
49. Zhang, Y.; Zheng, Y.; Hu, J.; Du, N.; Chen, F. Functional diversity of the microbial community in healthy subjects and periodontitis patients based on sole carbon source utilization. *PLoS ONE* **2014**, *9*, e91977.
50. Liu, W.; Pan, N.; Chen, W.; Jiao, W.T.; Wang, M.E. Effect of veterinary oxytetracycline on functional diversity of soil microbial community. *Plant Soil Environ.* **2012**, *58*, 295–301. [CrossRef]
51. Xu, X.H.; Guo, Y.; Sun, H.; Li, F.; Yang, S.; Gao, R.; Lu, X. Effects of phytase transgenic maize on the physiological and biochemical responses and the gut microflora functional diversity of *Ostrinia furnacalis*. *Sci. Rep.* **2018**, *8*, 4413. [CrossRef]
52. Riley, D.; Barber, S.D. Salt accumulation at the soybean (*Glycine max* (L.) Merr.) root-soil interface. *Soil Sci. Soc. Am. J.* **1970**, *34*, 154–155. [CrossRef]
53. Hong, S.; Gan, P.; Chen, A. Environmental controls on soil pH in planted forest and its response to nitrogen deposition. *Environ. Res.* **2019**, *172*, 159–165. [CrossRef] [PubMed]
54. Chen, S.; Rong, Y.; Su, Y.; Du, Z. Characteristics of soil respiration in farmland of Hexi oasis and its response to long-term fertilization. *J. Desert Res.* **2022**, *42*, 178–186.
55. Hughes, J.B.; Bohannan, B.J.M. Section 7 update: Application of ecological diversity statistics in microbial ecology. *Mol. Microb. Ecol. Man.* **2004**, *7*, 1321–1344.
56. Preston-Mafham, J.; Boddy, L.; Randerson, P.F. Analysis of microbial community functional diversity usingsole-carbon-source utilisation profiles—A critique. *FEMS Microbiol. Ecol.* **2002**, *42*, 1–14.

plants
MDPI

Article

Impact of Insect-Resistant Transgenic Maize 2A-7 on Diversity and Dynamics of Bacterial Communities in Rhizosphere Soil

Xiaohui Xu [1,2], Xin Liu [1,2,3], Fan Li [1,2], Chaofeng Hao [1,2], Hongwei Sun [1,2], Shuke Yang [1,2], Yue Jiao [4,*] and Xingbo Lu [1,2,3,*]

1 Shandong Key Laboratory of Plant Virology, Institute of Plant Protection, Shandong Academy of Agricultural Sciences, Jinan 250100, China; xuxiaohui1023@163.com (X.X.); liu18753136362@163.com (X.L.); lfzjnd@163.com (F.L.); chaofenghao2015@163.com (C.H.); hongweisun@126.com (H.S.); yangshuke426@126.com (S.Y.)
2 College of Life Sciences, Shandong Normal University, Jinan 250014, China
3 Development Center for Science and Technology, Ministry of Agriculture and Rural Affairs, Beijing 100176, China
4 Key Laboratory for Safety Assessment (Environment) of Agricultural Genetically Modified Organisms, Ministry of Agriculture and Rural Affairs, Jinan 250100, China
* Correspondence: jiaoyue@agri.gov.cn (Y.J.); luxb99@sina.com (X.L.)

Abstract: Artificial modification of *Bacillus thuringiensis* (Bt) proteins can effectively improve their resistance to target pests, but the effect of such modification on the diversity of rhizosphere microorganisms remains unclear. Transgenic maize 2A-7 contains two artificially modified Bt proteins, mCry1Ab and mCry2Ab. These proteins can enter soil and pose a potential threat to soil microbial diversity. To assess their impacts on rhizosphere bacteria communities, the contents of the two Bt proteins and changes in bacterial community diversity in the rhizosphere soils of transgenic maize 2A-7 and its control variety were analyzed at different growth stages in 2020. The results showed that the two Bt proteins were detected at low levels in the rhizosphere soils of 2A-7 plants. No significant differences in soil bacterial diversity were detected between 2A-7 and its control variety at any of the growth stages. Bioinformatics analysis indicated that the growth stage, rather than the cultivar, was the main factor causing changes in bacterial communities. This research provides valuable data for understanding the impact of Bt crops on the soil microbiome, and establishes a theoretical basis for evaluation of their safety.

Keywords: insect-resistant transgenic maize; artificially modified Bt protein; bacterial community diversity; rhizosphere soil; risk assessment

Citation: Xu, X.; Liu, X.; Li, F.; Hao, C.; Sun, H.; Yang, S.; Jiao, Y.; Lu, X. Impact of Insect-Resistant Transgenic Maize 2A-7 on Diversity and Dynamics of Bacterial Communities in Rhizosphere Soil. *Plants* **2023**, *12*, 2046. https://doi.org/10.3390/plants12102046

Academic Editors: C. Neal Stewart Jr. and Wei Wei

Received: 9 April 2023
Revised: 16 May 2023
Accepted: 17 May 2023
Published: 21 May 2023

1. Introduction

Since the approval of genetically modified (GM) crops for commercialization in the United States in 1996, the global area of GM crops has expanded by more than 110 times, reaching 190.4 million hectares in 2019 [1]. GM crops have been planted or imported in more than 70 countries [2]. The rapid adoption of GM crops has brought enormous economic, social, and environmental benefits [3]. Currently, approximately 50% of commercialized GM crops contain insect-resistant traits. These insect-resistant GM crops can produce at least one *Bacillus thuringiensis* (Bt) toxin protein, which can directly kill target pests without the need for pesticide spraying throughout the whole growth period. The application of insect-resistant GM crops can not only improve the yield and safety of crops, but it can also save on labor input, promote the development of low-or no-tillage systems, and reduce greenhouse gas emissions. Thus, it is beneficial for the sustainable development of society and for the environment [4,5].

Previous studies have shown that the natural Bt toxin produced by *Bacillus thuringiensis* has no adverse impact on the environment [6]. GM crops expressing artificially modified

Bt toxin proteins have strong toxicity and a broad insect resistance spectrum. In general, they target insects belong to the Lepidoptera, Diptera, and Coleoptera families; however, they also exhibit a certain level of toxicity to some non-target organisms, such as peach aphids [7]. During different growth stages, and after returning straw to the field, Bt proteins in GM crops enter the soil through exudates, pollen, and residues, which can be adsorbed by active particles in the soil and remain in the soil for a long time [8,9]. In soil, most Bt proteins form complexes with other soil factors, and their content levels are closely related to soil conditions such as type, temperature, pH, water content, and active particles [10]. During the adsorption process, protein concentration in soil usually exhibits a trend of first increasing, then decreasing, and finally reaching a steady state [11]. The content levels of Bt proteins and the activities of microbial communities are both elevated in rhizosphere soil compared with bulk soil. For this reason, changes in the structure and diversity of the rhizosphere microorganisms are often used as an early and effective indicators for assessing the effects of GM crops on soil ecology [12–14].

In the past ten years, a number of studies have been carried out to assess the biosafety of Bt crops with respect to soil microbial communities; however, their conclusions have been controversial. Two studies found that the planting of two Bt maize crops (Bt11 and Mon810) significantly altered the compositon, diversity and richness of fungal communities in rhizosphere soils [15,16]. In addition, transgenic cotton expressing the Cry1Ac protein was found to increase the functional diversity of soil microbial communities [17]. However, most Bt crops have been found to have no significant effects on soil microbial communities; these include Cry1Ac transgenic eggplant [18], transgenic maize Mon810 [19], Bt rice lines Huachi B6 and TT51 [20], and Cry3Bb maize MON863 [21]. Thus, studies on the impacts of GM crops on the soil ecosystem should fully consider the specificity of each transgenic plant, and be carried out on a "case-by-case" basis.

Various methods have been adopted for assessing the impact of GM crops on soil microbial diversity, such as plate colony-counting, Biolog, denaturing gradient gel electrophoresis (DGGE), and high-throughput sequencing [22–24]. The first three methods have obvious limitations, including high workloads, heavy time consumption and an inability to comprehensively reflect the composition, structure, and abundance of the microbial communities in the samples [25,26]. With the rapid development of high-throughput sequencing technology, this method is now widely used in environmental microbial diversity analyses [27,28]. Compared with traditional methods, high-throughput sequencing has a number of advantages, including low detection limits, high sensitivity, the absence of any culturing requirements, and the ability to systematically and rapidly evaluate the composition of samples (including known and unknown species), as well as structural variations in microbial communities. The content levels of microorganisms in rhizosphere soil are several times or even dozens of times higher than those in root-zone soil; among these microorganisms, bacteria, especially Gram-negative bacteria, are the most prevalent of all. Therefore, evaluating changes in the structure and diversity of the bacterial communities in rhizosphere soils can reflect, to a great extent, changes in microbial communities. The relative molecular weight of 16S rDNA is moderate, and the probability of its mutation is low. It consists of ten conservative regions and nine hypervariable regions [29]. Sequencing analysis of the hypervariable regions can easily distinguish different bacterial species. Therefore, 16S rDNA sequencing is widely used to evaluate microbial phylogeny, classification, and diversity [30–32].

In this study, to clarify the impact of transgenic maize 2A-7 expressing mCry1Ab and mCry2Ab genes on soil bacterial communities, the content levels of the two Bt proteins were measured in the rhizosphere soil of transgenic maize 2A-7 at different development stages in the summer of 2020. Then, the abundance, composition, and diversity of bacterial communities in the rhizosphere soil of transgenic maize 2A-7 and its control varieties were analyzed using 16s rDNA sequencing. A correlation analysis was conducted between changes in the bacterial community and changes in Bt protein content levels and the physicochemical properties [33] of rhizosphere soils, and the main factors leading to changes in bacterial communities were then identified. The results from the one-year field trial can provide necessary information for analyzing the impact of transgenic insect-resistant maize on soil bacterial communities, and provide a theoretical foundation for the biosafety regulation and commercialization of transgenic maize 2A-7.

2. Results

2.1. Changes of Bt Protein Contents in Rhizosphere Soil of Transgenic Maize 2A-7

To investigate the accumulation characteristics of the two modified Bt proteins (mCry1Ab and mCry2Ab) in the rhizosphere soil of transgenic maize 2A-7, the content levels of the two Bt proteins were measured with Cry1Ab and Cry2A Elisa kits, respectively. As shown in Figure 1, the two Bt proteins were detected in the rhizosphere soil of transgenic maize 2A-7 at relatively low levels, ranging from 0.15 ng/g to 1.16 ng/g (Figure 1). These values were less than one-thousandth of those of the two proteins expressed in the plant. Compared with the mCry2Ab protein, the content of mCry1Ab protein fluctuated greatly among different growth stages. The mCry1Ab content was highest at the seedling and silking stages, and lowest at the full ripening stage (Figure 1). The content of mCry2Ab was relatively stable at each growth stage, showing a slightly decreasing trend as the developmental process proceeded (Figure 1).

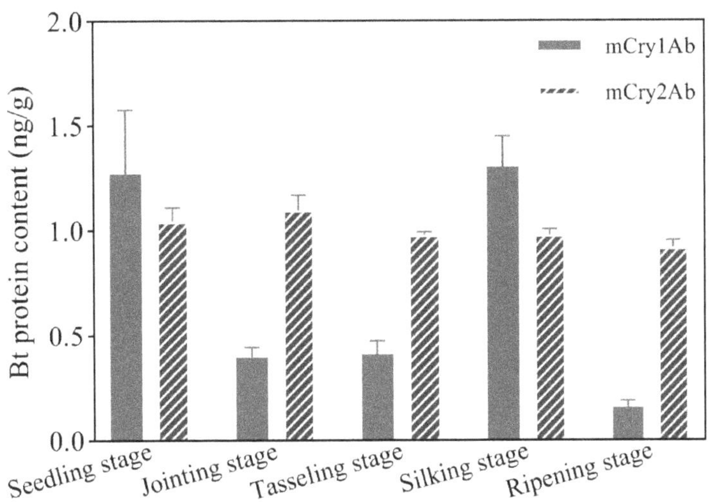

Figure 1. Content levels of mCry1Ab and mCry2Ab proteins in the rhizosphere soil of transgenic maize 2A-7 at different growth stages. Data are presented as the means ± standard deviations of three biological replicates.

2.2. Alpha Diversity of Rhizosphere Bacterial Community of Transgenic Maize 2A-7 and Its Control

Alpha diversity reflects the richness and diversity of species in microbial communities. In the current study, three indices (ACE, Simpson and Shannon) were used to identify differences in the alpha diversity of rhizosphere bacterial communities between transgenic maize 2A-7 and its control. As shown in Figure 2, the richness estimator ACE and the diversity indices Simpson and Shannon were not significantly different between the transgenic maize 2A-7 and its control at the same developmental stages. These results suggest that the artificial modification of Bt proteins in maize does not affect the diversity of rhizosphere communities.

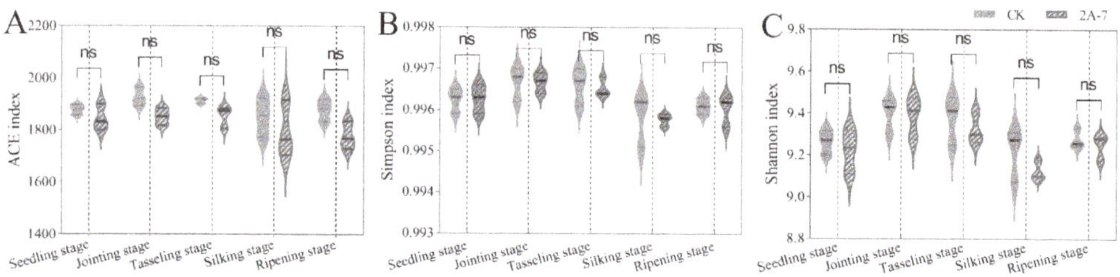

Figure 2. Comparison of three alpha diversity indices of rhizosphere bacterial communities from transgenic maize 2A-7 and its control. (**A**) ACE index; (**B**) Simpson index; (**C**) Shannon index. Significant difference analyses were carried out using a one way *t*-test and ANOVA. "ns" indicates no significant difference ($p > 0.05$). CK, non-transgenic control Dongdan 6531.

2.3. Beta Diversity of Rhizosphere Bacterial Community of the Two Maize Lines

To further analyze the impact of transgenic maize 2A-7 on the bacterial communities in rhizosphere soil, a principal co-ordinate analysis (PCoA) was conducted. Using different growth stages and maize inbred lines as two variables, the analysis revealed that bacterial communities in the rhizosphere soils of different maize varieties at different growth stages were significantly different ($p = 0.001$) (Figure 3A). To explain the key factors causing these differences, we separately analyzed the impact of developmental periods and different maize varieties on the diversity of bacterial communities. The results showed that the difference in maize varieties did not cause significant change in bacterial communities (Figure 3B). However, when the developmental stage was taken as the only variable, significant differences were observed in the bacterial communities among different development stages ($p = 0.001$) (Figure 3C). Specifically, PCoA revealed that there was no significant difference in the beta diversity of bacterial communities in the rhizosphere soil of the two maize varieties during the five growth stages (Figure 3D–H). Therefore, we suggest that the main factor affecting the changes in the diversity of rhizosphere bacterial communities is the growth stage rather than the variety of maize.

2.4. Bacterial Community Composition

The bacterial communities in the rhizosphere soil of transgenic maize 2A-7 and the control maize Dongdan 6531 are primarily clustered in 38 phyla, 93 classes, 186 orders, 272 families, 430 genera, and 528 species. The predominant bacteria at the phylum and genus levels were selected for further analysis based on a threshold of relative abundance > 1%. We found that Proteobacteria, Acidobacteria, Bacteroides, Gemmatimonadetes, Planctomycetes, and Verrucomicrobia were the dominant bacteria in the rhizosphere soils of the two maize varieties, accounting for over 80% of the relative bacterial abundance in total (Figure 4A). On the genus level, most of the bacteria could not be cultured; however, Nitrospira, Sphingomonas, MND1, and RB41 were the main known bacterial genera for the two maize varieties (Figure 4B). Overall, we found no significant difference in the bacterial community's composition between the two maize species at either the phylum or genus level (Figure 4).

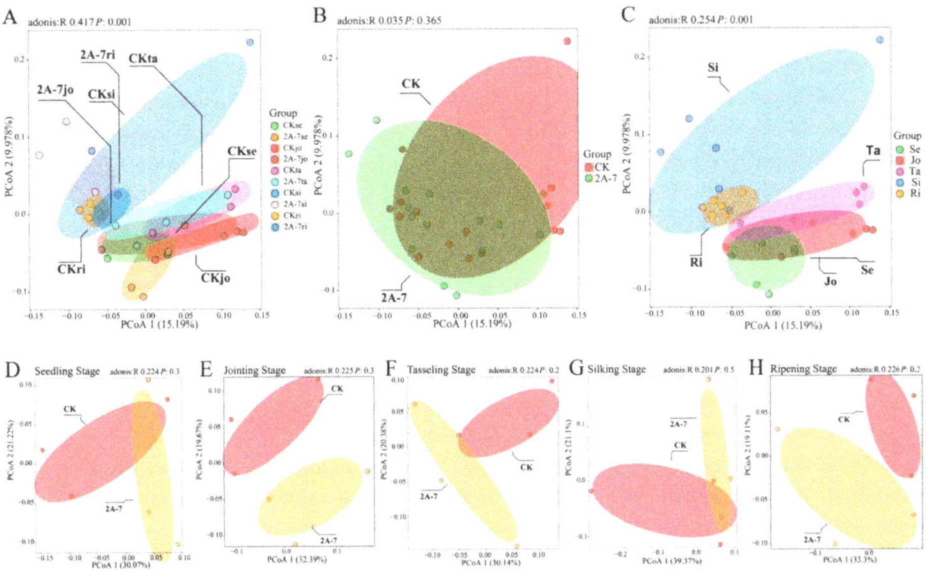

Figure 3. Principal co-ordinate analysis (PCoA) based on UniFrac distances of bacterial communities in the rhizospheres of transgenic maize 2A-7 and the control maize cultivar at different growth stages. (**A**) PCoA using two variables: different growth stages and maize inbred lines; (**B**,**C**) PCoA using one variable: different maize inbred lines (**B**) or different growth stages (**C**); (**D**–**H**) PCoA using different maize inbred lines as a variable at seedling stage (**D**), jointing stage (**E**), tasseling stage (**F**), silking stage (**G**) and ripening stage (**H**). The adonis R value represents the overall variation that can be explained by a certain grouping pattern in the PCoA analysis. The *p* value represents the significant difference based on Student's *t*-test ($p < 0.05$). CK, non-transgenic control Dongdan 6531. Se—seedling stage; Jo—jointing stage; Ta—tasseling stage; Si—silking stage; Ri—ripening stage.

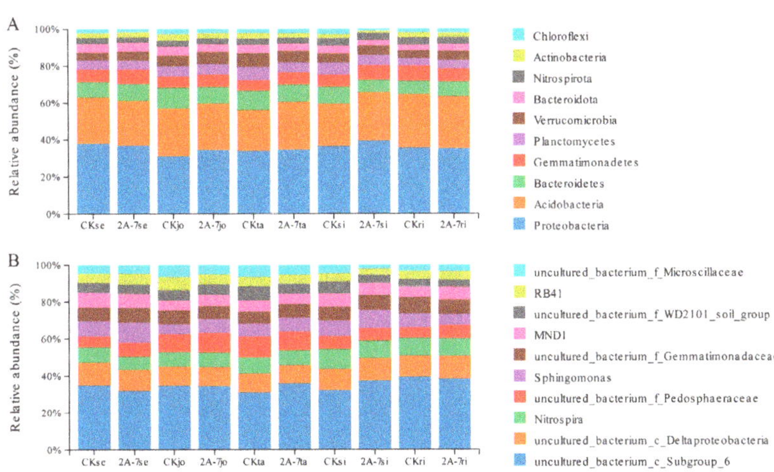

Figure 4. The composition of bacterial communities in the rhizosphere soils of transgenic maize 2A-7 and non-transgenic control Dongdan 6531. (**A**) The relative abundance of the top 10 phyla; (**B**) the relative abundance of the top 10 genera. CK, non-transgenic control Dongdan 6531. Se—seedling stage; Jo—jointing stage; Ta—tasseling stage; Si—silking stage; Ri—ripening stage.

2.5. Co-occurrence Network Analysis of Maize Rhizosphere Microbiome

To analyze the impact of different developmental stages on bacterial communities in maize rhizosphere soil, we constructed five co-occurrence networks based on OTU data from different growth stages of the transgenic maize 2A-7 and its control Dongdan 6531 (Figure 5). In general, the complex of the rhizosphere bacterial communities was similar at different stages, as networks possessed similar numbers of nodes and edges (Figure 5). As OTUs belonging to the same phylum were marked with the same color, we could easily observe dynamic changes in the dominant bacteria in rhizosphere soils at different developmental stages (Figure 5). The co-occurrence was most stable at the jointing stage, as reflected by the largest number of nodes and edges, as well as minimum values for connectance, clustering coefficients, and degree centralization (Table 1). The microbiome network had the smallest diameter and betweenness centralization at the seedling stage, while the highest values for degree centralization were recorded at the silking and ripening stages (Table 1).

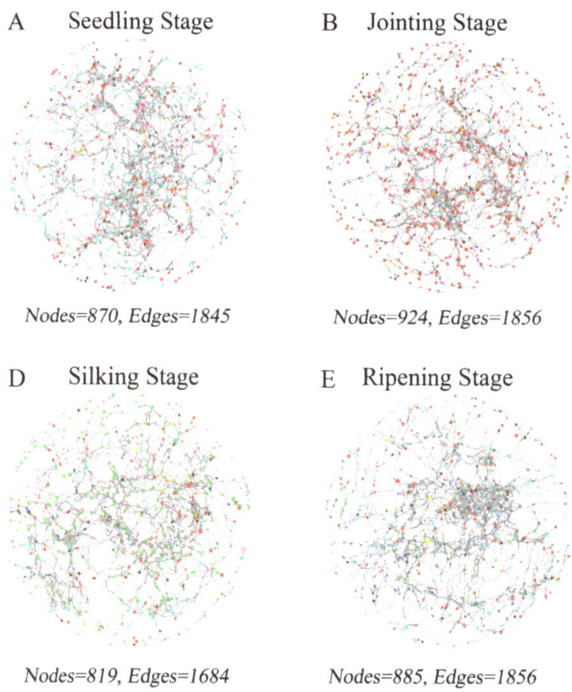

Figure 5. Networks of bacterial OTUs at different developmental stages in the two maize cultivars. OTUs in the same phylum are marked in the same color. (**A**) Seedling stage; (**B**) Jointing stage; (**C**) Tasseling stage; (**D**) Silking stage; (**E**) Ripening stage.

Table 1. The parameters of the bacterial OTU networks at five different stages.

Period	Connectance	Average Degree	Average Path Length	Diameter	Clustering Coefficient	Betweenness Centralization	Degree Centralization
Seedling Stage	0.0049	4.24	9.61	23	0.43	0.053	0.014
Jointing Stage	0.0044	4.02	10.43	33	0.39	0.085	0.010
Tasseling Stage	0.0043	3.84	10.30	28	0.40	0.073	0.008
Silking Stage	0.0050	4.11	10.61	30	0.43	0.079	0.018
Ripening Stage	0.0047	4.19	11.00	39	0.41	0.080	0.016

2.6. Linear Discriminate Analysis Effect Size (LEfSE) Analysis of Maize Rhizosphere Microbiome

LEFSE analysis (LDA > 3) was conducted on the genus level to identify the bacterial biomarkers causing the composition changes in the rhizosphere bacteria of transgenic maize 2A-7 and its near-isogenic control Dongdan 6531 at different developmental stages. Comparing the jointing and seedling stages, we found that Sphingomonas and Lysobacter were the dominant bacteria in the seedling stage, while the genus Luteolibacter was significantly enriched at the jointing stage (Figure 6A). In the comparison between the tasseling and silking stages, the dominant bacteria belonged to the RB41, Gemmata and Luteolibacterde genera, respectively (Figure 6B). Comparing the tasseling and silking stages, we found that the dominant bacteria in the tasseling stage were of the RB41, Bacillus and Bryobacter genera, while bacteria belonging to the Ramlibacter, Elin 6067, MND1, Sphingomonas, and Neurospora genera were most abundant at the silking stage (Figure 6C). A comparison between the silking and ripening stages revealed that bacteria in the Sphingomonas, Gemmata, Luteolibacter, Lysobacter and Massilia genera were enriched at the silking stage, while the predominant bacteria at the ripening stage were Actinobacteria, Holophagae, and Acidimicrobiia (Figure 6D).

2.7. Redundancy Analysis (RDA) of Bacteria Community Diversity with the Environmental Factors

To identify the relationships between the diversity of bacterial communities and other environmental factors, an RDA analysis was performed. The changes in two environmental factors, as well as the content levels of two modified Bt proteins and certain physicochemical properties [33], were used in this analysis. The results of the RDA analysis revealed that the pH level and available phosphorus content were the two most important environmental factors affecting the diversity of bacterial communities in the rhizosphere soils of the two maize varieties (Figure 7).

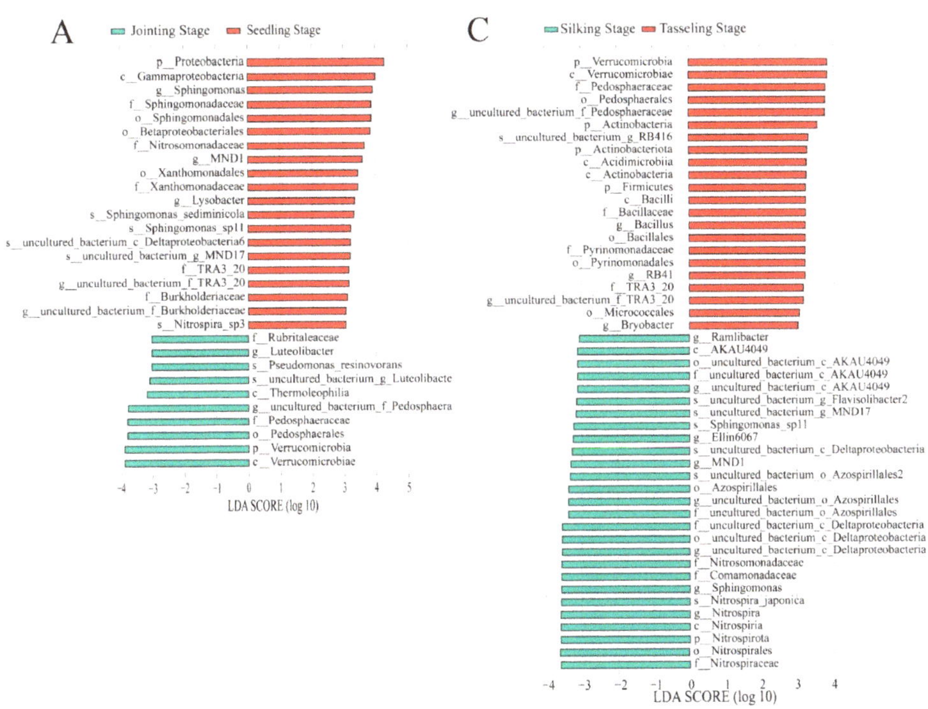

Figure 6. *Cont.*

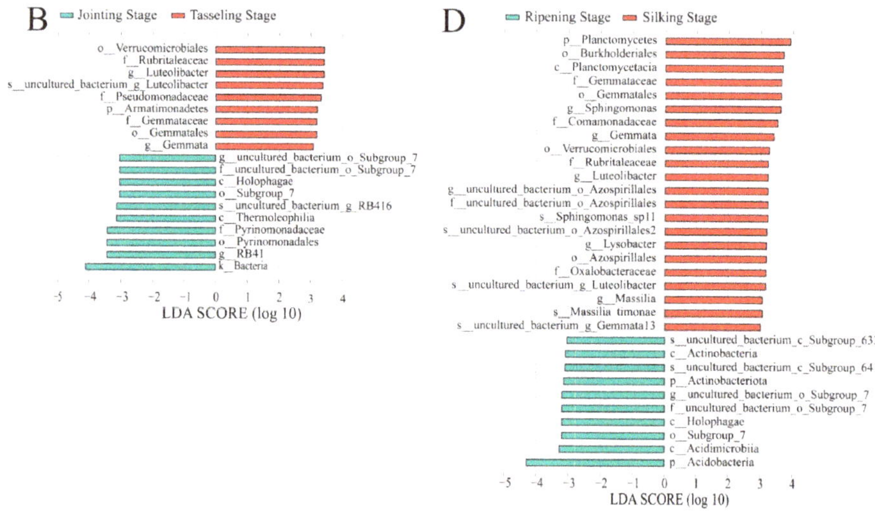

Figure 6. LDA scores of discriminative bacteria in the rhizosphere samples between two adjacent growth periods. (**A**) Jointing stage vs. seedling stage; (**B**) Tasseling stage vs. jointing stage; (**C**) Silking stage vs. tasseling stage; (**D**) Ripening stage vs. silking stage.

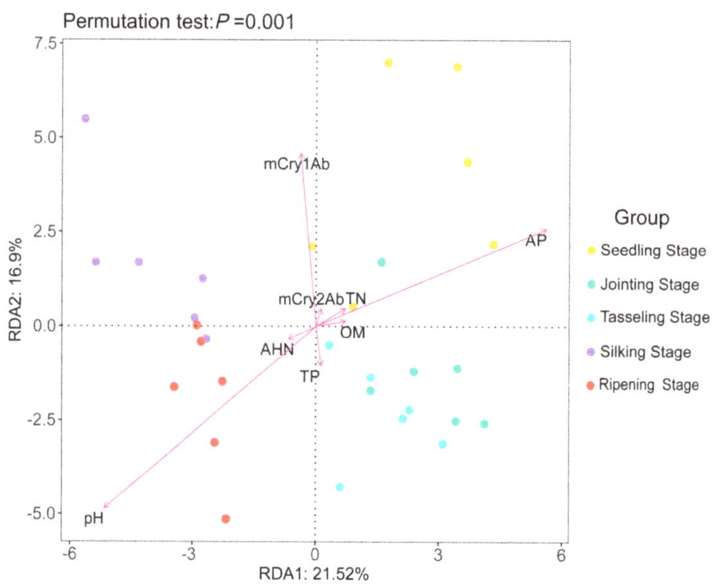

Figure 7. Redundancy analysis (RDA) of dynamic changes in bacteria communities in response to environmental factors. The arrows represent the environmental factors. TN—total nitrogen; AP—available phosphorus; TP—total phosphorus; AHN—alkaline hydrolyzable nitrogen; OM—organic matter. The length of the arrow line represents the degree of correlation; the longer the line, the greater the correlation.

3. Discussion

Rhizosphere soil microorganisms, consisting mainly of bacteria, are microorganisms which closely attach to rhizosphere soil particles. The amounts of bacteria in rhizosphere soil are much higher than in bulk soil, sometimes ten or more times higher. Such microorganisms establish a balanced relationship that is interdependent and mutually reinforcing with the roots of their host plant. Rhizosphere microorganisms can convert inorganic substances into organic substances and secrete growth-stimulating factors, thereby providing sufficient nutrients for the growth and development of plants. In turn, the plant roots provide suitable conditions and sources of nutrients and energy for rhizosphere microorganisms [34,35]. The microbial communities of rhizosphere soils can be influenced by various factors, including species, genotype, pH, and others [36,37]. The genomes of transgenic crops contains exogenous insertion fragments capable of producing proteins from other species, and these proteins can be released into the soil through root exudates, pollen, and residues [38–40]. Different transgenic crops exhibit different accumulation patterns of exogenous proteins in the soil during different growth stages [8,41–44]. In this study, the two modified Bt proteins mCry1Ab and mCry2Ab were artificially synthesized by optimizing the codons of the original proteins Cry1Ab and Cry2Ab, respectively [45]. mCry1Ab contains three classic structural domains of Cry1Ab protein, which are closely related to its toxicity, but missing the C-terminal fragment with 531 amino acid residues. The mCry2Ab protein showed much higher similarity to the Cry2Ab protein (99.68%), which was obtained only through codon optimization at two specific sites [45]. The contents of the two modified Bt proteins in the rhizosphere soil of transgenic maize 2A-7 were detected at different development stages in the Summer of 2020. A slight difference between the change trends of the two Bt proteins was found, i.e., the content of mCry1Ab was highest in the rhizosphere soil at the seedling and silking stages, while the content of mCry2Ab remained relatively stable during all growth periods (Figure 1). This difference may be related to the characteristics (stability and water-solubility, etc.) of the two Bt proteins, and the soil conditions (type, pH, temperature, water content, and other soil physicochemical properties, etc.), as has been observed by other researchers [8,10,43,44].

As concerns over the ecological safety of GM crops continue to grow, more studies have been conducted on the impact of GM crops on soil ecosystems. Most studies have shown that changes in the microbial communities in rhizosphere soil are closely related to environmental factors and seasonal changes, but not to the planting of transgenic GM crops [18,28,30,39,46]. Compared with the effects of seasonal transition, the planting of Cry1Ac transgenic eggplant produces only a minor impact on soil bacterial communities [18]. The field residue of Cry3Bb transgenic corn MON863 has no impact on changes in bacterial communities in soil, but does have a small impact on fungal communities, mainly due to changes in environmental factors, which are not related to differences in the variety of corn [39]. In other studies, researchers have found that the planting of Cry1Ah transgenic maize HGK60 produces no significant differences in the composition and diversity of bacterial communities compared to controls at the same growth stage, but differences have been identified among different growth periods [30,47]. In the current study, a one-year field trial was adopted to investigate the impact of the transgenic maize 2A-7 on the composition and diversity of microbial communities. High-throughput 16S rDNA sequencing was used to compare the microbiomes of rhizosphere soils from transgenic maize 2A-7 and its control line. The alpha and beta diversity analyses of the bacterial communities in the rhizosphere soils of transgenic maize 2A-7 and its control lines revealed no significant differences between the two maize cultivars during the same growth periods (Figures 2 and 3). However, samples at different developmental stages were distinct (Figure 3C). Therefore, developmental stages, rather than the cultivars, are the most important factor in determining the diversity of bacteria communities in rhizosphere soils. This finding is consistent with the results of most previous studies. Composition analysis at the phylum and genus level also revealed that the rhizosphere soils of the two maize cultivars in the current study had similar bacterial compositions at the same developmental stages (Figure 4), while soils at different stages had dominant bacterial species which were distinct (Figure 6).

Co-occurrence network analysis can reveal non-random associations between soil microorganisms. Ecological rules for microbial community structure and assembly have been gradually applied to the diversity analysis of various soil microorganisms [48]. Chen et al. found that with increases in pH value, the relationship between microbial communities in soil becomes closer [49]. Using co-occurrence analysis, Fang et al. found that microbial communities have a more complex and stable structure in summer compared with winter [50]. In the current study, the distribution patterns of bacteria in the rhizosphere soils of two maize cultivars at different stages were analyzed using a co-occurrence network analysis. It was found that the complexity of bacterial communities was similar at each developmental stage, but the centralization of bacterial communities was much higher at later developmental stages (Figure 5). Further LEFSE analysis showed that microbial community markers were different at adjacent developmental stages (Figure 6).

In a previous study, we found that there were no significant differences in the physicochemical properties and key soil enzyme activities of the rhizosphere soils of transgenic maize 2A-7 and its control line at the same developmental stage [33]. Compared to environmental factors, the growth period seemed to have the most significant impact on the bacterial communities, and materials from different growth periods distributed distinctly in different quadrants (Figure 7). In the present work, a redundancy analysis of dynamic changes in bacterial communities with environmental factors showed that changes in bacterial communities were significantly correlated with the pH value and the available phosphorus content (Figure 7). This finding is consistent with that of Li et al., who found that bacterial community composition was more closely correlated with soil pH values than with content levels of Bt protein [37]. In addition, we also found that mCry1Ab seemed to have a correlation with bacterial community changes (Figure 7). As mCry1Ab protein content has a strong positive correlation with the seedling and silking stages (Figures 1 and 7), we speculate that the impact of mCry1Ab protein on the rhizosphere soil bacterial communities partially stems from the growth period. Taking into account other environmental and growth period factors, mCry1Ab's impact on the soil bacterial community is not prominent; it is, at least, smaller than the impact of the growth period, pH and available phosphorus content.

Overall, we may state that there were no significant differences in the composition and diversity of bacterial communities in the rhizosphere soils of transgenic insect-resistant maize 2A-7 and its control Dongdan 6531 observed in the one-year field trial. The growth stage, rather than the Bt protein, is the main factor affecting changes in bacterial communities.

4. Materials and Methods

4.1. Plant and Soil Materials

Insect-resistant maize 2A-7 and the near-isogenic non-transgenic maize Dongdan 6531 were provided by China Agricultural University [33,45]. The two maize cultivars were planted in Jinan, Shandong, China (N 36°41′50″, E 117°04′33″) in the summer of 2020, using a randomized block design. Three replicates were given for each maize cultivar. The experimental site has cinnamon soil, which is neutral to weakly alkaline with middle level fertilizer. Seven days before planting, 3000 kg/acre of organic fertilizer and 30 kg/acre of compound fertilizer containing nitrogen (N), phosphorus (P), and potassium (K) were buried in the soil at a depth of 18 cm. During the entire growth period, no insecticides and herbicides were sprayed, but artificial weeding was carried out when necessary. Samples were collected from plant rhizosphere soils at five developmental stages, as described in our previous study [33], and then quickly transferred to liquid nitrogen prior to storage at −80 °C in a refrigerator in the laboratory.

4.2. Bt protein Content Determination

Cry1Ab was tested using the Cry1Ab/Cry1Ac protein ELISA kit (AP 003 CRBS, Envirologix, Portland, USA). Cry2Ab was tested using the Cry2A protein ELISA kit (AP 005 CRBS, Envirologix, Portland, USA). Standard curves were drawn based on the OD values of standard proteins at different concentrations to calculate the content of Bt proteins in each sample.

4.3. DNA Extraction and Sequencing

The DNA of the rhizosphere soil samples was extracted using a Soil Genomic DNA Extraction Kit (DP336, Tiangen, Beijing, China) following the manufacturer's instructions. The DNA purity and concentration of all the samples was measured with a Qubit dsDNA HS Assay Kit and Qubit 4.0 Fluorometer (Thermo Fisher Scientific, Waltham, MA, USA).

The universal primers (27F: AGRGTTTGATYNTGGCTCAG and 1492R: TASGGHTACCTTGTTASGACTT) were used to amplify the full-length 16S rRNA gene from the genomic DNA of each sample. To meet the multiplex sequencing requirement, both the forward and reverse 16S primers were tailed with sample-specific PacBio barcode sequences. We used a KOD One^{TM} PCR Master Mix (KMM-101, Toyobo, Osaka, Japan) for PCR amplification with the following program: 2 min at 95 °C for initial denaturation, followed by 25 cycles of 10 s at 98 °C for denaturation, 30 s at 55 °C for annealing, and, finally, 2 min at 72 °C for extension. All the PCR productions were purified using Agencourt AMPure XP Beads (Beckman Coulter, Indianapolis, IN, USA) and quantified with a Qubit dsDNA HS Assay Kit and Qubit 4.0 Fluorometer (Thermo Fisher Scientific, Oregon, OR, USA). SMRTbell libraries were established from the amplified DNA with a SMRTbell Express Template Prep Kit 2.0, following the constructions of the manufacturer. Purified SMRTbell libraries were then sequenced on a single PacBio Sequel II system.

4.4. Bioinformatic Analysis

The BMK Cloud (Biomarker Technologies, Beijing, China) was used to carry out most of the bioinformatic analysis. Raw data were filtered and demultiplexed using the SMRT Link software with the minPredicted Accuracy $\geq$ 0.9 and the minPasses $\geq$ 5 as the criteria to obtain the circular consensus sequencing (CSS) reads. After low-quality reads were eliminated through filtering, trimming and duplication deletion, unique reads were obtained. Sequences with a similarity of $\geq$97% were clustered into one operational taxonomic unit (OTU). Based on the naive Bayes classifier in QIIME2 [51], taxonomy annotations of the OTUs were assigned with a confidence threshold of 70%. The alpha diversity was calculated and graphed using QIIME2 and R software, respectively. Based on the OTU table, the UniFrac distances among different samples were calculated. PCoA analysis was used to analyze the beta diversity using the R package vegan and ggplot2. The bacterial co-occurrence networks were constructed with the R packages igraph, psych, Hmisc and WGCNA (Spearman $|\rho| > 0.7$ and $p < 0.01$) based on the OTU matrixes of individual growth stages. Linear discriminate analysis effect size (LEfSe) with the PERMANOVA based on 9,99 permutations was then employed between two adjacent growth stages using a web tool (https://bioincloud.tech/standalone-task-ui/lefse, accessed on 16 May 2023) to find the biomarkers (LDA > 3). The redundancy analysis (RDA) was carried out using the R packages vegan and ggplot2 with the OTU table.

4.5. Statistical Analysis

All the statistical analyses were performed using the R software (version 4.1.0). Student's *t*-test was used to calculate the significant differences between two groups based on *p*-values. A permutational multivariate analysis of variance (PERMANOVA) was used to examine significant differences among more than three groups. The LDA values in the linear discriminate analysis effect size (LEfSE) part were set to 3.0 to search the microbial biomarkers.

Author Contributions: Conceptualization, X.X. and X.L. (Xingbo Lu); methodology, X.L. (Xin Liu) and C.H.; investigation, S.Y., H.S. and F.L.; data curation, X.L. (Xin Liu) and H.S.; writing—original draft preparation, X.L. (Xin Liu) and Y.J.; writing—review and editing, X.X., Y.J. and X.L. (Xingbo Lu); funding acquisition, X.X. All authors have read and agreed to the published version of the manuscript.

Funding: This research was funded by the Agricultural Science and Technology Innovation Project of Shandong Academy of Agricultural Sciences (No. CXGC2022C02), and the Genetically Modified Organisms Breeding Major Projects of China (No. 2019ZX08013009-002-005).

Data Availability Statement: The Sequence Read Archive (SRA) accession number for 16S rDNA sequencing clean data is PRJNA943098. The data presented in this study are available in the graphs and tables provided in the manuscript.

Acknowledgments: We thank Xinyao Xia (Institute of Plant Protection, Chinese Academy of Agricultural Sciences) for bioediting the language of a draft of this manuscript.

Conflicts of Interest: The authors declare no conflict of interest.

References

1. ISAAA. *Global Status of Commercialized Biotech/GM Crops in 2019 (ISAAA Brief No. 55)*; ISAAA: Ithaca, NY, USA, 2019.
2. Anderson, J.A.; Ellsworth, P.C.; Faria, J.C.; Head, G.P.; Owen, M.D.K.; Pilcher, C.D.; Shelton, A.M.; Meissle, M. Genetically Engineered Crops: Importance of Diversified Integrated Pest Management for Agricultural Sustainability. *Front. Bioeng. Biotechnol.* **2019**, *7*, 24. [CrossRef] [PubMed]
3. Brookes, G. Genetically Modified (GM) Crop Use 1996–2020: Environmental Impacts Associated with Pesticide Use CHANGE. *GM Crops Food* **2022**, *13*, 262–289. [CrossRef] [PubMed]
4. Brookes, G. Farm income and production impacts from the use of genetically modified (GM) crop technology 1996–2020. *GM Crops Food* **2022**, *13*, 171–195. [CrossRef]
5. Brookes, G. Genetically Modified (GM) Crop Use 1996–2020: Impacts on Carbon Emissions. *GM Crops Food* **2022**, *13*, 242–261. [CrossRef] [PubMed]
6. Malovichko, Y.V.; Nizhnikov, A.A.; Antonets, K.S. Repertoire of the Bacillus thuringiensis Virulence Factors Unrelated to Major Classes of Protein Toxins and Its Role in Specificity of Host-Pathogen Interactions. *Toxins* **2019**, *11*, 347. [CrossRef]
7. Torres-Quintero, M.C.; Arenas-Sosa, I.; Zuñiga-Navarrete, F.; Hernández-Velázquez, V.M.; Alvear-Garcia, A.; Peña-Chora, G. Characterization of insecticidal Cry1Cb2 protein from Bacillus thuringiensis toxic to Myzus persicae (Sulzer). *J. Invertebr. Pathol.* **2022**, *189*, 107731. [CrossRef]
8. Liu, Y.; Li, J.; Luo, Z.; Wang, H.; Liu, F. The fate of fusion Cry1Ab/1Ac proteins from Bt-transgenic rice in soil and water. *Ecotoxicol. Environ. Saf.* **2016**, *124*, 455–459. [CrossRef]
9. Singh, A.K.; Dubey, S.K. Current trends in Bt crops and their fate on associated microbial community dynamics: A review. *Protoplasma* **2016**, *253*, 663–681. [CrossRef]
10. Li, Y.; Wang, C.; Ge, L.; Hu, C.; Wu, G.; Sun, Y.; Song, L.; Wu, X.; Pan, A.; Xu, Q.; et al. Environmental Behaviors of Bacillus thuringiensis (Bt) Insecticidal Proteins and Their Effects on Microbial Ecology. *Plants* **2022**, *11*, 1212. [CrossRef]
11. Tapp, H.; Stotzky, G. Monitoring the insecticidal toxins from Bacillus thuringiensis in soil with flow cytometry. *Can. J. Microbiol.* **1997**, *43*, 1074–1078. [CrossRef]
12. Rout, M.E.; Southworth, D. The root microbiome influences scales from molecules to ecosystems: The unseen majority. *Am. J. Bot.* **2013**, *100*, 1689–1691. [CrossRef] [PubMed]
13. Cordovez, V.; Dini-Andreote, F.; Carrión, V.J.; Raaijmakers, J.M. Ecology and Evolution of Plant Microbiomes. *Annu. Rev. Microbiol.* **2019**, *73*, 69–88. [CrossRef]
14. Bai, B.; Liu, W.; Qiu, X.; Zhang, J.; Zhang, J.; Bai, Y. The root microbiome: Community assembly and its contributions to plant fitness. *J. Integr. Plant Biol.* **2022**, *64*, 230–243. [CrossRef] [PubMed]
15. Castaldini, M.; Turrini, A.; Sbrana, C.; Benedetti, A.; Marchionni, M.; Mocali, S.; Fabiani, A.; Landi, S.; Santomassimo, F.; Pietrangeli, B.; et al. Impact of Bt corn on rhizospheric and soil eubacterial communities and on beneficial mycorrhizal symbiosis in experimental microcosms. *Appl. Environ. Microbiol.* **2005**, *71*, 6719–6729. [CrossRef] [PubMed]
16. Zhong, W.; Zeng, H.; Wang, J. Effect of Bt gene insertion on growth, physiology and gene expression of phosphorus transporter gene of corn after arbuscular mycorrhizal fungi colonization. *Chin. J. Eco-Agric.* **2017**, *25*, 1198–1205.
17. Zhang, W.; Cao, Z.; Wang, M.; Chen, X.; Wang, B. Absorption, translocation, and effects of Bt Cry1Ac peptides from transgenic cotton to the intercrops and soil functional bacteria. *Sci. Rep.* **2020**, *10*, 17294. [CrossRef] [PubMed]
18. Singh, A.K.; Rai, G.K.; Singh, M.; Dubey, S.K. Bacterial community structure in the rhizosphere of a Cry1Ac Bt-brinjal crop and comparison to its non-transgenic counterpart in the tropical soil. *Microb. Ecol.* **2013**, *66*, 927–939. [CrossRef]
19. Barriuso, J.; Valverde, J.R.; Mellado, R.P. Effect of Cry1Ab protein on rhizobacterial communities of Bt-maize over a four-year cultivation period. *PLoS ONE* **2012**, *7*, e35481. [CrossRef]

20. Fang, H.; Dong, B.; Yan, H.; Tang, F.; Wang, B.; Yu, Y. Effect of vegetation of transgenic Bt rice lines and their straw amendment on soil enzymes, respiration, functional diversity and community structure of soil microorganisms under field conditions. *J. Environ. Sci.* **2012**, *24*, 1259–1269. [CrossRef]
21. Xue, K.; Serohijos, R.C.; Devare, M.; Thies, J.E. Decomposition rates and residue-colonizing microbial communities of Bacillus thuringiensis insecticidal protein Cry3Bb-expressing (Bt) and non-Bt corn hybrids in the field. *Appl. Environ. Microbiol.* **2011**, *77*, 839–846. [CrossRef]
22. Shahmoradi, Z.S.; Tohidfar, M.; Marashi, H.; Malekzadeh-Shafaroudi, S.; Karimi, E. Cultivation Effect of Chitinase-Transgenic Cotton on Functional Bacteria and Fungi in Rhizosphere and Bulk Soil. *Iran. J. Biotechnol.* **2019**, *17*, e1982. [CrossRef] [PubMed]
23. Lv, N.; Liu, Y.; Guo, T.; Liang, P.; Li, R.; Liang, P.; Gao, X. The influence of Bt cotton cultivation on the structure and functions of the soil bacterial community by soil metagenomics. *Ecotoxicol. Environ. Saf.* **2022**, *236*, 113452. [CrossRef] [PubMed]
24. Griffiths, B.S.; Heckmann, L.H.; Caul, S.; Thompson, J.; Scrimgeour, C.; Krogh, P.H. Varietal effects of eight paired lines of transgenic Bt maize and near-isogenic non-Bt maize on soil microbial and nematode community structure. *Plant Biotechnol. J.* **2007**, *5*, 60–68. [CrossRef] [PubMed]
25. Mulder, C.; Wouterse, M.; Raubuch, M.; Roelofs, W.; Rutgers, M. Can transgenic maize affect soil microbial communities? *PLoS Comput. Biol.* **2006**, *2*, e128. [CrossRef] [PubMed]
26. Hallas, G.; Monis, P. Evaluation of heterotrophic plate and chromogenic agar colony counting in water quality laboratories. *MethodsX* **2015**, *2*, 415–422. [CrossRef] [PubMed]
27. Arshad, M.; Naqqash, T.; Tahir, M.; Leveau, J.H.; Zaheer, A.; Tahira, S.A.; Saeed, N.A.; Asad, S.; Sajid, M. Comparison of bacterial diversity, root exudates and soil enzymatic activities in the rhizosphere of AVP1-transgenic and nontransgenic wheat (Triticum aestivum L.). *J. Appl. Microbiol.* **2022**, *133*, 3094–3112. [CrossRef]
28. Wu, N.; Shi, W.; Liu, W.; Gao, Z.; Han, L.; Wang, X. Differential impact of Bt-transgenic rice plantings on bacterial community in three niches over consecutive years. *Ecotoxicol. Environ. Saf.* **2021**, *223*, 112569. [CrossRef]
29. Yang, B.; Wang, Y.; Qian, P.Y. Sensitivity and correlation of hypervariable regions in 16S rRNA genes in phylogenetic analysis. *BMC Bioinform.* **2016**, *17*, 135. [CrossRef]
30. Chen, Y.; Pan, L.; Ren, M.; Li, J.; Guan, X.; Tao, J. Comparison of genetically modified insect-resistant maize and non-transgenic maize revealed changes in soil metabolomes but not in rhizosphere bacterial community. *GM Crops Food* **2022**, *13*, 1–14. [CrossRef]
31. Singh, A.K.; Singh, M.; Dubey, S.K. Changes in Actinomycetes community structure under the influence of Bt transgenic brinjal crop in a tropical agroecosystem. *BMC Microbiol.* **2013**, *13*, 122. [CrossRef]
32. Liang, J.; Jiao, Y.; Luan, Y.; Sun, S.; Wu, C.; Wu, H.; Zhang, M.; Zhang, H.; Zheng, X.; Zhang, Z. A 2-year field trial reveals no significant effects of GM high-methionine soybean on the rhizosphere bacterial communities. *World J. Microbiol. Biotechnol.* **2018**, *34*, 113. [CrossRef] [PubMed]
33. Yang, S.; Liu, X.; Xu, X.; Sun, H.; Li, F.; Hao, C.; Lu, X. Effects of Insect-Resistant Maize 2A-7 Expressing mCry1Ab and mCry2Ab on the Soil Ecosystem. *Plants* **2022**, *11*, 2218. [CrossRef] [PubMed]
34. Mendes, R.; Garbeva, P.; Raaijmakers, J.M. The rhizosphere microbiome: Significance of plant beneficial, plant pathogenic, and human pathogenic microorganisms. *FEMS Microbiol. Rev.* **2013**, *37*, 634–663. [CrossRef]
35. Trivedi, P.; Leach, J.E.; Tringe, S.G.; Sa, T.; Singh, B.K. Plant-microbiome interactions: From community assembly to plant health. *Nat. Rev. Microbiol.* **2020**, *18*, 607–621. [CrossRef] [PubMed]
36. van Wyk, D.A.B.; Adeleke, R.; Rhode, O.H.J.; Bezuidenhout, C.C.; Mienie, C. Ecological guild and enzyme activities of rhizosphere soil microbial communities associated with Bt-maize cultivation under field conditions in North West Province of South Africa. *J. Basic Microbiol.* **2017**, *57*, 781–792. [CrossRef] [PubMed]
37. Li, P.; Li, Y.; Shi, J.; Yu, Z.; Pan, A.; Tang, X.; Ming, F. Impact of transgenic Cry1Ac + CpTI cotton on diversity and dynamics of rhizosphere bacterial community of different root environments. *Sci. Total Environ.* **2018**, *637–638*, 233–243. [CrossRef]
38. Zhou, X.Y.; Liu, N.; Zhao, M.; Li, H.; Zhou, L.; Tang, Z.W.; Cao, F.; Li, W. Advances in effects of insecticidal crystal proteins released from transgenic Bt crops on soil ecology. *Yi Chuan* **2011**, *33*, 443–448. [CrossRef]
39. Wang, Y.M.; Deng, J.X.; Zhang, J.X.; Jia, R.Z. Accumulation of Cry proteins in soil released from Bt rice after planting for multiple years. *Ying Yong Sheng Tai Xue Bao* **2022**, *33*, 119–125.
40. Meihls, L.N.; Higdon, M.L.; Siegfried, B.D.; Miller, N.J.; Sappington, T.W.; Ellersieck, M.R.; Spencer, T.A.; Hibbard, B.E. Increased survival of western corn rootworm on transgenic corn within three generations of on-plant greenhouse selection. *Proc. Natl. Acad. Sci. USA* **2008**, *105*, 19177–19182. [CrossRef]
41. Wang, H.; Ye, Q.; Wang, W.; Wu, L.; Wu, W. Cry1Ab protein from Bt transgenic rice does not residue in rhizosphere soil. *Environ. Pollut.* **2006**, *143*, 449–455. [CrossRef]
42. Li, F.; Wang, M.; Sun, H.W.; Yang, S.K.; Lu, X.B. Dynamics of Cry1ab protein content in the rhizosphere soil and straw debris of transgenic Bt corn. *Ying Yong Sheng Tai Xue Bao* **2013**, *24*, 1907–1913. [PubMed]
43. Wu, L.C.; Li, X.F.; Ye, Q.F.; Wang, H.Y. Expression and root exudation of Cry1Ab toxin protein in cry1Ab transgenic rice and its residue in rhizosphere soil. *Huan Jing Ke Xue* **2004**, *25*, 116–121. [PubMed]
44. Fu, Q.L.; Chen, S.W.; Hu, H.Q.; Li, Z.; Han, X.F.; Wang, P. Distribution of Bt protein in transgenic cotton soils. *Ying Yong Sheng Tai Xue Bao* **2011**, *22*, 1493–1498. [PubMed]
45. China Agricultural University. Corn Event 2A-7 and Its Identification Method. International Application No. PCT/CN2020/126452, 4 November 2020.

46. Baumgarte, S.; Tebbe, C.C. Field studies on the environmental fate of the Cry1Ab Bt-toxin produced by transgenic maize (MON810) and its effect on bacterial communities in the maize rhizosphere. *Mol. Ecol.* **2005**, *14*, 2539–2551. [CrossRef] [PubMed]
47. Wang, Y.; Zhang, M.; Li, S.; Li, P.; Lang, Z. Effects of Insect-Resistant Maize HGK60 on Community Diversity of Bacteria and Fungi in Rhizosphere Soil. *Plants* **2022**, *11*, 2824. [CrossRef] [PubMed]
48. Barberán, A.; Bates, S.T.; Casamayor, E.O.; Fierer, N. Using network analysis to explore co-occurrence patterns in soil microbial communities. *ISME J.* **2012**, *6*, 343–351. [CrossRef]
49. Chen, B.; Jiao, S.; Luo, S.; Ma, B.; Qi, W.; Cao, C.; Zhao, Z.; Du, G.; Ma, X. High soil pH enhances the network interactions among bacterial and archaeal microbiota in alpine grasslands of the Tibetan Plateau. *Environ. Microbiol.* **2021**, *23*, 464–477. [CrossRef]
50. Fang, W.; Fan, T.; Xu, L.; Wang, S.; Wang, X.; Lu, A.; Chen, Y. Seasonal succession of microbial community co-occurrence patterns and community assembly mechanism in coal mining subsidence lakes. *Front. Microbiol.* **2023**, *14*, 1098236. [CrossRef]
51. Bolyen, E.; Rideout, J.R.; Dillon, M.R.; Bokulich, N.A.; Abnet, C.C.; Al-Ghalith, G.A.; Alexander, H.; Alm, E.J.; Arumugam, M.; Asnicar, F.; et al. Reproducible, interactive, scalable and extensible microbiome data science using QIIME 2. *Nat. Biotechnol.* **2019**, *37*, 852–857. [CrossRef]

plants

MDPI

Article

Subchronic Toxicity of GmDREB3 Gene Modified Wheat in the Third Generation Wistar Rats

Jie Tian [1,†], Xiang-Hong Ke [1,†], Yuan Yuan [1], Wen-Xiang Yang [1], Xiao-Qiao Tang [1], Lan-Jie Pei [1], Jun Fan [1], Qin Zhuo [2], Xiao-Guang Yang [2], Jia-Fa Liu [1] and Bo-Lin Fan [1,*]

1 Hubei Provincial Key Laboratory for Applied Toxicology, Hubei Provincial Center for Disease Control and Prevention, Wuhan 430079, China; jiessie_tian@163.com (J.T.); cilan-724@163.com (X.-H.K.); emily_fighting@163.com (Y.Y.); ywx_21@163.com (W.-X.Y.); melon_qiao@163.com (X.-Q.T.); peilanjie123@163.com (L.-J.P.); u571_f16@163.com (J.F.); jfllionm@sohu.com (J.-F.L.)

2 Key Laboratory of Trace Element Nutrition of National Health Commission, Chinese Center for Disease Control and Prevention, Beijing 100050, China; zhuoqin@ninh.chinacdc.cn (Q.Z.); xgyangcdc@163.com (X.-G.Y.)

* Correspondence: vanbolin@163.com

† These authors contributed equally to this work.

Abstract: The aim of the current study was to evaluate the subchronic toxicity of GmDREB3 gene modified wheat in the third generation rats. SPF Wistar rats were fed with transgenic wheat diet (Gm), parental wheat diet (Jimai22) and AIN-93 rodent diet (Control), respectively, for two generations, to produce the third generation rats which were used for this study. The selected fresh weaned offspring rats (20/sex/group) were given the same diet as their parents for 13 weeks. No toxicity-related changes were observed in rats fed with Gm diet in the following respects: clinical signs, body weights, body weight gains, food consumption, food utilization rate, urinalysis, hematology, serum biochemistry and histopathology. The results from the present study demonstrated that 13 weeks consumption of Gm wheat did not cause any adverse effects in the third generation rats when compared with the corresponding Jimai22 wheat.

Keywords: subchronic toxicity; GmDREB3; genetically modified wheat; rats

Citation: Tian, J.; Ke, X.-H.; Yuan, Y.; Yang, W.-X.; Tang, X.-Q.; Pei, L.-J.; Fan, J.; Zhuo, Q.; Yang, X.-G.; Liu, J.-F.; et al. Subchronic Toxicity of GmDREB3 Gene Modified Wheat in the Third Generation Wistar Rats. *Plants* **2022**, *11*, 1823. https://doi.org/10.3390/plants11141823

Academic Editors: Wei Wei and C. Neal Stewart

Received: 10 June 2022
Accepted: 5 July 2022
Published: 12 July 2022

1. Introduction

Wheat, as one of the main food crops (maize, rice, wheat), is widely planted on about 220 million ha area of at least 43 countries in the world [1,2]; the global wheat output is estimated to increase by 1.0% every year [3]. Wheat is rich in starch, fat, protein and microelements and provides total food calories for about 40% of the world's population [4,5]. However, the output of wheat is seriously influenced by climatic changes and water scarcity in the environment. In the future, with the global temperature increasing, the frequency of drought occurrence will be greater than ever [6]. Jimai 22 wheat is a native crop mainly cultivated in northern China [7], but although it possesses some clear advantages of high yield, strong environment adaption and strong anti-reversion, it fails to be well promoted due to its poor resistance to drought. Consequently, to improve the drought tolerance of wheat is an urgent task in China and also around the world.

For many years, scientists have experimented with different ways to enhance drought resistance in crops, such as traditional hybridization methods and genetic engineering technology. Conventional hybridization has been quite effective to address the complex quality traits such as adaption to changing climate conditions, which are usually mediated by multiple genetic factors; however, it requires a huge commitment of time and human resources [8,9]. Genetic engineering technology overcomes some disadvantages of traditional breeding and selection methods, and therefore it is also applied as a promising way to develop new cultivars in modern agriculture to transfer beneficial genes from one organism to another to improve a certain trait of crops [10]. In this research, the GmDREB3 gene from soybean was inserted into the wild wheat (Jimai 22) to enhance its tolerance to drought. This Gm wheat was demonstrated to have drought resistance and can be planted in semi-arid and arid areas, but as a new transgenic crop it still faces the challenges connected with assessment in food safety for the reason of exogenous gene insertion. The unexpected toxic ingredients, reduction of the nutritional value and other unintended effects of this Gm wheat are still public concerns. Consequently, toxicity assessment is necessary to ensure safety before Gm Wheat can be accepted as food or feed into market [11–13].

Based on the previous two-generation reproduction studies of this Gm wheat [14,15], we performed this 13-week subchronic toxicity study in SPF grade Wistar rats, whose ancestors were fed with Gm wheat for their entire life cycle over two generations, to evaluate the potential adverse effects. This research followed OECD Guideline 408 for repeated dose 90-day toxicity study in rodents [16] and good laboratory practice standards (GLP standard).

2. Results

2.1. Clinical Observations, Body Weights and Food Consumption

Throughout the experiment, there were no toxicity-related signs observed in any group. The increase trends of body weights in the three groups were similar. Compared with the Jimai22 female group, the body weights of the Gm female rats showed no significant differences except during weeks 2, 3, 4 and 6 ($p < 0.05$ or $p < 0.01$). The body weights exhibited no difference between Gm male rats and Jimai22 male rats. In addition, no significant differences were observed in weekly food consumption of rats in the two wheat groups. The results are shown in Figures 1 and 2.

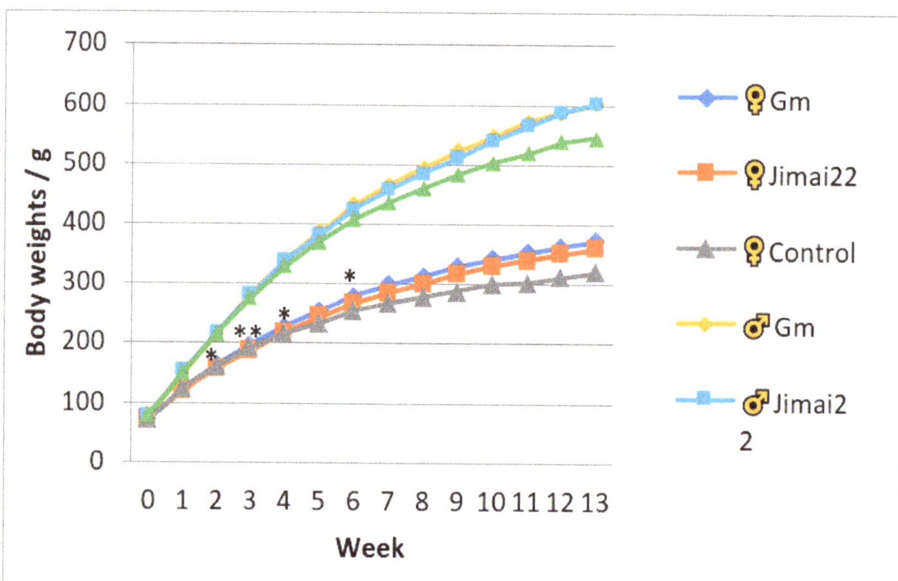

Figure 1. Weekly body weights of female and male animals in three groups. * represents $p < 0.05$ compared with Jimai22 group; ** represents $p < 0.01$ compared with Jimai22 group.

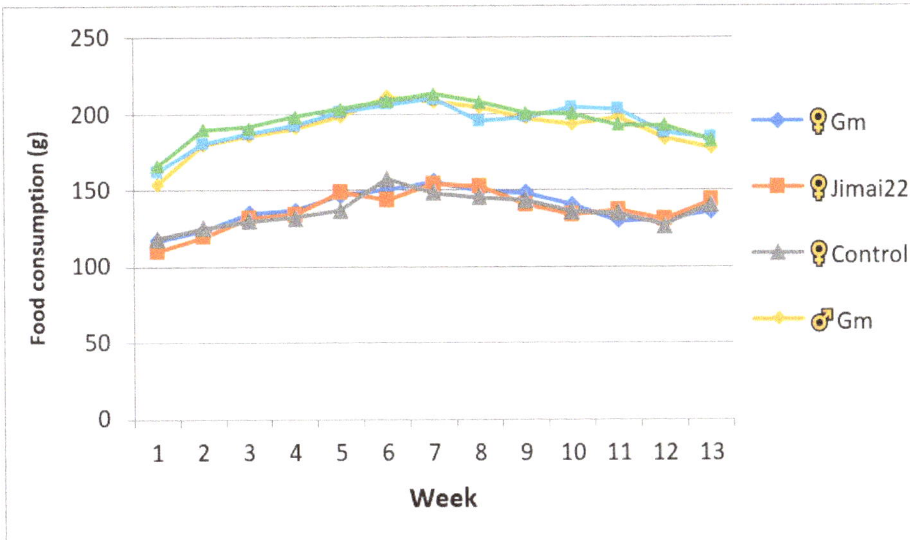

Figure 2. Weekly food consumption of female and male animals in three groups.

There were no significant differences between the Gm group and Jimai22 group ($p > 0.05$) in total body weight gains, total food consumption or total food utilization rate. Compared with the AIN-93 control group, the total body weight gains and total food utilization of the Gm group (females and males) were both significantly higher ($p < 0.01$, $p < 0.01$), but total food intake of Gm females and Gm males displayed no difference ($p > 0.05$). The results are shown in Figures 3–5.

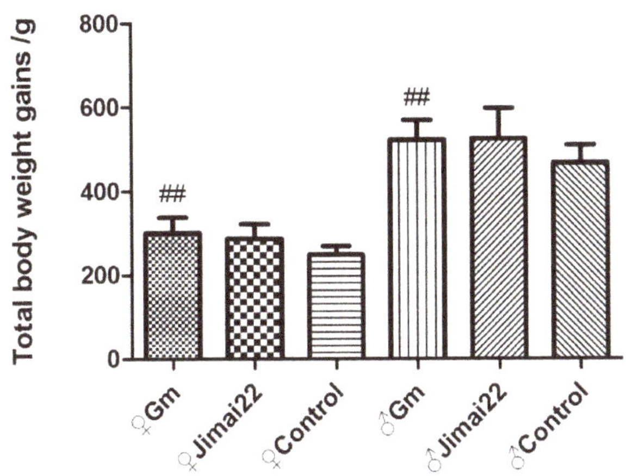

Figure 3. Total body weight gains of females and males during 13 weeks. ## represents $p < 0.01$ compared with control group.

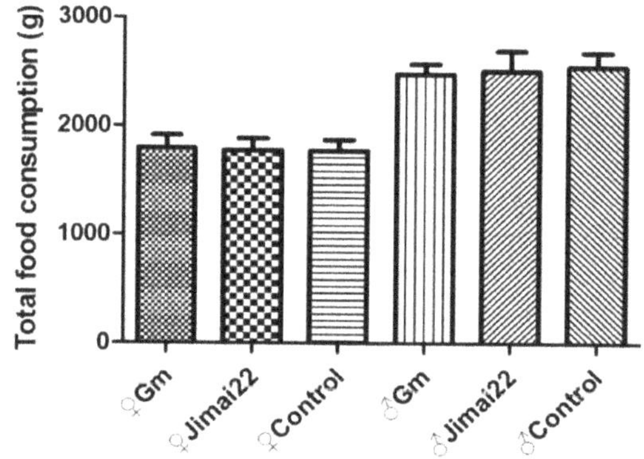

Figure 4. Total food consumption of females and males during 13 weeks.

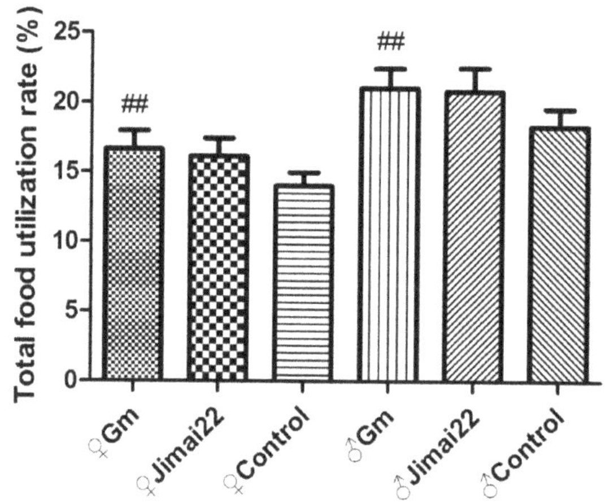

Figure 5. Total food utilization of females and males during 13 weeks. ## represents $p < 0.01$ compared with control group.

2.2. Urine Analysis

There were no statistically significant differences in any parameters of Gm female and Gm male rats compared with the Jimai22 or AIN-93 control groups (data not shown).

2.3. Hematology and Serum Biochemistry

Gm males had lower Hb, HCT % and a higher CHOL ($p < 0.05$, $p < 0.01$) and Gm females had a lower BUN and a higher CHOL ($p < 0.05$, $p < 0.01$), when compared with the Jimai22 group. When compared to the control group, Gm females had lower HCT%, BUN, ALT and AST, and higher Glu, CHOL, TP and Ca^{2+}; Gm males had lower HCT%, AST, Na^{+} and Cl^{-}, and a higher Glu ($p < 0.01$ or $p < 0.05$); the other serum biochemistry indices presented no significant differences among the three groups. The results are shown in Tables 1 and 2.

Table 1. Hematological parameters of male and female rats ($\bar{x} \pm s$, $n = 20$).

Sex	Group	WBC ($\times 10^9$/L)	RBC ($\times 10^{12}$/L)	Hb (g/L)	LYMP (%)	MONO (%)	NEUT (%)	EO (%)	BAS (%)	HCT (%)	PLT (%)	APTT (s)	PT (s)
Female	Gm	2.6 ± 1.1	7.66 ± 0.51	134 ± 32	70.0 ± 6.1	2.3 ± 0.9	25.8 ± 6.1	1.9 ± 0.9	0.0 ± 0.0	40.4 ± 2.2 ##	817 ± 227	11.8 ± 3.7	12.9 ± 0.8
	Jimai22	2.7 ± 1.2	7.75 ± 0.48	144 ± 9	68.7 ± 4.9	2.1 ± 0.8	26.6 ± 4.3	2.7 ± 2.0	0.0 ± 0.0	41.3 ± 2.2	792 ± 161	12.5 ± 5.9	12.3 ± 0.5
	Control	2.8 ± 1.2	7.91 ± 0.54	149 ± 10	67.2 ± 8.0	2.0 ± 0.8	29.0 ± 8.6	1.9 ± 0.8	0.0 ± 0.0	43.7 ± 2.8	846 ± 109	12.2 ± 0.8	12.8 ± 0.5
Male	Gm	4.9 ± 1.6	8.40 ± 0.28	146 ± 5 *	64.0 ± 4.8	2.4 ± 0.6	31.2 ± 4.6	2.4 ± 0.7	0.0 ± 0.0	41.7 ± 1.2 **##	894 ± 90	10.9 ± 0.8	12.8 ± 2.0
	Jimai22	4.9 ± 1.4	8.68 ± 0.49	151 ± 7	60.6 ± 6.9	2.8 ± 0.7	34.1 ± 6.7	2.5 ± 0.8	0.0 ± 0.0	43.4 ± 1.5	881 ± 105	11.5 ± 1.1	13.0 ± 0.41
	Control	5.5 ± 1.3	8.36 ± 0.34	148 ± 6	63.3 ± 4.5	2.2 ± 0.5	32.1 ± 4.4	2.1 ± 0.8	0.0 ± 0.0	44.3 ± 1.8	863 ± 99	12.5 ± 1.6	12.9 ± 0.57

* represents $p < 0.05$ compared with Jimai22 group; ** represents $p < 0.01$ compared with Jimai22 group; ## represents $p < 0.01$ compared with control group.

Table 2. Serum chemistry parameters of male and female rats ($\bar{x} \pm s$, $n = 20$).

Sex	Group	Glu (mmol/L)	BUN (mmol/L)	Cr (μmol/L)	CHOL (mmol/L)	TG (mmol/L)	ALT (U/L)	AST (U/L)	TP (g/L)	ALB (g/L)	ALB/GLO	K^+ (mmol/L)	Na^+ (mmol/L)	Cl^- (mmol/L)	Ca^{2+} (mmol/L)
Female	Gm	6.03 ± 0.98 #	4.12 ± 0.57 *##	64.9 ± 5.5	2.04 ± 0.48 **##	0.50 ± 0.16 a	31 ± 12 #	85 ± 18 ##	60.8 ± 2.6 #	32.3 ± 1.5	1.13 ± 0.05 a	4.43 ± 1.57	140.2 ± 1.4	103.1 ± 1.5	2.54 ± 0.09 ##
	Jimai22	6.33 ± 0.95	4.72 ± 0.76	67.4 ± 7.4	1.64 ± 0.35	0.57 ± 0.18 a	29 ± 6	86 ± 18	61.6 ± 3.0	33.1 ± 1.6	1.19 ± 0.08 a	4.00 ± 0.32	139.9 ± 1.7	103.3 ± 1.4	2.53 ± 0.07
	Control	5.21 ± 0.96	5.22 ± 1.03	68.2 ± 7.9	1.59 ± 0.35	0.52 ± 0.16	39 ± 8	113 ± 12	58.2 ± 3.6	31.3 ± 1.9	1.17 ± 0.07	4.53 ± 0.52	140.0 ± 2.6	103.4 ± 1.6	2.44 ± 0.08
Male	Gm	6.85 ± 0.86 ##a	4.89 ± 0.55	61.4 ± 6.3	1.88 ± 0.56 *	0.60 ± 0.24	34 ± 6 a	85 ± 9 #a	57.4 ± 2.7	28.7 ± 1.2	1.00 ± 0.06	4.39 ± 0.68	139.7 ± 0.8 ##	100.8 ± 1.5 ##	2.53 ± 0.08
	Jimai22	6.62 ± 0.93 a	4.71 ± 0.57	60.7 ± 5.1	1.52 ± 0.35	0.74 ± 0.46	32 ± 5 a	90 ± 13 a	58.2 ± 2.5	29.2 ± 1.4	1.01 ± 0.09	4.39 ± 0.47	140.2 ± 1.0	101.2 ± 1.6	2.52 ± 0.07
	Control	5.85 ± 0.84	4.51 ± 0.59	58.0 ± 4.5	1.76 ± 0.46	0.83 ± 0.38	38 ± 7	94 ± 12	57.7 ± 2.3	28.5 ± 1.01	0.98 ± 0.05	4.37 ± 0.34	143.2 ± 1.1	103.0 ± 1.2	2.50 ± 0.04

* represents $p < 0.05$ compared with Jimai22 group; ** represents $p < 0.01$ compared with Jimai22 group; # represents $p < 0.05$ compared with control group; ## represents $p < 0.01$ compared with control group; [a] represents this data was cited in our last paper (Tian et al., 2021) [15].

2.4. Organ Weights and Relative Organ Weights

When compared with the Jimai22 group, Gm female rats had higher absolute weights and relative weights of their liver and adrenal glands ($p < 0.01$ or $p < 0.05$). When compared with the control group, Gm female rats had higher absolute weights of heart, liver, spleen, kidneys and thymus ($p < 0.05$ or $p < 0.01$), and lower relative weights of brain and ovaries; Gm male rats had lower relative weights of heart and brain ($p < 0.01$ or $p < 0.05$). The results are shown in Tables 3 and 4.

Table 3. Organ weights and relative organ weights of females ($\overline{x} \pm s$, $n = 20$).

Organ	Gm		Jimai22		Control	
	Weights (g)	Relative Weights (%)	Weights (g)	Relative Weights (%)	Weights (g)	Relative Weights (%)
Body weight	354.2 ± 36.4		348.4 ± 33.1		302.0 ± 28.2	
Heart	1.02 ± 0.13 #	0.288 ± 0.034	1.02 ± 0.11	0.293 ± 0.029	0.93 ± 0.11	0.310 ± 0.035
Liver	9.32 ± 1.14 **##	2.648 ± 0.378 **	8.29 ± 0.84	2.379 ± 0.099	7.91 ± 1.04	2.631 ± 0.370
Spleen	0.75 ± 0.10 #	0.213 ± 0.022	0.71 ± 0.09	0.205 ± 0.020	0.66 ± 0.10	0.220 ± 0.034
Kidneys	2.11 ± 0.018 ##	0.598 ± 0.056	2.06 ± 0.23	0.592 ± 0.059	1.89 ± 0.15	0.630 ± 0.061
Brain	1.89 ± 0.25	0.538 ± 0.088 ##	1.89 ± 0.06	0.546 ± 0.056	1.87 ± 0.08	0.623 ± 0.059
Uterus	0.62 ± 0.21	0.178 ± 0.068	0.70 ± 0.29	0.203 ± 0.084	0.61 ± 0.17	0.202 ± 0.056
Ovaries	0.21 ± 0.03	0.061 ± 0.009 #	0.21 ± 0.04	0.059 ± 0.009	0.22 ± 0.04	0.073 ± 0.017
Adrenals	0.103 ± 0.024 *	0.030 ± 0.007 *	0.085 ± 0.018	0.024 ± 0.005	0.094 ± 0.026	0.031 ± 0.009
Thymus	0.461 ± 0.078 ##	0.131 ± 0.023	0.424 ± 0.139	0.122 ± 0.037	0.353 ± 0.086	0.117 ± 0.029

* represents $p < 0.05$ compared with Jimai22 group; ** represents $p < 0.01$ compared with Jimai22 group; # represents $p < 0.05$ compared with control group; ## represents $p < 0.01$ compared with control group.

Table 4. Organ weights and relative organ weights of males ($\overline{x} \pm s$, $n = 20$).

Organ	Gm		Jimai22		Control	
	Weights (g)	Relative Weights (%)	Weights (g)	Relative Weights (%)	Weights (g)	Relative Weights (%)
Body weight	587.1 ± 44.1		585.7 ± 69.5		524.0 ± 43.6	
Heart	1.49 ± 0.14	0.254 ± 0.023 #	1.52 ± 0.17	0.261 ± 0.027	1.44 ± 0.17	0.275 ± 0.030
Liver	15.08 ± 1.80	2.577 ± 0.327 a	14.80 ± 2.64	2.518 ± 0.254 a	14.32 ± 1.31	2.740 ± 0.218
Spleen	1.11 ± 0.19	0.189 ± 0.030	1.05 ± 0.17	0.179 ± 0.021	1.17 ± 0.55	0.224 ± 0.103
Kidneys	3.46 ± 0.28	0.592 ± 0.055 a	3.38 ± 0.41	0.579 ± 0.048 a	3.25 ± 0.58	0.625 ± 0.122
Brain	2.10 ± 0.08	0.359 ± 0.033 #	2.09 ± 0.09	0.361 ± 0.046	2.05 ± 0.08	0.394 ± 0.039
Testis	4.03 ± 0.53	0.688 ± 0.092	3.93 ± 0.81	0.672 ± 0.109	4.06 ± 1.18	0.773 ± 0.199
Epididymides	1.67 ± 0.19	0.287 ± 0.040	1.61 ± 0.21	0.278 ± 0.051	1.72 ± 0.62	0.329 ± 0.118
Prostate	1.49 ± 0.25	0.253 ± 0.038	1.51 ± 0.26	0.262 ± 0.058	1.39 ± 0.25	0.265 ± 0.043
Adrenals	0.090 ± 0.021	0.015 ± 0.004	0.084 ± 0.020	0.014 ± 0.003	0.091 ± 0.025	0.018 ± 0.005
Thymus	0.594 ± 0.168	0.101 ± 0.028	0.571 ± 0.152	0.097 ± 0.023	0.503 ± 0.119	0.096 ± 0.024

represents $p < 0.05$ compared with control group; a represents this data was cited in our last paper (Tian et al., 2021) [15].

2.5. Histopathology

There were some histopathological changes in three group animals, including: slight interstitial inflammation in the lungs; slight stomach glandular dilatation; slight inflammatory cell infiltration and slight vacuolation in livers; slight interstitial inflammatory infiltrate, hyaline casts and calcium deposition in kidneys; testis atrophy and sperm loss in epididymis. Details are shown in Table 5 and Figure 6. Histopathological results in all tissues did not reveal any significant difference between the Gm and Jimai22 groups.

Table 5. Histopathological results of three groups at week 13.

Histopathological Lesions	Groups Gm Group		Jimai22 Group		Control Group	
Sex	♀	♂	♀	♂	♀	♂
Number	20	20	20	20	20	20
Lung						
Slight interstitial inflammation	0	0	1	0	0	0
Stomach						
Slight glandular dilatation	0	0	0	1	0	0
Liver						
Slight inflammatory cell infiltration	0	1	1	1	0	0
Slight vacuolation	1	0	0	1	0	0
Severe vacuolation	0	0	0	1	0	0
Kidney						
Slight interstitial inflammatory cell infiltration	0	1	1	0	0	1
Slight hyaline casts	2	2	2	1	1	1
Slight calcium deposition	2	0	1	0	1	0
Testis						
Severe atrophy	-	1	-	2	-	1
Epididymis						
Sperm loss	-	1	-	2	-	1

"-": Not applicable.

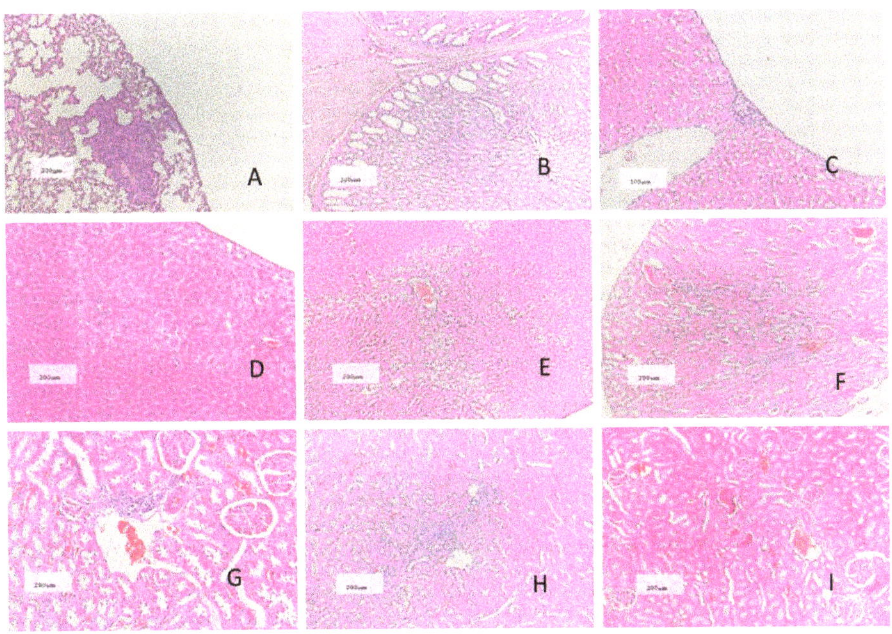

Figure 6. *Cont.*

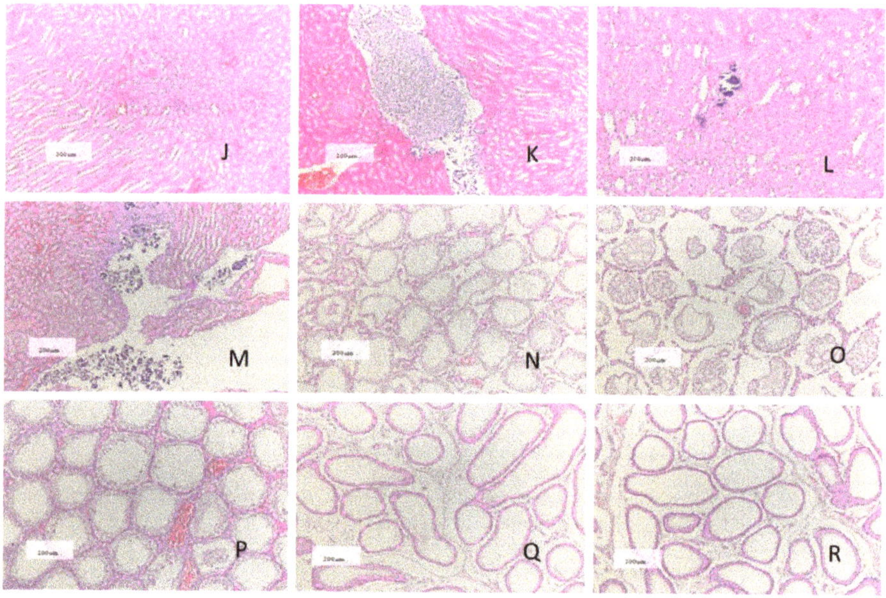

Figure 6. Non-specific histopathological changes in rats at week 13. (**A**): slight interstitial inflammation of lung in Jimai22 group; (**B**): Slight glandular dilatation of stomach in Jimai22 group; (**C**): slight inflammatory cells infiltration of liver in Jimai22 group; (**D**,**E**): slight hepatic vacuolation in Jimai22 and Gm group; (**F–H**): slight renal interstitial inflammatory cell infiltration in Control, Jimai22 and Gm group, respectively; (**I**,**J**): Slight renal hyaline casts in Jimai22 group and Gm group; (**K–M**): slight renal calcium deposition in Control, Jimai22 and Gm group, respectively; (**N–P**): testicular atrophy in Control, Jimai22 and Gm group, respectively; (**Q**,**R**): epididymal sperm loss in Control, Jimai22 group.

3. Discussion

Dehydration responsive element-binding proteins (DREBs) are members of a larger family of transcription factors (TFs), which are involved in the resistance to abiotic stress-induced oxidative damages by regulating the expression of genes in the stress defense pathway [17,18]. In addition, DREBs participate in the induction of salinity tolerance by acting on the downstream (auxin and ethylene signaling pathways) and upstream (ABA-independent signaling pathway) [19]. DREBs are also involved in immunity against several biotic stresses by modulating the expression of several downstream genes of the defense signaling cascade [20]. Therefore, DREB genes were used widely to genetically engineer plants for resistance to abiotic factors. The Gm wheat used in this study is a new variety produced by genetic engineering technology to introduce the GmDREB3 gene to the original wheat.

Previous nutritional analysis showed that this wheat was nutritionally equivalent to its non-Gm counterpart [21]; preceding studies also showed that this Gm wheat was safe in immunotoxicity studies [15,22], and in two-generation reproduction studies [14,15]. According to the OECD Test Guidelines and the Guidelines for Safety Assessment of Genetically Modified Plants of China, 90-day subchronic toxicity experiments in rodents are generally considered necessary for safety evaluation of all genetically modified foods [23–25], but routine 90-day subchronic toxicity assessment in transgenic foods were always conducted in single generation animals. Some studies suggested that long-term and multi-generational assessments are necessary in some cases to further assure the safety of Gm food/feed [26,27]. So far, no 90-day subchronic toxicity study has been performed to address the potential toxicological effects in animals exposed to GM feed over three generations. Consequently, a 13-week subchronic toxicity study was conducted in Wistar rats whose parental animals were pre-exposed to the Gm wheat for two generations.

In the present study, the Jimai22 group was used to test whether this Gm wheat would create any toxic effects due to the insertion of the GmDREB3 gene, while the AIN-93 basic diet was used as a nutritional control to analyze whether the wheat in the diet would affect the nutritional status of the rats. During the period of the 13-week study, compared to Jimai22 group, the body weights of the Gm female group were slightly higher in weeks 2, 3, 4 and 6, which was considered a normal body weight fluctuation. In addition, no remarkable difference was detected in the weekly food consumption, total body weight gains, total food consumption, or total food utilization rate, which also confirmed that the above-mentioned differences were of no toxic concern.

In hematology and serum biochemistry, compared with the Jimai22 group Gm male rats had lower HCT% and Hb and a higher CHOL, Gm females had a lower BUN and a higher CHOL. When compared with the control group: Gm females had lower HCT%, BUN, ALT and AST, higher Glu, CHOL, TP and Ca^{2+}; Gm males had lower HCT%, AST, Na^{+}, Cl^{-} and a higher Glu. All of the above-mentioned differences were considered to have no toxicological significance since these values were similar to those of the control group and the Jimai22 group; in addition, these observed differences were within the normal reference ranges for rats of this age in our laboratory. Serum total T4, T3 and TSH should be measured on samples obtained from each animal according to OECD 408 (2018 revised version), but these indexes was absent in this study because the investigation was performed in 2016. We acknowledge it was a limitation of the study. We will strictly refer to OECD guideline 408 (2018 version) in the future study of 90-day subchronic toxicity.

With respect to organ weights, significant increases were presented in absolute and relative weights of livers and adrenal glands in female rats of the Gm group, as compared with the Jimai22 group, which were considered to be not treatment-related since the respective findings were not supported by the biochemical or histopathological data generated in our study. Compared with the control group, Gm female rats had higher absolute weights of heart, liver, spleen, kidneys and thymus, and lower relative weights of brain and ovaries; also, Gm male rats had lower relative weights of heart and brain, which were attributed to increased body weights (females and males in Gm group) at the time of necropsy and were considered to have no toxicological significance.

In the histopathological analysis, some kinds of lesions were observed in the Gm group, but the severity and frequency of these lesions were comparable between the Gm and Jimai 22 group. Furthermore, the incidences of these findings were all within the normal reference range for such data established in our laboratory. Thus, all of the above-mentioned histopathological alterations in the Gm group were interpreted as incidental and were considered to be of no significance related to toxicity.

4. Materials and Methods

4.1. Plant and Diet

The Gm and Jimai22 wheats were provided by the Chinese Academy of Agricultural Sciences. Both seeds were processed into flour, then the two kind of wheat flour were formulated into rodent diets; both were at a percentage of 69.55% as previous diets of two generations, and meanwhile, the AIN-93 diet was set as a nutritional control [28]. The target genes in Gm, Jimai22 and control diets were detected using the PCR method [14].

4.2. Experiment Animals and Breeding Condition

In total, 120 specific pathogen-free (SPF) Wistar rats (60 males and 60 females) were obtained from previous two-generation reproductive experiments. The rats were housed in a good experimental environment and could eat and drink ad libitum. All experimental processes were approved and supervised by the Hubei Provincial Animal Management and Use Committee (NO. 2015022).

4.3. Study Design and Administration

The previous two generations of rats (F0 and F1 generation) were fed Gm diet, Jimai22 diet and AIN-93 control diet for their lifetime. A total of 20 males and 20 females of the third generation rats (F2 generation) were randomly selected from each group and were given the same diet as their parents for 13 weeks, to investigate the subchronic toxicity of this Gm wheat.

4.4. Clinical Observations, Body Weight and Food Consumption

All animals were observed daily for signs of toxicity and the availability of food. Animal body weights and food consumption of each group were recorded per week; total body weight gains, total food consumption and total food utilization rate for weeks 1–13 were calculated.

4.5. Urinalysis

Urine samples of all rats were collected by metabolism cages one day prior to necropsy. During the process of urine collection water was available to all rats freely. After observing and recording the urine appearance and volume, the urine indices, including BIL, UBG, KET, PRO, NIT, GLU, LEU, BLD, SG and PH, were determined using an automatic urine analyzer (Dirui; Changchun; China).

4.6. Hematology and Serum Biochemistry

At the end of week 13, all animals of the three groups were fasted overnight for 16 h and were given drinking water as before, then the blood samples were collected from abdominal aorta under anesthesia. RBC, WBC, PLT, HCT, Hb, NEUT%, LYMPH%, MONO%, EO% and BASO% were determined using an automatic blood analyzer XT2000iv (Sysmex Corporation; Godo; Japan). PT and APTT were measured using a coagulation function analyzer CA510 (Sysmex Corporation; Godo; Japan). GLU, BUN, CREA, CHOL, TG, ALT, AST, TP, ALB, ALB/GLO, Na^+, K^+, Cl^- and Ca^{2+} were measured using an automatic analyzer AU680 (Beckman Coulter, Inc.; California; America).

4.7. Histopathology

All animals were sacrificed and subsequently submitted to detailed necropsy and histopathology examination. Brain, heart, liver, spleen, kidneys, adrenal glands, thymus, ovaries (female), uterus (female), testes (male), epididymides (male) and prostate (male) were weighed, and the relative organ weights were calculated. The following organs were examined under a microscope: brain, pituitary, spinal cord, sciatic nerve, thyroid, parathyroid, salivary gland, sternum, thymus, oesophagus, trachea, heart, lung, stomach, spleen, liver, duodenum, jejunum, ileum, colon, rectum, pancreas, adrenals, kidneys, aorta, ovaries, uterus, mammary gland, testes, epididymides, prostate, urinary, bladder skeletal muscle, skin, eyes, and lymph node. In the histopathological analysis, the incidence and semi-quantitative score system used was recommended by Shackelford et al. [29]. The degree of lesions in each item was graded from one to five depending on severity: 1 = minimal (<1%); 2 = slight (1–25%); 3 = moderate (26–50%); 4 = moderate/severe (51–75%); 5 = severe/high (76–100%).

4.8. Statistical Analysis

Quantitative data in this study including animal body weights, food consumption, organ weight, hematology and biochemistry data were analyzed using one-way analysis of variance (ANOVA), then followed by Student's *t*-test when the variance of three group data was homogeneous. Urine data were analyzed using the Kruskal–Wallist test, and histopathological changes were analyzed using Chi-square test and Fisher's exact test. Statistically significant differences were considered if the *p*-value between the groups was less than 0.05.

5. Conclusions

The results of the 13-week subchronic toxicity study of the Gm wheat did not exhibit any toxic effects in Wistar rats, whose ancestors had been pre-exposed to the same wheat for two generations. This suggests that Gm wheat is as safe for food as its parental wheat in our current study.

Author Contributions: Writing—original draft: J.T.; investigation: J.T., X.-H.K., Y.Y., W.-X.Y., X.-Q.T., L.-J.P. and J.F.; methodology: B.-L.F., J.T., X.-G.Y. and Q.Z.; supervision: B.-L.F. and J.-F.L. All authors have read and agreed to the published version of the manuscript.

Funding: This study was supported by Ministry of Agriculture of the People's Republic of China (2016ZX08011005) and Hubei Provincial Public Health Academic Leader Program (No. 2021-73).

Institutional Review Board Statement: All experimental procedures were approved and supervised by the Hubei Provincial Animal Management and Use Committee (No. 2015022).

Informed Consent Statement: Not applicable.

Data Availability Statement: The data presented in this study are availability on request from the corresponding author. The data are not publicly available due to privacy.

Acknowledgments: We thank Ming Chen for kindly providing the samples of Gm wheat and Jimai22 control wheat.

Conflicts of Interest: The authors declare that they have no conflict of interest.

References

1. Curtis, T.; Halford, N.G. Food security: The challenge of increasing wheat yield and the importance of not compromising food safety. *Ann. Appl. Biol.* **2014**, *164*, 354–372. [CrossRef] [PubMed]
2. Rasheed, A.; Mujeeb-Kazi, A.; Ogbonnaya, F.C.; He, Z.H.; Rajaram, S. Wheat genetic resources in the post-genomics era: Promise and challenges. *Ann. Bot.* **2018**, *121*, 603–616. [CrossRef] [PubMed]
3. Ray, D.K.; Mueller, N.D.; West, P.C.; Foley, J.A. Yield trends are insufficient to double global crop production by 2050. *PLoS ONE* **2013**, *8*, e66428. [CrossRef] [PubMed]
4. Shiferaw, B.; Smale, M.; Braun, H.J.; Duveiller, E.; Reynolds, M.; Muricho, G. Crops that feed the world 10. Past successes and future challenges to the role played by wheat in global food security. *Food Secur.* **2013**, *5*, 291–317. [CrossRef]
5. De Oliveira Silva, A.; Ciampitti, I.A.; Slafer, G.A.; Lollato, R.P. Nitrogen utilization efficiency in wheat: A global perspective. *Eur. J. Agron.* **2020**, *114*, e126008. [CrossRef]
6. Anwaar, H.A.; Perveen, R.; Mansha, M.Z.; Abid, M.; Sarwar, Z.M.; Aatif, H.M.; Umar, U.; Sajid, M.; Aslam, H.M.U.; Alam, M.M.; et al. Assessment of grain yield indices in response to drought stress in wheat (*Triticum aestivum* L.). *Saudi J. Biol. Sci.* **2019**, *12*, 1818–1823. [CrossRef]
7. Li, H.J.; Zhou, Y.; Xin, W.L.; Wei, Y.Q.; Zhang, J.L.; Guo, L.L. Wheat breeding in northern China: Achievements and technical advances. *Crop J.* **2019**, *7*, 718–729. [CrossRef]
8. Ahmed, H.G.M.; khan, A.S.; Li, M.J.; Khan, S.H.; Kashif, M. Early selection of bread wheat genotypes using morphological and photosynthetic attributes conferring drought tolerance. *J. Integr. Agr.* **2019**, *18*, 2483–2491. [CrossRef]
9. Ayala, F.; Fedrigo, G.V.; Burachik, M.; Miranda, P.V. Compositional equivalence of event IND-OO412-7 to non-transgenic wheat. *Transgenic Res.* **2019**, *28*, 165–176. [CrossRef]
10. Nuccioa, M.L.; Paulb, M.; Batea, N.J.; Cohna, J.; Cutlerc, S.R. Where are the drought tolerant crops? An assessment of more than two decades of plant biotechnology effort in crop improvement. *Plant Sci.* **2018**, *273*, 110–119. [CrossRef]
11. Domingo, J.L. Safety assessment of GM plants: An updated review of the scientific literature. *Food Chem. Toxicol.* **2016**, *95*, 12–18. [CrossRef] [PubMed]
12. Fang, J.; Feng, Y.Q.; Zhi, Y.; Zhang, L.; Yu, Z.; Jia, X.D. A 90-day toxicity study of GmTMT transgenic maize in Sprague-Dawley rats. *Regul. Toxicol. Pharm.* **2017**, *85*, 48–54. [CrossRef] [PubMed]
13. Wang, E.H.; Yu, Z.; Jia, X.D.; Zhang, W.Z.; Xu, H.B. Effects of Parental Dietary Exposure to GM Rice TT51 on the Male Reproductive System of Rat Offspring. *Biomed. Environ. Sci.* **2016**, *29*, 267–274. [PubMed]
14. Tian, J.; Fan, B.L.; Ke, X.H.; Tian, M.X.; Yuan, Y.; Yang, W.X.; Tang, X.Q.; Zhuo, Q.; Yang, X.G.; Liu, J.F. Study on one-generation reproductive development toxicity of GmDREB3 gene modified wheat in Wistar rats. *Chin. J. Food Hyg.* **2019**, *31*, 313–319.
15. Tian, J.; Fan, B.L.; Ke, X.H.; Yuan, Y.; Yang, W.X.; Tang, X.Q.; Qu, J.J.; Qu, W.; Fu, S.H.; Zheng, Y.H.; et al. Two generation reproduction toxicity study of GmDREB3 gene modified wheat in Wistar rats. *Food Chem. Toxicol.* **2021**, *153*, 112310. [CrossRef] [PubMed]
16. OECD. Repeated Dose 90-day Oral Toxicity Study in Rodents. In *OECD Guidelines for the Testing of Chemicals*; OECD: Paris, France, 1998.

17. Rai, K.K.; Rai, N.; Rai, S.P. Prediction and validation of DREB transcription factors for salt tolerance in *Solanum lycopersicum* L.: An integrated experimental and computational approach. *Environ. Exp. Bot.* **2019**, *165*, 1–18. [CrossRef]
18. Pandey, B.; Sharma, P.; Saini, M.; Pandey, D.M.; Sharma, I. Isolation and characterization of dehydration responsive element-binding factor 2 (DREB2) from Indian wheat (*Triticum aestivum* L.) cultivars. *Aust. J. Crop Sci.* **2014**, *8*, 44–54.
19. Hichri, I.; Muhovski, Y.; Clippe, A.; Zizkova, E.; Dobrev, P.I.; Motyka, V.; Lutts, S. SlDREB2, a tomato dehydration-responsive element-binding 2 transcription factor, mediates salt stress tolerance in tomato and Arabidopsis. *Plant Cell Environ.* **2016**, *39*, 62–79. [CrossRef]
20. Muneer, S.; Jeong, B.R. Proteomic analysis of salt-stress responsive proteins in roots of tomato (*Lycopersiconesculentum* L.) plants towards silicon efficiency. *Plant Growth Regul.* **2019**, *77*, 133–146. [CrossRef]
21. Zhao, J.P.; Han, C.; Ren, S.; Wang, T.L.; Zhang, X.S.; Zhao, J.; Shen, Y.; Liu, T.T.; Zhuo, Q.; Li, Y. Comparison of nutritional component between transgenic wheat with drought-tolerance gene GmDREB3 and its non-transgenic counterpart. *Food Nutr. Chin.* **2017**, *23*, 21–24.
22. Li, Y.; Han, C.; Mao, H.M.; Shi, L.L.; Chen, C.; Zhao, J.P.; Zhuo, Q. Effect of GmDREB3 gene modified wheat on immune system of rats. *Food Nutr. Chin.* **2018**, *24*, 20–25.
23. Ozuna, C.V.; Barro, F. Safety evaluation of transgenic low-gliadin wheat in Sprague Dawley rats: An alternative to the gluten free diet with no subchronic adverse effects. *Food Chem. Toxicol.* **2017**, *107*, 176–185. [CrossRef] [PubMed]
24. Papineni, S.; Murray, J.A.; Ricardo, E.; Dunville, C.M.; Sura, R.K.; Thomas, J. Evaluation of the safety of a genetically modified DAS-444Ø6-6 soybean meal and hulls in a 90-day dietary toxicity study in rats. *Food Chem. Toxicol.* **2017**, *109*, 245–252. [CrossRef] [PubMed]
25. Papineni, S.; Passage, J.K.; Ekmay, R.D.; Thomas, J. Evaluation of 30% DAS-444Ø6-6 soybean meal in a subchronic rat toxicity study. *Regul. Toxicol. Pharmacol.* **2018**, *94*, 57–69. [CrossRef] [PubMed]
26. Wang, E.H.; Yu, Z.; Jia, X.D.; Xu, H.B. A two-generation reproduction study with transgenic Bt rice TT51 in Wistar rats. *Food Chem. Toxicol.* **2014**, *65*, 312–320. [CrossRef]
27. Snell, C.; Bernheim, A.; Berge, J.B.; Kuntz, M.; Pascal, G.; Paris, A.; Ricroch, A.E. Assessment of the health impact of GM plant diets in long-term and multigenerational animal feeding trials: A literature review. *Food Chem. Toxicol.* **2012**, *50*, 1134–1148. [CrossRef] [PubMed]
28. Reeves, P.G.; Nielsen, F.H.; Fahey, G.C. AIN-93 Purified Diets for Laboratory Rodents: Final Report of the American Institute of Nutrition Ad Hoc Writing Committee on the Reformulation of the AIN-76A Rodent Diet. *J. Nutr.* **1993**, *23*, 1939–1951. [CrossRef] [PubMed]
29. Shackelford, C.; Long, G.; Wolf, J.; Okerberg, C.; Herbert, R. Qualitative and quantitative analysis of nonneoplastic lesions in toxicology studies. *Toxicol. Pathol.* **2002**, *30*, 93–96. [CrossRef]

plants

MDPI

Article

Effects of Exogenous Salicylic Acid Application to Aboveground Part on the Defense Responses in Bt (*Bacillus thuringiensis*) and Non-Bt Corn (*Zea mays* L.) Seedlings

Yuanjiao Feng [1,2,3,4,†], Xiaoyi Wang [1,2,3,4,†], Tiantian Du [1,2,3,4], Yinghua Shu [1,2,3,4], Fengxiao Tan [1,2,3,4] and Jianwu Wang [1,2,3,4,*]

1 Key Laboratory of Agro-Environment in the Tropics, Ministry of Agriculture, South China Agricultural University, Guangzhou 510642, China
2 Guangdong Provincial Key Laboratory of Eco-Circular Agriculture, South China Agricultural University, Guangzhou 510642, China
3 Guangdong Engineering Research Center for Modern Eco-Agriculture and Circular Agriculture, Guangzhou 510642, China
4 Department of Ecology, College of Natural Resources and Environment, South China Agricultural University, Guangzhou 510642, China
* Correspondence: wangjw@scau.edu.cn; Tel.: +86-136-0286-3467
† These authors contributed equally to this work.

Abstract: Bt (*Bacillus thuringiensis*) corn is one of the top three large-scale commercialized anti-insect transgenic crops around the world. In the present study, we tested the Bt protein content, defense chemicals contents, and defense enzyme activities in both the leaves and roots of Bt corn varieties 5422Bt1 and 5422CBCL, as well as their conventional corn 5422 seedlings, with two fully expanded leaves which had been treated with 2.5 mM exogenous salicylic acid (SA) to the aboveground part for 24 h. The result showed that the SA treatment to the aboveground part could significantly increase the polyphenol oxidase activity of conventional corn 5422, the Bt protein content, and peroxidase activities of Bt corn 5422Bt1, as well as the polyphenol oxidase and peroxidase activity of Bt corn 5422CBCL in the leaves. In the roots, the polyphenol oxidase and peroxidase activity of conventional corn 5422, the polyphenol oxidase and superoxide dismutase activities of Bt corn 5422Bt1, the DIMBOA (2,4-dihydroxy-7-methoxy-2H, 1, 4-benzoxazin-3 (4H)-one) content, and four defense enzymes activities of Bt corn 5422CBCL were systematically increased. These findings suggest that the direct effect of SA application to aboveground part on the leaf defense responses in Bt corn 5422CBCL is stronger than that in non-Bt corn. Meanwhile, the systemic effect of SA on the root defense responses in Bt corn 5422CBCL is stronger than that in conventional corn 5422 and Bt corn 5422Bt1. It can be concluded that the Bt gene introduction and endogenous chemical defense responses of corns act synergistically during the SA-induced defense processes to the aboveground part. Different transformation events affected the root defense response when the SA treatment was applied to the aboveground part.

Keywords: Bt corn; salicylic acid; defense responses; Bt protein; defense chemicals; defense enzymes

Citation: Feng, Y.; Wang, X.; Du, T.; Shu, Y.; Tan, F.; Wang, J. Effects of Exogenous Salicylic Acid Application to Aboveground Part on the Defense Responses in Bt (*Bacillus thuringiensis*) and Non-Bt Corn (*Zea mays* L.) Seedlings. *Plants* **2022**, *11*, 2162. https://doi.org/10.3390/plants11162162

Academic Editors: Wei Wei and C. Neal Stewart, Jr.

Received: 25 July 2022
Accepted: 19 August 2022
Published: 20 August 2022

1. Introduction

Genetically engineered crops have been cultivated globally for 25 years over a continuously growing area. According to the ISAAA (International Service for the Acquisition of Agri-Biotech Applications), planting areas for transgenic crops have increased significantly, amounting to 190.4 million hectares in 2019 [1]. Meanwhile, the worldwide surface of insect-resistant Bt crops, alone or in combination with other traits such as herbicide tolerance, has exceeded 107 million hectares, among which Bt corn was planted over approximately 66 million hectares worldwide [1]. Corn (*Zea mays* L.) is a major food crop in China [2]. Bt corn is one of the top three large-scale commercialized anti-insect transgenic crops around the world, genetically modified to express insecticidal crystal protein from the bacterium *Bacillus thuringiensis*, which is toxic to some lepidopteran pests [3–6]. In addition to paying attention to the temporal and spatial expression of Bt protein in Bt corn [2,7,8] and the detection of insect resistance [5,9–13], the ecological risk assessment of Bt corn has also attracted the extensive attention of scholars at home and abroad [14–18]. To date, many studies have investigated the ecological risk assessment of Bt corn, mostly on the effect of Bt corn plantation on nontarget organisms on the ground [17,19,20]. In addition, studies have assessed the impact of Bt corn plantation and straw returning on the soil ecosystem [21–27].

Corn has evolved several strategies to avoid or defend the diseases and pests during their long coevolution processes, called "induced defense responses" [3,28–33]. The ways of inducing defense response usually include mechanical wounding, insect feeding, exogenous jasmonic acid (JA), and salicylic acid (SA) treatment, etc. Such defense responses can be studied by determining the contents of defense chemicals (such as DIMBOA (2,4-dihydroxy-7-methoxy-2H, 1, 4-benzoxazin-3 (4H)-one) and total phenolics) and the activities of some defensive enzymes (such as polyphenol oxidase, peroxidase, catalase, and superoxide dismutase), etc. [4,34]. As we know, the induced defense response in corn is characterized by a systematic response, and the treatment to the leaves may systematically induce the chemical defense response in the roots of corn [31,32]. It is believed that pest control can be achieved by combining the endogenous defense system with the introduced Bt genes, which is a promising alternative strategy for pest resistance management [3,4,35–37]. Nowadays, evaluating the difference in the induced defense responses between Bt corn and conventional corn is an important component to assess the ecological risk of transgenic crops [3,4,38–40]. So far, the main studies on the induced defense response of Bt corn include mechanical wounding, insect feeding, and exogenous JA treatment, and there are no reports on the direct and systematic effects of exogenous SA treatment on the defense response of Bt corn.

SA is a key signal compound in the induced defense response, hypersensitive response, and in systemic acquired resistance to pathogens in corn, and it can be transferred within the whole plant [31,41–43]. A mass of studies have shown that the treatment of exogenous SA to the aboveground part can affect the defense response [44–49]. For instance, foliar spraying of SA in *Thymus vulgaris* L. increased the contents of total phenolics and total flavonoids in the leaves after two months postapplication [44]. In the present study, after the aboveground part of Bt corns (5422Bt1 and 5422CBCL) and their conventional corn 5422 with two fully expanded leaves were treated with 2.5 mM exogenous SA for 24 h, we examined the content of Bt protein, DIMBOA, total phenolics, and the activities of polyphenol oxidase, peroxidase, catalase, and superoxide dismutase in the leaves and roots. It aimed to compare the direct induced defense response in the leaves and the systematic induced defense response in the roots of Bt corns and non-Bt corn with the application of SA to aboveground part. The findings of our study can be helpful in understanding the defense responses in Bt crops against diseases and pests, and also provide useful information for the assessment of the ecological risk of transgenic Bt crops.

2. Results

2.1. Direct and Systemic Effect of Exogenous Salicylic Acid Application to the Aboveground Part on the Content of Defense Chemicals in the Leaves and Roots of Bt and Non-Bt Corn Seedlings

2.1.1. Bt Protein Content

SA application to aboveground part significantly increased the content of Bt protein in the leaves of Bt corn 5422Bt1 by 59.12% ($t = 3.492$, $p = 0.013$), but had no obvious direct effect on the content of Bt protein in the leaves of another Bt corn, 5422CBCL (Figure 1).

The aboveground part treated with SA did not systematically effect the Bt protein content in the roots of the two Bt corns (5422Bt1 and 5422CBCL) (Figure 1).

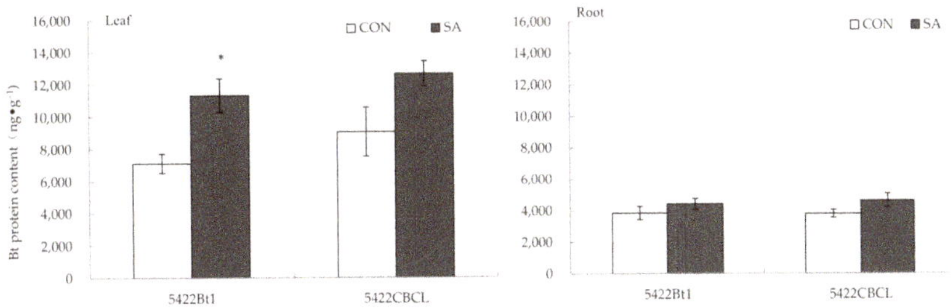

Figure 1. Direct and systemic effects of exogenous salicylic acid application to the aboveground part on the content of Bt protein in the leaves and roots of the Bt corns. The data in the figure are means ± standard errors. Significance among the data was determined by paired Student's *t*-test. "*" showed $p < 0.05$. CON: control; SA: salicylic acid.

2.1.2. DIMBOA Content

There was no difference in the DIMBOA content in the leaves of the conventional corn and the Bt corns (5422Bt1 and 5422CBCL) after the application of SA to the aboveground part (Figure 2).

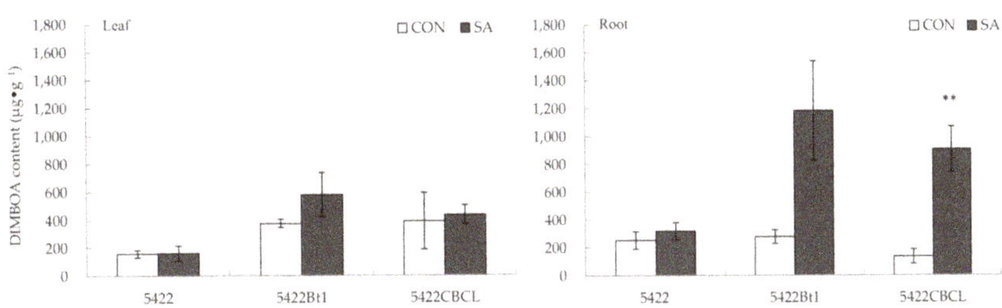

Figure 2. Direct and systemic effects of exogenous salicylic acid application to the aboveground part on the content of DIMBOA in leaves and roots of conventional and Bt corns. The data in the figure are means ± standard errors. Significance among the data was determined by paired Student's *t*-test. "**" showed $p < 0.01$. CON: control; SA: salicylic acid; DIMBOA: 2,4-dihydroxy-7-methoxy-2H, 1,4-benzoxazin-3 (4H)-one.

The SA application to aboveground part had no significant systemic effect on the DIMBOA content in the roots of the conventional corn 5422 and the Bt corn 5422Bt1 (Figure 2), but it increased systematically the DIMBOA content in the roots of another Bt corn, 5422CBCL, by 573.63% ($t = 4.521$, $p = 0.004$).

2.1.3. Total Phenolic Content

The SA application to aboveground part had no significant direct effect on the content of the total phenolics in the leaves of the two Bt corns and the conventional corn (Figure 3).

The SA treatment of the aboveground part had no significant systematic effect on the content of the total phenols in the roots of the two Bt corns (5422Bt1 and 5422CBCL) and the conventional corn 5422 (Figure 3).

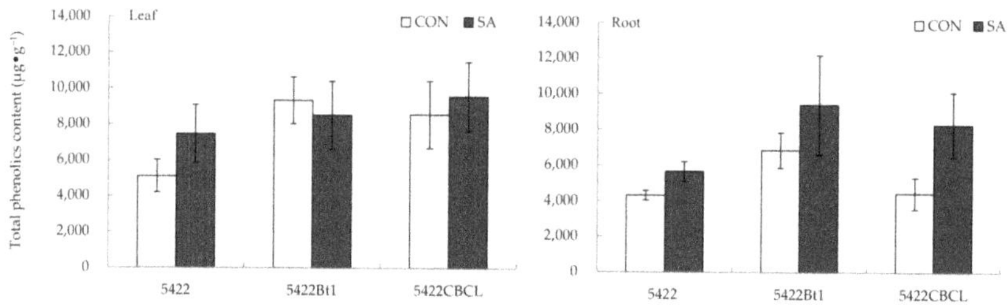

Figure 3. Direct and systemic effects of exogenous salicylic acid application to the aboveground part on the content of total phenolics in leaves and roots of conventional and Bt corns. The data in the figure are means ± standard errors. Significance among the data was determined by paired Student's *t*-test. CON: control; SA: salicylic acid.

2.2. *Direct and Systemic Effect of Exogenous Salicylic Acid Application to the Aboveground Part on the Activity of Defense Enzymes in the Leaves and Roots of Bt and Non-Bt Corn Seedlings*

2.2.1. Polyphenol Oxidase Activity

The direct effect of the SA treatment to the aboveground part on the activity of polyphenol oxidase of the conventional corn and the Bt corns in the leaves was different (Figure 4). The SA treatment of the aboveground part increased significantly the polyphenol oxidase activity in the leaves of the conventional corn 5422 by 74.28% ($t = 2.903$, $p = 0.027$), and that of the Bt corn 5422CBCL leaves by 135.58% ($t = 6.545$, $p = 0.001$), but it had no remarkable direct effect on the polyphenol oxidase activity in the leaves of the other Bt corn.

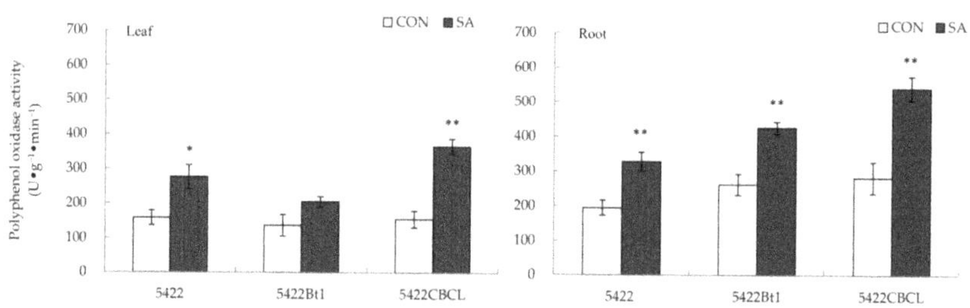

Figure 4. Direct and systemic effects of exogenous salicylic acid application to the aboveground part on the activity of polyphenol oxidase in the leaves and roots of conventional and Bt corns. The data in the figure are means ± standard errors. Significance among the data was determined by paired Student's *t*-test. "*" showed $p < 0.05$, "**" showed $p < 0.01$. CON: control; SA: salicylic acid.

The SA application to the aboveground part had a significant systematic effect on the activity of polyphenol oxidase in the roots of the conventional corn 5422 and the two Bt corns (Figure 4). After the SA was applied to the aboveground part, the polyphenol oxidase activity in the roots of the conventional corn and the Bt corns (5422Bt1 and 5422CBCL) was systematically increased by 68.33% ($t = 3.802$, $p = 0.009$), 62.59% ($t = 4.649$, $p = 0.004$), and 92.15% ($t = 4.543$, $p = 0.004$), respectively.

2.2.2. Peroxidase Activity

The SA application to the aboveground part increased tremendously the peroxidase activity in the leaves of the Bt corn 5422Bt1 and 5422CBCL by 94.06% ($t = 2.758$, $p = 0.033$) and 74.04% ($t = 2.629$, $p = 0.039$), respectively; however, it had no significant direct effect on the peroxidase activity in the leaves of the conventional corn 5422 (Figure 5).

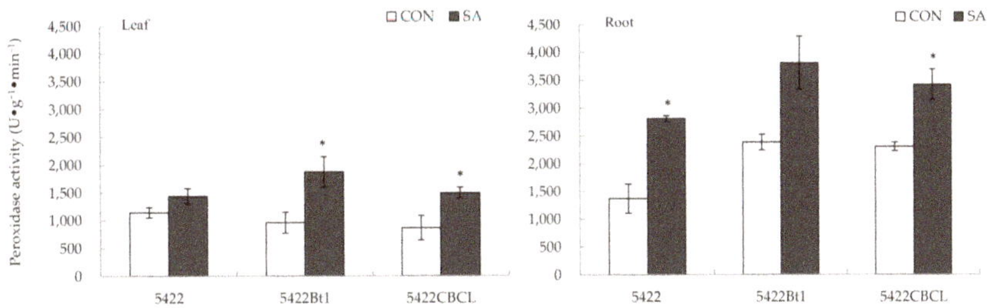

Figure 5. Direct and systemic effects of exogenous salicylic acid application to the aboveground part on the activity of peroxidase in the leaves and roots of conventional and Bt corns. The data in the figure are means ± standard errors. Significance among the data was determined by paired Student's *t*-test. "*" showed $p < 0.05$. CON: control; SA: salicylic acid.

The SA treatment to the aboveground part systematically increased the peroxidase activity in the roots of the conventional corn 5422 and Bt corn 5422CBCL by 105.11% ($t = 5.309$, $p = 0.010$) and 48.25% ($t = 3.955$, $p = 0.022$), respectively, but it had no obvious systematic effect on the peroxidase activity in the roots of another Bt corn, 5422Bt1 (Figure 5).

2.2.3. Catalase Activity

The SA treatment to the aboveground part had no obvious direct effect on the content of the catalase activity in the leaves of the two Bt corns (5422Bt1 and 5422CBCL) and the conventional corn (Figure 6).

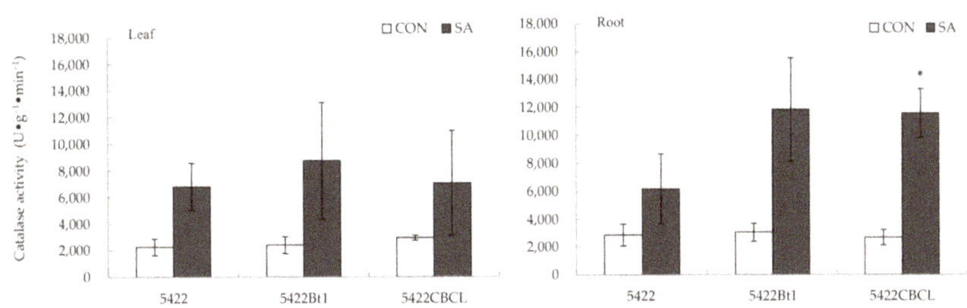

Figure 6. Direct and systemic effects of exogenous salicylic acid application to the aboveground part on the activity of catalase in the leaves and roots of conventional and Bt corns. The data in the figure are means ± standard errors. Significance among the data was determined by paired Student's *t*-test. "*" showed $p < 0.05$. CON: control; SA: salicylic acid.

The SA application to the aboveground part had no significant systematic effect on the activity of the catalase in the roots of the conventional corn 5422 and the Bt corn 5422Bt1 (Figure 6) but increased systematically the catalase activity in the roots of the Bt corn 5422CBCL by 335.54% ($t = 4.880$, $p = 0.011$).

2.2.4. Superoxide Dismutase Activity

There was no difference in the activity of superoxide dismutase in the leaves of the conventional corn and the Bt corns after the application of the SA to the aboveground part (Figure 7).

The salicylic acid treatment to the aboveground part had no obvious systematic effect on the superoxide dismutase activity in the roots of the conventional maize 5422 (Figure 7), but it increased systematically the superoxide dismutase activity in the roots of the Bt corns (5422Bt1 and 5422CBCL) by 129.63% ($t = 2.762$, $p = 0.033$) and 135.47% ($t = 3.206$, $p = 0.018$), respectively.

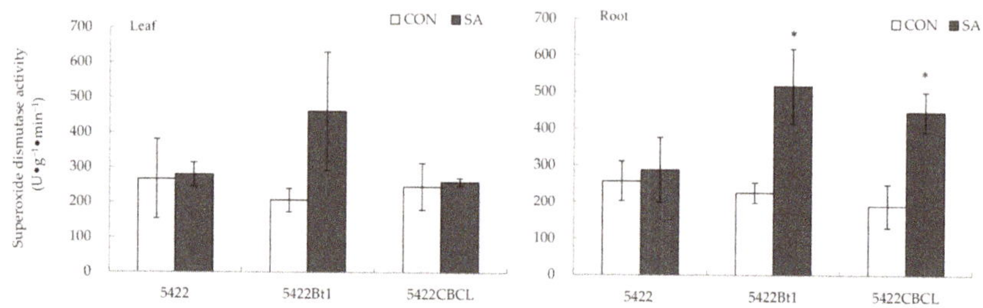

Figure 7. Direct and systemic effects of exogenous salicylic acid application to the aboveground part on the activity of superoxide dismutase in the leaves and roots of the conventional and Bt corns. The data in the figure are means ± standard errors. Significance among the data was determined by paired Student's t-test. "*" showed $p < 0.05$. CON: control; SA: salicylic acid.

3. Discussion

SA is a ubiquitous endogenous signal molecule in plants that can activate the defense and protection mechanisms for disease resistance, and it plays a critical role in plant allergic reactions to pathogens and systemic acquired resistance [31,43]. A mass of studies had proved that the treatment of exogenous SA to the aboveground part can affect directly the foliar defense response [31,44–49]. For example, SA foliage application further enhanced the activities of enzymes (SOD, POD, and CAT) and nonenzymatic antioxidants such as ascorbic acid and the phenolic contents [46]. Our research also showed that the SA treatment to the aboveground part could significantly increase the polyphenol oxidase activity of conventional corn 5422, the Bt protein content and peroxidase activities of Bt corn 5422Bt1, and the polyphenol oxidase and peroxidase activity of Bt corn 5422CBCL in the leaves.

Our study verified that SA application to the aboveground part significantly increased the content of Bt protein in the leaves of the Bt corn 5422Bt1, and this was similar to previous studies. For instance, the Bt protein content was increased in the first leaf of the Bt corn 34B24 by being treated with JA to the first leaf [37]. However, some studies showed that mechanical wounding, Asian corn borer (*Ostrinia furnacalis*) feeding, and exogenous JA treatment to the aboveground part had no significant effect on the Bt protein content in the leaves of the Bt corn 5422Bt1 and 5422CBCL [3,38,39]. In addition, our research showed that the direct effect of SA application to the aboveground part on the leaf defense responses in Bt corn 5422CBCL is stronger than that in non-Bt corn. For example, the SA application to the aboveground part increased tremendously the peroxidase activity in the leaves of the Bt corn 5422CBCL; however, it had no significant direct effect on the peroxidase activity in the leaves of the conventional corn 5422. It can be concluded that the Bt gene introduction and endogenous chemical direct defense responses in the leaves of corn act synergistically during the SA-induced defense processes to the aboveground part. These conclusions are consistent with the previous research results [3,38,39].

As we know, the induced defense response in plants is characterized by a systematic response [31,50,51]. Studies have reported that exogenous SA treatment to the aboveground part can affect systematically the defense response of roots [31,52–55]. For example, Khanna et al. [54] found that the root activities of CAT and SOD increased by 400.0 μM SA within 2 h of foliar treatment with SA in corn seedlings. Similarly, foliar application of 1.0 mM SA in the Oueslati olive variety (*Olea europeae* L.) resulted in increased root contents of total phenolics and flavonoids after 15 d [55]. Our results also proved the polyphenol oxidase and peroxidase activity of conventional corn 5422, the polyphenol oxidase and superoxide dismutase activities of Bt corn 5422Bt1, the DIMBOA content, and the four defense enzymes activities of Bt corn 5422CBCL in the roots were systematically increased after the SA treatment to the aboveground part.

In the present study, the aboveground part treated with SA did not systematically effect the Bt protein content in the roots of the two Bt corns (5422Bt1 and 5422CBCL). The previous studies also showed no significance in the Bt protein content in the roots of Bt corn 5422Bt1 and 5422CBCL after mechanical wounding and exogenous JA treatment to the aboveground part [3,38]. However, the Bt protein content was increased systematically in the roots of the Bt corn 5422CBCL which had been damaged by Asian corn borer in the first leaf [39]. Meanwhile, our study showed that the systemic effect of SA application to the aboveground part on the root defense responses in the Bt corn 5422CBCL is stronger than that in the non-Bt corn. For instance, the SA application to the aboveground part increased significantly the DIMBOA content, catalase, and superoxide dismutase activity in the roots of the Bt corn 5422CBCL; however, it had no significant systemic effect in the roots of the conventional corn 5422. It also can be concluded that the Bt gene introduction and the endogenous chemical systemic defense responses in the roots of the corns act synergistically during the SA-induced defense processes to the aboveground part. These conclusions are consistent with the previous research results [3,38]. However, the systematically induced chemical defense response was weaker in the roots for the Bt corns than the conventional corn after the first leaf was damaged by the Asian corn, which implies an antagonistic relationship between the Bt gene introduction and the chemical defense response in the roots [39].

Although the two Bt maize expressed the same insecticidal protein, their conversion events were different. The transformation event of Bt corn 5422Bt1 is Bt11, and the other Bt corn 5422CBCL is Mon810. Our results indicated that the SA treatment to the aboveground part could significantly increase the Bt protein content and peroxidase activities of Bt corn 5422Bt1, and the polyphenol oxidase and peroxidase activity of the Bt corn 5422CBCL in the leaves. However, in the roots, the polyphenol oxidase and superoxide dismutase activities of the Bt corn 5422Bt1, the DIMBOA content, and the four defense enzymes activities of the Bt corn 5422CBCL were systematically increased. These findings suggest that that the systemic effect of the SA application to the aboveground part on the root defense responses in the Bt corn 5422CBCL is stronger than that in the other Bt corn 5422Bt1. It can be concluded that different transformation events affected the root defense response when the SA treatment was applied to the aboveground part, and this was similar to previous studies [3,38,39]. For example, the gene expression of *Bx6*, *Bx9*, *PAL*, *PR-2a*, and *TPS* in the roots of the Bt corn 5422CBCL (Mon810) could be systematically induced by pest damage, whereas only the gene expressions of *Bx9*, *PAL*, and *TPS* in the roots of the Bt corn 5422Bt1 (Bt11) could be systematically induced by pest damage [39]. These results may be due to the content of the chemical defense substances of the corn itself, effected by different transformation events, which leads to the difference in their induced defense responses. The specific mechanism needs to be further studied.

It is well-known that the aboveground and belowground parts of plants are connected closely by the xylem and phloem vessels through which water, nutrients, photosynthates, and other plant compounds are transported [50,56]. Hence, the signal compounds and defense chemicals induced by disease and pests, or exogenous chemicals in the belowground parts may be transmitted to the aboveground parts, and accordingly, initiate the corresponding defense responses in the aboveground parts. Therefore, the research in regard to the induced defense responses of plants belowground has attracted wide attention from scholars [4,31–33,57–59]. Wang et al. [4] indicated that the Bt gene introduction affects the induced defense effects of the JA treatment to the belowground part of corn, leading to a stronger defense response in the roots of the Bt corns than in the conventional corn. Several studies were carried out to understand the systemic responses in nontreated leaves by exogenous SA treatment to the belowground part [31,57–59]. For instance, Song et al. [58] showed that the activities of POD, CAT, and SOD in the leaves were enhanced distinctly when the roots were treated with 0.1 mM SA after 21 d. As a result, it is still necessary to investigate the differences of the direct or systemically induced defense in the leaves and roots of Bt corn and non-Bt corn after the application of SA to the belowground parts.

4. Materials and Methods

4.1. Materials

Two varieties of Bt corn plants expressing insecticidal Cry1Ab protein, which were donated by Cindy Nakatsu (Department of Agriculture, Purdue University) and provided by Beck's Hybrids Company (Atlanta, IN, USA), were used for experiments. They were Bt corns 5422Bt1 (Bt11) and 5422CBCL (Mon810), and their conventional corn 5422. The transformation events of two of Bt corns were different. The transformation event of Bt corn 5422Bt1 is Bt11 and for Bt corn 5422CBCL it is Mon810. SA ((±)-Salicylic acid) was purchased from the Sigma-Aldrich company (St. Louis, MI, USA). The molecular weight of solid SA is 138.12 $g \cdot mol^{-1}$ (99%).

4.2. Experimental Design

Corn seeds were surface-sterilized in 5% hydrogen peroxide solution for 5 min and then germinated in wet cheese cloth at 25 ± 1 °C. Corn plants were grown in nutrient solution and kept at 22–28 °C under a light: dark 12:12 h photoperiod and 70% relative humidity in a growth chamber (Institute of Tropical and Subtropical Ecology, South China Agricultural University). The seedlings were then transferred to a 500 mL nutrient solution (5 mM KNO_3, 5 mM $Ca(NO_3)_2$, 2 mM $MgSO_4$, 1 mM KH_2PO_4, 46 μM H_3BO_3, 9 μM $MnCl_2$, 0.8 μM $ZnSO_4$, 0.3 μM $CuSO_4$, 0.1 μM $H_8MoN_2O_4$, 20 μM FeNaEDTA) with two seedlings per pot every other two days [4]. We have examined effects of the salicylic acid (SA) concentration (0.1, 0.5, 1.0, 2.5, 5.0 mM) and time (3, 12, 24, 48, 72 h) on the defense responses in corn after foliar application. The results showed that SA foliar application at 2.5 mM produces strong defense responses in corn, with the optimal induction time at 24 h. Treatments were applied to the corn seedlings with two fully expanded leaves. There were two treatments with four repeats: 2.5 mM salicylic acid treatment (SA) and the control (CON). In the 2.5 mM SA treatment, 100 μL SA solution containing 0.14% ethanol and 0.05% Tween-20 was spread uniformly on two fully expanded leaves of seedlings. The same volume of distilled water with 0.14% ethanol and 0.05% Tween-20 was used as control. After the leaves were treated for 24 h, the leaves (the treated part) and roots (the nontreated part) of each plant were collected separately to determine the content of Bt protein, DIMBOA, total phenolics, and the activities of polyphenol oxidase, peroxidase, catalase, and superoxide dismutase.

4.3. Analysis of Bt Protein

Bt protein in the leaf and root was quantified using commercial enzyme-linked immunosorbent assay (ELISA) kits according to the protocol of manufacturer (Agdia Company, Elkhart, IN, USA). Briefly [38], a 20 mg leaf sample or a 30 mg root sample was ground into powder in liquid nitrogen and immediately transferred into 2 mL centrifugal tube. The samples were then extracted by 1 mL PBST (Phosphate-Buffered Saline-Tween, provided with the kit). They were mixed thoroughly and centrifuged under 12,000 rpm at 4 °C for 10 min. The supernatant was diluted at a certain ratio (leaf: 1000:1; root: 600:1) with PBST for further detection. Diluted sample of 100 μL was added into each well of the plate (supplied with the kit), shaken for 15 min at 200 $r \cdot min^{-1}$ after covering preservative film, and followed by a 2 h incubation at room temperature. Then, the plate was washed 5 times with PBST and incubated for another 2 h after adding 100 μL enzyme conjugate in each well. The washing was repeated, and then 100 μL of TMB (Tetramethylbenzidine) substrate was added. After 15 min incubation, 50 μL 3 M H_2SO_4 was added to stop color development. Absorbance was measured at a wavelength 450 nm with a molecular devices microplate reader (Molecular Devices, San Jose, CA, USA) within 30 min. Bt protein concentration was measured by a five-point standard curve of purified Cry1Ab (supplied with the kit).

4.4. Analysis of DIMBOA

The procedure to prepare samples for the DIMBOA concentration was slightly modified from Ni and Quisenberry [60]. Fresh leaves and roots were weighed and ground into powder with a mortar in 10 mL distilled water. Aqueous extracts were incubated for 20 min, and samples were diluted with methyl alcohol in a ratio of 1:1. The methanol-diluted extract was centrifuged at 12,000 rpm for 15 min and filtered. The filtrate was evaporated to dryness under a vacuum. The residue was dissolved in 2 mL mixed solution (acetonitrile: 0.5% aqueous acetic acid, 1:1, *v*/*v*). Extracts were filtered through 0.45 μm membrane filters, and then the samples were stored at −20 °C for further measurement.

DIMBOA concentrations in the samples were quantified by high-performance liquid chromatography (HPLC) (Agilent 1100, USA) (column, Hypersil ODS C_{18} column (250 mm × 4 mm, 5 μm)) with a DAD detector by using external standard curves. Gradient elution was performed with a gradient of A (acetonitrile) and B (0.5% aqueous acetic acid), i.e., 25–45% of A from 0–10 min and 45–25% of A from 10–15 min. Solvent flow rate was set at 1 $mL \cdot min^{-1}$. The injection volume was 20 μL and the detection wavelength was 262 nm. DIMBOA contents in leaves and roots were determined according to the standard calibration curve obtained by peak area of a series of concentrations (0, 20, 40, 60, 80, 100 $\mu g \cdot mL^{-1}$) of DIMBOA standard samples. DIMBOA standard sample was purchased from the Shanghai ACMEC biochemical technology company. The purity of DIMBOA standard sample was 99%.

4.5. Analysis of Total Phenolics

Total phenolics contents were assayed according to Randhir and Shetty [61] and were determined as gallic acid equivalents (0, 25, 50, 100, 150, 200 $\mu g \cdot mL^{-1}$). Samples were weighed and ground into power in liquid nitrogen, soaked in 10 mL of 95% ethanol, and then kept in freezer for 48 h. The sample was centrifuged at 12,000 rpm for 10 min and filtered. Leaf: the filtrate of 0.5 mL was transferred into a test tube, and then 1.5 mL of 95% ethanol, 5 mL of distilled water, and 0.5 mL of Folin–Ciocalteu phenol reagent were added. Root: the filtrate of 1 mL was transferred into a test tube, and then 1 mL of 95% ethanol, 5 mL of distilled water, and 0.5 mL of Folin–Ciocalteu phenol reagent were added. After an incubation period of 5 min, 1 mL of 5% Na_2CO_3 was added, mixed well, and kept in dark for 1 h. The content of total phenolics was measured at 725 nm using UV–visible spectrophotometer (UV-2450 SHIMADZU, Kyoto, Japan).

4.6. Analysis of Polyphenol Oxidase Activity

Polyphenol oxidase (PPO) crude enzyme preparation was done according to Sivakumar and Sharma [62]. The samples were homogenized individually with 1 mL 0.1 M phosphate buffer (PH 6.5) in the ratio of 1:5 (*w*/*v*), centrifuged at 6000 rpm at 4 °C for 15 min, and the supernatant was used as crude enzyme for estimation.

The crude enzyme solution (10 μL), sample dilution (40 μL), and 6 concentrations (0, 15, 30, 60, 120, 180 $U \cdot L^{-1}$) of standard solutions were added to each well of microplate (Rapidbio Company, Plymouth, MI, USA), and thereafter the wells were incubated for 30 min at 37 °C. The microplate wells were washed with buffer for five times, and 50 μL HRP conjugate reagent was added and the wells, which were incubated and then washed five more times. An amount of 50 μL chromogen solution A and 50 μL chromogen solution B were added to each well and kept in dark for 15 min at 37 °C, then 50 μL stop solution was added to each well. The absorbance at 450 nm was measured by microplate reader. The polyphenol oxidase activity was calculated using a regression equation of the standard curve with the standard density and OD value.

4.7. Analysis of Peroxidase Activity

The activity of peroxidase (POD) in the leaf and root was quantified using the guaiacol colorimetric method described by Gao [63]. About 0.1 g of the sample was homogenized with 1 mL of 0.05 $mol \cdot L^{-1}$ PBS (phosphate-buffered saline) in an ice bath and centrifuged at 4 °C for 15 min at 8000 rpm. The supernatant was collected and used for the assay. For POD, the oxidation of guaiacol was measured by the increase in absorbance at 470 nm in every 30 s for 2 min. The assay contained 0.95 mL of 0.2% guaiacol, 1 mL phosphate buffer solution (pH 7.0), and 0.05 mL enzyme extract. The reaction was started with 1 mL of 0.3% H_2O_2.

4.8. Analysis of Catalase Activity

The activity of catalase (CAT) in the leaf and root was determined by hydrogen peroxide decomposition according to the method of Li [64]. An amount of 0.1 g of the sample was homogenized with 1 mL of 0.05 mol·L^{-1} PBS in an ice bath, then centrifuged at 4 °C for 15 min at 8000 rpm. The supernatant was collected and used for the assay. For CAT, the decomposition of H_2O_2 was followed by the decline in absorbance at 240 nm in every 30 s for 2 min. The 3 mL reaction mixture contained 1 mL phosphate buffer solution (pH 7.0), 1 mL of 0.3% H_2O_2, 0.95 mL of 0.2% guaiacol, and 0.05 mL enzyme extract. The reaction was initiated with enzyme extract.

4.9. Analysis of Superoxide Dismutase Activity

The activity of the superoxide dismutase (SOD) in the leaf and root was determined by measuring its ability to inhibit the photochemical reduction of nitroblue tetrazolium (NBT) according to the method of Gao [63]. An amount of 0.1 g of the sample was homogenized with 1 mL of 0.05 mol·L^{-1} PBS (pH 7.8) in an ice bath, then centrifuged at 4 °C for 15 min at 8000 rpm. The 3 mL reaction mixture contained 1.75 mL of 0.05 mol·L^{-1} PBS (pH 7.8), 0.3 mL of 130 mmol·L^{-1} methionine, 0.3 mL of 750 μmol·L^{-1} NBT, 0.3 mL of 100 μmol·L^{-1} EDTA-Na_2, and 0.05 mL of the enzyme extract. An amount of 0.3 mL of 20 μmol·L^{-1} riboflavin was added last. The test tubes containing the mixture were placed under two fluorescent lamps at 4000 lux. The reaction was started by switching on the light and was allowed to run for 20 min. The reaction was stopped by switching off the light and the absorbance at 560 nm was recorded. A nonirradiated reaction mixture that did not develop color was used as the control, and its absorbance was subtracted from A_{560}. The reaction mixture which lacked enzyme developed maximum color was the result of the maximum reduction of NBT. The reduction of NBT was inversely proportional to the enzyme activity. One unit of superoxide dismutase activity was defined as the amount of enzyme required to cause 50% inhibition of the rate of NBT reduction at 560 nm.

4.10. Statistical Analysis

The data were expressed as the means ± standard errors of four repeats. Analysis of group *t*-test was carried out using SPSS 13.0 (IBM, Armonk, MI, USA), and the significances were tested at 0.05 level using group *t*-test.

5. Conclusions

Our results indicated that direct effect of the SA application to the aboveground part on the leaf defense responses in the Bt corn 5422CBCL was stronger than that in conventional corn 5422. Meanwhile, the systemic effect of the SA application to the aboveground part on the root defense responses in the Bt corn 5422CBCL was stronger than that in the conventional corn 5422 and the Bt corn 5422Bt1. It can be concluded that the Bt gene introduction and the endogenous chemical defense responses of the corns act synergistically during the SA-induced defense processes to the aboveground part. Different transformation events affected the root defense response when the SA treatment was applied to the aboveground part.

Author Contributions: Writing—Original Draft Preparation, Y.F.; Methodology and formal analysis, X.W.; Investigation and resources, T.D.; Review and editing, Y.S.; Review and editing, F.T.; Supervision, J.W. All authors have read and agreed to the published version of the manuscript.

Funding: This research was financially supported by the National Natural Science Foundation of China (31971550, 32071619, 31370543, 31470574) and Guangdong Provincial Science and Technology Plan Project (2019B030301007, 2017A020216012).

Institutional Review Board Statement: Not applicable.

Informed Consent Statement: Not applicable.

Data Availability Statement: Data recorded in the current study are available in all Tables and Figures of the manuscript.

Acknowledgments: We thank Cindy Nakatsu (Purdue University, West Lafayette, IN) for providing us with corn seeds.

Conflicts of Interest: All the authors have no conflict of interest.

References

1. ISAAA. Global status of commercialized biotech/GM crops: 2019. *China Biotechnol.* **2021**, *411*, 114–119.
2. Nie, C.R.; Wang, J.W.; Luo, S.M.; Peng, Z.S.; Guo, J.W.; Xin, M.G. Spatial and temporal distribution of Bt proteins in Bt maize leaves. *Food Agric. Immunol.* **2021**, *32*, 450–459.
3. Feng, Y.J.; Jin, Q.; Tan, F.X.; Wang, J.W. Jasmonic acid induced defence responses in Bt (*Bacillus thuringiensis*) and non-Bt corn (*Zea mays* L.) seedlings. *Allelopath. J.* **2018**, *43*, 203–206. [CrossRef]
4. Wang, X.Y.; Feng, Y.J.; Suo, W.F.; Shao, Y.T.; Wang, J.W. Effects of jasmonic acid application to belowground part on the content of defence chemicals and activity of defence enzymes in Bt corn (*Zea mays* L.) seedlings. *Allelopath. J.* **2018**, *43*, 239–254. [CrossRef]
5. Horikoshi, R.J.; Vertuan, H.; Castro, A.A.; Morrell, K.; Griffith, C.; Evans, A.; Tan, J.; Asiimwe, P.; Anderson, H.; José, M.O.M.A.; et al. A new generation of Bt maize for control of fall armyworm (*Spodoptera frugiperda*). *Pest Manag. Sci.* **2021**, *77*, 3727–3736. [CrossRef] [PubMed]
6. Álvarez-Alfageme, F.; Devos, Y.; Camargo, A.M.; Arpaia, S.; Messéan, A. Managing resistance evolution to transgenic Bt maize in corn borers in Spain. *Crit. Rev. Biotechnol.* **2022**, *42*, 201–219. [CrossRef]
7. Wang, J.W.; Luo, S.M.; Feng, Y.J.; Cindy, N. The environmental distribution and ecological effects of Bt toxin in transgenic Crops. *Acta Ecol. Sin.* **2003**, *4*, 797–804.
8. Liang, J.G.; Zhang, D.D.; Li, D.Y.; Zhao, S.Y.; Wang, C.Y.; Xiao, Y.T.; Xu, D.; Yang, Y.Z.; Li, G.P.; Wang, L.L.; et al. Expression profiles of Cry1Ab protein and its insecticidal efficacy against the invasive fall armyworm for Chinese domestic GM maize DBN9936. *J. Integr. Agric.* **2021**, *20*, 792–803. [CrossRef]
9. Wang, Y.Q.; He, K.L.; Wang, Z.Y. Evolution of resistance to transgenic *Bacillus thuringiensis* maize in pest insects and a strategy for managing this. *Chin. J. Appl. Entomol.* **2019**, *56*, 12–23.
10. Bilbo, T.R.; Reay-Jones, F.P.F.; Greene, J.K. Evaluation of Insecticide thresholds in late-planted Bt and Non-Bt Corn for management of fall Armyworm (Lepidoptera: Noctuidae). *J. Econ. Entomol.* **2020**, *113*, 814–823. [CrossRef]
11. Li, G.P.; Feng, H.Q.; Ji, T.J.; Huang, J.R.; Tian, C.H. What type of Bt corn is suitable for a region with diverse lepidopteran pests: A laboratory evaluation. *GM Crops Food* **2020**, *12*, 115–124. [CrossRef] [PubMed]
12. Tavares, C.S.; Santos-Amaya, O.F.; Oliveira, E.E.; Paula-Moraes, S.V.; Pereira, E.J.G. Facing Bt toxins growing up: Developmental changes of susceptibility to Bt corn hybrids in fall armyworm populations and the implications for resistance management. *Crop Prot.* **2021**, *146*, 105664. [CrossRef]
13. Eghrari, K.; Oliveira, S.C.; Nascimento, A.M.; Queiroz, B.; Fatoretto, J.; de Souza, B.H.S.; Fernandes, O.A.; Môro, G.V. The implications of homozygous *vip3Aa20*- and *cry1Ab*-maize on *Spodoptera frugiperda* control. *J. Pest Sci.* **2022**, *95*, 115–127. [CrossRef]
14. Feng, Y.J.; Ling, L.; Fan, H.J.; Liu, Y.H.; Tan, F.X.; Shu, Y.H.; Wang, J.W. Effects of temperature, water content and pH on degradation of Cry1Ab protein released from Bt corn straw in soil. *Soil Biol. Biochem.* **2011**, *43*, 1600–1606. [CrossRef]
15. Lv, X.; Feng, Y.J.; Wang, X.Y.; Wang, J.W. Research progress on effects of Bt crop straw return on soil ecosystem. *Ecol. Sci.* **2019**, *38*, 235–240.
16. Shu, Y.H.; Du, Y.; Wang, J.W. Presence of Cry1Ab in the Bt maize–aphid (*Rhopalosiphum maidis*)–ladybeetle (*Propylea japonica*) system has no adverse effects on insect biological parameters. *Entomol. Exp. Appl.* **2019**, *167*, 553–560. [CrossRef]
17. Wang, B.F.; Wu, F.C.; Yin, J.Q.; Jiang, Z.L.; Song, X.Y.; Reddy, G.V.P. Use of taxonomic and trait-based approaches to evaluate the effect of Bt maize expressing Cry1Ie protein on non-target Collembola: A Case Study in Northeast China. *Insects* **2021**, *12*, 88. [CrossRef]
18. Yurchak, V.; Leslie, A.W.; Dively, G.P.; Lamp, W.O.; Hooks, C.R.R. Degradation of transgenic *Bacillus thuringiensis* proteins in corn tissue in response to post-harvest management practices. *Transgenic Res.* **2021**, *30*, 851–865. [CrossRef]
19. Xing, Y.J.; Qin, Z.F.; Feng, M.Y.; Li, A.M.; Zhang, L.; Wang, Y.; Dong, X.H.; Zhang, Y.X.; Tan, S.Q.; Shi, W.P. The impact of Bt maize expressing the Cry1Ac protein on non-target arthropods. *Environ. Sci. Pollut. Res.* **2019**, *26*, 5814–5819. [CrossRef]
20. Yin, Y.; Xu, Y.D.; Cao, K.L.; Qin, Z.F.; Zhao, X.X.; Dong, X.H.; Shi, W.P. Impact assessment of Bt maize expressing the Cry1Ab and Cry2Ab protein simultaneously on non-target arthropods. *Environ. Sci. Pollut. Res. Int.* **2020**, *27*, 21552–21559. [CrossRef]
21. Tan, F.X.; Wang, J.W.; Feng, Y.J.; Chi, G.L.; Kong, H.L.; Qiu, H.F.; Wei, S.L. Bt corn plants and their straw have no apparent impact on soil microbial communities. *Plant Soil* **2010**, *329*, 349–364. [CrossRef]
22. Liu, L.; Zhao, J.N.; Li, G.; Yang, D.L.; Wang, S.R.; Na, B.Q.; Na, R.S. Effects of transgenic Bt maize on soil available nutrients and enzyme activities. *Soils* **2012**, *44*, 167–171.
23. Li, F.; Wang, M.; Sun, H.W.; Yang, S.K.; Lu, X.B. Dynamics of Cry1Ab protein content in the rhizosphere soil and straw debris of transgenic Bt corn. *Chin. J. Appl. Ecol.* **2013**, *24*, 1907–1913.

24. Sun, H.W.; Li, F.; Yang, S.K.; Wang, M.; Lu, X.B. Effects of Bt transgenic maize on genetic diversity of rhizosphere soil bacteria. *J. Maize Sci.* **2013**, *21*, 10–16.
25. Zeng, P.; Feng, Y.J.; Zhang, W.C.; Zhang, Y.F.; Dong, W.C.; Wang, J.W. Change of Bt protein in soil after growing Bt corns and returning corn straws to soil and its effects on soil nutrients. *Chin. J. Appl. Ecol.* **2014**, *25*, 1997–2003.
26. Shu, Y.H.; Zhang, Y.Y.; Cheng, M.M.; Zeng, H.L.; Wang, J.W. Multilevel assessment of Cry1Ab *Bt*-maize straw return affecting the earthworm *Eisenia Fetida*. *Chemosphere* **2015**, *137*, 59–69. [CrossRef]
27. Wang, X.Y.; Feng, Y.J.; Yan, S.; Zeng, P.; Suo, W.F.; Wang, J.W. Effects of continuous planting Bt corn on Bt protein content and microbial quantity in soil. *Ecol. Environ. Sci.* **2016**, *25*, 1945–1952.
28. Karban, R.; Baldwin, I.T. *Induced Responses to Herbivory*; The University of Chicago Press: Chicago, IL, USA, 1997; pp. 1–11.
29. Heil, M. Indirect defence via tritrophic interactions. *New Phytol.* **2008**, *178*, 41–61. [CrossRef]
30. Feng, Y.J.; Wang, J.W.; Su, Y.J.; Luo, S.M. The role of jasmonic acid in the systemic induced defense response of aboveground and underground in corn (*Zea mays*). *Sci. Agric. Sin.* **2009**, *42*, 2726–2735.
31. Feng, Y.J.; Wang, J.W.; Luo, S.M.; Jin, Q.; Fan, H.Z.; Su, Y.J.; Liu, Y.H. Effects of exogenous application of jasmomic acid and salicylic acid on the leaf and root induction of chemical defence in maize (*Zea mays* L.). *Allelopath. J.* **2010**, *25*, 133–146.
32. Feng, Y.J.; Wang, J.W.; Luo, S.M. Review on the induced defense response between plant aboveground and belowground parts. *Ecol. Sci.* **2010**, *29*, 292–297.
33. Feng, Y.J.; Wang, J.W.; Luo, S.M.; Fan, H.Z.; Jin, Q. Costs of jasmonic acid induced defense in aboveground and belowground parts of corn (*Zea mays* L.). *J. Chem. Ecol.* **2012**, *38*, 984–991. [CrossRef] [PubMed]
34. Hemanth Kumar, N.K.; Kumari, M.S.; Singh, M.V.; Jagannath, S. Oxidative stress and antioxidant metabolic enzymes response of maize (*Zea mays* L.) seedlings to a biotic stress (Alachlor) condition. *Int. J. Agric.* **2016**, *8*, 2227–2231.
35. Bates, S.L.; Zhao, J.Z.; Roush, R.T.; Shelton, A.M. Insect resistance management in GM crops: Past, present and future. *Nat. Biotechnol.* **2005**, *23*, 57–62. [CrossRef]
36. Ferry, N.; Edwards, M.G.; Gatehouse, J.; Capell, T.; Christou, P.; Gatehouse, A.M. Transgenic plants for insect pest control: A forward looking scientific perspective. *Transgenic Res.* **2006**, *15*, 13–19. [CrossRef] [PubMed]
37. Feng, Y.J.; Wang, J.W.; Luo, S.M. Effects of exogenous jasmonic acid on concentrations of direct-defense chemicals and expression of related genes in Bt (*Bacillus thuringiensis*) corn (*Zea mays*). *Agric. Sci. China* **2007**, *6*, 1456–1462. [CrossRef]
38. Feng, Y.J.; Jin, Q.; Wang, J.W. Systemic induced effects of mechanical wounding on the chemical defense of Bt corn (*Zea mays*). *Chin. J. Plant Ecol.* **2010**, *34*, 695–703.
39. Feng, Y.J.; Wang, J.W.; Jin, Q. Asian corn borer (*Ostrinia furnacalis*) damage induced systemic response in chemical defence in Bt corn (*Zea mays* L.). *Allelopath. J.* **2010**, *26*, 101–112.
40. Xu, H.; Wang, X.Y.; Chi, G.L.; Tan, B.C.; Wang, J.W. Effects of *Bacillus thuringiensis* genetic engineering on induced volatile organic compounds emission in maize and the attractiveness to a parasitic wasp. *Front. Bioeng. Biotechnol.* **2019**, *7*, 160. [CrossRef]
41. Klessig, D.F.; Malam, Y.J. The salicylic acid signal in plants. *Plant Mol.* **1994**, *26*, 1439–1458. [CrossRef]
42. Cueto-Ginzo, I.A.; Serrano, L.; Sin, E.; Rodríguez, R.; Morales, J.G.; Lade, S.B.; Medina, V.; Achon, M.A. Exogenous salicylic acid treatment delays initial infection and counteracts alterations induced by maize dwarf mosaic virus in the maize proteome. *Physiol. Mol. Plant Pathol.* **2016**, *96*, 47–59. [CrossRef]
43. Belt, K.; Huang, S.; Thatcher, L.F.; Casarotto, H.; Singh, K.; Aken, O.V. Salicylic acid-dependent plant stress signalling via mitochondrial succinate dehydrogenase. *Plant Physiol.* **2017**, *173*, 2029–2040. [CrossRef] [PubMed]
44. Khalil, N.; Fekry, M.; Bishr, M.; Zalabani, S.E.; Salama, O. Foliar spraying of salicylic acid induced accumulation of phenolics, increased radical scavenging activity and modified the composition of the essential oil of water stressed *Thymus vulgaris* L. *Plant Physiol. Biochem.* **2018**, *123*, 65–74. [CrossRef]
45. Naz, R.; Sarfraz, A.; Anwar, Z.; Yasmin, H.; Nosheen, A.; Keyani, R.; Roberts, T.H. Combined ability of salicylic acid and spermidine to mitigate the individual and interactive effects of drought and chromium stress in maize (*Zea mays* L.). *Plant Physiol. Biochem.* **2021**, *159*, 285–300. [CrossRef]
46. Parveen, A.; Arslan, A.M.; Hussain, I.; Perveen, S.; Rasheed, R.; Mahmood, Q.; Hussain, S.; Ditta, A.; Hashem, A.; Al-Arjani, A.F.; et al. Promotion of growth and physiological characteristics in water-stressed *Triticum aestivum* in relation to foliar-application of salicylic acid. *Water* **2021**, *13*, 1316. [CrossRef]
47. Sultan, I.; Khan, I.; Chattha, M.U.; Hassan, U.M.; Barbanti, L.; Calone, R.; Ali, M.; Majeed, S.; Ghani, M.A.; Batool, M.; et al. Improved salinity tolerance in early growth stage of maize through salicylic acid foliar application. *Ital. J. Agron.* **2021**, *16*, 1810. [CrossRef]
48. Hasanuzzaman, M.; Ahmed, N.; Saha, T.; Rahman, M.; Rahman, K.; Alam, M.M.; Rohman, M.M.; Nahar, K. Exogenous salicylic acid and kinetin modulate reactive oxygen species metabolism and glyoxalase system to confer waterlogging stress tolerance in soybean (*Glycine max* L.). *Plant Stress* **2022**, *3*, 100057. [CrossRef]
49. Pan, L.W.; Xiang, C.Y.; Ding, J.W.; Liu, J.B.; Du, J.; Cao, G.Y. Physiological response of salicylic acid on physiological characteristics of rape seeding under salt stress. *J. Tianjin Agric. Univ.* **2022**, *29*, 22–26.
50. Van Dam, N.M.; Harvey, J.A.; Wäckers, F.L.; Bezemer, T.M.; Van der Putten, W.H.; Vet, L.E.M. Interactions between aboveground and belowground induced responses against phytophages. *Basic Appl. Ecol.* **2003**, *4*, 63–77. [CrossRef]
51. Xu, T.; Wang, J.W.; Luo, S.M. Cloning of the key genes in maize oxylipins pathways and their roles in herbivore induced defence. *Chin. Sci.* **2005**, *50*, 2457–2466.

52. Ludwig-Müller, J.; Schubert, B.; Pieper, K.; Ihmig, S.; Hilgenberg, W. Glucosinolate content in susceptible and resistant Chinese cabbage varieties during development of clubroot disease. *Phytochemistry* **1997**, *44*, 407–417. [CrossRef]
53. Du, Y.P.; Ji, X.L.; Jiang, E.S.; Cui, L.J.; Zhai, H. Phylloxera resistance induced by salicylic and jasmonic acids in Kyoho grapevine. *Acta Entomol. Sin.* **2014**, *47*, 443–448.
54. Khanna, P.; Kaur, K.; Gupta, A.K. Salicylic acid induces differential antioxidant response in spring maize under high temperature stress. *Indian J. Exp. Biol.* **2016**, *54*, 386–393. [PubMed]
55. Methenni, K.; Abdallah, M.B.; Nouairi, I.; Smaoui, A.; Ammar, W.B.; Zarrouk, M. Salicylic acid and calcium pretreatments alleviate the toxic effect of salinity in the Oueslati olive variety. *Sci. Hortic.* **2018**, *233*, 349–358. [CrossRef]
56. Taiz, L.; Zeiger, E. *Plant Physiology*; Sinauer Associates, Inc.: Sunderland, MA, USA, 1998.
57. Kiddie, G.A.; Doughty, K.J.; Wallsgrove, R.M. Salicylic acid-induced accumulation of glucosinolates in oilseed rape (*Brassica napus* L.) leaves. *J. Exp. Bot.* **1994**, *45*, 1343–1346. [CrossRef]
58. Song, Y.L.; Dong, Y.J.; Kong, J.; Tian, X.Y.; Bai, X.Y.; Xu, L.L. Effects of root addition and foliar application of nitric oxide and salicylic acid in alleviating iron deficiency induced chlorosis of peanut seedlings. *J. Plant Nutr.* **2017**, *40*, 63–81. [CrossRef]
59. Colak, N.; Kurt-Celebi, A.; Fauzan, R.; Torun, H.; Ayaz, F.A. The protective effect of exogenous salicylic and gallic acids ameliorates the adverse effects of ionizing radiation stress in wheat seedlings by modulating the antioxidant defence system. *Plant Physiol. Biochem.* **2021**, *168*, 526–545. [CrossRef]
60. Ni, X.; Quisenberry, S.S. Comparison of DIMBOA concentrations among wheat isolines and corresponding plant introduction lines. *Entomol. Appl.* **2000**, *96*, 275–279. [CrossRef]
61. Randhir, R.; Shetty, K. Developmental stimulation of total phenolics and related antioxidant activity in light-and dark-germinated corn by natural elicitors. *Process Biochem.* **2005**, *40*, 1724–1732. [CrossRef]
62. Sivakumar, G.; Sharma, R.C. Induced biochemical changes due to seed bacterization by Pseudomonas fluorescens in maize plants. *Indian Phytopathol.* **2003**, *56*, 134–137.
63. Gao, J.F. *Plant Physiology Experimental Guidance*; Higher Education Press: Beijing, China, 2006.
64. Li, H.S. *Experiments Principle and Technology of Plant Physiology and Biochemistry*; Higher Education Press: Beijing, China, 2003.

plants MDPI

Article

Proteomic Analysis of *Bt cry1Ac* Transgenic Oilseed Rape (*Brassica napus* L.)

Zheng-Jun Guan [1,2,†], Min Zheng [2,3,†], Zhi-Xi Tang [2], Wei Wei [2,*] and C. Neal Stewart, Jr. [4]

1 Department of Life Sciences, Yuncheng University, Yuncheng 044000, China; zhengjunguan@126.com
2 State Key Laboratory of Vegetation and Climate Change, Institute of Botany, Chinese Academy of Sciences, Beijing 100093, China; zhengmin0816@126.com (M.Z.); tangzhixi@ibcas.ac.cn (Z.-X.T.)
3 Department of Hotel Management, Linyi Technician Institute, Linyi 276005, China
4 Department of Plant Sciences and Center for Agricultural Synthetic Biology, University of Tennessee, 2505 EJ Chapman Drive, Knoxville, TN 37996-4561, USA; nealstewart@utk.edu
* Correspondence: weiwei@ibcas.ac.cn
† These authors contributed equally to this work.

Abstract: Oilseed rape (*Brassica napus* L.) is an important cash crop, but transgenic oilseed rape has not been grown on a commercial scale in China. It is necessary to analyze the characteristics of transgenic oilseed rape before commercial cultivation. In our study, differential expression of total protein from the leaves in two transgenic lines of oilseed rape expressing foreign Bt Cry1Ac insecticidal toxin and their non-transgenic parent plant was analyzed using a proteomic approach. Only shared changes in both of the two transgenic lines were calculated. Fifteen differential protein spots were analyzed and identified, namely, twelve upregulated expressed protein spots and three downregulated protein spots. These proteins are involved in photosynthesis, transporter function, metabolism, protein synthesis, and cell growth and differentiation. The changes of these protein spots in transgenic oilseed rape may be attributable to the insertion of the foreign transgenes. However, the transgenic manipulation might not necessarily cause significant change in proteomes of the oilseed rape.

Keywords: *cry1Ac* gene; oilseed rape; proteomics; 2D gel electrophoresis

Citation: Guan, Z.-J.; Zheng, M.; Tang, Z.-X.; Wei, W.; Stewart, C.N., Jr. Proteomic Analysis of *Bt cry1Ac* Transgenic Oilseed Rape (*Brassica napus* L.). *Plants* **2023**, *12*, 2319. https://doi.org/10.3390/plants12122319

Academic Editor: Pavel Kerchev

Received: 26 May 2023
Revised: 9 June 2023
Accepted: 12 June 2023
Published: 15 June 2023

1. Introduction

With the rapid development of genetic engineering technology, the application and effects of transgenic plants have gradually attracted public attention worldwide [1,2]. Transgenic technology can produce novel foods more effectively and reduce the need for pesticides, leading to less environmental pollution [3,4]. In particular, Bt toxins isolated from *Bacillus thuringiensis* have been expressed in transgenic plants to confer inherent pest resistance. Bt crops have been overwhelmingly successful and beneficial with respect to increasing yields and reducing chemical pesticide use [5]. However, the technology might have unintended negative impacts on the environment and human health. Nutrition compounds in transgenic Bt crops may be changed, and may even generate toxic compounds due to of the unknown effects of exogenous genes [6–11]. Therefore, Bt plants are commonly selected as experimental systems in studies that focus on detecting potential effects of exogenous genes in transgenic plants.

Oilseed rape (*Brassica napus* L.) is an important oil and cash crop, and it is a major source of edible vegetable oil and proteins. As such, the improvement of rape quality and yield has received increased attention. Herbicide-resistant transgenic oilseed rape was one of the earliest biotech crops developed and has a large range of applications. At present, herbicide-tolerant transgenic oilseed rape has been commercialized in the United States, Canada, Australia and other countries [12]. Although insect-resistant Bt oilseed rape has not been commercialized, it has served as a model biotech crop for studying the effects of transgenic plants (e.g., Cao et al. 2014; Liu et al. 2018) [13,14]. Owing to current methods of transgene insertion, transgenic plants may exhibit random genomic side effects such as variable gene expression, mutations to endogenous loci at the points of insertion, homologous gene expression inhibition effects (i.e., silencing), activation effects and other unanticipated changes in crop phenotypes. Therefore, it is necessary to assess potential unexpected consequences of transgenic Bt oilseed rape before commercial cultivation.

Due to the potential unknown effects and unpredictability of exogenous genes, proteomic technology is an effective and direct method for detecting the unintended effects of transgenic plants [15,16]. Differential proteomics is a major research area that mainly aims to detect and confirm different proteins produced by special stimulation between two or more groups of samples [17–19]. Two-dimensional gel electrophoresis, a core technique of differential proteomic analysis, can isolate different proteins that exist in plants [20]. Previous studies have determined proteome changes in transgenic plants by using proteomics [21–26]. Up to now, the research on proteomic analysis of transgenic plants has not raised any new safety issues [25,26]. In our study, in order to better understand the influencing mechanism of the exogenous *Bt cry1Ac* gene on oilseed rape, a proteomic approach was used to analyze the differential expression of proteins in transgenic Bt oilseed rape.

2. Results

2.1. Comparative Proteomic Analysis of Transgenic and Non-Transgenic Bt Oilseed Rape Leaves

To evaluate the unintended effects of a foreign gene on the leaf expression profile of oilseed rape, differential proteomics between transgenic oilseed rape lines (GT1 and GT9) and the control plant ('Westar') were analyzed. Two-dimensional (2D) gel electrophoresis patterns of leaf proteins with high resolution and reproducibility were successfully obtained. After staining with Coomassie brilliant blue, results for 2D gel electrophoresis of total proteins from the leaves of two lines of transgenic oilseed rape and the non-transgenic control were obtained (Figure 1). Protein spots were detected in the leaves of transgenic oilseed rape, and the number and score of protein spots that matched between transgenic oilseed rape and its control were counted (Table 1). 30 differentially expressed protein spots were successfully detected and identified in the 2 transgenic lines, including 21 upregulated expressed protein spots and 9 downregulated protein spots. Protein spots that were changed in both transgenic lines were further analyzed.

Table 1. Distribution of protein spots in groups and their MS identification between two transgenic oilseed rape lines (GT1 and GT9) and the non-transgenic oilseed rape plants (Westar).

Spot Number [a]	Protein Name	Accession No. [b]	Score [c]	Theor. pI [d]	Exp. pI [d]	Theor. MW [e]	Exp. MW [e]	Ratio [f]
		Energy						
802 (↑)	Ribulose-1,5-bisphosphate carboxylase/oxygenase large subunit	gi | 11466371	86	5.06	6.14	52,405	53,186.5	1.07
65 (↑)	Transketolase	gi | 18411711	97	5.81	5.94	79,837	80,374.4	1.06
858 (↑)	Ribulose bisphosphate carboxylase small chain 1B	gi | 15240912	68	6.27/	7.59	20,155	20,558.2	1.06
367 (↑)	Ribulose bisphosphate carboxylase/oxygenase activase	gi | 297612474	82	4.98	7.56	38,775	39,108.5	1.08
699 (↑)	Ribulose-1,5-bisphosphate carboxylase/oxygenase large subunit	gi | 11466371	134	5.06	6.14	52,405	53,186.5	1.12
844 (↑)	Ribulose bisphosphate carboxylase small chain F1	gi | 132091	390	6.27	8.23	14,358	20,455.2	1.06
476 (↑)	Ribulose bisphosphate carboxylase large chain	gi | 2500677	78	6.28	6.39	48,953	49,069.4	1.08
872 (↑)	Ribulose bisphosphate carboxylase small chain F1	gi | 132091	103	6.27	8.23	14,358	20,455.2	1.06
		Transporters						
586 (↓)	Calcium-transporting ATPase	gi | 302756809	63	8.07	6	112,406	113,321.0	0.93
87 (↓)	V-type proton ATPase catalytic subunit A	gi | 15219234	96	4.53	5.11	68,682	69,111.0	0.93

Table 1. *Cont.*

Spot Number [a]	Protein Name	Accession No. [b]	Score [c]	Theor. pI [d]	Exp. pI [d]	Theor. MW [e]	Exp. MW [e]	Ratio [f]
		Metabolism						
651 (↓)	Alanine aminotransferase 2-like	gi | 30698866	59	4.95	5.95	59,380	59,986.2	0.85
325 (↑)	Glutamine synthetase	gi | 12643761	114	4.24	6.16	47,214	47,714.0	1.09
		Protein synthesis						
513 (↑)	Ribosomal protein L11	gi | 56404772	66	9.60	9.27	14,973	15,151.0	1.10
		Cell growth/division						
319 (↑)	Cis-zeatin O-glucosyltransferase	gi | 242093988	63	4.79	6.06	50,372	50,813.9	1.05

a: Assigned spot numbers as indicated in Figure 1; b: Accession numbers according to NCBInr; c: The Mascot searched score (M. S.) against the database NCBInr; d: The experimental and theoretical pI of the identified proteins; e: The experimental and theoretical mass (Da) of the identified proteins; f: The normalized spot volume in GT1/GT9 leaves divided by the normalized volume in Westar leaves. The upwards arrow '↑' and the downwards arrow '↓' stands for upregulated and downregulated protein spot, respectively.

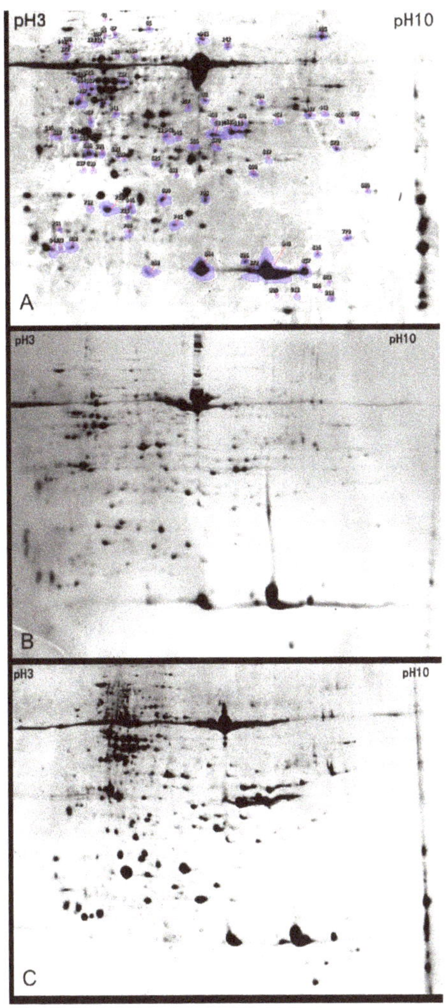

Figure 1. Two-dimensional gel electrophoresis of non-transgenic Westar (**A**), transgenic GT1 (**B**), and GT9 (**C**) oilseed rape. Potentially differential protein spots were identified and marked in blue for further analysis (**A**).

2.2. Identification and Functional Evaluation of the Differentially Expressed Proteins

The detected spots of differentially expressed proteins were excised from the 2D gels and identified using MALDI-TOF/TOF and database searches (such as NCBInr and KEGG). Fifteen differentially expressed protein spots were successfully identified, including twelve upregulated expressed protein spots and three downregulated protein spots. Table 1 provides information such as the Mascot score, NCBI accession ID and name and molecular weight of the successfully identified proteins.

To evaluate the characteristics of the identified proteins, the theoretical and experimental ratios of molecular weight (MW) and isoelectric point (pI) were determined, respectively (Table 1). The closer the theoretical and experimental values of the identified proteins are, the greater the certainty that the identification made by means of Mass Spectrometry (MS) database searching will be the MS identification obtained. About 80% of the theoretical and experimental molecular weight values of the identified proteins were similar, but their pI values were different, indicating that the identified proteins had different characteristics and possible isoforms of the MS data.

The identified proteins were classified into six categories on the basis of their biological activities (Table 1 and Figure 2). In this study, 53.33% (eight spots) of the identified proteins were related to energy functions, 13.33% (two spots) to transporters, 13.33% (two spots) to metabolism, 6.67% (one spot) to protein synthesis, 6.67% (one spot) to cell growth/division, and 6.67% (one spot) presented an unclear classification.

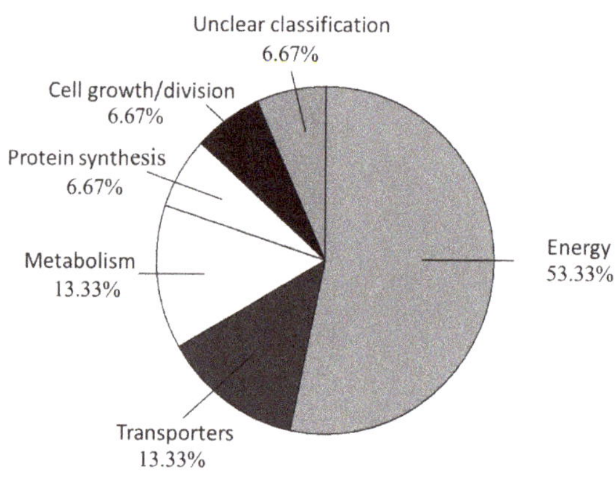

Figure 2. Graphic representation of the functional distribution of differentially abundant protein species (%) identified in non-transgenic Westar and transgenic GT1 and GT9.

In this study, several differentially expressed proteins—including ribulose-1,5-bisphosphate carboxylase/oxygenase large subunit, transketolase, ribulose bisphosphate carboxylase small chain 1B, ribulose bisphosphate carboxylase/oxygenase activase, ribulose bisphosphate carboxylase small chain F1 and ribulose bisphosphate carboxylase large chain—were found to be upregulated and involved in the energy metabolism pathway. Furthermore, some proteins that showed downregulated expression, such as calcium-transporting ATPase and V-type proton ATPase catalytic subunit A, may be related to transporters (Table 1).

3. Discussion

To investigate and characterize the different proteins between transgenic and non-transgenic oilseed rape, proteomic detecting tools (2DE and MS) were applied to identify the differentially expressed proteins between two transgenic Bt oileed rape lines (GT1 and GT9) and its non-transgenic parent plant (Westar) to evaluate proteomic changes between transgenic and non-transgenic plants in our study. Only the protein spots that changed in both transgenic lines in comparison to the non-transgenic plants were further analyzed in order to detect changes that were likely caused by the insertion of foreign transgenes and genetic manipulation. Fifteen protein spots were detected and identified, although the difference seems to not be biologically significant, with a ratio ranging from 0.85 to 1.12. The results suggested that transgenic manipulation might not cause differences in *B. napus* plant proteomes, while such a change could be considered significant when comparing a single transgenic line to its parent plant, e.g., Liu et al. [23]. In this study, these detected proteins of slight difference were mainly related to energy functions, indicating that proteins involved in energy were likely affected by the transgenic manipulation. Yang et al. [27] investigated different proteins between transgenic and non-transgenic rice plants (*Oryza sativa* L.) via comparative proteomic analysis, and a similar result was obtained, that the most abundant category was energy-related proteins among those identified. Liu et al. [23] employed proteomic approaches to study protein abundance changes in seeds from the Bt transgenic line GT1 of oilseed rape (*B. napus*), and eight proteins were more abundant in transgenic oilseed rape seeds than in non-transgenic seeds. The current work studied the proteomic change in the leaf of oilseed rape caused by gene transformation, and no protein with the previous study was detected on seed proteomes. It was suggested that those detected proteins in seeds [23] and in leaves (current study) could be tissue-specific, and the impact of inserted transgenes could be also different between seeds and leaves.

In this study, the highest proportion of detected proteins was related to photosynthesis. Further works are needed to elucidate the potential impact on plant productivity. Ribulose-1,5-bisphosphate carboxylase/oxygenase (rubisco), which is the most abundant protein in plants, is widely found in organelles with photosynthetic functions. Rubisco, composed of one larger and one smaller subunit, is a key enzyme that affects the carbon assimilation rate during photosynthesis. Unlike other enzymes, rubisco is a bifunctional enzyme that catalyzes carboxylation and oxygenation, an intersection of recycling reactions of both photosynthetic carbon reduction and oxidation [28]. In our study, seven upregulated protein spots were related to rubisco subunits. In particular, upregulated ribulose bisphosphate carboxylase/oxygenase activase (protein spot 367) in the leaf of transgenic oilseed rape, which was identified in this study, may change the conformation of the rubisco active site, helping to accelerate the carboxylation process and avoid digestion [29,30]. The expression levels of rubisco in both transgenic GT1 and GT9 were higher than in the non-transgenic parent plant, likely because insertion of the Bt gene required numerous rubisco molecules in order to express the inserted transgene in oilseed rape. The increased abundance of rubisco is likely able to help plants adapt to their environment by conferring a higher carbon assimilation rate [31]. This could be a plant strategy to deal with potential adverse impacts caused by transgene insertion. The co-expressed rubisco gene in three transgenic rice lines was remarkably upregulated under salt stress [32]. It implicated that rubsico could be susceptive to transgene manipulation and environmental growth stress.

Transketolase plays a key role in the Calvin cycle of photosynthesis and is involved in the synthesis of nucleic acids, carbohydrates, amino acids and lipids. Transketolase is identified as a target of herbicidal substance α-terthienyl, revealed by a proteomics study [33]. Overexpression of the transketolase gene promotes chilling tolerance by increasing the activities of photosynthetic enzymes, alleviating oxidative damage and stabilizing cell structure in *Cucumis sativus* L. [34]. Transgenic *Chlamydomonas reinhardtii* cells that overexpressed transketolase of *Pyropia haitanensis* grew better than wild-type cells in response to osmotic stress [35]. Compared to wild-type cotton, five upregulated transketolase protein spots were identified in a transgenic cotton line with a *crylAc* gene from *Bacillus thuringiensis* (BT) [36]. In our study, it can be inferred that the increase of transketolase in transgenic oilseed rape is likely caused by Bt gene insertion.

V-type proton ATPase catalytic subunit A (V-H+-ATPase), a kind of H^+-ATPase, is mainly responsible for catalyzing the hydrolyzation of ATP. This enzyme plays a key role in ion balance within plant cells, and it may modulate the stress resistance of plants (including salt, drought, cold and excessive heavy metal stresses) [37,38]. For example, salt stress reduced the V-H^+-ATPase and the V-H^+-PPase activity in potato cultivars [39].

Calcium-transporting ATPase is a Ca^{2+} transportation system, and it is crucial in regulating intracellular or extracellular Ca^{2+} concentration and signal transduction [40]. V-H^+-ATPase may be related to calcium-transporting ATPase [41]. The results showed that the expression of V-H^+-ATPase has the same decreasing tendency as that of calcium-transporting ATPase in transgenic oilseed rape (GT1 or GT9). Therefore, it could be assumed that despite the insect-resistant features, decreased expression of the two enzymes may weaken the abiotic stress resistance of transgenic plants. To test this hypothesis, future experiments could be conducted to assess the performance of the transgenic plant under abiotic stress conditions.

Alanine aminotransferase 2-like belongs to the pyridoxal phosphate multigene family, and it may help in regulating carbon and nitrogen metabolism in plant cells [42]. A decrease in the expression of alanine aminotransferase 2-like in transgenic oilseed rape (GT1 or GT9) may impact the relevant physiological metabolism and reduce resistance to anti-anoxia and pathogens. Glutamine synthetase (GS) is an important enzyme involved in the assimilation of inorganic nitrogen into organic forms in higher plants [43]. Salt stress induced glutamine synthetase activity in the roots and the leaves of *Trigonella foenum-graecum* L. plants. [44]. GS2-cosuppressed rice plants exhibited a poor plant growth phenotype and a poor nitrogen transport ability [45]. Six genes encoding GS-protein were found and identified from the transcriptome data of the asparagus (*Gracilaria lemaneiformis* L.) genome [46]. In this study, increased expression of GS in transgenic oilseed rape (GT1 or GT9) could enhance the regulation of nitrogen metabolism and improve salt tolerance in transgenic plants.

Ribosomal protein L11 is a highly conserved protein located at the base of the L7/L12 stalk of the ribosome, and is mainly involved in promoting ribosomal RNA folding during protein synthesis [47]. Expression of ribosomal protein L11 in transgenic oilseed rape (GT1 and GT9) increased, likely due to the synthesis of the Bt protein, which requires the participation of a large number of ribosomes. Cis-zeatin-O- glucosyltransferase may stimulate the activity of cytokinins in plants, thereby regulating a series of physiological and biochemical processes (e.g., stimulating growth, retarding senescence and plant stress resistance) [48]. Upregulated expression of cis-zeatin-O- glucosyltransferase in the leaves of transgenic oilseed rape could delay plant senescence to a certain degree and improve the production of transgenic oilseed rape.

In this study, owing to the insertion of the Bt gene, changes in the transgenic oilseed rape leaf proteome were detected, and these changes played important roles in processes such as energy conversion, protein transport and metabolism. These results provided useful information for further illuminating the potential effects of transgenic oilseed rape on human health and environment. Our study inferred that exogenous DNA in a host oilseed rape genome might affect plant photosynthesis, which requires further study. Incidental differences among differences in transgenic-line-associated, photosynthesis-related proteins may have effects on other plant traits, such as biomass production and chlorophyll concentrations, even though no differences were observed here (data not shown). Although there were some unintended protein variations in transgenic oilseed rape leaves, there were no obvious functional proteomic changes produced in the oilseed rape leaf proteome. This study presented a well-established relationship between the identified proteins in transgenic oilseed rape via MS/MS and the databases. The role of each identified protein was curated. The proteins identified in this study were not unique to oilseed rape, but are common among plants. Hence, the identified differences are deemed to not be novel or hazardous, and no significant change in proteomes was found. As such, these transgenic oilseed rape lines likely have no non-target effects in the proteomes. Further works should focus on the safety issues that were solely caused by the expression of Bt Cry1Ac toxin.

In summary, the differential expression of total protein in transgenic oilseed rape was compared using a proteomic approach. Twelve upregulated expressed protein spots and three downregulated protein spots were analyzed and identified. Those protein spots in oilseed rape leaves that related to energy conversion, protein transport, and metabolism may be affected by the transgenic procedure. The results showed that some unintended protein variations, due to the transformation of foreign transgenes in oilseed rape leaves, may not be biologically significant. However, further confirmation of the safety implications of the changes must be considered in risk assessment, especially under environmental change conditions.

4. Materials and Methods

4.1. Plant Materials and Planting

The following three plant types were used in this study: *Brassica napus* 'Westar' (non-transgenic maternal parent) and GT1 and GT9 (two transgenic *B. napus lines*). *B. napus* Westar, a spring-type oilseed rape, was transformed with genetically linked GFP and the Bt (Cry1Ac) gene regulated by independent CaMV 35S promoters in the pSAM12 plasmid [49], and two of the GFP/Bt transformed lines (GT1 and GT9) were used in this current study.

Greenhouse experiments of three biological replicates were conducted at the Institute of Botany, Chinese Academy of Sciences, Beijing, China. The three types of plants were grown in plastic basins in the greenhouse. The potting mix consisted of vermiculite, peat moss and clay soil in the proportion 1:1:1 (*v:v*). The plants were cultured for 16 h of supplemental light per day at a temperature ranging from 18 to 25 °C. The leaves in four to five-leaf stages were selected as experimental materials.

4.2. Protein Extraction and Quantification

Proteins were extracted from the top expanded leaves of the two transgenic lines and the non-transgenic *B. napus* using the method described by Joosen et al. [50]. The presence of transgenes was confirmed by PCR with specific primers (Cry1Ac transgene: 5′-ATTTGGGG-AATCTTTGGTCC-3′ and 5′-ACAGTACGGATT-GGGTAGCG-3′; GFP gene: 5′-TACCCAGATCATATGAAGCGG-3′ and 5′-TTGGGATCTTTCGAAA GGG-3′) for Bt and GFP transgenes at the 4–5 leaf stage at 4 weeks after seed germination [49]. After 1 g of the leaf was ground up in a mortar with liquid nitrogen, proteins were extracted using 10 mL of extraction buffer (10% *w:v* trichloroacetic acid/acetone, 0.07% *w:v* DTT), depolymerized using lysis buffer (7 M urea, 2 M thiourea, 4% *w:v* CHAPS, 1 mM PMSF, 50 mM DTT, 0.5% *w:v* Triton X-100, and 0.5% *v:v* IPG-buffer) and subsequently measured using the Protein Quantification Kit by Bradford method (Beijing Boling Kewei Bio-Technique Co., LTD, Beijing) using BSA (- Sigma-Aldrich. Inc., Beijing) as the standard [51].

4.3. 2D Gel Electrophoresis

The 10 mg protein sample was mixed with 200 μL of isoelectric focusing buffer (7 M urea, 2 M thiourea, 4% *w:v* CHAPS, 50 mM DTT, 0.5% *v:v* IPG-buffer and 0.001% *w:v* bromophenol blue) and loaded onto 18 cm IPG linear dry strips (pH 3–10). After passive rehydration for 14 h, the strips were focused using the Protean Isoelectric Focusing System (Bio-Rad), and the following program was used: 2 h at 50 V, 1 h at 100 V, 1 h at 200 V, 1 h at 500 V, 1 h at 1000 V, 5 h of a linear gradient to 8000 V and 5 h at 8000 V [50].

The strips were equilibrated at room temperature for 15 min in the equilibration solution (6 M urea, 0.375 M Tris-HCl [pH 8.8], glycerol, 2% SDS with 1% DTT, and 0.001% bromophenol blue, followed by carboxymethylation with 2.5% iodoacetamide). The equilibrated strips were run on 12.5% SDS polyacrylamide gels at 5 mA/gel for 45 min and 20 mA/gel for 4–6 h until the dye front reached the bottom of the gel. Proteins were visualized using Coomassie Brilliant Blue G-250 staining after 1 h of protein fixation in a solution containing 40% ethanol and 10% acetic acid. Destaining was performed with the same fixing solution for 1 h, followed by 5 washes with water.

4.4. Image and MS Analysis

Well-separated gels of the three independent biological replicates were used for proteomic comparisons. The gels were scanned with the UMAX Power Looker 2100XL Scanner (Shiqun International Trading Co., Ltd., Shanghai, China) and analyzed for proteome differences. Progenesis Samespots DIGE enable (v4.5) visual tools (CloudScientific Technology

Co., Ltd., Shanghai, China) were used for image analysis, spot detection, matching between gels and normalization. Three biological repeats for each sample were examined, and the results were shown in average ± SD (n = 3). Spots of interest were manually excised from the Gel Analysis Program (GAP) stained 2DE gels.

Protein spots with significant changes were analyzed using matrix-assisted laser desorption/ionization-time of flight mass spectrometry (MALDI-TOF-MS) [52]. Spots were considered reproducible if they were detected in all the biological replicates. Protein spots were considered to be differentially accumulated when the change was more than 1.5-fold with statistically significant differences ($p < 0.05$) [23].

Protein identification was performed by searching for MS and MS/MS data in the National Center for Biotechnology Information (NCBI) databases using a built-in Mascot server (V2.1, Matrix Science, London, UK). Proteins were identified using a minimum of two MS/MS spectra matching the databank sequence. All identifications were manually validated. The search parameters were used as follows: trypsin was selected as the digestive enzyme, carbamidomethylation of cysteine as a fixed modification, oxidation of methionine as a variable modification, 100 ppm mass tolerance for precursor ions, 0.2 Da of peptide and fragment mass tolerance and one missed cleavage. The proteins for which the Mascot scores were more than threshold score 55 were considered to be reliably identified ($p < 0.05$).

Author Contributions: Conceptualization, W.W.; formal analysis, M.Z. and Z.-J.G.; resources, C.N.S.J.; investigation, M.Z. and Z.-X.T.; writing, Z.-J.G., M.Z., C.N.S.J. and W.W.; supervision, W.W. and C.N.S.J.; funding acquisition, W.W. All authors have read and agreed to the published version of the manuscript.

Funding: This research was funded by "National Natural Science Foundation of China (NSFC) grant (grant number: 31370357)", "The Biodiversity Investigation, Observation and Assessment Program (2019-2023) of the Ministry of Ecology and Environment of China", "The Natural Science Foundation of Shanxi Province, China, grant number: 202203021211111" and "The Yuncheng University Scientific Research Projects, grant number: XK-2021006, XKX-202206, CXY-202212E278".

Data Availability Statement: The data presented in this study are available on request from the corresponding author.

Acknowledgments: We thank the three anonymous reviewers whose comments/suggestions helped improve and clarify this manuscript.

Conflicts of Interest: The authors declare no conflict of interest. The funders had no role in the design of the study; in the collection, analyses, or interpretation of data; in the writing of the manuscript, or in the decision to publish the results.

References

1. Fu, X.; Ying, Y. Food safety evaluation based on near infrared spectroscopy and imaging: A Review. *Crit. Rev. Food Sci.* **2016**, *56*, 1913–1924. [CrossRef] [PubMed]
2. Boqvist, S.; Söderqvist, K.; Vågsholm, I. Food safety challenges and one health within Europe. *Acta Vet. Scand.* **2018**, *60*, 1. [CrossRef] [PubMed]
3. Satoh, R.; Nakamura, R.; Komatsu, A.; Oshima, M.; Teshima, R. Proteomic analysis of known and candidate rice allergens between non-transgenic and transgenic plants. *Regul. Toxicol. Pharm.* **2011**, *59*, 437–444. [CrossRef] [PubMed]
4. Oboa, B.; Doicd, F.; Jop, E. Transgenic plant-mediated phytoremediation: Applications, challenges, and prospects. In *Assisted Phytoremediation*; Vimal, P., Ed.; Elsevier Inc.: Amsterdam, The Netherlands, 2021; pp. 179–202.
5. Kumar, K.; Gambhir, G.; Dass, A.; Tripathi, A.K.; Singh, A.; Jha, A.K.; Yadava, P.; Choudhary, M.; Rakshit, S. Genetically modified crops: Current status and future prospects. *Planta* **2020**, *251*, 91–118. [CrossRef]
6. Sanchis, V. From microbial sprays to insect-resistant transgenic plants: History of the biospesticide *Bacillus thuringiensis*. A review. *Agron. Sustain. Dev.* **2010**, *1*, 217–231. [CrossRef]
7. Ricroch, A.E.; Bergé, J.B.; Kuntz, M. Evaluation of genetically engineered crops using transcriptomic, proteomic, and metabolomic profiling techniques. *Plant Physiol.* **2011**, *155*, 1752–1761. [CrossRef]
8. Herman, R.A. Unintended compositional changes in genetically modified (GM) crops, 20 years of research. *J. Agric. Food Chem.* **2013**, *61*, 11695–11701. [CrossRef]

9. Kim, J.-K.; Park, S.-Y.; Lee, S.-M.; Lim, S.-H.; Kim, H.-J.; Oh, S.-D.; Yeo, Y.; Cho, H.-S.; Ha, S.-H. Unintended polar metabolite profiling of carotenoid-biofortified transgenic rice reveals substantial equivalence to its non-transgenic counterpart. *Plant Biotechnol. Rep.* **2013**, *7*, 121–128. [CrossRef]
10. De Schrijver, A.; De Clercq, P.; Maagd, R.A.; van Frankenhuyzen, K. Relevance of Bt toxin interaction studies for environmental risk assessment of genetically modified crops. *Plant Biotechnol. J.* **2016**, *13*, 1221–1223. [CrossRef]
11. Narva, K.E.; Wang, N.X.; Herman, R. Safety considerations derived from Cry34Ab1/Cry35Ab1 structure and function. *J. Invertebr Pathol.* **2017**, *142*, 27–33. [CrossRef]
12. ISAAA. *Global Status of Commercialized Biotech/GM Crops in 2019: Biotech Crops Drive Socio-Economic Development and Sustainable Environment in the New Frontier*; ISAAA Brief No. 55; ISAAA: Ithaca, NY, USA, 2019.
13. Cao, D.; Stewart, C.N., Jr.; Zheng, M.; Guan, Z.-J.; Tang, Z.-X.; Wei, W.; Ma, K.-P. Stable *Bacillus thuringiensis* transgene introgression from *Brassica napus* to wild mustard *B. Juncea*. *Plant Sci.* **2014**, *227*, 45–50. [CrossRef] [PubMed]
14. Liu, Y.-B.; Stewart, C.N., Jr.; Li, J.-S.; Wei, W. One species to another: Sympatric Bt transgene gene flow from *Brassica napus* alters the reproductive strategy of wild relative *Brassica juncea* under herbivore treatment. *Ann. Bot.* **2018**, *122*, 617. [CrossRef] [PubMed]
15. Coll, A.; Nadal, A.; Rossignol, M.; Puigdomènech, P.; Pla, M. Proteomic analysis of MON810 and comparable non-GM maize varieties grown in agricultural fields. *Transgenic Res.* **2011**, *20*, 939–949. [CrossRef] [PubMed]
16. Wang, Y.; Xu, W.; Zhao, W.; Hao, J.; Luo, Y.; Tang, X.; Zhang, Y.; Huang, K. Comparative analysis of the proteomic and nutritional composition of transgenic rice seeds with *Cry1ab/ac* genes and their non-transgenic counterparts. *J. Cereal Sci.* **2012**, *55*, 226–233. [CrossRef]
17. Gong, C.-Y.; Wang, T. Proteomic evaluation of genetically modified crops: Current status and challenges. *Front. Plant Sci.* **2013**, *4*, 41. [CrossRef]
18. Agapito-Tenfen, S.Z.; Vilperte, V.; Benevenuto, R.F.; Rover, C.M.; Traavik, T.I.; Nodari, R.O. Effect of stacking insecticidal cry and herbicide tolerance epsps transgenes on transgenic maize proteome. *BMC Plant Biol.* **2014**, *14*, 346–352. [CrossRef]
19. Shukla, P.; Gautam, R.; Singh, N.K. A proteomic study of cysteine protease induced cell death in anthers of male sterile tobacco transgenic plants. *Physiol. Mol. Biol. Plants* **2019**, *25*, 1073–1082. [CrossRef]
20. Jiang, Q.; Li, X.; Niu, F.; Sun, X.; Hu, Z.; Zhang, H. iTRAQ-based quantitative proteomic analysis of wheat roots in response to salt stress. *Proteomics* **2017**, *17*, 1600265. [CrossRef]
21. Xue, K.; Yang, J.; Liu, B.; Xue, D. The integrated risk assessment of transgenic rice *Oryza sativa*: A comparative proteomics approach. *Food Chem.* **2012**, *135*, 314–318. [CrossRef]
22. Sestili, F.; Paoletti, F.; Botticella, E.; Masci, S.; Saletti, R.; Muccilli, V.; Lafiandra, D. Comparative proteomic analysis of kernel proteins of two high amylose transgenic durum wheat lines obtained by biolistic and *Agrobacterium*-mediated transformations. *J. Cereal Sci.* **2013**, *58*, 15–22. [CrossRef]
23. Liu, Y.-B.; Zhang, Y.-X.; Song, S.-Q.; Li, J.-S.; Stewart, C.N., Jr.; Wei, W.; Zhao, Y.-J.; Wang, W.-Q. A proteomic analysis of seeds from *Bt*-transgenic *Brassica napus* and hybrids with wild *B. Juncea*. *Sci. Rep.-UK* **2015**, *5*, 15480. [CrossRef]
24. Vidal, N.; Barbosa, H.; Jacob, S.; Arruda, M. Comparative study of transgenic and non-transgenic maize (*Zea mays*) flours commercialized in Brazil, focusing on proteomic analyses. *Food Chem.* **2015**, *180*, 288–294. [CrossRef] [PubMed]
25. García-Molina, M.D.; Muccilli, V.; Saletti, R.; Foti, S.; Masci, S.; Barro, F. Comparative proteomic analysis of two transgenic low-gliadin wheat lines and non-transgenic wheat control. *J. Proteom.* **2017**, *165*, 102–112. [CrossRef] [PubMed]
26. Liu, W.; Liang, L.; Zhang, Z.; Dong, M.; Jin, W. iTRAQ-based quantitative proteomic analysis of transgenic and non-transgenic maize seeds. *J. Food Compos. Anal.* **2020**, *92*, 103564. [CrossRef]
27. Yang, Y.; Dai, L.; Zhu, K.; Xia, H.; Chen, L.; Liu, H.; Chen, K. Foreign protein detection in transgenic rice revealed by comparative proteomic analysis. *Crop. Sci.* **2015**, *552*, 2225–2233. [CrossRef]
28. Bathellier, C.; Tcherkez, G.; Lorimer, G.H.; Farquhar, G.D. Rubisco is not really so bad. *Plant Cell Environ.* **2018**, *41*, 705–716. [CrossRef] [PubMed]
29. Gulfishan, M.; Jahan, A.; Bhat, T.A.; Sahab, D. Plant senescence and organ abscission. In *Senescence Signalling and Control in Plants*; Sarwat, M., Narendra Tuteja, N., Eds.; Academic Press: Cambridge, MA, USA, 2019; pp. 255–272.
30. Tommasi, I.C. The mechanism of Rubisco catalyzed carboxylation reaction: Chemical aspects involving acid-base chemistry and functioning of the molecular machine. *Catalysts* **2021**, *11*, 813. [CrossRef]
31. Salesse-Smith, C.E.; Sharwood, R.E.; Busch, F.A.; Kromdijk, J.; Bardal, V.; Stern, D.B. Overexpression of Rubisco subunits with RAF1 increases Rubisco content in maize. *Nat. Plants* **2018**, *4*, 802–810. [CrossRef]
32. Li, W.; Qiang, X.-J.; Han, X.-R.; Jiang, L.-L.; Zhang, S.H.; Han, J.; He, R.; Cheng, X.G. Ectopic Expression of a *Thellungiella salsuginea* aquaporin gene, TsPIP1;1, increased the salt tolerance of rice. *Int. J. Mol. Sci.* **2018**, *19*, 2229–2235. [CrossRef]
33. Zhao, B.; Huo, J.; Liu, N.; Zhang, J.; Dong, J. Transketolase is identified as a target of herbicidal substance α-terthienyl by proteomics. *Toxins* **2018**, *10*, 41–43. [CrossRef]
34. Bi, H.; Li, F.; Wang, H.; Ai, X. Overexpression of transketolase gene promotes chilling tolerance by increasing the activities of photosynthetic enzymes, alleviating oxidative damage and stabilizing cell structure in *Cucumis sativus* L. *Physiol. Plant.* **2018**, *167*, 502–515. [CrossRef] [PubMed]
35. Shi, J.; Wang, W.; Lin, Y.; Xu, K.; Xu, Y.; Ji, D.; Chen, C.; Xie, C. Insight into transketolase of *Pyropia haitanensis* under desiccation stress based on integrative analysis of omics and transformation. *BMC Plant Biol.* **2019**, *19*, 475–491. [CrossRef] [PubMed]

36. Wang, L.; Wang, X.; Jia, X.; Huang, Q.; Tan, Y.; Guo, A. Comparative proteomics of *Bt*-transgenic and non-transgenic cotton leaves. *Proteome Sci.* **2015**, *13*, 15–30. [CrossRef] [PubMed]
37. Hisabori, T. Regulation machineries of ATP synthase from phototroph. In *Advances in Botanical Research*; Hisabori, T., Ed.; Academic Press: Cambridge, MA, USA, 2020; Volume 96, pp. 1–26.
38. Wang, F.-W.; Wang, C.; Sun, Y.; Wang, N.; Li, X.-W.; Dong, Y.-Y.; Yao, N.; Liu, X.M.; Chen, H.; Chen, X.-F.; et al. Overexpression of vacuolar proton pump ATPase (V-H+-ATPase) subunits B, C and H confers tolerance to salt and saline-alkali stresses in transgenic alfalfa (*Medicago sativa* L.). *J. Integr. Agric.* **2016**, *15*, 2279–2289. [CrossRef]
39. Jaarsma, R.; de Boer, A.H. Salinity tolerance of two potato cultivars (*Solanum tuberosum*) correlates with differences in vacuolar transport activity. *Front. Plant Sci.* **2018**, *9*, 737–741. [CrossRef]
40. Costa, A.; Luoni, L.; Marrano, C.A.; Hashimoto, K.; Köster, P.; Giacometti, S.; De Michelis, M.I.; Kudla, J.; Bonza, M.C. Ca^{2+}-dependent phosphoregulation of the plasma membrane Ca^{2+}-ATPase ACA8 modulates stimulus-induced calcium signatures. *J. Exp. Bot.* **2017**, *68*, 3215–3230. [CrossRef]
41. Yadav, A.K. Role of plant Ca^{2+}-ATPase in calcium homeostasis during development and stresses. In *Calcium Transport Elements in Plants*; Upadhyay, S.K., Ed.; Academic Press: Cambridge, MA, USA, 2021; pp. 103–128.
42. Zhong, M.; Liu, X.; Liu, F.; Ren, Y.; Wang, Y.; Zhu, J.; Teng, X.; Duan, E.; Wang, F.; Zhang, H.; et al. FLOURY ENDOSPERM12 encoding alanine aminotransferase 1 regulates carbon and nitrogen metabolism in rice. *J. Plant Biol.* **2019**, *62*, 61–73. [CrossRef]
43. Wang, X.; Wei, Y.; Shi, L.; Ma, X.; Theg, S.-M. New isoforms and assembly of glutamine synthetase in the leaf of wheat (*Triticum aestivum* L.). *J. Exp. Bot.* **2015**, *66*, 6827–6834. [CrossRef]
44. Ouerghi, Z.; Lachaâl, M.; Chebbi, M.; Ben Abdallah, S.; Amdouni, T.; Msilini, N. Does the source of nitrogen affect the response of fenugreek plants to saline stress? In *Agrochimica: International Journal of Plant Chemistry, Soil Science and Plant Nutrition of the University of Pisa*; Pisa University Press: Pisa, Italy, 2017; Volume 61, pp. 2283–5431.
45. Bao, A.; Zhao, Z.; Ding, G.; Shi, L.; Xu, F.; Cai, F. The stable level of glutamine synthetase 2 plays an important role in rice growth and in carbon-nitrogen metabolic balance. *Int. J. Mol Sci.* **2015**, *16*, 12713–12736. [CrossRef]
46. Liu, X.; Zhang, Q.; Huan, Z.; Zhong, M.; Chen, W.; Du, H. Identification and characterization of glutamine synthetase isozymes in *Gracilaria lemaneiformis*. *Aquat. Bot.* **2018**, *146*, 23–30. [CrossRef]
47. Hinck, A.P.; Markus, M.A.; Huang, S.R.; Grzesiek, S.; Kustonovich, I.; Draper, D.E.; Torchia, D.A. The RNA binding domain of ribosomal protein L11: Three- dimensional structure of the RNA-bound form of the protein and its interaction with 23 S rRNA. *J. Mol. Biol.* **2015**, *274*, 101–113. [CrossRef] [PubMed]
48. Shang, X.-L.; Xie, R.-R.; Tian, H.; Wang, Q.-L.; Guo, F.-Q. Putative zeatin O-glucosyltransferase OscZOG1 regulates root and shoot development and formation of agronomic traits in rice. *J. Integr. Plant Biol.* **2016**, *58*, 627–641. [CrossRef] [PubMed]
49. Halfhill, M.D.; Richards, H.A.; Mabon, S.A.; Stewart, C.N., Jr. Expression of GFP and Bt transgenes in *Brassica napus* and hybridization with *Brassica rapa*. *Theor. Appl. Genet.* **2001**, *103*, 659–667. [CrossRef]
50. Joosen, R.; Cordewener, J.; Supena, E.D.J.; Vorst, O.; Lammers, M.; Maliepaard, C.; Zeilmaker, T.; Miki, B.; America, T.; Custers, J.; et al. Combined transcriptome and proteome analysis identifies pathways and markers associated with the establishment of rapeseed microspore-derived embryo development. *Plant Physiol.* **2007**, *144*, 155–172. [CrossRef]
51. Ramagli, L.S. Quantifying protein in 2-D PAGE solubilization buffers. In *2-D Proteome Analysis Protocols. Methods in Molecular Biology*; Link, A.J., Ed.; Humana Press: Totowa, NJ, USA, 1999; Volume 112, pp. 99–103.
52. Li, X.-H.; Wu, X.-F.; Yue, W.-F.; Liu, J.-M.; Li, G.-L.; Miao, Y.-G. Proteomic analysis of the silkworm (*Bombyx mori* L.) hemolymph during developmental stage. *J. Proteome Res.* **2006**, *5*, 2809–2814. [CrossRef] [PubMed]

plants

Review

Transgene Bioconfinement: Don't Flow There

Jessica N. Stockdale and Reginald J. Millwood *

Department of Plant Sciences, University of Tennessee (UTK), Knoxville, TN 37996, USA
* Correspondence: rmillwood@utk.edu

Abstract: The adoption of genetically engineered (GE) crops has led to economic and environmental benefits. However, there are regulatory and environmental concerns regarding the potential movement of transgenes beyond cultivation. These concerns are greater for GE crops with high outcrossing frequencies to sexually compatible wild relatives and those grown in their native region. Newer GE crops may also confer traits that enhance fitness, and introgression of these traits could negatively impact natural populations. Transgene flow could be lessened or prevented altogether through the addition of a bioconfinement system during transgenic plant production. Several bioconfinement approaches have been designed and tested and a few show promise for transgene flow prevention. However, no system has been widely adopted despite nearly three decades of GE crop cultivation. Nonetheless, it may be necessary to implement a bioconfinement system in new GE crops or in those where the potential of transgene flow is high. Here, we survey such systems that focus on male and seed sterility, transgene excision, delayed flowering, as well as the potential of CRISPR/Cas9 to reduce or eliminate transgene flow. We discuss system utility and efficacy, as well as necessary features for commercial adoption.

Keywords: bioconfinement; transgenic plants; GE crops; gene flow; male sterility; transgene excision; delayed flowering

Citation: Stockdale, J.N.; Millwood, R.J. Transgene Bioconfinement: Don't Flow There. *Plants* **2023**, *12*, 1099. https://doi.org/10.3390/plants12051099

Academic Editor: Wei Wei

Received: 6 February 2023
Revised: 23 February 2023
Accepted: 24 February 2023
Published: 1 March 2023

1. Introduction

In the US, more than 90% of commercially cultivated crops are genetically engineered (GE) [1]. These crops confer well-characterized transgenic traits, primarily insect and herbicide resistance, that have been engineered in a few crop species with low potential for gene flow [1]. Over the last three decades, transgene movement beyond cultivation has been a concern associated with GE crops [2–4]. Transgene flow may occur via the dispersal of pollen, seed, or vegetative-propagules, although pollen-mediated gene flow is commonly described as the major route of concern [4,5]. Indeed, much research has focused on potential environmental impacts of transgene movement beyond cultivation [6–8] as well as the potential for the establishment of free-living transgenic populations [9–12]. Transgene flow has been documented for a few crops, yet few free-living transgenic populations have been identified and there have been no reports of measurable negative environmental impacts [13–18]. This may be due to the reproductive biology of current GE crops and the area of cultivation. For example, many of these crops are highly domesticated, primarily self-pollinate, and are cultivated in regions with few to no sexually compatible wild relatives [19].

Gene flow is more likely to occur in crop species with high outcrossing frequencies often associated with wind-pollinated crops and obligate outcrossing species, such as switchgrass (*Panicum virgatum* L.). The probability of interspecific hybridization is also higher if there are wild relatives native to the region of cultivation [20]. Additionally, certain transgenes and their conferred traits may also raise biosafety concerns in regard to environmental and ecological risk if they were to escape to unmanaged wild relative populations [19]. For example, certain transgenes may provide advantageous traits, such as improved tolerance to abiotic stress conditions including drought, salinity, or low and high temperatures [20,21]. There is concern that GE trait introgression into wild populations could provide a competitive advantage to transgenic hybrids under stress conditions. This may lead to potential ecological consequences such as decreased biodiversity or possibly local population extinction [22,23].

To address these concerns several transgene confinement approaches have been proposed. The implementation of physical or biological confinement (bioconfinement) strategies could prevent transgene flow. Physical confinement approaches utilize physical barriers such as enclosures, fencing, flower removal, and geographic isolation that provide spatial separation from sexually compatible crops and wild relatives. These approaches are often not practical solutions for large-scale commercialization in terms of physical labor and facility requirements [24]. Alternatively, biological confinement approaches offer feasible and sustainable strategies to contain transgenes. Since the early 1990s, bioconfinement approaches have been studied for their potential to limit or eliminate transgene flow. To date, none are commonly used in commercial crops. The adoption of a bioconfinement system would allow new GE crops to reach the market that may not otherwise. Numerous technologies have been proposed through various approaches including maternal inheritance, genome incompatibility [25], apomixis [26], cleistogamy [27,28], and total sterility, but their application is limited. In this review, we survey efficient and widely applicable bioconfinement technologies such as male and seed sterility, transgene excision, and delayed flowering. These approaches have been achieved through genetic engineering, including type II clustered regularly interspaced short palindromic repeat/ CRISPR-associated protein 9 (CRISPR/Cas9) gene editing. We also discuss the most effective bioconfinement methods developed to date and compare their potential to eliminate transgene flow. Lastly, perspectives are provided for features we believe necessary for commercial adoption of a bioconfinement system.

2. Engineered Reproductive Sterility

2.1. Male Sterility

Engineered male sterility is a bioconfinement approach that can be designed to yield pollenless flowers or flowers with sterile transgenic pollen. Generally, GE male sterile phenotypes can be produced through the tightly-controlled and targeted expression of genes that yield products cytotoxic to reproductive tissues [21,25,29]. To date, genes used for engineered male sterility include *barnase, Diphtheria toxin A-chain*, and the endonuclease *Eco*RI. Each have been expressed in several plant species with varying results which are described below.

2.1.1. Barnase-Induced Male Sterility

The bacterial gene *barnase* (from *Bacillus amyloliquefaciens*) has been used to generate male sterile plants. Barnase is a ribonuclease that, when expressed in plant cells, leads to RNA degradation and subsequent cell death [30]. When utilized for engineered male sterility, barnase expression has been targeted to tapetal cells which form the nutritive cell layer surrounding microspore mother cells [31,32]. These cells are necessary for viable pollen development [31]. When barnase is expressed in this manner, tapetal cell death prevents microspore development and confers total pollen ablation (Figure 1a) [30]. Fertility can be restored by crossing with restoration lines genetically engineered to express barstar (also from *Bacillus amyloliquefaciens*) which acts as a repressor of barnase [33,34].

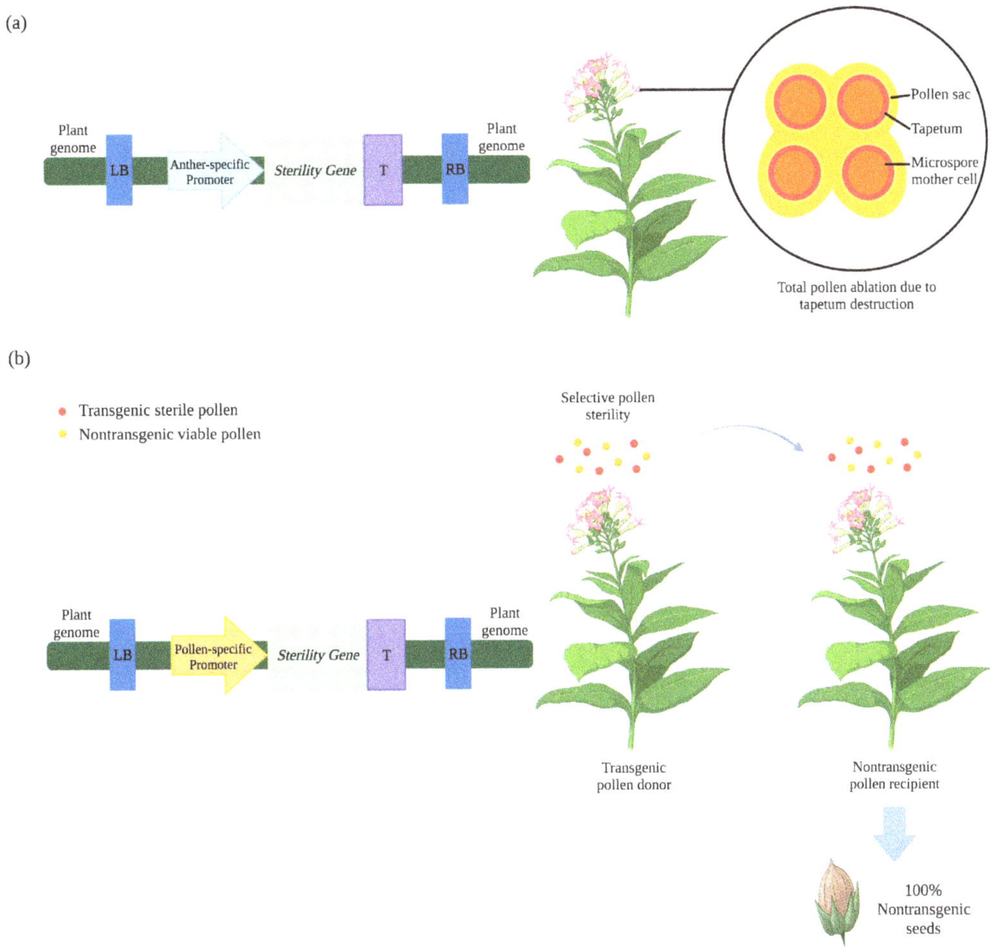

Figure 1. Engineered male-sterility approaches. (**a**) A sterility gene, such as *Barnase*, is located between an anther-specific promoter and a terminator (T) within the left (LB) and right borders (RB) on T-DNA inserted into the plant genome. *Barnase* expression prevents pollen formation by degrading RNA in the tapetal cells, the nutritive cells surrounding microspore mother cells. (**b**) A sterility gene, such as *Eco*RI endonuclease, is located between a pollen-specific promoter and a terminator within the left and right borders on the T-DNA, and in this example is inserted once in the plant genome. In this theoretical scenario, *Eco*RI expression leads to the death of transgenic pollen grains and nontransgenic pollen remains viable. Pollen movement from transgenic to nontransgenic plants yields 100% nontransgenic seeds. Figure created with Biorender.com, accessed on 22 February 2023.

Experimental results in tobacco (*Nicotiana tabacum* L.) and canola (*Brassica napus* L.) under the control of the tapetum-specific promoter TA29 (derived from tobacco) revealed barnase expression conferred completely pollenless plants in 92% and 77% of transgenic events, respectively [30]. Except for the absence of pollen, these transgenic plants were indistinguishable from nontransgenic tobacco controls in vegetative and floral morphology and displayed a similar growth rate and height through all developmental phases.

Using this approach, barnase has been used for male sterile plant production in several monocot species including wheat (*Triticum aestivum* L.) [35], creeping bentgrass (*Agrostis stolonifera* L.) [36], and rice (*Oryza sativa* L.) [37]. In wheat, tapetum-specific promoters ca55 (derived from maize (*Zea mays* L.)), and T72 and E1 (derived from rice) were used to drive barnase expression [35]. Out of 29 transgenic events, 11 were chosen for phenotypic analysis and crossing experiments. Two events contained barnase under the control of the promoter E1, six were under the control of the T72 promoter, and the remaining three events utilized the ca55 promoter. All transgenic events displayed a completely pollenless phenotype that was stably inherited across two generations of backcrossing except for one event (T72) which produced fertile transgenic progeny in cross-pollination experiments. These transgenic plants displayed normal phenotypes and were comparable to wild-type plants in height, leaf size, and tiller count. When expressed in creeping bentgrass using the tapetum-specific promoter TAP (derived from rice), barnase expression led to significantly reduced pollen production in 90% of transgenic lines [36]. For plants that produced pollen, viability was measured and all pollen grains tested were non-viable. No off-target expression was reported, and the male sterile phenotype was stably inherited in progeny when backcrossed to nontransgenic creeping bentgrass. When barnase was expressed in rice using the pollen-specific promoter Zm13 (derived from maize) [37], transgenic events were crossed to nontransgenic rice. The resulting progeny produced 50% viable nontransgenic pollen. Transgenic pollen expressing barnase was nonviable which is expected for transformants with a single T-DNA copy. The cross-pollination experiments also revealed that pollen-specific expression of barnase had no effect on seed yield, and plants displayed normal growth and development.

Barnase has also been expressed in male-specific tissues of GE trees such as pine (*Pinus radiata* D. Don) [38], eucalyptus (*Eucalyptus globulus* Labill.) [38], and hybrid poplar (*Populus tremula* × *Populus tremuloides*) [39]. In pine and eucalyptus, a male cone-specific promoter PrMC2 (derived from pine) was used to express the modified barnase coding sequence barnaseH102E [38]. BarnaseH10E2 contained a single amino acid substitution which was expected to reduce RNAse activity and prevent off-target expression. When expressed in pine, 97% of transgenic events did not produce pollen. Nine of these transgenic events were studied under field conditions, where pollenless cones were observed over a 2-year study. Similar results were observed in eucalyptus where 95% of transgenic events analyzed were pollenless. Under field conditions, 16 of 17 transgenic events produced pollenless cones over a four-year study. Cone dissection revealed that one transgenic event produced pollen, but the pollen was non-viable. Additionally, vegetative growth and development were normal for both transgenic pine and eucalyptus. Barnase was also expressed in poplar hybrids (*Populus tremula* × *Populus tremuloides*) under the control of the tapetum-specific promoter TA29 [39]. Throughout a four-year field study, two of six transgenic trees were completely pollenless and four transgenic trees released approximately six pollen grains per catkin (a flower cluster found in certain tree species) during the first two years of the study. Transgenic pollen production was significantly less compared to nontransgenic trees which produced approximately 73,000 to 85,000 pollen grains per catkin during the same period. In the final two years of the study, no pollen was produced in transgenic trees whereas 775,000 to 2,700,000 pollen grains per catkin were produced in nontransgenic controls. Unfortunately, plant growth was negatively affected as determined by stem volume (height × diameter^2) measurements. Of 18 transgenic poplar events, 17 were significantly smaller compared to nontransgenic controls. Because growth of transgenic poplar expressing the β-glucuronidase (GUS) reporter gene was similar to that of nontransgenic controls, this phenotype was attributed to off-target barnase expression and possibly worsened by field conditions.

In summary, the targeted expression of barnase yields pollenless flowers or sterile pollen production and may be an effective strategy to prevent transgene flow with little to no negative effects on plant growth and development. Although examples of off-target expression were observed in few transgenic events, the use of barnase for bioconfinement is applicable across many crop species, stable across several generations, and can effectively confer male sterility. Nonetheless, this bioconfinement system can be further optimized through the use of synthetic promoters that provide increased sterility gene expression to targeted tissues for enhanced efficacy in a wide array of plant species.

2.1.2. Diphtheria Toxin A-Chain-Induced Male Sterility

Another well-characterized cytotoxic gene used for engineered male sterility is *Diphtheria toxin A-chain* (*DTA*) (from *Corynebacterium diphtheriae*). When expressed in cells, DTA catalyzes ADP-ribosylation and inactivation of elongation factor 2 [40]. This action inhibits protein synthesis resulting in cell death. Targeted *DTA* expression has been used for cell–cell interaction and functional analysis studies [41,42]. The characterization of tapetum- and pollen-specific promoters through the expression of DTA in transgenic plants was found to yield pollenless and pollen-sterile plants with minimal to no effects on plant growth [41,42]. For this reason, targeted DTA expression has been adopted as a bioconfinement strategy through male sterile plant production. Male fertility can be restored by crossing with transgenic lines engineered to express a diphtheria toxin repressor (DTxR) (also from *Corynebacterium diphtheriae*) [43].

Tapetum-specific expression of DTA by the promoter TA29 was analyzed in seven transgenic tobacco plants, all of which contained a single T-DNA copy [42]. From these transformants, anther development was identical to that of nontransgenic control plants except for the destruction of the tapetum and collapse of the pollen sac resulting in pollenless plants. No off-target expression or negative effects on growth were reported suggesting TA29 promoter activity is localized to tapetum cells.

Another study used the pollen-specific promoter Lat52 derived from tomato (*Solanum lycopersicum* L.) to drive the expression of DTA in 20 transgenic tobacco events [41]. Of the 20 events, 16 were male-sterile and displayed one of two phenotypes. Half of the events displayed a pollenless phenotype, whereas the other half had a significant reduction in pollen grains of which 50% were nonviable. Two transgenic events that produced 46.3% and 50.9% nonviable pollen were selected for further analyses, and the pollen-sterile phenotype was stably inherited through four generations of self-pollination. Additionally, two of the 20 transgenic events displayed off-target expression which resulted in decreased flower size.

Both tapetum- or pollen-DTA expressing plants are phenotypically similar to barnase expressing male-sterile plants. Tapetum- and pollen-specific expression of DTA yielded male-sterile tobacco with few to no off-target affects [41,42]. However, concerns may arise in regard to the use of DTA in crops. Although a single molecule of DTA can induce cell death, diphtheria toxin B chain is required to transport DTA into cells [40,44]. The use of DTA in this manner may raise concerns regarding potential pollen-feeding insects and human health effects even though DTA is not highly toxic alone. Despite its potential, these concerns may create roadblocks when used as a bioconfinement strategy.

2.1.3. *Eco*RI Endonuclease-Induced Male Sterility

Another gene studied for bioconfinement purposes is the *Eco*RI endonuclease (from *Escherichia coli*) which has been expressed in pollen to confer selective male sterility in tobacco (Figure 1b) [45]. *Eco*RI endonuclease is a well-characterized type II restriction endonuclease that creates a double-stranded break at the 5′-GAATTC-3′ recognition sequence [46]. The expression of *Eco*RI in pollen causes repeated double-stranded breaks in the pollen nuclear genome which induces cell death [45].

Pollen-specific expression of *Eco*RI endonuclease by the Lat52 promoter conferred selective male sterility in tobacco, and all transformants displayed over 99% and up to 100% bioconfinement efficacy when used as the pollen donor source and crossed to nontransgenic tobacco plants under greenhouse and field conditions [45]. Additionally, bioconfinement efficacy was stable across generations and transgenic plants were indistinguishable from nontransgenic plants in regard to growth and vegetative morphology. Thus far, this approach has been characterized in tobacco, but more research is needed in other species since recognition-site frequency of restriction endonucleases and promoter activity can vary. Furthermore, this approach is advantageous for pollen-feeding insects since it permits pollen production, whereas other male-sterility approaches may be food-source limiting.

2.2. Conditional Seed Sterility

Conditional seed sterility is another promising bioconfinement approach that was originally designed to protect intellectual property in genetically engineered plants [47–49]. Such approaches are referred to as genetic use restriction technologies (GURTs). GURTs are often designed as an inducible system that activates sterility gene expression under specific conditions [19,50]. One of the earliest GURTs, termed Genesafe technology, was originally developed in the 1990's to prevent the unauthorized use of transgenic seeds by inducing seed sterility in the next generation [47–49,51]. Although promising, Genesafe was never commercialized. This was partially due to public opposition since growers could not save and reuse seeds [19,50,52]. However, this approach holds great potential as a bioconfinement system specifically for vegetatively or clonally propagated crops such as switchgrass, certain hemp cultivars (*Cannabis sativa* L.), and poplar.

GURT designs often require activation by an external stimulus, such as a chemical or temperature treatment, which would activate an engineered genetic circuit and inhibit seed germination in the subsequent generation (Figure 2). These GURTs are comprised of four components: (i) a genetic switch which is inducible and can activate or repress the trait switch; (ii) a trait switch which is typically a recombinase gene responsible for removing a blocker sequence; (iii) a blocker sequence flanked by recombinase recognition sites (located between the seed-specific promoter and sterility gene); and (iv) a sterility gene controlled by a seed-specific promoter [50,53]. Altogether, the application of a stimulus activates the genetic switch, which amplifies the trait switch and removes a blocker sequence to bring the seed-specific promoter and sterility gene in close proximity [50,52]. Seed-specific expression of the sterility gene renders transgenic seeds sterile.

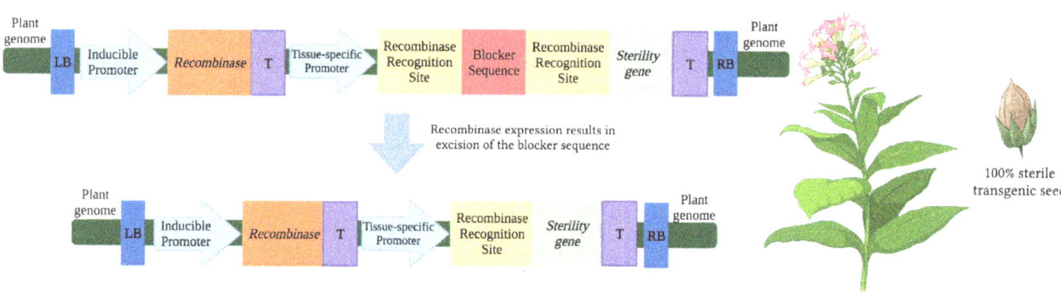

Figure 2. Conditional sterility via inducible system activation of sterility genes. Within the left (LB) and right borders (RB) of the T-DNA located in the plant genome, an inducible promoter is located upstream of a recombinase gene and terminator (T). An external stimulus is applied to induce recombinase gene expression. The recombinase enzyme subsequently excises the blocker sequence by cutting DNA segments at flanking recognition sites. The tissue-specific promoter (ex. seed-specific) is moved into close proximity of the sterility gene and transcription can occur. The sterility gene product leads to cell death. In this theoretical example, a homozygous tobacco line with a seed-specific promoter driving sterility gene expression leads to 100% sterile transgenic seeds. Figure created with Biorender.com, accessed on 18 January 2023.

The original Genesafe prototype was a repressible system and required hybridization of two parent plants (P1 and P2) to produce inactive, inducible seeds carrying the entire Genesafe system [47,53]. P1 carried the seed-specific promoter and sterility gene divided by the blocker sequence that produces a repressor protein. P2 carried a recombinase gene that, without application of a stimulus, was repressible by the blocker sequence in P1 when hybridized. After hybridization, a stimulus treatment would release the repressor, and the recombinase would excise the blocker sequence. As a result, the sterility gene would be expressed in the next generation of seeds. This design was tested in tobacco using late embryo-specific promoters LEA4A and LEA14 derived from cotton (*Gossypium herbaceum* L.) to drive expression of the sterility gene *saporin* (from *Saponaria officinalis*), a ribosome-inactivating protein that inhibits protein synthesis when expressed in plant cells [54]. The blocker sequence was successfully excised using the recombinase system Cre/*loxP*; however, seed-sterility was not observed [55]. This was attributed to the inefficient activity of the cotton-derived promoter in tobacco, as well as insufficient saporin production. This design was not further tested because a new and simpler design was created which used a single parent to carry the Genesafe system. The new design was chemically inducible, where application of a stimulus would permit expression of the recombinase gene and excise the blocker sequence [47,53]. As a result, the seed-specific promoter would express the sterility gene and was expected to produce non-germinable seeds in the following generation. This design can be altered to use other components to make up the chemically-inducible system but was described with the Cre/*loxP* recombinase system, LEA promoters, and the *saporin* gene. However, no data were provided for seed sterility.

Although Genesafe was not previously commercialized for its original purpose, it could be immensely useful for bioconfinement if used in clonally or vegetatively propagated crops. Improved tools and system components are available that can be used to optimize the system. Specifically, efficient recombinase systems can be employed such as gene-deletor technology (described in Section 3) which has displayed up to 100% transgene excision from pollen and/or seeds [56]. Improved inducible- and tissue-specific synthetic promoters are available which have been optimized for strong inducible and highly specific spaciotemporal expression [57–60]. With optimized components and new designs, Genesafe holds promise as a bioconfinement system that can be incorporated into a wide array of crops.

3. Transgene Excision

Transgene excision has been proposed as a strategy to eliminate transgene flow. This strategy utilizes inducible- or tissue-specific promoters to express a site-specific recombinase. The DNA targeted for excision is flanked by site-specific recombinase recognition sites during design and cloning of the transformation construct. Depending on the placement of the recognition sites, this strategy can be used to remove transgenes such as selectable marker genes (SMGs) or entire T-DNA cassettes. The excised DNA is degraded by non-specific nucleases within the cell, and components that were not located between the recognition sites would remain in the genome including the left and right T-DNA borders and one recombinase recognition site (Figure 3a) [61,62].

3.1. Transgene Excision via Cre/loxP Recombinase

The most studied and best characterized bidirectional recombinase systems include Cre-*loxP* (from bacteriophage P1), FLP-*FRT* (from *Saccharomyces cerevisiae*), and R-*RS* (from *Zygosaccharomyces rouxii*) [63–65]. Tyrosine recombinases that recognize relatively short sequences (34–58 base pairs) are used in these systems to mediate DNA cleavage [63–65]. Dependent on the orientation of the recognition sites relative to each other, integration, excision, or inversion of targeted sequences of DNA can occur [66]. Although all of these outcomes have been observed, excision and degradation of the targeted DNA is observed most often since it is kinetically favored [66].

The Cre-*loxP* recombinase system has been studied in Arabidopsis (*Arabidopsis thaliana* L.) [67] and tobacco [67–70] studies, where pollen was targeted for transgene excision to prevent pollen-mediated gene flow [67,70]. In Arabidopsis and tobacco greenhouse experiments, the microspore-specific promoter NTM19 (derived from tobacco) was used to drive Cre expression and produce transgene-free pollen [67]. Transgenic Arabidopsis and tobacco plants were used in self- and cross-pollination experiments, followed by backcross experiments. Backcross experiments displayed 100% transgene excision in three Arabidopsis lines. In transgenic tobacco, over 99% transgene excision was observed in five lines. In addition to high transgene excision efficiency, plants displayed normal growth and development. Another study used the Cre/*loxP* recombinase system to excise the SMG from transgenic tobacco through the use of the pollen- and embryo-specific promoter DLL (derived from Arabidopsis) to drive Cre expression [70]. Eight transgenic tobacco lines were self-pollinated and 99% of the progeny displayed partial or complete excision of the SMG. However, there was notable variation across transgenic lines, ranging from 3.8% to 81.5% excision efficiency. Four of these lines were used in cross-pollination experiments as pollen donors to nontransgenic tobacco, where complete or partial excision of the SMG was observed in 99% of plants. These experiments also displayed variation across transgenic lines which ranged from 40% up to 96.2% excision efficiency. High excision efficiency has been observed and supports the potential of the recombinase system for transgene excision; however, the variation among lines should be addressed by utilizing promoters that provide consistent and targeted high expression levels.

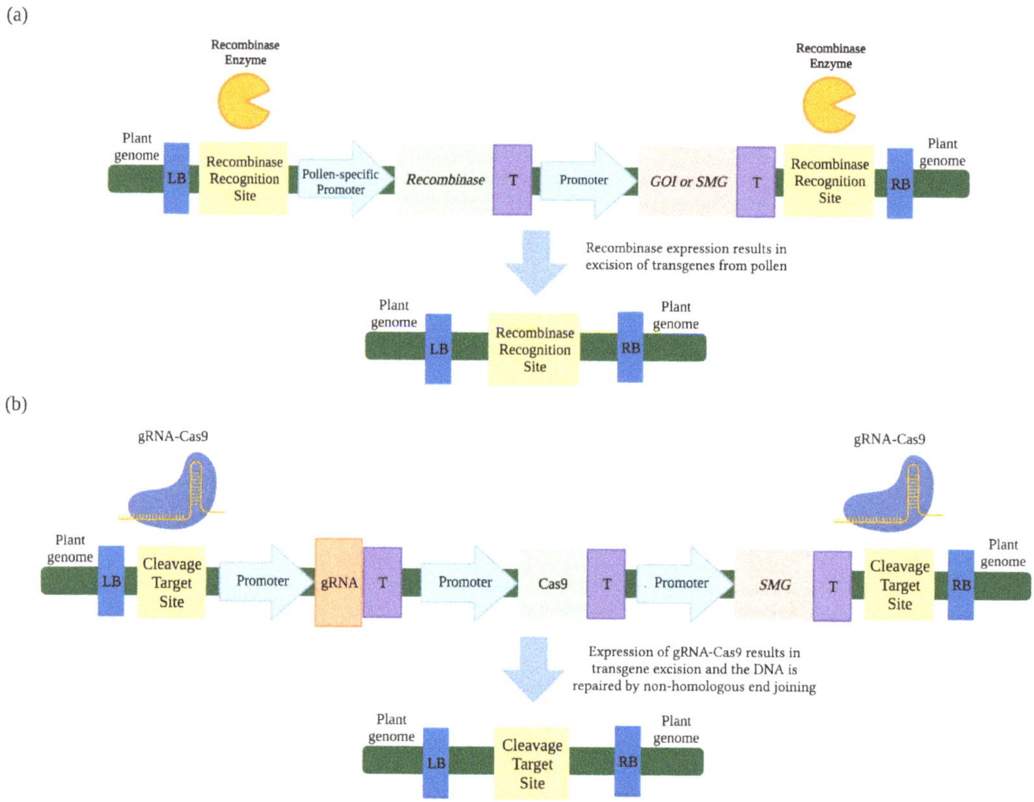

Figure 3. Transgene excision by site-specific recombinase and CRISPR/Cas9 gene editing. (**a**) Recombinase expression is controlled by a pollen-specific promoter. In pollen, the recombinase recognizes sites flanking the gene of interest (GOI) or selectable marker gene (SMG) and the recombinase expression cassettes and removes the DNA between the recognition sites. One recognition site and left (LB) and right (RB) borders remains while the DNA removed is degraded within the plant cell. (**b**) CRISPR/Cas9 excision utilize cleavage target sites (CTSs) that flank the guideRNA (gRNA), Cas9 enzyme, and SMG expression cassettes. The CTSs are comprised of identical DNA sequences. Expression of a single gRNA and Cas9 enzyme excises the T-DNA between the CTSs. In this example, cleaved DNA ends are repaired by non-homologous end joining, leaving behind one CTS and left and right borders and the DNA removed is degraded within the plant cell. Figure created with Biorender.com, accessed on 22 February 2023.

Seed-specific expression of the Cre/*loxP* recombinase system was also studied to prevent seed-mediated transgene flow. In tobacco (*Nicotiana benthamiana* L.), the seed-specific promoter napin (derived from canola) was used to drive Cre expression and determine the efficiency of SMG excision from seeds [69]. Tobacco was co-transformed with two plasmids, the first of which contained the SMG flanked by the *loxP* recognition sites and the second construct contained Cre driven by the napin promoter. A total of 10 lines were co-transformed, self-pollinated, and the progeny were subjected to analysis. Excision of the SMG occurred in all 10 lines, eight of which displayed 100% excision efficiency and two lines displayed 76.8% and 93.1% excision efficiency. Additionally, all transgenic lines were normal in regard to vegetative and reproductive morphology. The expression of the Cre/*loxP* recombinase system was also studied in tobacco (*Nicotiana tabacum* L.) under the control of the seed-specific promoter BcNA1 (derived from canola) to remove the SMG [68]. Eight transgenic lines were obtained and self-pollinated, and progeny were subjected to a germination assay on selection media to determine whether excision had occurred. In four of the transgenic lines, 100% excision was observed. In the remaining transgenic lines, excision occurred at a rate of 25.4%, 29.9%, 59.6%, and 71.8%. Additionally, all transgenic lines appeared normal in regard to plant growth and fertility. In transgenic canola, Cre expression was driven by the embryo-specific promoter USP88 (derived from *Vicia faba* L.) to excise the SMG and produce SMG-free progeny [71]. Progeny from eight transgenic lines were analyzed and an average of 32.7% excision efficiency was observed, none of which were reported to have adverse morphological features. Although these lines were able to produce SMG-free progeny, the average excision rates were relatively low compared to other studies. To improve this system, stronger and more tightly-controlled promoters should be selected to drive Cre expression.

Both pollen- and seed-specific expression of the Cre/*loxP* recombinase system have displayed up to 100% transgene excision in multiple transgenic lines. However, significantly lower efficiency has been observed in several lines. To reduce variability among lines, it is apparent that design changes, specifically for promoters that drive recombinase expression, as well as further testing and optimization are needed to ensure high and consistent levels of transgene excision in all transgenic lines. If consistency is achieved, the Cre/*loxP* recombinase system may be a useful bioconfinement tool.

3.2. Transgene Excision from Pollen via CinH-RS2

Other well-characterized recombinase systems are CinH-*RS2* (from *Acinetobacter* plasmids pKLH2 and pKLH204) and ParA-*MRS* (from the broad host range plasmid RK2), which are unidirectional recombinase systems and use serine recombinases to irreversibly excise targeted sequences of DNA [72–74]. Recombinases of the CinH-*RS2* and ParA-*MRS* recombinase systems recognize 113 and 133 base pair sequences, respectively, which are much longer than the tyrosine-derived recombinase systems previously described and are expected to decrease potential off-target DNA cleavage [75,76].

Transgene excision from pollen has been studied in tobacco using the CinH-*RS2* recombinase system [61]. In transgenic tobacco, the pollen specific Lat52 promoter (derived from tomato) was used to drive expression of CinH recombinase. The *RS2* recognition sites flanked all transgenes in the construct and were within the T-DNA borders. A green fluorescent protein (GFP) visual marker gene was located in this region and expression was driven by the Lat59 pollen-specific promoter (derived from tomato). In this design, successful excision would prevent expression of GFP in pollen. Pollen grains from five transgenic lines with a single T-DNA copy were collected and analyzed, three of which displayed less than 1% GFP-expressing pollen and the other two lines displayed 1.94% and 17.48% GFP-expressing pollen. Furthermore, transgenic plants were phenotypically indistinguishable from nontransgenic controls in regard to plant growth and reproduction. Thus, the authors concluded that highly efficient transgene excision is possible, and this approach holds promise for transgene bioconfinement.

Based on these results, the CinH-*RS2* recombinase system can effectively excise transgenes from pollen without negatively impacting plant growth. The pollen analysis showed promising results. However, cross-pollination experiments should be performed to confirm that transgenes are not present in the progeny. Further research is also needed to evaluate design and system components for improvements to achieve a 100% excision efficiency as observed in other designs.

3.3. Transgene Excision from Pollen and/or Seeds via Fused loxP-FRT Recombinase System

Gene-deletor technology is a term used to describe a site-specific recombinase system that uses fused *loxP-FRT* recognition sites on either side of the transgene (s), along with expression of either Cre or FLP recombinase genes [56]. Cre and/or FLP recombinases were driven by pollen-specific promoters Lat52 (derived from tomato) or BGP1 (derived from *Brassica campestris*), or pollen- and seed-specific promoter PAB5 (derived from Arabidopsis) to remove transgenes from pollen and/or seeds of GE tobacco. Gene-deletor technology displayed 100% transgene excision in pollen and seeds from multiple transgenic tobacco events. Specifically, analysis of progeny produced from self- and cross-pollination experiments revealed that constructs with fused *loxP-FRT* recognition sequences and promoter:recombinase combinations BGP1:FLP, PAB5:FLP, and Lat52:Cre displayed 100% excision efficiency in multiple lines. Additionally, no deleterious effects on vegetative or reproductive growth were reported. In the same study, the promoter:recombinase combination BGP1:FLP was used with only the *FRT* recognition site and displayed 99% excision efficiency, a slightly lower rate than when both *loxP-FRT* recognition sites were present (100%). The study also examined how the simultaneous expression of Cre and FLP recombinases driven by Lat52 affected transgene excision in the presence of fused *loxP-FRT* recognition sites. This design produced up to 45% transgene excision, a significantly lower rate from Cre expression alone.

These studies suggest gene-deletor technology increased excision efficiency and can effectively eliminate transgene flow in tobacco without negatively impacting plant growth and development. The promoter:recombinase combinations with high performance should be tested in other crop species to determine if efficacy is comparable.

3.4. CRISPR/Cas9-Mediated Transgene Excision

In recent years, type II clustered regularly interspaced short palindromic repeat/CRISPR-associated protein 9 (CRISPR/Cas9) has been utilized for transgene excision. It requires two Cas9 cleavage target sites (CTS) on either side of the target DNA and a guide RNA (gRNA)-Cas9 assembly that recognizes and cleaves the CTSs. The excised DNA is degraded by non-specific nucleases in the cell. Plasmid construct components that were not located between the CTSs would remain in the genome (Figure 3b) [77].

In one study, CRISPR/Cas9 transgene excision efficacy was examined for the excision of SMGs in transgenic rice [78]. To do so, callus of transgenic rice that constitutively expressed GUS was transformed with a CRISPR/Cas9 construct. The cleavage target sites were located at the ends of the 1.6 kb *GUS* gene. Since the *GUS* gene is expected to be removed from the genome, successful excision would be determined by lack of GUS expression. In this experiment, a total of 34 plants were produced from 12 transgenic events. Three of these events displayed 100% transgene excision. Despite high excision efficiency, the authors report that normal vegetative growth was observed in four of 34 transgenic plants and only two plants produced a small quantity of seeds. No specific details were provided for the adverse phenotypes and off-target Cas9 activity. While this study displayed 100% transgene excision from multiple transgenic events, reports of no phenotypic and off-target effects are necessary to eliminate concerns of potential negative outcomes.

Another study also aimed to excise SMGs from rice through CRISPR/Cas9-mediated transgene excision [79]. The promoter Pssi (derived from rice) was expected to express strongly in stem, shoot tip, and inflorescence tissue, and was therefore selected to drive Cas9 nuclease expression. Transgenic rice was generated to constitutively express GUS; however, the *GUS* gene was separated into two segments denoted as 'GU' and 'US', interrupted by a CRISPR/Cas9 cassette and a selectable marker gene that acted as a blocker sequence flanked by CTSs. Additionally, both 'GU' and 'US' carried a 1027 bp 'U' sequence as a homologous region, inducing CRISPR/Cas9-mediated homology-directed repair following excision of the blocker sequence resulting in a functional *GUS* reporter gene if excision was successful. Transgene excision occurred in 11 of 15 transgenic lines. In lines where excision was observed, T_1 lines were produced and three were randomly selected for further analysis. In one line, 6.7% homozygous excision and 66.7% heterozygous excision was observed. Another displayed 73.3% homozygous excision and 13.3% heterozygous excision, and the last line displayed 86.7% homozygous excision. In contrast to the study previously described, no adverse effects were observed in regard to plant growth and reproduction.

CRISPR/Cas9-mediated transgene excision displayed high excision efficiency in rice. However, there are limited studies that have used CRISPR/Cas9 for transgene excision, one of which resulted detrimental effects on plant morphology and reproduction. The use of tissue-specific promoters to drive Cas9 expression, such as those that target pollen or seeds, may help reduce potential off-target effects in vegetative tissues. We believe CRISPR/Cas9-mediated transgene excision can be a useful bioconfinement system, but further optimization is needed.

4. Delayed Flowering

The transition from vegetative to flowering phase is a complex regulatory system controlled by microRNAs (miRNAs), transcription factor (TF) genes, and floral identity genes. miRNAs are a class of small regulatory RNAs that regulate gene expression and RNA silencing through RNA degradation or translational inhibition [80,81]. The overexpression of certain miRNAs can silence floral identity genes that may result in shifts in flowering time or reduced flower production. This approach has the potential to shift flowering time to prevent complete or partial overlap with the flowering time of sexually compatible plants, thus preventing or limiting transgene flow [82,83].

Overexpression of miR156 Delayed Flowering

miRNA 156 (miR156) is one of several miRNAs responsible for regulating the floral transition in the shoot apical meristem [84]. miR156 expression is regulated by the age-dependent pathway, and expression is high in seedlings and declines as the plant matures. It has been characterized to repress transcripts of the *SQUAMOSA PROMOTER BINDING PROTEIN-LIKE* (*SPL*) TF family that upregulate floral identity genes *SUPPRESSOR OF OVEREXPRESSION OF CONSTANS1* (*SOC1*) and *LEAFY* (*LFY*) [84]. Thus, the overexpression of miR156 is expected to prevent the upregulation of *SOC1* and *LFY* as the plant matures, which consequently inhibits/delays the transition from the vegetative to flowering phase. For this reason, the overexpression of miR156 has been studied in switchgrass under greenhouse and field conditions to determine its potential for transgene bioconfinement via delayed flowering [85,86].

In a greenhouse study, switchgrass plants engineered to overexpress miR156 (from rice) were categorized as low, moderate, and high based on transgene expression levels [85]. Transgenic switchgrass with low expression of miR156 flowered at the same time and displayed normal reproductive growth compared to nontransgenic controls. These plants also displayed an increase in tiller number and vegetative biomass. Plants considered to have moderate expression of miR156 did not flower after 360 days and were also observed to have an increase in tiller number and vegetative biomass. However, these lines had several notable defects when compared to nontransgenic controls, including reduced plant height, decreased internode diameter, leaf blade width, and leaf sheath length. Similar to the plants categorized as moderate expression, transgenic switchgrass with high expression of miR156 did not flower after 360 days. Additionally, high expression of miR156 led to an increase in tiller number and severe defects including a reduction in plant height and biomass, as well as significantly decreased internode diameter and length, leaf sheath length, and leaf blade length and width compared to nontransgenic switchgrass [85].

Subsequently, a two-year field study was conducted using low and moderate miR156 overexpression lines to determine flowering time and biomass production under field conditions [86]. Compared to nontransgenic controls, low overexpression lines were highly variable in flowering time during the first year of establishment post-transplanting. Once these lines were established, there was no delay in flowering during the second year. Reduced biomass was observed in some plants, and fewer flowers and seeds were observed in all low expression lines. However, when flowering was monitored for moderate overexpression lines, many plants did not flower in either year of the field study. Of the plants that did flower, timing was delayed by 12 weeks in the first year and two weeks in the second year. Also, some plants displayed reduced biomass production whereas others were unaffected. Additionally, decreased flower and seed production was observed in all moderate overexpression plants.

The overexpression of miR156 in switchgrass results in delayed and decreased flowering. Several phenotypes observed in some of the low and moderate overexpression lines are desirable, including significantly delayed flowering. To optimize this system, miR156 expression levels may be fine-tuned through promoter selection that yields expression levels high enough to delay flowering without a negative impact on plant fitness. Delayed or decreased flowering could also be achieved by downregulating the *SPL* gene (s) that are targets of miR156, including *SPL 2/10* and *SPL 3/4/5* [86]. Temporal downregulation of these genes individually or in combination prior to floral transition could be achieved through RNA interference (RNAi) or CRISPR interference (CRISPRi) and result in delayed flowering without negatively affecting plant biomass production. Additionally, expression of other miRNAs known to regulate floral development can be altered to confer delayed flowering [87]. For example, miR172 expression is negatively correlated with miR156 expression, and overexpression of miR172 has been observed to accelerate flowering time in Arabidopsis [88]. Downregulation of miR172 could also confer similar phenotypes to those of miR156 overexpression and be useful for bioconfinement purposes.

5. Perspectives

GE crops that have been approved for commercial release are considered low risk of transgene flow and generally safe [17]. It is likely that regulatory bodies will require new GE crops with a high risk of gene flow or those that contain novel transgenic traits be put through lengthy and costly biosafety assessments before nonregulated status is approved. These crops would benefit from the inclusion of a bioconfinement system that significantly limits or eliminates transgene flow. In this review, we have highlighted the most prospective bioconfinement designs and many of these approaches have been tested in several plant species. Several designs have shown promise with nearly 100% bioconfinement efficacy. Should a system be adopted, the design must meet several key criteria. First, the system should be highly reliable, consistently reaching near 100% efficiency to reduce transgene exposure to unintended hosts. In addition, system components should not pose biosafety concerns regarding ecological risks, human health, or non-target organisms that may be exposed such as pollen-feeding insects. Lastly, it is vital that the bioconfinement system have no negative impact on crop yield, whether that is vegetative biomass, seed, or fruit yield. Many of the strategies reviewed here meet most of these criteria yet further improvements are needed for more robust and targeted gene expression. Indeed, the future is bright for this area of research and we expect advances in synthetic biology to be leveraged so new tools, such as inducible synthetic promoters, can be fine-tuned for robust gene expression in target tissue as well as improved efficiency of recombinases and CRISPR/Cas9-based methods. Although no technology is without risks, we are optimistic that several bioconfinement systems will soon be adopted that reduce biosafety concerns of stakeholders, regulators, and the public.

Author Contributions: Conceived and conceptualization the manuscript, R.J.M. and J.N.S.; writing—original draft preparation, J.N.S.; and writing, review and editing, R.J.M. and J.N.S. All authors have read and agreed to the published version of the manuscript.

Funding: This work was supported through funding from the USDA NIFA Biotechnology Risk Assessment Research Grant Program (2019-33522-29989) and the University of Tennessee.

Data Availability Statement: Data sharing not applicable.

Acknowledgments: We would like to thank Neal Stewart for his mentorship. We also thank Rob Sears and Bryce Trull for their suggestions and support.

Conflicts of Interest: The authors declare no conflict of interest.

References

1. ISAAA. *Global Status of Commercialized Biotech/GM Crops in 2019: Biotech Crops Drive SocioEconomic Development and Sustainable Environment in the New Frontier*; ISAAA Brief No. 55; ISAAA: Ithaca, NY, USA, 2019.
2. Chandler, S.; Dunwell, J.M. Gene flow, risk assessment and the environmental release of transgenic plants. *Crit. Rev. Plant Sci.* **2008**, *27*, 25–49. [CrossRef]
3. Ellstrand, N.C. Current knowledge of gene flow in plants: Implications for transgene flow. *Philos. Trans. Biol. Sci.* **2003**, *358*, 1163–1170. [CrossRef]
4. Rizwan, M.; Hussain, M.; Shimelis, H.; Hameed, M.U.; Atif, R.M.; Azhar, M.T.; Asif, M. Gene flow from major genetically modified crops and strategies for containment and mitigation of transgene escape: A review. *Appl. Ecol. Environ. Res.* **2019**, *17*, 11191–11208. [CrossRef]
5. Lu, B.R. Transgene escape from GM crops and potential biosafety consequences: An environmental perspective. *Collect. Biosaf. Rev.* **2008**, *4*, 66–141.
6. Chang, H.; Snow, A.A.; Mutegi, E.; Lewis, E.M.; Heaton, E.A.; Miriti, M.N. Extent of pollen-mediated gene flow and seed longevity in switchgrass (*Panicum virgatum* L.): Implications for biosafety procedures. *Biomass Bioenergy* **2018**, *109*, 114–124. [CrossRef]
7. Kesoju, S.R.; Kramer, M.; Brunet, J.; Greene, S.L.; Jordan, A.; Martin, R.C. Gene flow in commercial alfalfa (*Medicago sativa* subsp. *sativa* L.) seed production fields: Distance is the primary but not the sole influence on adventitious presence. *PLoS ONE* **2021**, *16*, e0248746. [CrossRef]
8. Millwood, R.; Nageswara-Rao, M.; Ye, R.; Terry-Emert, E.; Johnson, C.R.; Hanson, M.; Burris, J.N.; Kwit, C.; Stewart, C.N., Jr. Pollen-mediated gene flow from transgenic to non-transgenic switchgrass (*Panicum virgatum* L.) in the field. *BMC Biotechnol.* **2017**, *17*, 40. [CrossRef] [PubMed]
9. Arriola, P.E.; Ellstrand, N.C. Crop-to-weed gene flow in the genus *Sorghum* (Poaceae): Spontaneous interspecific hybridization between johnsongrass, *Sorghum halepense*, and crop sorghum, *S. bicolor*. *Am. J. Bot.* **1996**, *83*, 1153–1159. [CrossRef]
10. Cao, Q.J.; Xia, H.; Yang, X.; Lu, B.R. Performance of hybrids between weedy rice and insect-resistant transgenic rice under field experiments: Implication for environmental biosafety assessment. *J. Integr. Plant Biol.* **2009**, *51*, 1138–1148. [CrossRef] [PubMed]
11. Ellstrand, N.C. "Born to run"? Not necessarily: Species and trait bias in persistent free-living transgenic plants. *Front. Bioeng. Biotechnol.* **2018**, *6*, 88. [CrossRef]
12. Bauer-Panskus, A.; Miyazaki, J.; Kawall, K.; Then, C. Risk assessment of genetically engineered plants that can persist and propagate in the environment. *Environ. Sci. Eur.* **2020**, *32*, 32. [CrossRef]
13. Zapiola, M.L.; Campbell, C.K.; Butler, M.D.; Mallory-Smith, C.A. Escape and establishment of transgenic glyphosate-resistant creeping bentgrass *Agrostis stolonifera* in Oregon, USA: A 4-year study. *J. Appl. Ecol.* **2007**, *45*, 486–494. [CrossRef]
14. Greene, S.L.; Kesoju, S.R.; Martin, R.C.; Kramer, M. Occurrence of transgenic feral alfalfa (*Medicago sativa* subsp. *sativa* L.) in alfalfa seed production areas in the United States. *PLoS ONE* **2015**, *10*, e0143296. [CrossRef]
15. Reichman, J.R.; Watrud, L.S.; Lee, E.H.; Burdick, C.A.; Bollman, M.A.; Storm, M.J.; King, G.A.; Mallory-Smith, C. Establishment of transgenic herbicide-resistant creeping bentgrass (*Agrostis stolonifera* L.) in nonagronomic habitats. *Mol. Ecol.* **2006**, *15*, 4243–4255. [CrossRef] [PubMed]
16. Schafer, M.G.; Ross, A.A.; Londo, J.P.; Burdick, C.A.; Lee, E.H.; Travers, S.E.; Van de Water, P.K.; Sagers, C.L. The establishment of genetically engineered canola populations in the US. *PLoS ONE* **2011**, *6*, e25736. [CrossRef] [PubMed]
17. NASEM. *Genetically Engineered Crops: Experiences and Prospects*; NASEM: Washington, DC, USA, 2016.
18. Ellstrand, N.C.; Rieseberg, L.H. When gene flow really matters: Gene flow in applied evolutionary biology. *Evol. Appl.* **2016**, *9*, 833–836. [CrossRef]
19. Stewart, C.N., Jr. Controlling transgene flow from engineered crops to unintended hosts by molecular approaches. In *Gene Flow: Monitoring, Modeling and Mitigation*; Wei, W., Stewart, C.N., Eds.; CABI: Wallingford, UK, 2021; pp. 118–124.
20. Warwick, S.I.; Beckie, H.J.; Hall, L.M. Gene flow, invasiveness, and ecological impact of genetically modified crops. *Annu. N. Y. Acad. Sci.* **2009**, *1168*, 72–99. [CrossRef]
21. Ding, J.; Duan, H.; Deng, Z.; Zhao, D.; Yi, G.; McAvoy, R.; Li, Y. Molecular strategies for addressing gene flow problems and their potential applications in abiotic stress tolerant transgenic plants. *Crit. Rev. Plant Sci.* **2014**, *33*, 190–204. [CrossRef]

22. Pilson, D.; Prendeville, H.R. Ecological effects of transgenic crops and the escape of transgenes into wild populations. *Annu. Rev. Ecol. Evol. Syst.* **2004**, *35*, 149–174. [CrossRef]
23. Haygood, R.; Ives, A.R.; Andow, D.A. Consequences of recurrent gene flow from crops to wild relatives. *Proc. R. Soc. London. B Biol. Sci.* **2003**, *270*, 1879–1886. [CrossRef]
24. Kausch, A.P.; Hague, J.; Oliver, M.J.; Li, Y.; Daniell, H.; Mascia, P.; Watrud, L.S.; Stewart, C.N., Jr. Transgenic perennial biofuel feedstocks and strategies for bioconfinement. *Biofuels* **2010**, *1*, 163–176. [CrossRef]
25. Daniell, H. Molecular strategies for gene containment in transgenic crops. *Nat. Biotechnol.* **2002**, *20*, 581–586. [CrossRef]
26. Sandhu, S.; Blount, A.R.; Quesenberry, K.H.; Altpeter, F. Apomixis and ploidy barrier suppress pollen-mediated gene flow in field grown transgenic turf and forage grass (*Paspalum notatum* Flüggé). *Theor. Appl. Genet.* **2010**, *121*, 919–929. [CrossRef]
27. Faisal, S.; Guo, Y.; Zang, S.; Cao, B.; Qu, G.; Hu, S. Morphological and genetic analysis of a cleistogamous mutant in rapeseed (*Brassica napus* L.). *Genet. Resour. Crop Evol.* **2018**, *65*, 397–403. [CrossRef]
28. Fargue, A.; Colbach, N.; Pierre, J.; Picault, H.; Renard, M.; Meynard, J.M. Predictive study of the advantages of cleistogamy in oilseed rape in limiting unwanted gene flow. *Euphytica* **2006**, *151*, 1–13. [CrossRef]
29. Klocko, A.L. Strategies to facilitate containment of genetically engineered crops. *CABI Rev.* **2020**, *15*, 1–11. [CrossRef]
30. Mariani, C.; Beuckeleer, M.D.; Truettner, J.; Leemans, J.; Goldberg, R.B. Induction of male sterility in plants by a chimaeric ribonuclease gene. *Nature* **1990**, *347*, 737–741. [CrossRef]
31. Echlin, P. The role of the tapetum during microsporogenesis of angiosperms. *Pollen* **1971**, 41–61. [CrossRef]
32. Goldberg, R.B.; Beals, T.P.; Sanders, P.M. Anther development: Basic principles and practical applications. *Plant Cell* **1993**, *5*, 1217–1229. [CrossRef] [PubMed]
33. Hartley, R.W. Barnase and barstar: Expression of its cloned inhibitor permits expression of a cloned ribonuclease. *J. Mol. Biol.* **1988**, *202*, 913–915. [CrossRef]
34. Mariani, C.; Gossele, V.; Beuckeleer, M.D.; Block, M.D.; Goldberg, R.B.; Greef, W.D.; Leemans, J. A chimaeric ribonuclease-inhibitor gene restores fertility to male sterile plants. *Nature* **1992**, *357*, 384–387. [CrossRef]
35. Block, M.D.; Debrouwer, D.; Moens, T. The development of a nuclear male sterility system in wheat. Expression of the barnase gene under the control of tapetum-specific promoters. *Theor. Appl. Genet.* **1997**, *95*, 125–131. [CrossRef]
36. Luo, H.; Kausch, A.P.; Hu, Q.; Nelson, K.; Wipff, J.K.; Fricker, C.C.R.; Owen, T.P.; Moreno, M.A.; Lee, J.Y.; Hodges, T.K. Controlling transgene escape in GM creeping bentgrass. *Mol. Breed.* **2005**, *16*, 185–188. [CrossRef]
37. Hague, J.P.; Dellaporta, S.L.; Moreno, M.A.; Longo, C.; Nelson, K.; Kausch, A.P. Pollen sterility—A promising approach to gene confinement and breeding for genetically modified bioenergy crops. *Agriculture* **2012**, *2*, 295–315. [CrossRef]
38. Zhang, C.; Norris-Caneda, K.H.; Rottmann, W.H.; Gulledge, J.E.; Chang, S.; Kwan, B.Y.H.; Thomas, A.M.; Mandel, L.C.; Kothera, R.T.; Victor, A.D.; et al. Control of pollen-mediated gene flow in transgenic trees. *Plant Physiol.* **2012**, *159*, 1319–1334. [CrossRef]
39. Elorriaga, E.; Meilan, R.; Ma, C.; Skinner, J.S.; Etherington, E.; Brunner, A.; Strauss, S.H. A tapetal ablation transgene induces stable male sterility and slows field growth in *Populus*. *Tree Genet. Genomes* **2014**, *10*, 1583–1593. [CrossRef]
40. Pappenheimer, A.M., Jr. Diphtheria toxin. *Annu. Rev. Biochem.* **1977**, *46*, 69–94. [CrossRef]
41. Twell, D. Diphtheria toxin-mediated cell ablation in developing pollen: Vegetative cell ablation blocks generative cell migration. *Protoplasma* **1995**, *187*, 144–154. [CrossRef]
42. Koltunow, A.M.; Truettner, J.; Cox, K.H.; Wallroth, M.; Goldberg, R.B. Different temporal and spatial gene expression patterns occur during anther development. *Plant Cell* **1990**, *2*, 1201–1224. [CrossRef]
43. Fourel, G.; Phalipon, A.; Kaczorek, M. Evidence for direct regulation of diphtheria toxin gene transcription by an Fe^{2+}-dependent DNA-binding repressor, *DtoxR*, in *Corynebacterium diphtheriae*. *Infect. Immun.* **1989**, *57*, 3221–3225. [CrossRef]
44. Yamaizumi, M.; Mekada, E.; Uchida, T.; Okada, Y. One molecule of diphtheria toxin fragment A introduced into a cell can kill the cell. *Cell* **1978**, *15*, 245–250. [CrossRef] [PubMed]
45. Millwood, R.; Moon, H.S.; Poovaiah, C.R.; Muthukumar, B.; Rice, J.H.; Abercrombie, J.M.; Abercrombie, L.L.; Green, W.D.; Stewart, C.N., Jr. Engineered selective plant male sterility through pollen-specific expression of the *Eco*RI restriction endonuclease. *Plant Biotechnol. J.* **2016**, *14*, 1281–1290. [CrossRef] [PubMed]
46. Wilson, G.G. Type II restriction—Modification systems. *Trends Genet.* **1988**, *4*, 314–318. [CrossRef]
47. Oliver, M.J.; Quisenberry, J.E.; Trolinder, N.; Keim, D.L. Control of Plant Gene Expression. U.S. Patent No. 5,723,765, 3 March 1998.
48. Oliver, M.J.; Quisenberry, J.E.; Trolinder, N.; Keim, D.L. Control of Gene Expression. U.S. Patent No. 5,925,808, 20 July 1999.
49. Oliver, M.J.; Quisenberry, J.E.; Trolinder, N.; Keim, D.L. Control of Gene Expression. U.S. Patent No. 5,977,441, 2 November 1999.
50. Sang, Y.; Millwood, R.; Stewart, C.N., Jr. Gene use restriction technologies for transgenic plant bioconfinement. *Plant Biotechnol. J.* **2013**, *11*, 649–658. [CrossRef] [PubMed]
51. FAO. *Potential Impacts of Genetic Use Restriction Technologies (GURTs) on Agricultural Biodiversity and Agricultural Production Systems*; FAO: Rome, Italy, 2001.
52. Lombardo, L. Genetic use restriction technologies: A review. *Plant Biotechnol. J.* **2014**, *12*, 995–1005. [CrossRef]
53. Oliver, M.J.; Hake, K. Seed-based containment strategies. In *Plant Gene Containment*; Oliver, M.J., Li, Y., Eds.; John Wiley & Sons, Inc.: Hoboken, NJ, USA, 2013; pp. 113–124.

54. Ferreras, J.; Barbieri, L.; Girbés, T.; Battelli, M.G.; Rojo, M.A.; Arias, F.J.; Rocher, M.A.; Soriano, F.; Mendéz, E.; Stirpe, F. Distribution and properties of major ribosome-inactivating proteins (28 S rRNA N-glycosidases) of the plant *Saponaria officinalis* L. (Caryophyllaceae). *Biochim. Biophys. Acta (BBA)-Gene Struct. Expr.* **1993**, *1216*, 31–42. [CrossRef]
55. Oliver, M.J.; Luo, H.; Kausch, A.P.; Collins, H. Seed-based strategies for transgene containment. In Proceedings of the 8th International Symposium on the Biosafety of Genetically Modified Organisms, Montpellier, France, 26–30 September 2004; pp. 154–161.
56. Luo, K.; Duan, H.; Zhao, D.; Zheng, X.; Deng, W.; Chen, Y.; Stewart, C.N., Jr.; McAvoy, R.; Jiang, X.; Wu, Y.; et al. 'GM-gene-deletor': Fused *loxP-FRT* recognition sequences dramatically improve the efficiency of FLP or CRE recombinase on transgene excision from pollen and seed of tobacco plants. *Plant Biotechnol. J.* **2007**, *5*, 263–374. [CrossRef]
57. Liu, W.; Stewart, C.N., Jr. Plant synthetic promoters and transcription factors. *Curr. Opin. Biotechnol.* **2016**, *37*, 36–44. [CrossRef] [PubMed]
58. Huang, D.; Kosentka, P.Z.; Liu, W. Synthetic biology approaches in regulation of targeted gene expression. *Curr. Opin. Plant Biol.* **2021**, *63*, 102036. [CrossRef]
59. Yang, Y.; Chaffin, T.A.; Ahkami, A.H.; Blumwald, E.; Stewart, C.N., Jr. Plant synthetic biology innovations for biofuels and bioproducts. *Trends Biotechnol.* **2022**, *40*, 1454–1468. [CrossRef]
60. Ali, S.; Kim, W.C. A fruitful decade using synthetic promoters in the improvement of transgenic plants. *Front. Plant Sci.* **2019**, *10*, 1433. [CrossRef] [PubMed]
61. Moon, H.S.; Abercrombie, L.L.; Eda, S.; Blanvillain, R.; Thomson, J.G.; Ow, D.W.; Stewart, C.N., Jr. Transgene excision in pollen using a codon optimized serine resolvase CinH-*RS2* site-specific recombination system. *Plant Mol. Biol.* **2011**, *75*, 621–631. [CrossRef]
62. Yau, Y.Y.; Stewart, C.N., Jr. Less is more: Strategies to remove marker genes from transgenic plants. *BMC Biotechnol.* **2013**, *13*, 36. [CrossRef]
63. Hoess, R.H.; Abremski, K. The Cre-*lox* recombination system. In *Nucleic Acids and Molecular Biology*; Eckstein, F., Lilley, D.M.J., Eds.; Springer: Berlin/Heidelberg, Germany, 1990; Volume 4, pp. 99–109.
64. Senecoff, J.F.; Bruckner, R.C.; Cox, M.M. The FLP recombinase of the yeast 2-micron plasmid: Characterization of its recombination site. *Proc. Natl. Acad. Sci. USA* **1985**, *82*, 7270–7274. [CrossRef]
65. Araki, H.; Jearnpipatkul, A.; Tatsumi, H.; Sakurai, T.; Ushio, K.; Muta, T.; Oshima, Y. Molecular and functional organization of yeast plasmid pSR1. *J. Mol. Biol.* **1985**, *182*, 191–203. [CrossRef]
66. Gidoni, D.; Srivastava, V.; Carmi, N. Site-specific excisional recombination strategies for elimination of undesirable transgenes from crop plants. *Vitr. Cell. Dev. Biol.- Plant* **2008**, *44*, 457–467. [CrossRef]
67. Mlynárová, L.; Conner, A.J.; Nap, J.P. Directed microspore-specific recombination of transgenic alleles to prevent pollen-mediated transmission of transgenes. *Plant Biotechnol. J.* **2006**, *4*, 445–452. [CrossRef] [PubMed]
68. Hamzeh, S.; Motallebi, M.; Zamani, M.R. Efficient seed-specifically regulated autoexcision of marker gene (*nptII*) with inducible expression of interest gene in transgenic *Nicotiana tabacum*. *Turk. J. Biol.* **2016**, *40*, 1–11. [CrossRef]
69. Kopertekh, L.; Schulze, K.; Frolov, A.; Strack, D.; Broer, I.; Schiemann, J. Cre-mediated seed-specific transgene excision in tobacco. *Plant Mol. Biol.* **2010**, *72*, 597–605. [CrossRef]
70. Polóniová, Z.; Jopčík, M.; Matušíková, I.; Libantová, J.; Moravčíková, J. The pollen- and embryo- specific Arabidopsis DLL promoter bears good potential for application in marker-free Cre/*loxP* self-excision strategy. *Plant Cell Rep.* **2015**, *34*, 469–481. [CrossRef] [PubMed]
71. Ye, X.; Vaghchhipawala, Z.; Williams, E.J.; Fu, C.; Liu, J.; Lu, F.; Hall, E.L.; Guo, S.X.; Frank, L.; Gilbertson, L.A. Cre-mediated autoexcision of selectable marker genes in soybean, cotton, canola and maize transgenic plants. *Plant Cell Rep.* **2022**, *42*, 45–55. [CrossRef]
72. Liu, W.; Yuan, J.S.; Stewart, C.N., Jr. Advanced genetic tools for plant biotechnology. *Nat. Rev. Genet.* **2013**, *14*, 781–793. [CrossRef]
73. Sobecky, P.A.; Easter, C.L.; Bear, P.D.; Helinski, D.R. Characterization of the stable maintenance properties of the par region of broad-host-range plasmid RK2. *J. Bacteriol.* **1996**, *178*, 2086–2093. [CrossRef] [PubMed]
74. Kholodii, G.; Mindlin, S.; Gorlenko, Z.; Petrova, M.; Hobman, J.; Nikiforov, V. Translocation of transposition-deficient (TndPKLH2-like) transposons in the natural environment: Mechanistic insights from the study of adjacent DNA sequences. *Microbiology* **2004**, *150*, 979–992. [CrossRef] [PubMed]
75. Shao, M.; Blechl, A.; Thomson, J.G. Small serine recombination systems ParA-*MRS* and CinH-*RS2* perform precise excision of plastid DNA. *Plant Biotechnol. J.* **2017**, *15*, 1577–1589. [CrossRef] [PubMed]
76. Thomson, J.G.; Ow, D.W. Site-specific recombination systems for the genetic manipulation of eukaryotic genomes. *Genesis* **2006**, *44*, 465–476. [CrossRef]
77. He, Y.; Mudgett, M.; Zhao, Y. Advances in gene editing without residual transgenes in plants. *Plant Physiol.* **2022**, *188*, 1757–1768. [CrossRef]
78. Srivastava, V.; Underwood, J.L.; Zhao, S. Dual-targeting by CRISPR/Cas9 for precise excision of transgenes from rice genome. *Plant Cell Tissue Organ Cult.* **2017**, *129*, 153–160. [CrossRef]
79. Tan, J.; Wang, Y.; Chen, S.; Lin, Z.; Zhao, Y.; Xue, Y.; Luo, Y.; Liu, Y.G.; Zhu, Q. An efficient marker gene excision strategy based on CRISPR/Cas9-mediated homology-directed repair in rice. *Int. J. Mol. Sci.* **2022**, *23*, 1588. [CrossRef]

80. Brodersen, P.; Sakvarelidze-Achard, L.; Bruun-Rasmussen, M.; Dunoyer, P.; Yamamoto, Y.Y.; Sieburth, L.; Voinnet, O. Widespread translational inhibition by plant miRNAs and siRNAs. *Science* **2008**, *320*, 1185–1190. [CrossRef]
81. Jonas, S.; Izaurralde, E. Towards a molecular understanding of microRNA-mediated gene silencing. *Nat. Rev. Genet.* **2015**, *16*, 421–433. [CrossRef]
82. Brookes, G.; Barfoot, P.; Melé, E.; Messeguer, J.; Bénétrix, F.; Bloc, D.; Foueillassar, X.; Fabié, A.; Poeydomenge, C. *Genetically Modified Maize: Pollen Movement and Crop Co-Existence*; PG Economics Ltd.: Dorchester, UK, 2004; pp. 3–20.
83. Ortega-Molina, J. Results of the studies into the coexistence of genetically modified and conventional maize. In Proceedings of the COPA-COGECA Colloquy on the Co-Existence and Thresholds of Adventitious Presence on GMOs in Conventional Seeds, Brussels, Belgium, 2004.
84. Luo, Y.; Guo, Z.; Li, L. Evolutionary conservation of microRNA regulatory programs in plant flower development. *Dev. Biol.* **2013**, *380*, 133–144. [CrossRef] [PubMed]
85. Fu, C.; Sunkar, R.; Zhou, C.; Shen, H.; Zhang, J.Y.; Matts, J.; Wolf, J.; Mann, D.G.J.; Stewart, C.N., Jr.; Tang, Y.; et al. Overexpression of miR156 in switchgrass (*Panicum virgatum* L.) results in various morphological alterations and leads to improved biomass production. *Plant Biotechnol. J.* **2012**, *10*, 443–452. [CrossRef] [PubMed]
86. Johnson, C.R.; Millwood, R.; Tang, Y.; Gou, J.; Sykes, R.W.; Turner, G.B.; Davis, M.F.; Sang, Y.; Wang, Z.Y.; Stewart, C.N., Jr. Field-grown miR156 transgenic switchgrass reproduction, yield, global gene expression analysis, and bioconfinement. *Biotechnol. Biofuels* **2017**, *10*, 255. [CrossRef] [PubMed]
87. Trumbo, J.L.; Zhang, B.; Stewart, C.N., Jr. Manipulating microRNAs for improved biomass and biofuels from plant feedstocks. *Plant Biotechnol. J.* **2015**, *13*, 337–354. [CrossRef] [PubMed]
88. Aukerman, M.J.; Sakai, H. Regulation of flowering time and floral organ identity by a microRNA and its *APETALA2*-like target genes. *Plant Cell* **2003**, *15*, 2730–2741. [CrossRef] [PubMed]

plants

MDPI

Review

Enabling Trade in Gene-Edited Produce in Asia and Australasia: The Developing Regulatory Landscape and Future Perspectives

Michael G. K. Jones [1,*], John Fosu-Nyarko [1], Sadia Iqbal [1], Muhammad Adeel [1], Rhodora Romero-Aldemita [2], Mahaletchumy Arujanan [3], Mieko Kasai [4], Xun Wei [5], Bambang Prasetya [6], Satya Nugroho [7], Osman Mewett [8], Shahid Mansoor [9], Muhammad J. A. Awan [9], Reynante L. Ordonio [10], S. R. Rao [11], Abhijit Poddar [12], Penny Hundleby [13], Nipon Iamsupasit [14] and Kay Khoo [15]

1 Crop Biotechnology Research Group, Centre for Crop and Food Innovation, Food Futures Institute, Murdoch University, Perth, WA 6150, Australia
2 ISAAA—BioTrust Global Knowledge Center on Biotechnology, International Service for the Acquisition of Agri-Biotech Applications (ISAAA), IRRI, Los Banos 4031, Philippines
3 Malaysian Biotechnology Information Centre, Monash University Malaysia, Jalan Lagoon Selatan, Bandar Sunway 47500, Malaysia
4 Japan Plant Factory Association, 6-2-1 Kashiwanoha Kashiwa, Chiba 277-0012, Japan
5 Zhongzhi International Institute of Agricultural Biosciences, Research Center of Biology and Agriculture, Shunde Graduate School, University of Science and Technology Beijing, Beijing 100024, China
6 National Biosafety Committee of Genetically Engineered Products (KKH-PRG), Research Center for Testing Technology and Standards, National Research and Innovation Agency (BRIN), Central Jakarta 10340, Indonesia
7 Research Center for Genetic Engineering, Research Organization for Life Sciences and Environment, National Research and Innovation Agency (BRIN), Central Jakarta 10340, Indonesia
8 Australian Seed Federation, 20 Napier Cl, Deakin, Canberra, ACT 2600, Australia
9 National Institute for Biotechnology and Genetic Engineering (NIBGE), Faisalabad 44000, Pakistan
10 Crop Biotech Center, Philippine Rice Research Institute, Munoz 3119, Philippines
11 Sri Balaji Vidyapeeth University, Pondicherry 607402, India
12 MGM Advanced Research Institute, Pondicherry 607402, India
13 John Innes Centre, Norwich, Research Park, Norwich NR4 7UH, UK
14 Biotechnology Alliance Association, Bangkok 10900, Thailand
15 Regulatory Affairs Manager, Seeds Asia-Pacific, BASF Australia Ltd., 12/28 Freshwater Pl, Southbank, VIC 3006, Australia
* Correspondence: m.jones@murdoch.edu.au; Tel.: +61-(0)4-1423-9428

Citation: Jones, M.G.K.; Fosu-Nyarko, J.; Iqbal, S.; Adeel, M.; Romero-Aldemita, R.; Arujanan, M.; Kasai, M.; Wei, X.; Prasetya, B.; Nugroho, S.; et al. Enabling Trade in Gene-Edited Produce in Asia and Australasia: The Developing Regulatory Landscape and Future Perspectives. *Plants* **2022**, *11*, 2538. https://doi.org/10.3390/plants11192538

Academic Editors: C. Neal Stewart, Jr. and Wei Wei

Received: 2 August 2022
Accepted: 13 September 2022
Published: 27 September 2022

Abstract: Genome- or gene-editing (abbreviated here as 'GEd') presents great opportunities for crop improvement. This is especially so for the countries in the Asia-Pacific region, which is home to more than half of the world's growing population. A brief description of the science of gene-editing is provided with examples of GEd products. For the benefits of GEd technologies to be realized, international policy and regulatory environments must be clarified, otherwise non-tariff trade barriers will result. The status of regulations that relate to GEd crop products in Asian countries and Australasia are described, together with relevant definitions and responsible regulatory bodies. The regulatory landscape is changing rapidly: in some countries, the regulations are clear, in others they are developing, and some countries have yet to develop appropriate policies. There is clearly a need for the harmonization or alignment of GEd regulations in the region: this will promote the path-to-market and enable the benefits of GEd technologies to reach the end-users.

Keywords: gene editing; genome editing; Cas9; GEd; biosafety; Asia; Australasia; Asia-Pacific; regulations; crops; path-to-market; trade; harmonization; science diplomacy

1. Introduction

Long before the development of Mendelian genetics, farmers selected superior plants from natural populations. Since the early 1900s, the principles of Mendelian genetics have been applied increasingly to improve crop yields, eventually resulting in the 'green revolution'. More recently, applied crop researchers and plant breeders have sought to improve crops by developing better combinations of genetic material by making use of the increasingly available genomic information, bioinformatics, and molecular biology to identify genes that underlie useful traits and using marker-assisted selection to follow them to develop better combinations of genes for improved traits.

In parallel with conventional breeding approaches, gene transfer methods have enabled researchers to transfer genes between related or unrelated species to deliver genetically modified (GM) plants or crops, sometimes referred to genetically modified organisms (GMOs) or living modified organisms (LMOs). GM crops now provide more than 10% of the world's food, but unlike conventional or mutation breeding, both of which are forms of genetic manipulation, the commercial growth of GM crops has been highly regulated, and this has increased the costs of their development. The lack of an international harmonization of regulations for GM crops has been a major factor in holding back their wider implementation [1].

Over the last ten years, so-called 'New Breeding Technologies' (NBTs) (also called New Genetic Technologies—NGTs) have emerged, based on the development of targeted mutations using a combination of a double-stranded (ds) nucleases such as Cas9, directed to DNA sequence sites to be cut by guide RNA, making use of the Clustered Regularly Interspaced Short Palindromic Repeats (CRISPR) system employed by bacteria in virus—defense interactions (e.g., CRISPR/Cas9). Amongst the crop science researchers, there is now an air of excitement about the potential applications of genome-, or preferably, gene-editing (GEd), for crop improvement. In some cases, GEd technology is also described as 'Precision Breeding' or 'Plant Breeding Innovations'. There is now a whole new toolbox of promising applications of GEd with great potential for plant breeders to use and therefore achieve many of their desired outcomes more rapidly. GEd methods are predicted to be game-changing technologies that are poised to revolutionize both basic research and plant breeding [2]. However, it is relevant to note that GEd-based technologies cannot be used to achieve all of the results that can be achieved using GM technologies, because in GEd, current regulations may limit the accessible gene pool. Nevertheless, there are some clear advantages of GEd such as developing non-GM improvements for yield, quality, and abiotic/biotic stress tolerance [3,4]. GEd technology is of particular interest in the developing world where food and nutritional security in a climate change scenario is a real challenge, and GEd can be used efficiently to develop desirable traits in important staple foods [5].

As is often the case, the speed of new scientific developments such as GEd have outpaced policy and regulatory aspects, and this is challenging regulators worldwide. It has become evident that the beneficial applications of NBTs will not be achieved without sensible science-based policies and regulations. In particular, without international harmonization of policies on gene-edited products, the benefits they can confer to society will not be achieved. Scientists may not be in a position to make policies and regulations, but they can influence the process through engagement with policy-makers, politicians, and the public to promote international harmonization of GEd policies and regulations. Only by ensuring internationally compatible, science-based policies for gene-edited crops can the world benefit fully from these exciting new technologies.

The issue of policies that relate to NBTs, and how they could become non-tariff trade barriers, has been recognized with relevant publications that have focused on policies in North and South America, and Europe, but not on the Asia-Pacific region (used here as excluding the Russian Federation, Central Asian republics, Bhutan, Nepal, and the Middle East) including Australasia. As an example, Australia is a major food exporter to Asia, with the top eight grain-importing countries being in Asia, and 70% of exported horticultural produce going to Asia. Without an understanding of international policies on GEd and their harmonization, the trade in products developed using GEd technologies will be severely curtailed.

Here, we provide an overview of the science that underlies GEd and discuss the current regulatory status for gen-edited products, focusing on countries in Asia and Australasia. We discuss the path-to-commercialization and the need to promote the harmonization of national policies, which is vital to enable the future trade of GE products in the region.

1.1. Gene-Editing Technology

Much has been written about the science that underlies GEd, the pace of publications about gene-edited crops is accelerating, and the processes involved have been well-described. In this section, we outline some of the technologies used or available in the GEd toolbox, present some examples of their applications, and provide references to more detailed reviews on this subject.

Although GEd evolved through a variety of approaches such as oligonuceotide directed mutagenesis (ODM), editing using zinc finger nucleases (ZFNs) and transcription activator-like effector nucleases (TALENs), the most widely used GE technology is based on the CRISPR/Cas9 system, which was developed from a bacterial immunity mechanism [6]. The CRISPR-associated endonuclease (Cas9, a dsDNAse) enables what is essentially targeted mutagenesis: the site of the targeted cleavage of DNA is determined by the associated short 19–21 nucleotide single guide RNA (sgRNA). CRISPR/Cas9 facilitates the targeted modification of genetic information at a specific genomic location, enabling the alteration, deletion, or addition of DNA bases at a specific site, which results from DNA repair mechanisms. The cut ends of the DNA sequence may be repaired, most commonly with the loss or addition of a small number of bases (indels) by non-homologous end joining (NHEJ). With the discovery of novel tools of CRISPR and variants of bacterial dsDNA cleaving enzymes, there are recent exciting developments where the PAM sequence requirement for the efficient cleavage of DNA is minimized [7]. Alternatively, with the addition of a donor nucleotide template, nucleotides can be added at the site of DNA cleavage by a process of homologous recombination (HR). The advantage of the CRISPR/Cas9 and associated GEd systems is their relative simplicity and adaptability. A comparison of tools used for GEd (ZFNs, TALENS and CRISPR/Cas9) is provided in [6].

Additional techniques in the GE toolbox include base-editing to convert one base into another, modifying the expression levels of target genes by editing promoter sequences, editing genes for transcription factors for multiplexed outcomes, epigenetic changes such as the patterns of DNA methylation or acetylation, and simultaneous targeting of multiple genes to modify multiple or polygenic traits using several sgRNAs at once [8]. The latter enables all the homologous alleles in a polyploid plant, or in a multi-gene family, to be targeted at the same time. Alternative DNAses and gRNAs are also available, some with increased specificity [8].

The application of GEd technology may involve the introduction of a plasmid encoding a dsDNAse such as Cas9 and sgRNA in a T-DNA into the cells of explants or protoplasts (enabling the cells to make their own Cas9 and gRNAs) together with a selectable gene for more efficient selection of potentially edited lines—in this case, selfing of the T0 plants and the selection of segregants null for the editing cassette are required for commercial development. Alternatively, there are methods that do not involve the introduction of T-DNA into the target plant cells, in particular, the introduction of the Cas9 protein and gRNA as a ribonucleoprotein (RNP) complex by particle bombardment, followed by plant regeneration and screening for edited plants. Cas9 is a large protein to deliver into a cell, and its expression via T-DNA is generally more stable and increases the number of edited cells. The use of carbon nanotubes with agrobacterium only for the translocation of GEd components without transformation provides alternatives to the longer process of transgene segregation, particularly in vegetatively propagated crops [9–11].

It takes more time to identify edited plants after RNP delivery, but this is the only approach that can be used to edit heterozygous clonally propagated species such as potato without losing the original variety. In this case, detecting the presence of targeted mutations in regenerated plants can be achieved by polymerase chain reaction (PCR) and restriction enzyme analysis, targeted Sanger sequencing or genome sequencing.

GEd technology can also be used in gene drive technology (GDT) where there is a biased inheritance of a gene or genetic element from parent to offspring, resulting in an increase in the frequency of the element until most members of a population contain that genetic element. The potential for GDT in agriculture, driven by mutations created using GE technology, is likely to be developed to eliminate pests, diseases, or weeds from a population, but its discussion was outside the scope of this review.

In summary, GEd technologies can be described as a collection of advanced molecular biology techniques that facilitate precise, efficient, and targeted modifications at genomic loci [12].

1.2. Summary of Research on GEd Products

It was beyond the scope of this paper to provide a complete list of all published work on GEd for all plants including staple crops such as grains and important horticultural crops. However, the 'European Sustainable Agriculture Through Genome Editing' organization (EU-SAGE) has established a database of the latest evidence on GEd applications in plants and many crops. This is a searchable database that describes the plant species, traits, the GEd techniques employed, countries where the work was undertaken and applications covering studies of any crop developed for market-oriented agricultural production as a result of GEd. The database is regularly updated and can be accessed at genome search | EU-SAGE. The traits described are related to plant yield and growth, improved food/feed quality, industrial uses, abiotic and biotic stress tolerances, herbicide tolerance, color, flavor, and storage performance. A summary of the current data on GEd crops in the EU-SAGE database is provided in Figure 1.

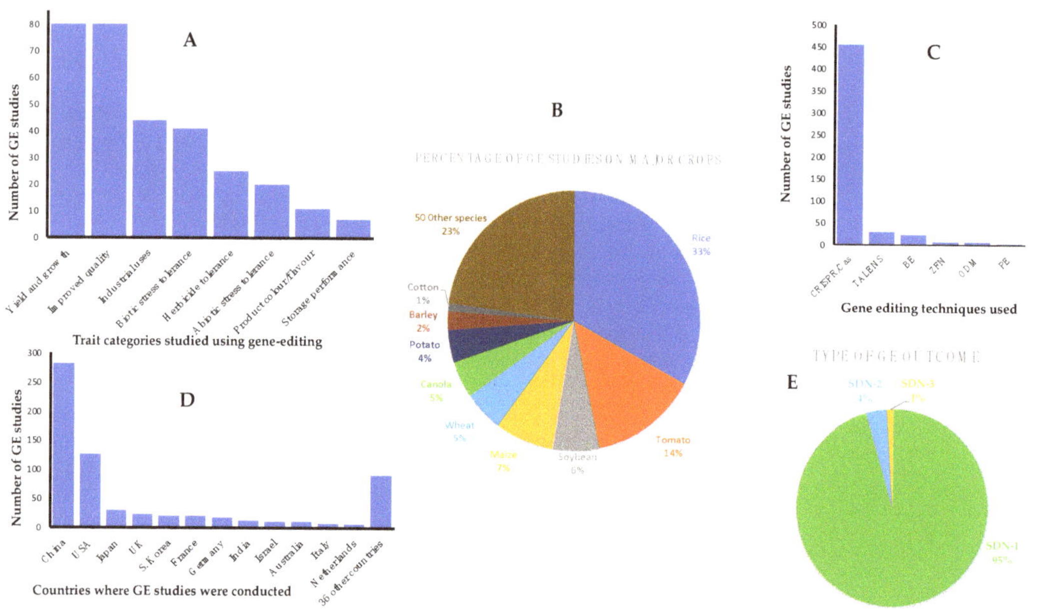

Figure 1. A summary of the published work on GEd to improve traits of plant/crop species. (**A**) The number of GEd studies targeting different traits of crop plants. (**B**) The percentage of GE studies on different species of plants. (**C**) The frequency of use of five GEd techniques in current plant research. BE and PE indicate base-editing and prime-editing. (**D**) Countries where GEd of plants has been undertaken. (**E**) The outcome of GEd type in plant research. (Source: EU-SAGE database; https://www.eu-sage.eu (accessed on 2 July 2022).

From the summary provided in Figure 1, we highlight a few examples of the applications of GEd technology to crop improvement, indicating that GEd, in one form or another, has been applied to all major crop plants including the cereals maize, rice, wheat, barley, sorghum, the staple potato, and industrial crops such as canola and cotton. Other major food crops with data in the EU-SAGE database include soybean, brassicas, tomato, oranges, grapefruit, cassava, flax, cucumber, watermelon and mushrooms, sugarcane, and sugarbeet [6,12,13]. The breeding targets include biotic factors such as resistance to diseases and pests (e.g., resistance to powdery mildew, rice blast, bacterial blight, citrus canker, viruses); quality traits such as the amylose:amylopectin ratio to reduce the Glycemic Index, high oleic acid content, flavor, reduced browning, reduced anti-nutritional factors, improved nutrition such as vitamins A, C, and D; herbicide tolerance; hybrid/breeding systems and maturity dates; grain size, grain number, number of tillers, protein quality, reduced pre-harvest sprouting, reduced allergenicity; improved stress tolerance (e.g., to drought, heat and cold stress), and trait stacking.

1.3. Definitions of Gene-Editing—Site Directed Nucleases (SDN)

Gene-editing, in which there is a spontaneous repair at a ds-break (dsB) site whose repair does not introduce external DNA, is usually referred to as Site-Directed Nuclease 1 (SDN-1). If a repair oligonucleotide is incorporated at the dsB, this is referred to as Site-Directed Nuclease 2 (SDN-2), and if a completely new gene cassette is inserted at the dsB, then it is Site-Directed Nuclease-3 (SDN-3). There is general agreement on the definitions of SDN-1 and SDN-3, but SDN-2 can be interpreted in several ways. To date, an SDN-3 product is regarded as a GMO in almost all jurisdictions.

In its broadest sense, SDN-2 is applied to either the insertion of one or a few bases from an HDR oligonucleotide, or the insertion of a complete allele from within the plant's gene pool, which could have been introduced by conventional breeding. The latter includes allele swopping, which is a standard aim in conventional breeding. The broader definition is much more useful because the question arises on how many nucleotides can be added before an SDN-2 event becomes defined as SDN-3? This aspect is discussed in more detail in Section 3.3.1.

The other question relates to whether recombinant DNA, usually in the form of a Cas9/gRNA cassette with a selectable marker gene, is used initially to improve the efficiency of identifying edited plants amongst the regenerants, followed by selfing and identification of the edited progeny, which do not contain the editing cassette (i.e., do not contain any externally derived DNA).

In considering the current GEd policies of the countries in Asia and Australasia, these issues become apparent, and the current status of GEd regulations in these countries is discussed below.

2. The Policies and Regulation of Gene-Editing in Asia and Australasia

Here, we discuss the policies and regulations that relate to the products of GEd technology on a country basis. The aim is to provide an actual or potential pathway-to-market of GEd products where possible in each country: at present, data are not available for Myanmar, Laos, Cambodia, or Vietnam.

2.1. GEd Regulations in Australia

The regulation of gene technology in Australia is governed by the Commonwealth Government *Gene Technology Act 2000*, with corresponding State and Territory laws and an Intergovernmental Gene Technology Agreement (https:/www.genetechnology.gov.au, accessed on 2 July 2022). The Act establishes the statutory office holder, the Gene Technology Regulator, to administer the Act and corresponding State and Territory legislation. State and Territory governments take part in governing the scheme through the Gene Technology Minister's Meeting and Gene Technology Standing Committee, which support a nationally consistent regulatory system for gene technology. How the Act is implemented is described in the *Gene Technology Regulations 2001* including the duties of the Office of the Gene Technology Regulator (OGTR).

An overview of the scheme is provided in Figure 2, which summarizes how the scheme is governed, the groups that provide advice, and who takes part in the consultations.

Figure 2. An overview of the bodies that administer or provide information or advice to the Gene Technology Regulator in Australia (https://www.ogtr.gov.au/; accessed on 2 July 2022).

To undertake work with GMOs, the Gene Technology Regulator and staff at the OGTR administer the Act and regulate the use of GM organisms. The Office is based in the Commonwealth Government Department of Health, with the remit to ensure that dealings are safe and well-managed to protect human health and the environment. It covers all aspects of gene regulation including health, the environment, industrial applications, and agriculture.

The duties of the Regulator include administering licenses, developing policy principles and guidelines, codes of practice, ensuring compliance with the legislation and providing advice on gene technology to Government ministers, other agencies, and the public (https://www.ogtr.gov.au/; accessed on 2 July 2022).

The *Gene Technology Act 2000* is restricted to governing living organisms: the standards for the safety, content, and labelling of food is the responsibility of Food Standards Australia New Zealand (FSANZ). Although the scheme has withstood the test of time, the OGTR recognized that new technologies were outpacing the regulations, and to ensure the legislation was fit-for-purpose, a series of reviews of the scheme were undertaken, with the aim of future-proofing it for current and future scientific developments.

The Gene Technology Regulations can designate organisms that are not GMOs (Schedule 1). Natural mutations and mutations induced by chemical or irradiation treatments are classified as 'not GMOs', but the status of SDN-1 (targeted changes with unguided repair), template guided repair with oligonucleotides (SDN-2), or with longer templates (SDN-3) was not clear. Following the review, the outcome was that only SDN-1 products were excluded from the regulations, but organisms edited by SDN-2 or ODM and SDN-3 were regulated as GMOs. If no external DNA is used to generate the edited organism (e.g., RNP process or transient expression of the editing cassette), the organisms are not GMOs. Similarly, if an integrated editing expression cassette is no longer present in null segregants, then the resultant organism is not a GMO. Table 1 summarizes the current status of the regulation of GEd organisms in Australia.

Table 1. The status of organisms with SDN-1 modifications in Australia by method of SDN application.

	SDN Protein Applied (with or without sgRNA)	**SDN Expressed from a Transgene That Is Only Transiently Present in the Organism**	**SDN Expressed from Transgene Integrated in the Genome**
Status of the initial organism modified by SDN-1	Not a GMO (Schedule 1 item 4)	GMO while transgene or its expressed products are present Not a GMO when transgene and expressed products have degraded (Schedule 1 items 4 + 10)	GMO
Status of offspring inheriting the SDN-1 modification	Not a GMO (Schedule 1 item 9(a))	Not a GMO (Schedule 1 item 9(b))	GMO if SDN transgene also inherited Not a GMO if no SDN transgene inherited (Schedule 1 item 9(b))

What this means is that organisms developed from SDN-2 and SDN-3 treatments are still regulated as GMOs in Australia. This result puts Australia at odds with some other developed countries in which SDN-2 edits are not classified as GMOs if they introduce sequences from plants in the same gene pool, and which could have been developed by conventional breeding techniques. In effect, the valuable opportunity for allele swopping in breeding cannot be conducted without a product being designated as a GMO.

The overall scheme describing the pathway for the deregulation of GEd products in Australia is shown in Figure 3.

2.1.1. Modernizing and Future-Proofing Gene Technology Regulatory Schemes in Australia

As previously described, the importation and cultivation of GMOs in Australia is regulated through a nationally consistent legal scheme, the *Gene Technology Act 2000* and the *Gene Technology Regulations 2001*. The Act is administered by the Gene Technology Regulator, who is responsible for making decisions on whether to approve field trials and the commercial release of GM crops. GM products are regulated by a number of authorities with specific areas of responsibility (Figure 3) such as FSANZ, which sets the standards for the safety, content, and labelling of food.

Following a Technical Review of the Gene Technology Regulations, the regulatory status of some GEd techniques in Australia was clarified (Table 1) (i.e., organisms modified using site-directed nucleases without templates to guide genome repair (i.e., SDN-1) are not regulated as GMOs). These organisms are treated the same as those resulting from conventional breeding process, and no consultation with the Regulator is required. If a template is used to guide genome repair (i.e., SDN-2 and SDN-3), the resulting organisms are GMOs, as are organisms modified using ODM. However, there remains a lack of regulatory clarity when it comes to foods derived from gene technology. For example, a product developed using SDN-1 is not a GMO for cultivation purposes; however, whether it is regulated as a GM food is subject to the outcomes of the ongoing FSANZ review of food derived from new breeding techniques.

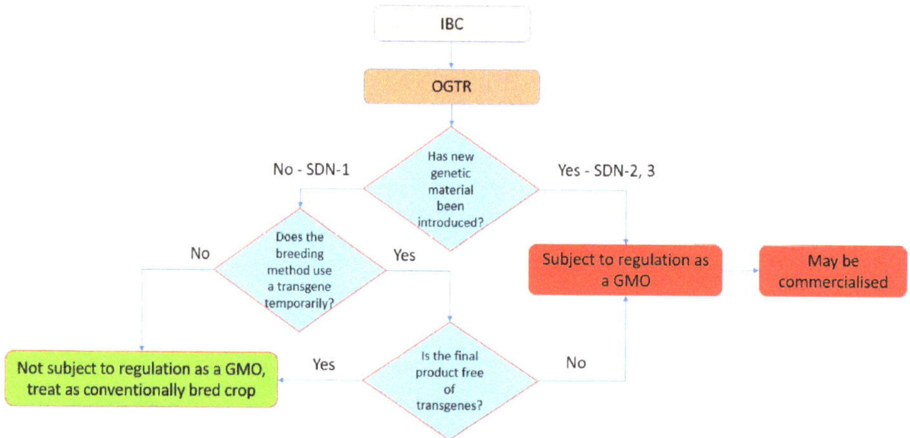

Figure 3. A summary of the pathways to the deregulation of SDN-1 GEd products in Australia.

In December 2020, the Australian Department of Health launched a review paper on 'Modernizing and Future-Proofing Australia's Gene Technology Regulatory Scheme' (https://www.genetechnology.gov.au/resources/publications/2017-review-consultation-regulation-impact-statement-modernising-and-future-proofing-national-gene-technology-scheme; accessed on 2 July 2022). This paper presented three options for consultation:

Option A: Status quo—no changes.
Option B: Risk-tiering model—dealings classified according to their indicative risk.
Option C: Matrix model—the nature of the dealing determines its classification.

Option B, a risk-tiering approach, was the preferred option of most responders and was recommended for adoption in the subsequent Decision Regulatory Impact Statement. Option B enables dealings with GMOs to be distinguished based on indicative risk (i.e., enabling a proportionate risk response). This approach is summarized in Figure 4 below:

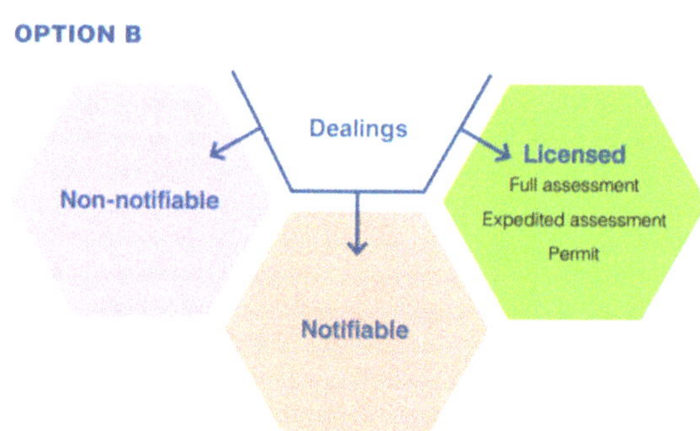

Figure 4. Preferred Australian model for assessing the risk of GEd foods. Option B: Risk-tiering model—where dealings are classified according to their indicative risk (https://www.genetechnology.gov.au/resources/publications/2017-review-consultation-regulation-impact-statement-modernising-and-future-proofing-national-gene-technology-scheme; accessed on 2 July 2022).

For example, the gene technology used to create the GMO would be a relevant consideration. If a specific gene technology (i.e., some types of GEd) presents a very low risk and a case-by-case assessment is not required, then these dealings could be eligible for one of the 'lighter-touch' pathways. However, under this option, even those dealings classified as non-notifiable are still considered as a 'GMO' and are not 'excluded' from regulation. This is significant compared to the SDN-1 exclusion described earlier.

2.1.2. Food Derived from New Breeding Techniques

FSANZ is currently undertaking a review on its regulatory approach to food derived from new breeding techniques (NBT food) (FSANZ 2019, Review of food derived using new breeding techniques, Final Report). The starting point for the review is that the need for the pre-market assessment of food derived from NBT food is essentially a question about risk, and how NBT food compares to conventional food. If it can be demonstrated that NBT food is equivalent in risk to conventional food, then it may be argued that a pre-market safety assessment is unnecessary. FSANZ suggests that when assessing the risk(s) from NBT food, the size of genetic change, whether it was intended or not, and the method used to effect the genetic change are irrelevant considerations (FSANZ 2019).

According to FSANZ (2019), the crucial factor from a food safety perspective when any genetic change is made is the impact of that change on the food. If a genetic change is made using a new breeding technique, and the introduced change has not resulted in new or altered product characteristics compared to conventional food, then FSANZ concludes "the NBT food will carry the same risk as the equivalent conventional food".

This provides a clear basis for excluding these foods from a requirement for the pre-market safety assessment as a GM food.

2.2. GE Regulations in New Zealand

Agricultural exports constitute a large part of New Zealand's (NZ) economy, and at present, no GM crops are grown. GM organisms are regulated by the *Hazardous Substances and New Organisms (HSNO) Act 1996*. In Australia, this Act defines what is a GMO, where GMO means, unless expressly provided otherwise by regulations, any organism in which any of the genes or other genetic material (a) have been modified by in vitro [not defined] techniques; or (b) are inherited or otherwise derived, through any number of replications, from any genes or other genetic material which has been modified by in vitro techniques, and organisms that are not genetically modified.

In 2014, the NZ Environmental Protection Authority ruled that plants produced using GEd methods, where no foreign DNA remained in the edited plant, would not be regulated as GMOs. However, following a challenge in the High Court, this decision was overturned so that NZ currently regards all products of GEd as GMOs [14].

2.3. The Regulatory Status of GEd Produce in Japan

The Japanese government positions GEd as a key innovative technology. In 2014, the Government started funding research with the goal of bringing products developed through GEd to market, under the 'Cross-ministerial Strategic Innovation Promotion Program' (SIP Program) [15]. Discussions on the related policy were then initiated since the Government recognized that a science-based and applicable policy was vital for the commercialization of GEd products. After five years of discussion, which involved inviting many expert opinions and holding public hearings, the Government finally published the handling policy in 2019. Since GM technology may be used during the early stages of GEd techniques, and GEd can result in different types of products, depending on whether or not a repair template is used as well as the nature of the template used, the Government discussed each case separately, and determined what type of products were or were not subject to existing GMO regulations. Table 2 summarizes the GEd handling policy under the *Food Sanitation Law*, the *Feed Safety Law*, and the *Cartagena Law* (the environmental safety law) and each responsible Ministry. There is a discrepancy in the handling policy resulting from different definitions of a GMO under each relevant law.

Table 2. A summary of the policies for handling food products derived from GEd in Japan.

Type of GE Outcome	Food Sanitation Law (MHLW)	Feed Safety Law (MAFF)	Cartagena Law (MOE/MAFF)
SDN-1: Deletion/insertion, no template	Non-GMO	Non-GMO	Non-GMO
SDN-2 Change with template	Non-GMO (1-few bp)	Non-GMO (1-few bp)	GMO/Non-GMO *
SDN-3 Addition of new genetic material	GMO	GMO	GMO/Non-GMO *

* Non-GMO: when the template DNA is from the same species (self-cloning/intragenic) or from sexually compatible species ('natural' occurrence). MHLW: *Ministry of Health, Labor and Welfare*, MOE: *Ministry of the Environment*, MAFF: *Ministry of Agriculture, Forestry and Fisheries*.

The definition of a GMO under the *Food Sanitation Law* and the *Feed Safety Law* can be described as 'an organism obtained through recombinant DNA techniques'. Although the definition of a GMO is process-based, produce-based risk assessments have also been undertaken for GMOs in Japan. Therefore, in determining the handling policy for GEd products, the Ministry of Health, Labor, and Welfare (MHLW) considers whether or not the GEd product can be distinguished from a product obtained through spontaneous or induced mutation: if they are indistinguishable, the risk of a GEd product should be within that of the product obtained through conventional breeding. Hence, if the genetic change induced by GEd is either a nucleotide deletion, substitution, or insertion resulting from the repair of a double-strand break (i.e., SDN-1), such a product is determined as not being subject to regulation as a GMO (Figure 5). In addition, if the genetic change as a result of GEd is up to a 'few base pairs' (i.e., SDN-2), the product is not subject to regulation as a GMO. However, if the genetic change is larger than a few base pairs, the product is regulated as a GMO regardless of the source of the template DNA (i.e., SDN-2 and 3).

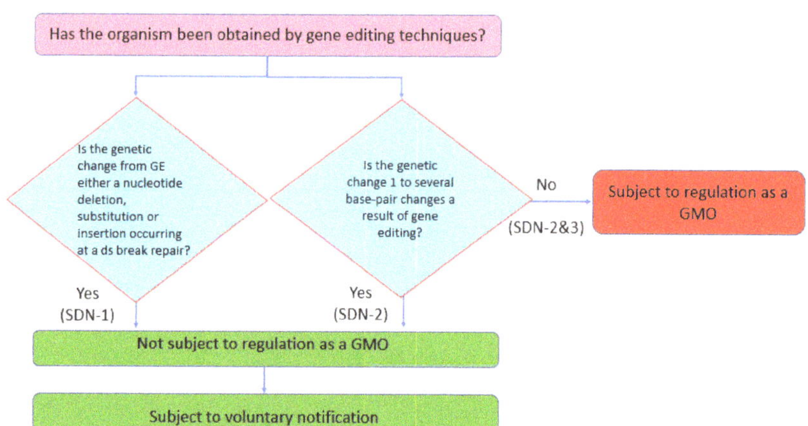

Figure 5. Policy decisions for GEd/GM products in Japan by the *Food Sanitation Law* and *Feed Safety Law*.

The definition of a GMO under the environmental safety regulation (*Cartagena Law*) is that as a 'living modified organism is an organism that possesses extracellularly processed nucleic acids or its replicate excluding the case that the nucleic acids are from the same species (self-cloning) or from sexually compatible species (natural occurrence)'. Consequently, if the final product does not possess nucleic acids that have been processed outside the cell, the product is not subject to regulation as a GMO (Figure 6). On the other hand, if a template DNA is used so that the final product contains extracellularly processed nucleic acids, the product is subject to regulation as a GMO. However, since the *Cartagena Law* does not regulate a product, if the extracellularly processed nucleic acids are from the same species or from sexually compatible species, the same rule applies as for a GEd product. This means that GMO/non-GMO determination of SDN-2 and 3 type products is made on a case-by-case basis.

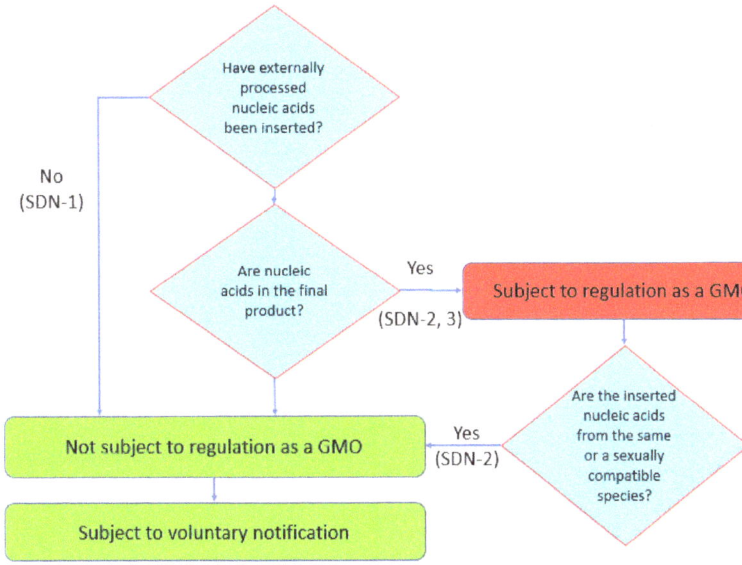

Figure 6. Pathway to the deregulation of GEd products under the Cartegena Law (Environmental Safety) in Japan.

Because GEd is a new technology and is evolving, the Japanese Government still requests developers to submit 'voluntary' pre-marketing notification documents, even if the product is not subject to GM regulation. This notification allows the Government to further confirm that products derived from GEd are as safe as products obtained through conventional breeding, and to gather information on what products are on the market so that the Government can take additional action if needed. The request for voluntary notification also reflects the consumers' concerns that GEd products would be commercialized without any checking or tracking system.

The information that should be included in notification documents to MHLW are as follows:

- Crop, variety, use of the product, and purpose of use.
- The GEd method used, and details of the modification.
- Absence of foreign DNA (using appropriate methods, including Southern blot, Next Generation Sequencing, and PCR.).
- Confirmation that the change (including off-target changes) in DNA does not produce new allergens or increase known toxins (i.e., MHLW currently expects a developer to conduct (i) analysis of off-target edits, and (ii) homology searches to known allergens and toxins not only at the targeted site, but also an off-target edit site, if identified).
- Any change of metabolites relating to a targeted metabolic pathway.
- Year and month of launch (after commercialization)

The information required by the MOE and MAFF is basically the same, except that discussion of any possible influence on biological diversity is required instead of a discussion on allergenicity and toxicity. The notification document is posted on each relevant Ministry's website immediately after submission. By May 2022, three products had undergone the voluntary pre-marketing notification process in Japan: these were high-GABA tomato, sea bream with more meat and better feed efficiency, and tiger pufferfish with faster growth and better feed efficiency. Several public research projects are in progress, with more products expected to be notified soon.

2.4. The Regulatory Status of GEd Produce in China

Chinese researchers have a strong track record of research on GEd crops, as evidenced by the fact that China holds more patents on plants than any other country [16], but to date, no GEd crop products have been commercialized [17]. The safety evaluation of GEd crops has not been subject to the same regulations as GMOs since January 2022, when the Guidelines for Safety Evaluation of Agricultural Gene Editing Plants (Trial Edition) was issued by the Ministry of Agriculture and Rural Affairs (MARA, formerly MOA). However, in the four following regulatory steps, which include crop variety registration, seed production evaluation, seed business evaluation and processing evaluation, GEd crops are still regulated as GMOs.

2.4.1. Current GEd Regulatory Status in China

There are five steps for a GEd product to proceed from the laboratory to the market. These are: safety evaluation, crop variety registration, seed production evaluation, seed business evaluation, and processing evaluation. Except for the first step, Chinese oversight of GEd plants is based mainly on the *Regulations on the Safety Management of Agricultural Genetically Modified Organisms* (referred to as the '*Regulations*') issued by the State Council in 2001. Article 3 states 'that an agricultural GMO denoted in these Regulations relates to animals, plants, micro-organisms, and their products in which genetic engineering technology was used to change the genome composition for agricultural production or agricultural product processing'. This means that all crops and their products obtained through GEd technology are classified as agricultural GMOs, and are included in the safety management of agricultural GMOs according to law.

There are three specific administrative measures and related guidelines that support the *Regulations*. In 2002, the then Ministry of Agriculture (MOA) issued the *Administrative Measures for the Safety Evaluation of Agricultural Genetically Modified Organisms*, which have since been revised three times, in 2004, 2016, and 2017. The *Administrative Measures for the Import Safety of Agricultural Genetically Modified Organisms* have been revised, twice in 2004 and 2017. For the *Administrative Measures for the labelling ('marks') of Agricultural Genetically Modified Organisms*, a revision was made in 2004. These revised versions made more specific provisions on the safety evaluation, import safety approval, and identity management system for the safe management of agricultural GMOs. In January 2022, MARA formulated and published new *Guidelines for Safety Evaluation of Agricultural Gene Editing Plants* (Trial Edition). It further standardized the safety evaluation management of agricultural GEd plants. This is a milestone for R&D using NBTs and industrial promotion in China. As a result, the current path-to-commercial growth of crop plants described in Figure 7 is expected to pave the way for more crops with advanced traits to be commercialized.

2.4.2. China's Government Organizations Responsible for GEd Monitoring

The main supervising authorities of GEd/GM procedures in China are MARA, together with its subordinate agricultural departments, which are specifically responsible for the supervision and management of the safety of agricultural GEd products and GMOs. MARA is also responsible for reviewing and issuing Safety Certificates, Variety Certificates, Seed Production Permits, Seed Business Permits, and Processing Permits for agricultural GMOs. In addition, other departments including the Science and Technology Departments, Development and Reform Commission, Ministry of Commerce, General Administration of Customs, and Market Supervision Departments of China, are also responsible for the R&D investment, market access, domestic circulation, import and export, and labelling management. For other organisms (such as trees), authorities such as Ministry of Forestry have the power of oversight for relevant GM products in their areas.

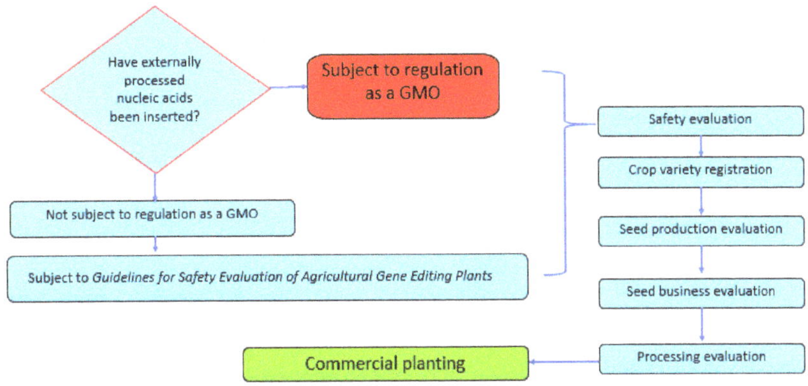

Figure 7. The path to commercialization for GM and GEd crop plants in China.

Although the Ministry of Ecology and Environment (MEE) appears not to be concerned with the domestic safety regulation of GEd products, it is responsible for China's national responses to the secretariat of the Convention on Biodiversity Diversity (CBD), and is active in environmental risk assessment and management in China.

2.4.3. Progress toward Deregulation of GEd Products in China

In China, there have been rapid developments in GEd technology applied to crops, especially based on the CRISPR/Cas9 system [12,18]. There is currently an ongoing discussion on the regulation of GEd produce, which has focused mainly on the supervision of their safe use to support the current increasing trend of applications for GEd technologies for crop improvement [19]. Based on the rapid developments of GEd technologies, there is an urgent need to improve public trust and establish a science-based regulatory system for GEd R&D and products. The current desire by the Chinese scientific community is for regulatory bodies to implement a consultation system that will assess GEd products on a case-by-case basis, especially for products of SDN-1 and SDN-2 technology, bringing them closer in line with international standards [17].

2.5. The Regulatory Landscape for GEd in India

In 1989, well before there was global consensus on regulating biotechnology through the *Cartagena Protocol*, India enacted its domestic biotechnology regulatory framework through the '*Rules for the Manufacture, Use, Import, Export, and Storage of Hazardous micro-organisms/Genetically engineered organisms or cells*' (*Rules* 1989) under the *Environment (Protection) Act of* 1986. The *Rules* regulate activities involving genetic engineering from the bench to market through six competent agencies at the *process*, *product*, and *processed* '*product thereof*' levels. '*Product thereof*' includes any item that either contains GMOs or are derived from a GMO, but may not contain GMO in the final product (e.g., oil derived from a GMO is known as 'product thereof'). These agencies are either regulatory or advisory and operate at different levels, starting from institutions to the central government agencies. Research activities including the import, export, and transfer of materials and contained experiments are monitored by *Institutional Biosafety Committees* (IBSC) at the institutional level and regulated by the *Review Committee on Genetic Manipulations* (RCGM) at the federal level, and operate from the Department of Biotechnology (DBT). In contrast, the *Genetic Engineering Appraisal Committee* (GEAC), the apex regulatory body, operates from the Ministry of Environment, Forest and Climate Change (MoEF&CC), and approves confined field trials, the environmental release of GM products, and the large-scale commercial import and export of GM materials.

With the introduction of various forms of GEd, in which processes such as SDN-1 and SDN-2 involve deletion or minor edits in the genome, similar to products that have mutations identical to those found in nature, regulators have debated whether these changes should be regulated, and if so, how. In India, there has been a clear picture where *Rule* 3 of the *Rules* 1989 defined genetic engineering more broadly to cover the deletion or insertion of a nucleotide or a DNA sequence coding for a gene or regulatory element, which may result in the development of a trait that may or may not exist in nature. As a result, emerging technologies of GEd were essentially regulated under *Rules* 1989 at the process and product levels.

However, following inter-ministerial and stakeholder consultations, and considering the available options, government actions were consolidated under *Rules* 1989, which led to a more sensible regulatory approach to GEd technologies. Several guidelines and office memoranda were then issued under the *Rules* 1989, which required Indian regulators to adopt a process-weighted regulatory mechanism where regulatory decentralization and deregulation were introduced, based on the process of modification and the product, and depending on the presence of exogenous DNA. As a result, laboratory experiments involving SDN-1 modifications now require no approval, except that the IBSC needs to be informed. In contrast, SDN-2 and SDN-3 experiments require the approval of the IBSC and RCGM, respectively.

Recently, the MoEF and CC have deregulated all SDN-1 and SDN-2 forms of GEd products if they are free from exogenous DNA. However, experimental plant work on GEd is still regulated by the IBSC and must be reported to the RCGM, and be undertaken under containment conditions until the plant is free from exogenous DNA. To obtain the deregulated status, an applicant needs to submit information to the IBSC for appraisal and inform the RCGM. Significantly, all SDN-1 and SDN-2 plants are exempt from further biosafety trials under field conditions as well as from food and feed safety assessments, and require no approval from GEAC for commercialization. In contrast, editing technologies with a footprint of exogenous DNA (i.e., SDN-3 is still regulated at the process and product levels as for 'traditional' GMOs).

To facilitate assessment of the type of SDN modification in a GEd product, the DBT has guidelines for the type of required Risk Assessment and Risk Management. The assessment is conducted by the IBSC and RCGM based on information submitted on the plant biology, programmable nuclease, template nucleotide sequence(s), molecular characterization, editing and selection methods, possible off-target mutations, and stability of edits over generations. Although the guidelines give a much clearer roadmap involving experimentation with GEd plants, they do not specify the number of base changes allowed for SDN-2 GEd, that is, editing having detectable edit footprints and the source of the template DNA. Furthermore, in parallel with the deregulation of SDN-2 and its definitions in the guidelines, altered expression profiles due to defined footprint edits are considered as an allelic form comparable to those available in a primary/secondary gene pool. It is possible that the latter has made biosafety assessments quite questionable, especially when the source of the template DNA crosses species barriers. Thus it has not specified the regulatory status for plants with inserted detectable nucleotides obtained through SDN-2 methods, or imported GEd seeds for experimentation, cultivation, or food/feed use. The regulatory pathway for GEd in India is shown in Figure 8.

Despite some aspects that still need clarity, good progress has been made, especially because India has not yet assessed or provided any commercial approval for GEd plants. However, it is expected that the current roadmap will facilitate R&D being undertaken on GEd in India (Table 3) and those planning to use the technology to improve crop plants.

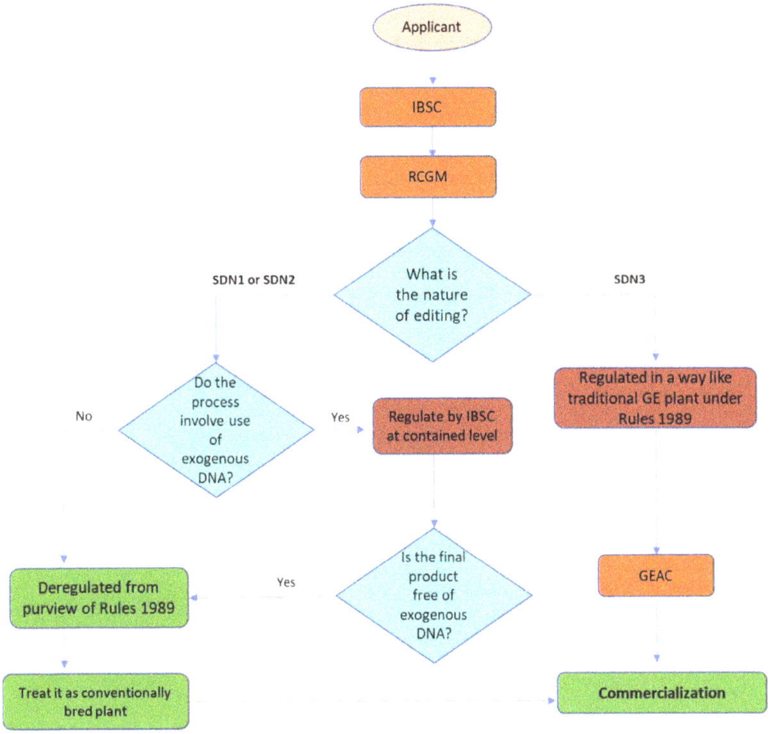

Figure 8. The regulatory pathway for gene-edited plants in India.

Table 3. Some ongoing R&D on GEd at Indian institutions.

Institutes	Crop	Trait
National Institute of Plant Genome Research, New Delhi	Indian Mustard Rice Chickpea Rice/Maize-	Glufosinate alkaloid reduction to tolerant level Disease resistance and herbicide tolerance Seed size and quality Improvement of root architecture and stress/nutrient response/abiotic stress tolerance
Bose Institute	Tomato	Adjusting complex traits
Junagadh Agricultural University, Gujarat	Groundnut	High oleic acid and low linoleic acid
Indian Agricultural Research Institute, New Delhi	Rice	Yield, nitrogen use efficiency, water use efficiency, abiotic and biotic stress tolerance
International Center for Genetic Engineering and Biotechnology, New Delhi	Rice	Low phytate; nutrient use efficiency; herbicide tolerance
National Research Center on Plant Biotechnology	Indian Mustard	Seed meal quality
Tamil Nadu Agricultural University, Coimbatore	Rice	Disease resistance and nutritional quality
Institute of Life Sciences (ILS), Bhubaneswar	Bhimkol (*Musa balbisiana*)	Seedless
National Agri-Food Biotechnology, Mohali	Banana	Increase levels of beta carotene

2.6. The Regulatory Status of GEd Produce in Pakistan

Pakistan has a population of over 220 million. With one of the highest population growth rates, the country will require technologies such as GEd to enhance the yield, quality, and nutritional value of food and fiber crops, which are the mainstay of its economy. To date, different institutions in Pakistan are working in collaboration with researchers in other countries to improve important crops such as wheat, rice, cotton, potato, oilseed brassicas, soybean, and tomato using various GEd tools (Table 4).

Table 4. The current R&D on the GEd of crops in Pakistan.

Institute	Crop	Targeted Trait/s
National Institute for Biotechnology and Genetic Engineering (NIBGE)	Wheat	Yield improvement, disease resistance, nutritional enhancement
	Potato	Disease resistance and quality improvement
	Cotton	Quality improvement, biotic and abiotic stress tolerance
	Rice	Yield improvement, disease resistance, and herbicide tolerance
	Brassica	Edible oil quality improvement
Centre for Excellence in Molecular Biology (CEMB)	Tomato	Virus resistance
	Potato	Reduction in cold-induced sweetening, scab and blight resistance
	Cotton	Induction of male sterility and virus resistance
	Corn	Herbicide tolerance
Forman Christian College University (FCCU)	Wheat	Quality improvement
	Cotton	Heat/drought resistance
National Institute of Genomics and Advanced Biotechnology (NIGAB)	Potato	Reduction in potato browning
	Wheat	Yield improvement and root growth improvement
	Tomato	Enhancing shelf life
National Center for Genome Editing (NCGE), University of Agriculture	Wheat	Yield improvement and quality improvement
	Brassica	Edible oil quality improvement
	Cotton	Disease resistance

Being a signatory of the *Convention of Biological Diversity* (CBD) and *Cartagena Protocols*, Pakistan is obliged to address the potential risks or hazards posed by genetically modified organisms (GMOs) before releasing them into the environment. For this purpose, the *Ministry of Environment (now named Ministry of Climate Change)* constituted a *National Biosafety Committee* (NBC) to consider issues regarding GMOs and their products before conducting controlled laboratory research, field studies, and commercial release. To date, gene-edited organisms (if a foreign gene is present) are treated as GMOs in accordance with the guidelines provided in the *Pakistan Biosafety Rules*, 2005 issued by the *Ministry of Environment* under the umbrella of the *Pakistan Environmental Protection Act*, 1997. In contrast, if a foreign gene is absent or segregated out, then the edited plant will not be subject to GMO regulations. A meeting of all stakeholders was organized by the National Institute for Biotechnology and Genetic Engineering (NIBGE) to discuss regulations for the release of GEd products in Pakistan. The consensus was that the GEd policies of Japan and Australia will be adopted. The *Institutional Biosafety Committee* (IBC) considers all GEd cases and submits recommendations on the status of projects or research outcomes as GMO or not to the NBC. This follows very much a policy that SDN-1 and SDN-2 products will not be regulated as GMOs, whereas when a foreign gene is present, the product will be treated as a GMO (SDN-3).

According to the proposed guidelines, three tiers have been established for monitoring and implementation, namely the IBC, the Technical Advisory Committee (TAC), and the NBC. First, the Principal Investigator (PI) and researchers of a GEd project are responsible for the safety, hazards, and risks to themselves and the community. The IBC undertakes the initial risks assessment and later monitors and inspects the processes involved in the project. After evaluating the application and the available resources, keeping in mind any associated risk or qualification of the responsible PI, the IBC will forward the application to the TAC.

The TAC plays a major role in evaluating and reviewing all the applications more technically for licensing. The TAC ensures that the GMOs or any product under consideration go through the necessary risk assessment according to the guidelines provided in the *Pakistan Biosafety Rules*, 2005. Third, the NBC, operating directly under the Ministry of Environment, oversees the overall monitoring, safety, and risk assessment/management of the laboratory, fieldwork, and commercial release of GMOs and their products (Figure 9).

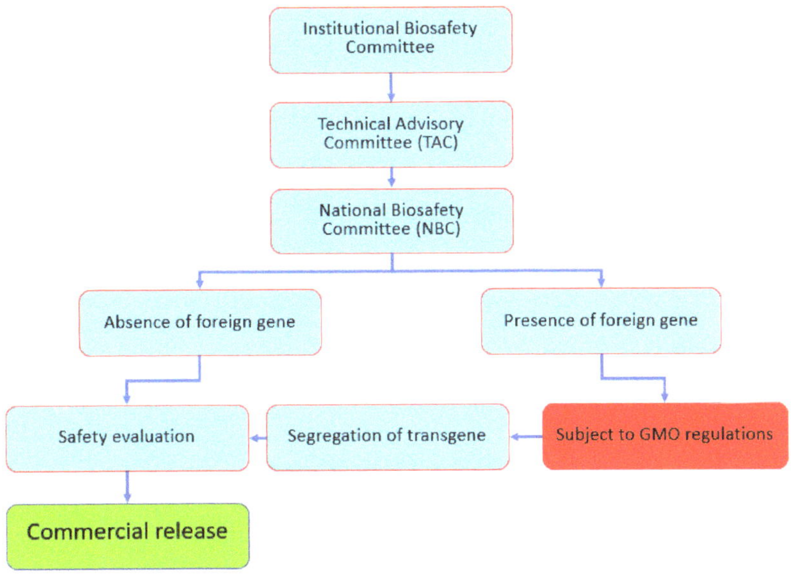

Figure 9. The administrative framework for commercial release of GEd or GM crops in Pakistan.

Because of the flexible regulatory procedures for GEd crops in Pakistan, academic and research institutes have reported the use of the CRISPR-Cas systems in crops such as rice, potato, wheat, and cotton (Table 4). In basmati rice, for example, knockout mutants have been developed to counter bacterial blight disease, caused by *Xanthomonas aryzae pv. oryzae* (Xoo) by disturbing effector binding elements (EBEs) in the promoter region of the OsSWEET14 gene [20]. Similarly, a CRISPR-Cas13-mediated multiplexing approach has been used in potatoes to confer resistance to multiple strains of potato virus Y (PVY) [21]. For wheat, the phytic acid content has been decreased to biofortify the crop by enhancing the accumulation of iron and zinc in mature grains [22]. NIBGE is carrying out broad-spectrum R&D work on major crops such as wheat, rice, soybean, and cotton to improve the yield, quality, and develop resistance to biotic/abiotic stresses through CRISPR-Cas systems in commercially grown cultivars. Despite all the work in progress, as no GEd crop has yet progressed through the system or gone to national trials. However, the pace of ongoing initiatives by the Government shows the seriousness of promoting and accepting GEd crops to combat adverse challenges posed by climate change and reduced arable land.

2.7. The Regulatory Status of GEd Produce in Thailand

At present, Thailand does not have specific laws to regulate GM crops. To undertake such GM studies, researchers need to follow the *Biosafety Guidelines for Work Related to Modern Biotechnology* (2016). The guidelines embrace all work related to gene manipulation employing recombinant DNA (rDNA) technology for all purposes including the development of transgenic plants, animals, and micro-organisms, production of vaccines, commercial and industrial manufacturing of rDNA-derived products, and the release of transgenic materials and products into the environment.

The importation of GM crops and their products is evaluated under the *Plant Quarantine Act* (1964) (amended in 1994): in 1994, the *Department of Agriculture, Ministry of Agriculture and Cooperatives*, made a "Ministerial Declaration" prohibiting the import and transit of all transgenic plants, unless permission is granted by the Director-General of the *Department of Agriculture*, and only for experimental purposes. For foods derived from GM crops, the Ministerial Order (2002) under the *Food Act* 1965 (Amendment 2002) requires the labelling of GM products.

Additional guidelines were developed later including the Biosafety Guidelines for Contained Used of GM Micro-organisms at Pilot and Industrial Scales, Biosafety Guidelines for Plants Carrying Stacked Genes and Their Derivatives and Guideline on Plant Biosafety in Research Greenhouses, while the precautionary principle was adopted for biosafety actions (https://www.tei.or.th/file/library/2020-Progress-Biodiversity-eng_30.pdf; accessed 15 July 2022).

New regulations are being considered at present. These are:

1. The *Draft Genetically Modified Foods Regulation*, which includes the regulation lists of those GM plants that have passed food safety assessment by the risk evaluation agencies designated by the Thai Food and Drug Administration. Food products consisting of or produced from GM food (GMF) ingredients are prohibited from production/processing, retail or importation into Thailand unless they have undergone food safety evaluation. New GMFs must be assessed by designated risk evaluation agencies.
2. The *Draft Biodiversity Law*, which includes regulating living modified organisms ("LMOs") that are alive and capable of transferring genetic material to subsequent generations. Under this draft, LMOs are strictly prohibited from being released into the environment unless they are on the Release List, and such a release is handled according to the requirements stipulated in the draft and its secondary legislation. To be announced on the Release List, the applicant must submit an application to the Competent Authority, together with a biosafety risk assessment report (https://www.lexology.com/library/detail.aspx?g=767d7504-faae-42d1-b0e1-a4b7854b296c; accessed 15 July 2022).
3. The *Draft Biosafety Assessment Guidelines for Genome Editing Technology*, which includes three types of GEd technology, and details of the minimum requirements for a food and feed safety assessment (https://www.isaaa.org/webinars/2021/gethai/ppt/Policy%20Considerations%20of%20Genome%20Editing%20in%20Thailand%20-%20Chalinee%20Kongsawat.pdf). The draft *Biodiversity Law*, which covers research, field trials, and the commercialization of GM plants, animals, and micro-organisms, will be sent to the Cabinet for approval by the end of 2022 (https://apps.fas.usda.gov/newgainapi/api/Report/DownloadReportByFileName?fileName=Agricultural%20Biotechnology%20Annual_Bangkok_Thailand_10-20-2020).

2.7.1. Government Organizations Responsible for GM/GEd Regulations in Thailand

There are four main government agencies involved in the regulation of research and products of agricultural biotechnology. These are:

Department of Agriculture (DOA), Ministry of Agriculture, and Cooperatives (MOAC), responsible for regulating imported GM and GEd seed for planting, conducting GM/GEd research and development, and risk assessment

The National Center for Genetic Engineering and Biotechnology (BIOTEC), Ministry of Higher Education, Science, Research and Innovation (MHESI), is responsible for developing technical guidelines and providing technical advice under the Biosafety Program

The Ministry of Natural Resources and Environment (MONRE) is responsible for drafting the National Biosafety Law (Biodiversity Law) and is the national focal point for the Convention on Biological Diversity (CBD) and Cartagena Protocol on Biosafety (CPB) and The Thai Food and Drug Administration (TFDA), Ministry of Public Health

(MOPH) is responsible for regulating and monitoring the use of GM/GEd food including labelling and regulating the imports of GM/GEd-contained food products.

2.7.2. Examples of GM/Ed Crop Products Being Commercialized in Thailand

Although no GM or GEd crops have been approved for field trials or commercial cultivation in Thailand, the importation of GM crops does take place, and is limited to corn, soybean, and cotton for feed and industrial use.

GEd technology in Thailand is defined as a type of genetic engineering in which a nucleotide sequence is inserted, deleted, or replaced in the genome of a living organism using engineered nucleases, thus, it is regulated in the same manner as GM crops. GM/GEd products must undergo risk assessment before approval for commercial release. However, for SDN-1, there are only three components that need to be considered: a comparison of the nucleotide sequence difference between a GEd product and its unmodified counterpart, the product specification, and an off-target analysis. According to the *Draft Biodiversity Law*, to take GM/GEd crops to the market, they must be approved after risk assessment by the relevant authority and then included on the Release List. According to the draft GEd regulations, GEd products need to be assessed on a case-by-case basis. Minimum assessment requirements are imposed on SDN-1 products whereas SDN-3 products are assessed rigorously.

Since the *Draft Biodiversity Law* is still under consideration by the Cabinet, and the secondary laws are under development, the picture is not clear enough to provide a certain path-to-market for GEd crops in Thailand.

2.8. The Regulatory Status of GEd Produce in the Philippines

The Philippines was the first ASEAN country (1990) to initiate a biotech regulatory system for GMOs. It recognized the value of modern biotechnology and the need to develop policies that could evolve with technological developments. To develop policies for the expected NBT products (including GEd), imported or from local research, in January 2018, the Department of Agriculture—Biotech Program Office (DA-BPO) initiated a Study Group to review the status of eight different NBTs (SDNs, ODMs, Cisgenesis, Intragenesis, RNA-dependent DNA methylation (RdDM), grafting with GM material, reverse breeding, agro-infiltration and synthetic genomics) to provide science-based policy recommendations on how such products should be treated. The existing GMO regulations (Joint Department Circular no. 1, series of 2021 or JDC1, s2021) were the defining departure point. The Study Group findings and consultation meetings (February–September 2018) generated "A Review of the New Plant Breeding Techniques (NBTs) from the Viewpoint of Regulation", which was submitted to the DA-BPO (in November 2018) and to the National Committee on Biosafety of the Philippines (NCBP) in May 2019. The NCBP then commissioned an Ad Hoc Technical Working Group (TWG) (from June–December 2019) to resolve the regulation of such products and this resulted in a policy on NBTs, NCBP Resolution No. 1 series of 2020, also known as "*The Regulation of Plant and Plant Products Derived from the Use of Plant Breeding Innovations (PBIs) Or New Plant Breeding Techniques (NBTs)*", issued in April 2021.

The resolution synonymized NBTs with plant breeding innovations (PBIs). In summary, products of NBTs can be (a) GMO, if, as defined under Executive Order (EO) 514 (2006), they contain a novel combination of genetic materials obtained by modern biotechnology. "Novel combination" was defined by the Ad Hoc TWG as a "resultant genetic combination/change in a living organism that is not possible to obtain through conventional breeding", or (b) non-GMOs or conventional products, if they do not contain a novel combination of genetic material. Only GMOs will be regulated under JDC No. 1, whereas their non-GM counterparts are not regulated under this JDC, but are still subject to regulations normally applied to conventional plant products. To facilitate understanding of the techniques covered by the policy, a decision tree was created (Figure 10). In the tree, it is clear that GEd can lead to the formation of a non-GMO (PBI Case 1: SDN1, SDN2) or a GMO (PBI Case 2: SDN3), depending on the use of a repair template and the nature of the insert.

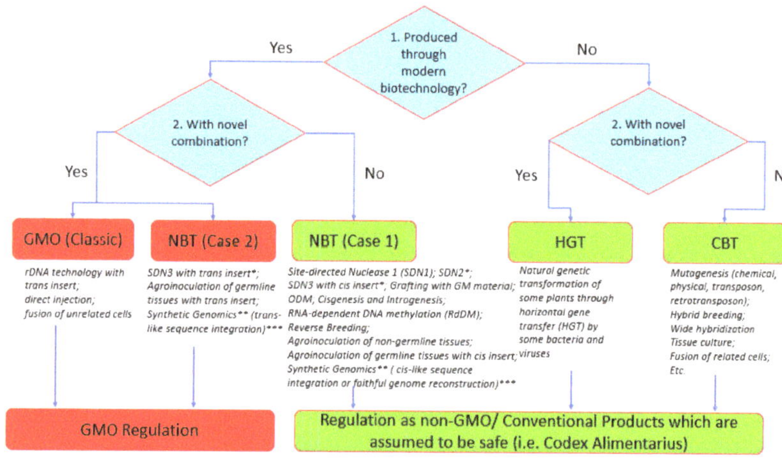

Figure 10. Decision tree for NBT products from the Ad Hoc Technical Working Group of the National Committee on Biosafety of Philippines (NCBP). * includes insertion using the CRISPR-CAS with Prime Editing; ** not to be confused with Synthetic Biology, which specializes on sequences/genetic elements (e.g., unnatural base pairs) in the genome that are not found in nature (beyond novel combination); *** relates to a largely synthetically assembled genome).

The decision tree followed a "20-bp rule"—this means that if an inserted sequence is 20-bp or more of a foreign DNA sequence, it is regarded as a GMO. However, an insertion of 19 bp or less of a foreign sequence is not regarded as a novel combination of genetic materials, hence the product is not subject to GM regulation. Additionally, GEd techniques that involve a repair template that inserts a 19-bp sequence and below will be considered as SDN-2, while those with an insertion of 20-bp or more will be considered as SDN-3. The SDN-2 edits will automatically be exempted from GMO regulation since the short insertion is not regarded as a novel combination of genetic material. In addition, other products of NBTs/PBIs that possess remnant border sequences from a transformation vector will be assessed based on the length of such sequences. The 20-bp rule described here was motivated by the report of the Joint Research Center of the European Commission (2011), and by the 20-bp guide sequence (crRNA) requirement in CRISPR-Cas9 systems that allow *Streptococcus pyogenes* Cas9 in nature to recognize and digest invading viral DNA.

After the issuance of the NCBP resolution, the Department of Agriculture commissioned a Technical Advisory Group on Modern Biotechnology and related innovations for agriculture and fisheries (April, 2021). This resulted in a draft memorandum entitled "Rules and Procedure to Evaluate and Determine When Products of Plant Breeding Innovations (PBIs) or New Plant Breeding Techniques (NBTs) Are Covered under the DOST-DA-DENR-DOH-DILG Joint Department Circular No. 1, series of 2016 (JDC1) based on the NCBP Resolution no. 1, series of 2021", which was finally published as Memorandum Circular No. 8 (MC8) in May 2022, ending a 4-year journey since the creation of the Study Group under DA-BPO.

As per MC8, there will be a Technical Consultation for Evaluation and Determination (TCED), which is a technical evaluation of a PBI product to determine whether or not the final product is or is not a GMO. The process will be carried out by members of the Bureau of Plant Industry (BPI) Biotechnology Core Team (BCT-PBI) and assigned external expert(s). When a product has been officially determined to be a non-GMO, the team will issue a "*Certificate of Non-Coverage from the JDC1,* s2021" to the product developer. The certificate essentially recognizes the PBI product as being a conventional product, indicating that there is no need to obtain a Commercial Propagation Permit under JDC1, s2021 to register it as a variety or to be commercialized. A locally developed GEd crop, which is expected imminently to undergo TCED evaluation is bacterial leaf blight-resistant rice being co-developed by the International Rice Research Institute (IRRI) and the Philippine Rice Research Institute (PhilRice).

2.9. The Regulatory Status of GEd Crops and Produce in Malaysia

The *Malaysian Biosafety Act* 2007 and the *Biosafety (Approval and Notifications) Regulations* 2010 served as the national guidelines for regulating LMOs in Malaysia. The former fulfils Malaysia's obligation as a party to the *Cartagena Protocol* on Biosafety. Before the Act was tabled and the regulations came into force, the lead government agency, the *National Resources and Environmental Ministry* (NRE), consulted various stakeholders including the Malaysian Biotechnology Corporation, the Malaysian Biotechnology Information Center, the Malaysian Manufacturers Association, Consumer's Association, and other Non-Governmental Organizations (NGOs).

The *Biosafety Act 2007* describes the framework of the LMO regulation system; its overall objective is to protect humans, plants, animal health, and the environment by regulating the release, importation, exportation, and contained use of LMOs. Among others, the Act prohibits LMO activities to be conducted unless proposers have received prior approval or provided notification and the possible risks have been assessed. On the other hand, the Biosafety (Approval and Notification) Regulations 2010 contain additional information on the Act including how an application for approval for release or import activity must be made and matters relating to the certification of approval.

In Malaysia, a LMO is defined as a living organism that possesses a novel combination of genetic material through the use of modern biotechnology (defined as the application of in vitro nucleic acid techniques including recombinant DNA or fusion of cells beyond a taxonomic family) excluding traditional breeding and selection techniques. Currently, only a few activities are exempt such as pharmaceutical products of LMOs with relevant international treaties, organizations, or under written laws such as genetically engineered vaccines. However, the products cannot be released intentionally into the environment. Field trials involving LMOs growing in an open environment are considered as a 'release activity', and approval must be obtained before the trials start.

If the product of an LMO is to be commercialized, it needs to be clearly identified and labelled. In 2010, the *Food (Amendment) Regulations* 2010 was passed, requiring all GM food and ingredients to be labelled before release into the market. However, a USDA Foreign Agricultural Service Report in 2018 noted that while Malaysia does not have any large-scale domestic production of GM crops, the country has expanded the number of GM products for commercial use, with more than 30 products being approved for import and market release (https://bch.cbd.int/en/countries/MY; Biosafety Clearing House, Malaysia, Conventional on Biological Diversity; accessed 30 July 2022). In addition, although mandatory labelling guidelines have been established, they have yet to be enforced (https://www.biosafety.gov.my/wp-content/uploads/2021/08/Garis-Panduan-Pelabelan-Makanan-dan-Bahan-Makanan-Melalui-Bioteknologi-Moden.pdf). According to the report, GEd plant products represent a very small fraction (if any) of the imported biotech produce and there is no commercialized production of GEd crops to date. Past projects include a confined field trial of GEd-papaya with delayed ripening. Interestingly, Malaysia is a significant importer of GM products for livestock feed such as soybean, demonstrating that the agricultural community supports the expanded use of GM products. A United Nation's survey in Malaysia indicated that while industry and academia had a reasonable knowledge of the biosafety laws and regulatory framework, domestic consumer awareness was low.

Recognizing the increasing importance of GEd products, the National Biosafety Board (NBB) has been actively engaging stakeholders to improve current regulations and to encourage research on GEd crops and produce in Malaysia. In reviewing the current regulations, the NBB has divided the regulatory scope into: from laboratory (notification) to market (approval) as well as into the three types of modifications, namely, SDN-1, SDN-2, and SDN-3. The division of regulations for each category is provided in Table 5.

Table 5. The regulatory scope of the Biosafety Act in Malaysia for LMOs.

Notification—Part IV of Act	Approval—Part III of Act	
	Developing LMOs—From Bench to Market	
R&D	**R&D**	**Commercialization**
Contained use of LMOs Import for contained use Export of LMOs	Field trials	Direct introduction to the environment. Commercialized planting. Placing in the market.
	Direct Commercial Use—No R&D	
Export of LMOs Contained use for industrial production	Importation of LMOs/products for placing in the market or release	

Islamic Perspective for GEd Regulation of GE Produce in Malaysia

Malaysia has a diverse population: several groups have advocated the use of GEd technologies from an Islamic Fiqhi perspective. They argue that the country should have a comprehensive Islamic framework since Islam is the official religion and there is a Muslim majority in Malaysia. While this perspective focuses mainly on GEd for medical treatment, it would not be surprising if religious approval or certification is sought by consumers before consuming GEd products, for example, whether the crops are halal. Although this has not been discussed by the NBB, as GEd products become more established in Malaysia, this aspect may become more important as it reflects Malaysia's culture and religion.

2.10. The Regulatory Status of GEd Produce in Indonesia

Indonesia, through the *National Biosafety Committee for Genetically Modified Organisms*, recognizes GEd as an important breakthrough that will play a significant role in its economy. As a megadiverse country with abundant bioresources, it is expected that the agriculture sectors (farming, forestry and fisheries) will benefit from the application of GEd technology as leading economic drivers, employing 29% of the total workforce [23].

In Indonesia, the earliest published reports on R&D on GEd in agriculture were in 2020 on the rice GA20ox-2 gene (cvs Kitaake and Mentong) encoding a gibberellin biosynthetic enzyme [24–26] and the OsCKX2 gene in the rice cv Mentik Wangi [27]. Several research institutes and universities are working on the GEd of different crops. For example, to improve the productivity of rice and soybean and cassava storage tolerance at the Research Center for Biotechnology-Indonesian Institute of Sciences, and improve the resistance to bacterial leaf blight in rice, resistance to geminivirus in chili pepper, increasing artemisinin content in artemisia, and resistance to citrus vein phloem degeneration (CVPD) in oranges at the Indonesian Center for Agricultural Biotechnology and Genetic Resources Research and Development (ICABIOGRAD) of the Ministry of Agriculture, and orchid flowering quality and disease resistance at the Universitas Gadjah Mada.

The *Indonesian Institute of Sciences* (LIPI) and ICABIOGRAD have been integrated into the newly established *Indonesian National Agency for Research and Innovation* (BRIN). The funding for GE R&D is obtained mainly from the government of Indonesia: some activities are supported by Japanese agencies (JICA/JST and JSPS). In Indonesian GEd crops, the produce and research have not been specifically regulated. The Ministry of Agriculture and the then LIPI in collaboration with the *Biosafety Commission for Genetically*

Modified Product (KKH-PRG) organized focused group discussions (FGD) attended by regulators and scientists from Ministries, Government agencies, and universities to try to reach a consensus on Indonesia's stance on GEd issues and regulations. The first FGD, conducted by ICABIOGRAD in February 2019, discussed the progress of GEd research in Indonesia and worldwide, and the state of regulation in Indonesia. The FGD noted that GEd was being undertaken at ICABIOGRAD, the Research Center for Biotechnology, LIPI, the Indonesian Research Institute for Biotechnology and Bioindustry (PPBI), and the Universitas Gadjah Mada (UGM) on diverse commodities (rice, orange, palm oil, artemisia, and orchids), and for different purposes such as increasing productivity and stress tolerance. The FGD acknowledged that, globally, GEd products were reaching the stage of commercial release. The FGD also acknowledged that GEd would play an increasingly important role, and because no regulations existed, there was the need to develop appropriate regulations to govern the conduct of GEd research, the process for the release of GEd products, and to resolve any potential confusion on the definitions of the GM and GEd processes and products.

The second FGD on GEd (20 February 2019), organized by the Directorate for Conservation of Biodiversity (Ministry of Environment and Forestry) and the *Indonesian Commission on Biosafety of Genetic Engineering Product* concluded that GEd products differed from GM products, and that there needed to be a clear definition of GEd to differentiate it from GM products. The FGD noted that since GM may be involved in the early development of some GEd products, some mechanisms of GM product regulation were in place, and could be applied for early assessment of the GEd products to determine whether the final products fall under GM product regulations.

The definition of GM/genetic engineering in Indonesia is stated in Government Regulation No. 21/2005 on the Biosafety of Genetically Modified Products as "any living organism, part of it and/or its processed products with new genetic make-ups resulting from the application of modern biotechnology". The definitions of LMOs and "modern biotechnology" were in line with the definitions under the *Cartagena Protocol* in which LMOs mean any living organism that possesses a novel combination of genetic material obtained through the use of modern biotechnology, and which involves the application of (a) in vitro nucleic acid techniques including recombinant deoxyribonucleic acid (DNA) and direct injection of nucleic acid into cells or organelles, or (b) the fusion of cells beyond their taxonomic family, which overcome natural physiological reproductive or recombination barriers, and that are not techniques used in traditional breeding and selection. In Indonesia, the decisions on GM products fall under the responsibility of the KKH-PRG, assisted by the *Technical Team for Biosafety for Genetically Modified Products* (TTKH-PRG). The TTKH-PRG consisted of TTKH-PRG for food (coordinated by the *Agency for Drug and Food Controls* (BPOM), TTKH-PRG for feed (coordinated by the Ministry of Agriculture), and TTKH-PRG for environmental safety (coordinated by the Ministry of Environment and Forestry). The flow of assessment is shown in Figure 11.

The definition of GMOs/LMOs was discussed at the third FGD (at the Indonesian Institute of Sciences, LIPI) in January 2020. This FGD, attended by members of the Indonesian Biosafety Committee, the Technical Team for Biosafety, the heads of biotechnology-based associations such as the Indonesia Biotechnology Consortium, Indonesia Biotechnology Society, and Indonesian Association for Agriculture Biotechnology, the Deputy for Life Sciences of LIPI, the Deputy under the Indonesian Agency for Drug and Food Control, the Director of the Research Center for Biotechnology-LIPI, the Head of BBBiogen-MoA, the Deputy Assistant under the Coordinating Ministry of Economy, and scientists from universities and research institutions, discussed decisions made at the previous FGD and shared experiences of regulators and scientists from three other countries. The FGD also discussed the need to reassess the definition of GMOs in Government Regulation No. 21/2005 to accommodate GEd products, which although developed mostly using modern biotechnology, could result in products that can be categorized as non-GMO. In other words, the FGD acknowledged that the process of producing a product via modern biotechnology could result in a non-GM product, depending on its genetic make-up (Table 5). The modern biotechnology products categorized as SDN-1 were determined to be non-GM because of their similarity with commonly occurring natural mutations. The policy on SDN-2 and SDN-3 products is still under discussion. However, the view was expressed that products of SDN-2 and SDN-3 could also be considered as non-GM when they fall into particular categories. For example, SDN-2 could be considered as non-GM as long as the template is from the same gene pool and no novel protein was produced. SDN-3, which is a knock-in mutation, could be considered as cisgenic, in which case the donor gene or DNA fragments come from the same gene-pool (Table 5). The decision on whether cisgenics are considered as non-transgenic, thus non-GM, are still under discussion.

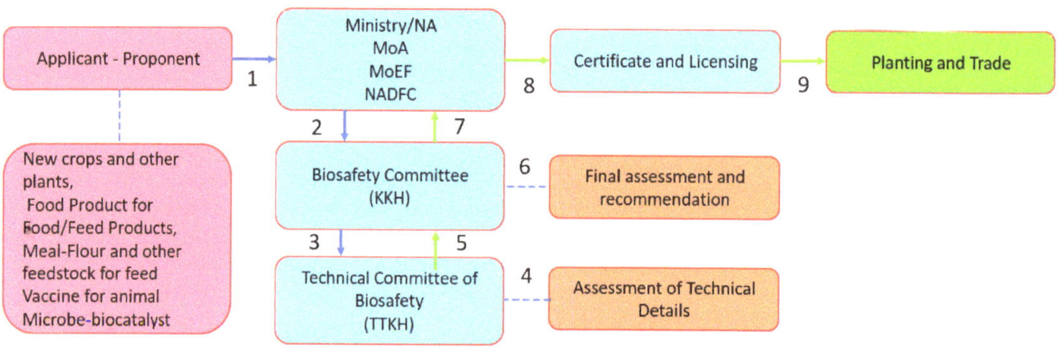

Figure 11. The procedures of the Biosafety Certification of new agricultural products in Indonesia. (The process follows the numbers as indicated, starting at the Ministerial level, proposal reviews by the Biosafety Committee and its Technical Committee, which assesses technical details, make a final assessment and recommendation for approval at the Ministry level, for issuing a certificate for licensing for planting or trade).

The FGD recognized the importance of GEd products to the Indonesian economy and recommended that a harmonization of the regulations/technical guidelines between GM and GEd products was needed, even before regulations on GEd were established, so that GEd products can be filed for release using the existing guidelines. This would require stating that the product is GEd and a non-GM, backed with scientific evidence. The FGD also recommended that a task force for the assessment, consisting of TTKH-PRG members and renowned scientists in the field, should be appointed to provide scientific recommendations to KKH-PRG. The scheme for handling the release of GEd products in Indonesia, also discussed during the third FGD, followed the available scheme for GM products, which requires the proponent to file a request for product assessment to the Agency for Drug and Food Control for food, to the Ministry of Agriculture for feed, and/or to the Ministry of Environment and Forestry for environmental safety. The responsible Ministry or Agency will request the KKH-PRG assess the status of the products for their GM status or if applicable, assess provided scientific data to support the non-GM status of the product. A recommendation would then be issued for the product to be released as for non-GM products. Since the third FGD, KKH-PRG has received one application for assessment of a GEd product, based on the SDN-1 approach. However, no decisions on GEd products have been made thus far. A flow diagram for the assessment process is provided in Figure 12.

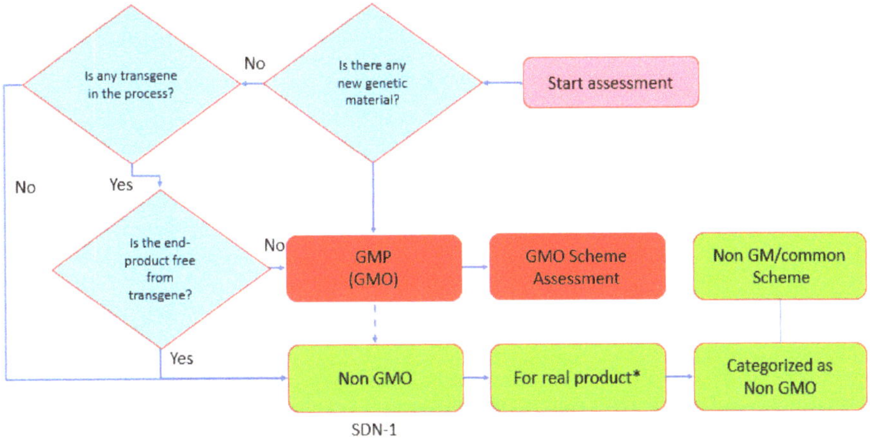

Figure 12. The mechanism of assessment of GEd products in Indonesia (* note for hypothetical products, there needs to be supporting data of molecular analysis and phenotype).

2.11. The Regulatory Status of GEd Produce in Taiwan

In Taiwan, the regulatory policy on GE products is being reviewed by an *ad hoc* expert committee convened by the Taiwan Food and Drug Administration. Currently, mandatory pre-consultation and notification is required for GE foods and basic/safety information (i.e., evidence of no foreign DNA, or adverse effects) and reference materials need to be submitted for review, based on the draft policy disclosed in 2021. The cultivation of GEd products in Taiwan will be controlled under the *Plant Variety and Plant Seed Act* administered by the Council of Agriculture. The definition of whether GEd products are captured under this regulation is also not clear, and any new policy will need to clarify whether there will be exemptions for any GEd products.

2.12. The Regulatory Status of GEd Produce in South Korea

The South Korean regulatory authorities have taken the view that GEd products are currently captured under the existing Transboundary Movement, etc. of Living Modified Organisms Act (commonly referred to as the *LMO Act*), which governs GM products. The *Act* is currently under review by five regulatory agencies. To exempt any GEd product from regulation, it would be necessary to introduce new amending regulations.

In May 2021, the Ministry of Trade, Industry, and Energy (MOTIE) released a draft revision of the *LMO Act*. The final decision for the exemption of GEd products from the regulations is presented after a pre-review by a Preliminary Review Committee, consisting of government regulatory agencies. Under Article 7–3 of the draft *Preliminary Review of New Living Modified Organism*, a new LMO may be exempt if:

- It has been made without introduction of foreign genes;
- Foreign genes are not retained in the final product; or
- The scientific evidence presented is credible enough to support that the final LMO developed by modern bioengineering technology, other than those set forth in Sub-paragraphs 1 and 2, can also be made by traditional breeding or natural mutation.

However, under the new Article 7–4 (Information collection for safety management of organisms), additional information may be requested for a preliminary review including the detection method and reference material.

The *Preliminary Review Committee* will consist of representatives of seven different Ministries. This new constitution of the committee could make assessments and the review process difficult (the current review system for GM products involves only five agencies).

In addition, the introduction of a new category of 'new genetically modified organisms' appears unhelpful in the growing GEd field. Consultations on the draft are ongoing.

3. The Commercialization of GE Crops in Asia and Australia

The adoption of GM crops, which still hold much promise for addressing production challenges, has been severely hampered by strong opposition from anti-GM groups, leading to a lack of public/consumer acceptance. This has meant that in Asia as well as elsewhere globally, the cultivation and commercialization of GM crops has been confined to a few crop types, mainly cotton, corn, and soybean, and a limited number of countries. In the Asia-Pacific region, there is only meaningful cultivation in China (GM cotton), India (GM cotton), Pakistan (GM cotton), Vietnam (GM corn), the Philippines (GM corn), and Australia (GM cotton and canola). Apart from Australia, these countries import large amounts of grain from GM sources, mainly for feed and oil, and are important global markets for GM crops produced in both North and South America. The global grain trade is a directional flow of GM exports from the Americas to the leading markets of China, Japan, and Korea. However, the markets of some South-East Asian countries have become increasingly significant and are now important grain destinations for GM soybean and corn.

When trading GM commodity crops, regulatory approvals are a necessity and are already in place for currently traded products in the importing countries. Before commercializing a GM commodity crop, technology provider companies analyze the trade patterns and identify key export countries to make sure that the approvals are in place before launching in the country of cultivation. A recent study determined that the costs of discovery, development, and authorization of a new GM trait is in the vicinity of USD 115 million and takes about 16.5 years [1]. This expense, both in terms of time and money, practically rules out the ability for academic institutions and small- to medium-sized companies to commercialize GM commodity crops.

In contrast, crop traits developed using GEd, if not subject to the same regulatory constraints as GM crops, have the potential to offer an alternative solution to the production problems faced by farmers in the Asia-Pacific region. Because of the high regulatory cost to launch GM crops, commercialized events have been mainly limited to broadacre commodity crops where the seed market is big enough to recoup the up-front investments. If the regulatory costs for GEd crops are much lower than for GM crops, investment costs will be lower, making the investment in minor crops such as vegetables a real possibility. This would also increase the opportunity for the commercialization of GEd crops developed by local research institutions in Asia-Pacific countries, and may be brought to market by small- or medium-sized companies for regional production. This is already happening in the examples of the GABA tomatoes from Sanatech Seed and the GEd Madai sea bream from the Regional Fish Company in Japan. There is major R&D on GEd crops in China and India, although research in South East Asia on GE crops is still at an early stage. However, there is recognition of the potential of the technology, with governmental and semi-governmental research organizations having active programs in crops of local importance such as palm oil, driven by the hope that the regulation of GEd crops will not mirror that of GM crops.

In short, how GEd products are regulated in the Asia-Pacific will determine how much of the current and future research pipelines can become a commercial reality. As described above, there is mixed news regarding the regulation of GEd crops in SE Asia and Australasian countries. While uncertainty still remains for many countries, it appears that most governments have recognized that GEd technology in agriculture can play a big part in sustainability and production goals in the face of climate change, and there have been some developments in ASEAN countries (Singapore, Malaysia, Philippines, Indonesia, Myanmar, Vietnam, Laos, Cambodia, Thailand, Brunei) that support the view that GEd products that are equivalent to what can be developed through more conventional means should not be additionally regulated. The final position taken by a large block such as the EU may also influence the final policy decisions in ASEAN countries.

3.1. An Overview of GEd Technology Regulations in Asia and Australasia

Figure 12 provides and overview of the regulatory status of countries in Asia and Australasia. In summary, there is not yet a harmonized or uniform regulatory system for GE products in these countries to support the smooth international trading of agricultural products. The regulatory status of SDN-1 crops appears settled for a few countries, namely, the Philippines, Japan and Australia, which exempt some GEd products, and New Zealand (regulates GEd products as GM) (Figure 13). Some countries such as Singapore and Indonesia indicate that they will exempt some GEd products from regulations, while Thailand may opt for reduced regulatory requirements.

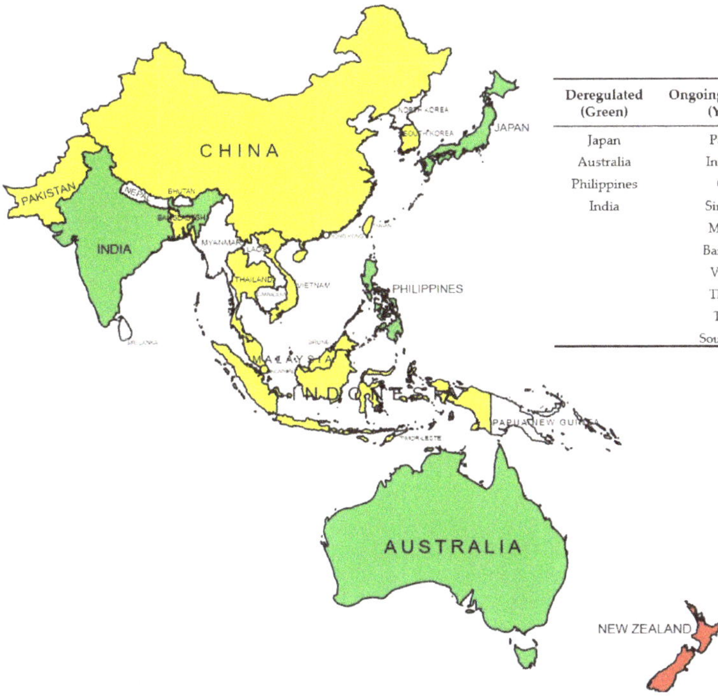

Deregulated (Green)	Ongoing discussions (Yellow)	Regulated (Red)	No regulations (White)
Japan	Pakistan	New Zealand	Myanmar
Australia	Indonesia		Laos
Philippines	China		Cambodia
India	Singapore		Papua New Guinea
	Malaysia		East Timor
	Bangladesh		Sri Lanka
	Vietnam		North Korea
	Thailand		Nepal
	Taiwan		Bhutan
	South Korea		

Figure 13. The regulatory status for GEd crops in countries in the Asia-Pacific region. It is based on the deregulation of SDN-1 crops (green), with some countries also deregulating SDN-1 and SDN-2 products, as described in the text. Countries with ongoing discussions (yellow) and regulated as GMOs (red). Note that regulation of GEd crops in China is under discussion, but does not use SDN terminology: at present GEd is still under GMO product safety management measures, but with less onerous requirements in the pathway to commercial approval.

The good news for advancing GEd technology is that the overall trend is clearly toward either exemption for some types of GEd products or toward fewer regulatory requirements compared to the GM products.

3.2. Lessons on GE Regulations, Trade and Harmonization from Other Parts of the World

For countries in Asia and Australasia keen to apply the GEd technology to improve crops, the goal is not only to provide food security for the nation, but also increases the quality and tonnage of their major export crops. To achieve that, the regulatory climate of GEd crops should be similar to those of the importing countries. Many countries have developed systems that make the current trading of GM crops possible, and as Asian and Australasian countries want to trade in the international market in the future when GEd crops become acceptable commodities, it will be imperative that lessons from successful countries are learnt and decisions on the deregulation of GEd crops are taken with a view to brevity and clarity. Examples from nations around the world are discussed below.

3.2.1. Using Established and Trusted Frameworks

The two major economies in North America, the United States of America (USA) and Canada, have moved swiftly to support the development, deregulation, and commercialization of GEd crops, mainly by employing the framework established successfully and implemented to regulate GM crops. As the largest grower of GM crops [28], the USA was the first country to approve a GEd product for commercial sale. This was Calyxt's GEd soybean with no trans-fats and lower saturated fat produced using TALENs [29]. The enthusiastic uptake of GEd technology in the USA perhaps reflects that even the introduction of GM technology in the 1990s did not trigger the need for new regulations, but instead relied on applying the existing regulatory frameworks to oversee these new crops. The adoption of such a strategy by Asian and Australasian countries could remove barriers and red tape, which has slowed decision-making on deregulation, even of SDN-1 crops, viewed as the least complicated of the outcomes of GEd technology.

In the USA, up to three different agencies can be involved in the regulatory oversight, depending on the final product and how the plant was produced. These are the U.S. Department of Agriculture's '*Animal and Plant Health Inspection Service*' (USDA-APHIS), responsible for protecting agriculture from pest and diseases, the *Environmental Protection Agency* (EPA), which regulates pesticides and therefore regulates biotech crops that have pesticide properties (e.g., insect resistant crops, where they are considered plant incorporated protectants (PIPs)), and the Food and Drug Administration (FDA), which oversees food safety. The FDA consultation process is voluntary, and the result of this consultation is an acknowledgement from the agency that there are "no further questions" concerning human or animal food derived from the GM plant based on the information provided by the applicant.

In 2020, the USDA-APHIS finalized its SECURE (Sustainable, Ecological, Consistent, Uniform, Responsible, Efficient) rule, which exempts certain GEd plants that otherwise could have been developed through conventional breeding. This focuses on regulating the properties of GEd plants, rather than the process used to create them. APHIS states that the exemptions are intended to bring the regulation of potential GEd plants more in line with the guidelines for conventionally bred crops. Therefore, GEd crops that do not contain foreign DNA are not regulated as GMOs if they pose no risk to other plants, and show no food safety attributes different from those of traditionally bred crops. It becomes the responsibility of the developer to assure that products to enter the market are safe for use and consumption (as in the case for conventional crops). The FDA and EPA are yet to announce whether their existing policies and regulations, related to GMOs, would be used to regulate GEd crops and food. Further guidelines for requesting a *Regulatory Status Review* (RSR) can be found on the APHIS-USDA website.

Like the USA, Canada has a well-established product-orientated approach to policy and regulatory oversight and regulates all plants with novel traits (PNTs), regardless of the technology used to create them. Health Canada considers that GE technologies do not present any unique or specifically identifiable food safety concerns compared to other technologies of plant development. Therefore, GEd plant products should be regulated like all other products of plant breeding within the Novel Food Regulations (i.e., by the novel traits and how these traits impact food safety). The health, product, and trait-based strategies used for the deregulation of GEd crops in the USA and Canada, if adopted, will help smaller economies in Asia and Australasia focus their research funds and time on traits important for their economies. GEd traits of such significance are likely to be accepted by consumers, possibly forcing governments to deregulate their use.

3.2.2. Modified and Progressive Assessment and Deregulation

The evidence that positive changes in GEd regulations can be achieved is exemplified by the UK. The 2001/2018 EU Court of Justice directive ruled that a GEd product was to be regulated as a GMO within the EU (which then included the United Kingdom, UK) and would not fall under the mutagenesis exemption of the directive as for older established mutagenesis technologies with a safe history of use. The UK formally left the EU in January 2020, and this gave the UK scope to move from the restrictive GMO EU Directive and set its own regulatory path. Within the UK, England, Scotland, Northern Ireland, and Wales have national laws that control the release of GMOs into the environment. In England, the Department of Environment, Food, and Rural Affairs (Defra) is responsible for the environmental release of GM plants, with all applications submitted to Defra also being passed on to the statutory Advisory Committee on Releases to the Environment (ACRE) appointed under Section 124 of the *UK Environmental Protection Act* 1990 (EPA) to provide advice to the Government regarding the release and marketing of GMOs. The committee works within the legislative framework set out by 'Part VI of the EPA', and in England, the *GMO Deliberate Release Regulations* 2002 *Act*, which together previously implemented EU Directive 2001/18/EC. The principal role of ACRE is to consider each application on a case-by-case basis and evaluate the risks to human health and the environment.

Defra launched a consultation exercise in early 2021, resulting in an announcement by the UK Government on Genetic Technologies (published 29 September 2021) for a two-step reform. The first step removed the regulatory burden for research groups by enabling the field trials of GEd crops (free from transgenes) to go ahead without being subject to existing GMO rules, under a new on-line 'notification' system to Defra. The second step was to bring forward primary legislation to amend the regulatory definitions of a GMO and to exclude organisms that have genetic changes that could have been achieved through traditional breeding or that could occur naturally [30]. These crops would then be regulated in line with conventional crops, with 'novel food' oversight [31] where appropriate. This would enable much easier trade relationships with counties that have adopted similar regulations. However, the impact on trade with the EU is still a hurdle, as the UK regulations differ from those in the EU. It should also be noted that currently, the changes to legislation apply only to England: how that may impact trade with Wales and Scotland is not yet clear.

Another example of recent progress in GEd policies and regulations is that two countries in Africa have now deregulated GEd crops: Nigeria and Kenya. As the first country in Africa, Nigeria has authorized guidelines on GEd in December 2020 through its National Biosafety Management Agency. Decisions will be made on a case-by-case basis: if edited lines do not contain a new combination of genetic material, they can be classified as conventional varieties or products. In February 2022, Kenya's National Biosafety Authority published guidelines that provide the framework for exemptions of GEd organisms and products from the Biosafety Act, enabling a case-by-case approval that would treat them as conventional varieties or breeds [32]. Malawi, Ethiopia, and Ghana are also currently developing their policies, while South Africa is currently in an on-going appeals process after a decision to consider all GEd plants as GMOs.

Similarly, for Norway, it has been proposed that GEd crops that do not contain DNA from another species be regulated as conventional plants, but would still require notification. In Israel, GEd crops that do not contain DNA from another species are also regulated as conventional plants. The status of GEd plants has yet to be defined in the Russian Federation [33]. A similar approach is adopted by most South and Central American countries where GEd crops are regulated as conventional plants unless they contain foreign DNA: in some countries, notification to the authorities is required to approve this exemption [34,35].

3.2.3. Avoiding the "Protection without Clarity Regulation"

It is to be hoped that the lessons from the European Union's (EU) approach to regulating GM and now GEd crops will be avoided by countries in the Asia-Pacific region. This relates to the isolated approach adopted by the EU to regulate GM/GEd crops, where crucial definitions leave room for diverse interpretations, while at the same time, scientists in the community and some member countries are not on board with the regulations.

Before 2018, several Member States (e.g., Sweden and the UK, then being part of the EU) had interpreted the mutagenesis exemption to the 2001/18 EU GMO directive to include precision mutagenesis applications such as GEd crops, that had been edited in a way that would result in a product indistinguishable from one obtained through traditional mutagenesis techniques (i.e., chemical or radiation induced mutagenesis).

In 2016, nine NGOs filed a case to the French Courts, (later referred to as the CJEU), to challenge the interpretation of the earlier exemption on GEd (EU GMO directive [36]) to allow 'GM through the backdoor'. Following the 2001/18 EU GMO Directive, over 117 research facilities signed a position paper urging the European Policy Makers to act to safeguard Europe's competitiveness on these new technologies [37]. For many, the ruling fell short of delivering clarity on the regulatory status of GEd, and how such crops would be monitored [38].

The EU Council later requested that the EU Commission look in more detail at the impact of the CJEU ruling. The results, published in April 2021, concluded that the current GMO legislation was 'not fit-for-purpose' for some NGTs and their products, and that it needed to be adapted to keep in line with scientific and technological progress [39]. The lack of clarity surrounding the future regulatory climate for GEd crops has resulted in several EU-based companies focusing on developing GEd crops for non-EU markets [40]. How the EU will be able to implement the regulation and traceability of such crops is also not clear. This debate will continue for the EU, and at the time of writing, a public consultation was open to gain feedback on a proposed new legal framework for plants obtained by targeted mutagenesis and cisgenesis, and for food and feed products. However, any new legislation for review is not expected before Q2 2023. The EU situation is a salutary lesson that Asia-Pacific countries would do well not to follow the EU example on regulating GEd crops.

3.3. Issues Relating to the Commercialization of GEd Plants

3.3.1. Differing Definitions of SDN-2 in Different Jurisdictions

From the information provided above, it is evident that although national definitions of SDN-1 and SDN-3 products of GEd are clear, those for SDN-2 products are not. An overview of SDN-2 definitions in Asia and Australia is provided below, together with comparisons from the EU and the UK (Table 6).

Table 6. Differing definitions of SDN-2 in different jurisdictions.

Country	SDN-2 Definition	Comments
Australia	An organism modified by the repair of single-strand or double-strand breaks of genomic DNA induced by a site-directed nuclease, if a nucleic acid template was added to guide homology-directed repair.	SDN-2 products are still regulated as GMOs (this remains a fundamental point of disagreement between industry and the Regulator, which prefers a decision based on what changes this makes to the final product)
China	GEd crops in China do not fall into the categories of SDN-1, SDN-2	No clear equivalent definition available
Japan	Change with template of 1 to a few bps (the definition is ambiguous to leave room to implement a flexible policy)	SDN-2 is not regarded as GMO if the template DNA is from the same species or from a sexually compatible species.
Philippines	A gene-editing technique that inserts a maximum of 19-bp DNA sequence (foreign or non-foreign) from a repair template, producing a non-GMO.	The definition is different from most others

Table 6. *Cont.*

Country	SDN-2 Definition	Comments
Thailand (draft definition)	A technique in which template DNA is used to modify a targeted DNA sequence to be an intended sequence modification as expected by homology-directed repair (HDR).	Legislation still under consideration. Maximum insert size proposed to be 10 kb.
India	In the recently issued 'Guidelines for Safety Assessment of Genome Edited Plants' 2022', SDN-2 has been defined as site-directed mutagenesis using a DNA sequence template. Further elaborated in the 'comments' column	SDN-2 involves a template-guided repair of a targeted DNA break using an externally supplied template sequence. The donor carries one or several small mutations flanked by twosequences matching both ends of the DNA break, and is thus recognized as a repair template, allowing the introduction of the mutation(s) at the target site. The resultant mutant carries modified sequence, leading to altered expression profile of the gene and/or altered activity of the encoded protein/RNA. Thus, the edited version could be regarded as an allelic form comparable to those available in primary/secondary gene pool'.
Pakistan	No official definition	An agreed definition of SDN-2 is still under discussion in Pakistan.
Bangladesh	No official definition	GEd policies under discussion.
Indonesia	Targeting a specific location Non-GM SDN-2 products: (a) Use own gene pool, no novel protein (b) Use other gene pool as repair template, no novel protein	SND-2 Classified as GM if a novel protein is produced. SDN-3—cisgenics using own gene pool—under discussion whether non-GM or GM Use of another gene pool or production of a novel protein—regulated as a GMO.
Vietnam	No definition at present	
EU	In SDN-2 applications, specific point mutations, small deletions/additions are generated as a result of the introduction into the cell of a repair DNA template (donor DNA) homologous to the targeted area. By means of homologous recombination (HR), precise and small genetic modification can be achieved.	Regulated as a GMO.
UK	SDN-1 and 2 combined—changes that could have been produced by traditional breeding. Genetic Technology (Precision Breeding) Bill—Parliamentary Bills—UK Parliament under discussion.	Allowed to take SDN-1/2 to field trials without a GMO license—but need to notify Defra (competent authority). The aim of current legislation is to allow edits that could have been obtained by traditional breeding or in nature not to be viewed as GMOs.

3.3.2. Proving the Absence of External DNA or New Allergenic Peptides in SDN-1 Plants

One issue relates to the method used to generate SDN-1 plants. Often, the editing machinery (Cas protein and guide RNAs) are introduced as a T-DNA. Once a successful edit is identified, the plant is selfed and selected for null segregants (i.e., plants containing the edit but no longer containing the T-DNA). However, the introduction of undesired DNA sequences from the plasmid backbone might still occur. The question follows, how much effort needs to be made to determine whether there is introduced DNA or not? SDN-1 plants with no additional T-DNA may legally be grown commercially, but if introduced DNA is subsequently found, the plant may become a GMO, and growing that plant without a license may then become illegal. It remains to be determined how much effort should be put into demonstrating a negative, that is, that there is no added DNA. At the very least, for the commercialization of SDN-1 edited plants, the plants would be sequenced at the site of editing to show that no vector DNA is present, and quantitative PCR used to identify the absence of any backbone DNA. In addition, the sequences surrounding the edit site should be checked using appropriate databases to show that no new small ORF formed (e.g., >30 amino acids) and that no new allergenic peptides resulted. These considerations emphasize the need to keep good records and the importance of the stewardship of relevant data.

3.3.3. Off-Target Edits

It is important to view any off-target edits of GEd in context. Spontaneous mutations occur naturally at rates of ~10^{-8} to 10^{-9} per site per generation, and pan-genome sequencing has revealed that many sequence differences exist within a species [41]. There are also more than 3200 varieties generated by induced mutagenesis, not counting the variation caused by somaclonal variation [42]. Off-target GEd may occur at sites similar to the target site, if present. Off-target edits can be minimized by checking the genomic sequence of the plant if available and choosing alternative guide RNA sequences. Recently developed CRISPR editing systems are also more specific, and various studies have indicated that the proper design of gRNAs leads to undetectable levels of off-target edits [43]. Studies on rice [44], cotton [45], and maize [43,46] attributed nearly all of the variation in re-sequenced GEd plants to tissue culture-induced somaclonal variation. This is similar to findings for transgenic plants [47]. The history of the safe consumption of foods from plants is based on the fact that many mutations, regardless of origin, have no phenotypic effect. In conventional breeding, these neutral genetic changes cannot be removed from plant populations. Clearly, possible off-target edits in plants present much-reduced safety concerns compared to those that might arise in medical applications of GEd [41], and off-target edits that result in an undesirable phenotype will routinely be eliminated in subsequent crop breeding.

3.3.4. Unintentional Low-Level Presence of Edited Seeds

An SDN-1 plant may be approved for growth in one country, but not in a country that imports such seeds. The low level presence (LLP) of GM or GEd products is more an issue of bulk grain transport, and of the asynchronous development of national regulations, but could disrupt trade. A group of countries have joined to form the Global Low-Level Presence Initiative (GLI: https://llp-gli.org/; accessed 2 July 2022). The GLI is composed of 15 importing and exporting countries committed to working collaboratively on approaches to minimizing unnecessary trade disruptions when addressing LLP, and includes some countries in SE Asia. These countries are Australia, Argentina, Brazil, Canada (member and Secretariat for the GLI), Colombia, Costa Rica, Indonesia, Mexico, Paraguay, Philippines, Russia, South Africa, United States, Uruguay, and Vietnam. Although formed to address the small amounts of commercialized GM grain, despite the best industry management practices, which might be present in shipments to countries where the use of the grain has not yet been authorized, it is also relevant to GEd grains for the same reasons.

3.3.5. Patents and Licensing

Any group considering the commercialization of a GEd crop or its products should seek professional and legal advice concerning the potential gene targets and the need for a license to use a particular GEd technology. It is beyond the scope of this review to provide such advice.

3.4. *Science Diplomacy—A Pathway toward Regulatory Harmonization?*

Science diplomacy is a policy discussion in which the aim is to combine the values of science and diplomacy to help provide technical solutions to global challenges such as food security [48]. Biotechnology provides an example where science diplomacy can be used to facilitate effective science communication between stakeholders and promote impactful scientific outcomes in multilateral negotiations including the harmonization of regulatory triggers for plant-based GEd [49].

The multilateral treaty system and on-going negotiations at the United Nations and other regional organizations is likely to have a direct impact on the regulations related to the commercialization of biotechnology products. There is a real need to include current scientific knowledge on GEd products in the context of the Cartagena Protocol on Biosafety, which deals with the environmental release of biotech crops. Without clarity on the regulatory status of GE crops in the international treaty process, efforts to harmonize regulations will be more difficult and can lead to institutional drift [50]. The benefits of international GEd policy harmonization may be achieved more rapidly through negotiation simulations, science policy/diplomacy education [51], policy advocacy platforms [52], and grant structures [53].

Science diplomacy can therefore help increase the communication between policy/diplomacy communities at the level of product (regulatory triggers), process (international conventions), and education (science communication), thus reducing institutional drift [54]. Various organizations and networks such as the International Rice Research Institute (IRRI), International Maize and Wheat Improvement Center (CIMMYT), and the Consultative Group on International Agricultural Research (CGIAR) are in key positions to increase the impact of science advice in international treaties to promote harmonization. Science diplomacy can enable cross-sectoral treaty discussions and national capacities in developing regulatory guidelines related to agricultural biotechnology, its commercialization, and trade

4. Conclusions and Prospects for the Future

The current population in Asia and Australasia is about 4.75 billion people (about 60% of the world's total population) and it is predicted to grow to 5.5 billion by 2050 (www.worldometers.info; accessed 2 July 2022). New GE technologies hold the promise to make a major contribution to increasing the world's food supply for both farmers and consumers, and for human health and the environment. The frontiers of plant breeding are moving from transgenesis as a potentially dominant form of plant breeding, mainly because of the onerous and prohibitive regulations surrounding GMOs/LMOs. In contrast, developing regulations are treating GEd plants more as mutagenized crops, in a targeted rather than random way [55]. Since many thousands of commercialized crops derived from mutagenic treatments are available without regulation as GMOs, and which can also be sold as 'organic', the developing regulatory landscape for GEd crop products could mark the end of pointless battles over GMOs. There is now an opportunity to fundamentally alter the 'risk–utility' balance for GEd products [55].

However, it is clear that regulatory harmonization, or at least alignment, of GEd crops and produce is crucial, and this is particularly the case for Asia and Australasia. Without the harmonization/alignment of GE regulations, crop industries may well face the same trade issues that have limited the wider commercialization of GM crops. From the narrative provided above, it appears that increasingly more of the world's nations are proceeding to a rational approach of regulating GEd crops, following the principle that like products should be regulated in the same way. Some have been influenced by EU policies, and some are swayed by NGOs and activists, rather than because they disagree with the science. Commonly accepted definitions of SDN-1, SDN-2, and SDN-3 would be a great help. With the aid of science diplomacy and meaningful international discussions, the harmonization or alignment of GEd regulations can be achieved, thus enabling the full benefits of GEd technologies to be realized.

Author Contributions: Conceptualization: M.G.K.J., M.A. (Muhammad Adeel), M.A. (Mahaletchumy Arujanan) and R.R.-A. Writing: all stated authors contributed sections of the manuscript on GEd regulations in their own regulatory jurisdictions, or on commercialization of GEd crops. Review and editing: M.G.K.J., S.I. and J.F.-N. All authors have read and agreed to the published version of the manuscript.

Funding: This project was funded by a grant to M.G.K.J. and M.A. (Muhammad Adeel) from the Department of Agriculture, Forestry, and Fisheries (formerly DAWE), Government of Australia, Project Assisting Small Exporters (PASE), Building Capacity for Small Exporters to Exploit 'new breeding technologies', Grant number 4-FA4N7WL. Author P.H. acknowledges funding support from the Biotechnology and Biological Sciences Research Council (BBSRC), the Institute Strategic Programs GRO [BB/J004588/1], and GEN [BB/P013511/1] to the John Innes Centre.

Data Availability Statement: Not applicable.

Acknowledgments: We thank the authors from Murdoch University, Perth; ISAAA, the Philippines and the Malaysian Biotechnology Information Center for helping to co-ordinate this review, and all authors for their contributions.

Conflicts of Interest: The authors declare no conflict of interest.

References

1. AgbioInvestor. Time and cost to develop a new GM trait. In *A Study on Behalf of Crop Life International: Cost and Time Required for the Discovery, Development and Authorisation of a New Plant Biotechnology-Derived Genetic Trait*; 2022; pp. 1–45. Available online: www.agbioinvestor.com (accessed on 17 July 2022).
2. Belhaj, K.; Chaparro-Garcia, A.; Kamoun, S.; Patron, N.J.; Nekrasov, V. Editing plant genomes with CRISPR/Cas9. *Curr. Opin. Biotechnol.* **2015**, *32*, 76–84. [CrossRef]
3. Zafar, S.A.; Zaidi, S.S.; Gaba, Y.; Singla-Pareek, S.L.; Dhankher, O.P.; Li, X.; Mansoor, S.; Pareek, A. Engineering abiotic stress tolerance via CRISPR/Cas-mediated genome editing. *J. Exp. Bot.* **2020**, *71*, 470–479. [CrossRef]
4. Zaidi, S.S.; Mukhtar, M.S.; Mansoor, S. Genome editing: Targeting susceptibility genes for plant disease resistance. *Trends Biotechnol.* **2018**, *36*, 898–906. [CrossRef]
5. Zaidi, S.S.; Vanderschuren, H.; Qaim, M.; Mahfouz, M.M.; Kohli, A.; Mansoor, S.; Tester, M. New plant breeding technologies for food security. *Science* **2019**, *363*, 1390–1391. [CrossRef]
6. Kouhen, M.; García-Caparrós, P.; Twyman, R.M.; Abdelly, C.; Mahmoudi, H.; Schillberg, S.; Debez, A. Improving environmental stress resilience in crops by genome editing: Insights from extremophile plants. *Crit. Rev. Biotechnol.* **2022**, 1–16. [CrossRef]
7. Awan, M.J.A.; Amin, I.; Mansoor, S. Mini CRISPR-Cas12f1: A new genome editing tool. *Trends Plant Sci.* **2022**, *27*, 110–112. [CrossRef]
8. Gaj, T.; Sirk, S.J.; Shui, S.-L.; Liu, J. Genome-editing technologies: Principles and applications. *Cold Spring Harb. Perspect. Biol.* **2016**, *8*, a023754. [CrossRef]
9. Ahmar, S.; Mahmood, T.; Fiaz, S.; Mora-Poblete, F.; Shafique, M.S.; Chattha, M.S.; Jung, K.-H. Advantage of nanotechnology-based genome editing system and its application in crop improvement. *Front. Plant Sci.* **2021**, *12*, 663849. [CrossRef]
10. Chandrasekaran, R.; Rajiv, P.; Abd-Elsalam, K.A. 14—Carbon nanotubes: Plant gene delivery and genome editing. In *Carbon Nanomaterials for Agri-Food and Environmental Applications*; Abd-Elsalam, K.A., Ed.; Elsevier: Amsterdam, The Netherlands, 2020; pp. 279–296.
11. Mujtaba, M.; Wang, D.; Carvalho, L.B.; Oliveira, J.L.; Espirito Santo Pereira, A.d.; Sharif, R.; Jogaiah, S.; Paidi, M.K.; Wang, L.; Ali, Q. Nanocarrier-mediated delivery of miRNA, RNAi, and CRISPR-Cas for plant protection: Current trends and future directions. *ACS Agric. Sci. Technol.* **2021**, *1*, 417–435. [CrossRef]
12. Zhang, Y.; Massel, K.; Godwin, I.D.; Gao, C. Applications and potential of genome editing in crop improvement. *Genome Biol.* **2018**, *19*, 1–11. [CrossRef]
13. EU-SAGE. European Sustainable Agriculture through Genome Editing. 2002. Available online: https://www.eu-sage.eu/genome-search (accessed on 2 July 2022).
14. Fritsche, S.; Poovaiah, C.; MacRae, E.; Thorlby, G. A New Zealand Perspective on the Application and Regulation of Gene Editing. *Front. Plant Sci.* **2018**, 9. [CrossRef] [PubMed]
15. Cross-Ministerial Strategic Innovation Promotion Program. Japan's Science and Technology Policy. Council for Science and Technology Policy Brochure about SIP. Pioneering the Future: Japanese Science, Technology and Innovation 2021. Available online: https://www8.cao.go.jp/cstp/panhu/sip_english/sip_en.html (accessed on 13 May 2022).
16. Martin-Laffon, J.; Kuntz, M.; Ricroch, A.E. Worldwide CRISPR patent landscape shows strong geographical biases. *Nat. Biotechnol.* **2019**, *37*, 613–620. [CrossRef] [PubMed]
17. Wei, X.; Pu, A.; Liu, Q.; Leng, Y.; Fu, Z.; Wu, F.; An, X.; Long, Y. Commercialization and Supervision Policies of Gene Edited Crops in China and Other Main Countries. *ACS Agric. Sci. Technol* **2022**, *2*, 167–173. [CrossRef]
18. Mao, Y.; Botella, J.R.; Liu, Y.; Zhu, J.-K. Gene editing in plants: Progress and challenges. *Nat. Sci. Rev.* **2019**, *6*, 421–437. [CrossRef]
19. Huang, S.; Weigel, D.; Beachy, R.N.; Li, J. A proposed regulatory framework for genome-edited crops. *Nat. Genet.* **2016**, *48*, 109–111. [CrossRef]
20. Zafar, K.; Noureen, A.; Awan, M.J.A.; Anjum, N.; Aslam, M.Q.; Khan, M.Z.; Amin, I.; Mansoor, S. Genome Editing to Develop Disease Resistance in Crops. In *Genome Engineering for Crop Improvement*; Upadhyay, S.K., Ed.; John Wiley & Sons: Hoboken, NJ, USA, 2021; pp. 224–252.

21. Noureen, A.; Zuhaib Khan, M.; Amin, I.; Zainab, T.; Ahmad, N.; Haider, S.; Mansoor, S. Broad-spectrum resistance against multiple PVY-strains by CRISPR/Cas13 system in *Solanum tuberosum* crop. *GM Crops Food* **2022**, *13*, 97–111. [CrossRef]
22. Ibrahim, S.; Saleem, B.; Rehman, N.; Zafar, S.A.; Naeem, M.K.; Khan, M.R. CRISPR/Cas9 mediated disruption of Inositol Pentakisphosphate 2-Kinase 1 (TaIPK1) reduces phytic acid and improves iron and zinc accumulation in wheat grains. *J. Adv. Res.* **2022**, *37*, 33–41. [CrossRef]
23. Hanri, M.; Sholihah, N.K. Labor Market Brief: Productivity and Innovation Research Group. LPEM FEB University of Indonesia. 2020. Available online: https://www.lpem.org/research-group/social-protection-and-labor-research-group/ (accessed on 2 June 2022).
24. Apriana, A.; Santoso, T.J.; Sisharmini, A.; Reflinur; Ambarwati, A.D.; Hadiarto, T.; Sustiprijatno; Nuryati. Phenotypic and genetic stability evaluation of the targeted GA20ox-2 gene mutation in CRISPR/Cas9 mutant rice derived from Mentong cultivar. In *AIP Conference Proceedings*; AIP Publishing: College Park, MD, USA, 2022; Volume 2462.
25. Mubarok, H.; Basunanda, P.; Santoso, T.J. Tolerance of T2 generation 'Kitaake'Rice (*Oryza sativa* L.) CRISPR/Cas9-OsGA20ox-2 mutant strains to drought condition. *Ilmu Pertan. (Agric. Sci.)* **2019**, *4*, 123–130. [CrossRef]
26. Santoso, T.; Trijatmiko, K.; Char, S.; Yang, B.; Wang, K. Targeted mutation of GA20ox-2 gene using CRISPR/Cas9 system generated semi-dwarf phenotype in rice. *IOP Conf. Ser. Earth Environ. Sci.* **2020**, *482*, 012027. [CrossRef]
27. Ruzyati, M.; Sisharmini, A.; Apriana, A.; Santoso, T.J.; Purwanto, E.; Samanhudi, S.; Yunus, A. Construction of CRISPR/Cas9_gRNA-OsCKX2 module cassette and its introduction into rice cv. Mentik Wangi mediated by *Agrobacterium tumefaciens*. *Biodiversitas J. Biol. Divers.* **2022**, *23*, 2679–2689. [CrossRef]
28. *ISAAA Brief 55–2019. Executive Summary: Biotech Crops Drive Socio-Economic Development and Sustainable Environment in the New Frontier. International Service for the Acquisition of Agri-Biotech Applications*; ISAAA: Ithaca, NY, USA, 2019. Available online: https://www.isaaa.org/resources/publications/briefs/55/executivesummary/default.asp (accessed on 2 June 2022).
29. Calyxt Press Release. *First Commercial Sale of Calyxt High Oleic Soybean Oil on the U.S.* Saint Paul, MN, USA, 2019. Available online: https://calyxt.com/first-commercial-sale-of-calyxt-high-oleic-soybean-oil-on-the-u-s-market/ (accessed on 2 June 2022).
30. *Department for Environment, Food and Rural Affairs (Defra) Consultation Outcome—Genetic Technologies Regulation*; Government Response: London, UK, 2021.
31. Food Standard Agency. *Comparing International Approaches to Food Safety Regulation of GM and Novel Foods*; F.S.A.: London, UK, 2021. [CrossRef]
32. Buchholzer, M.; Frommer, W.B. An increasing number of countries regulate genome editing in crops. *New Phytol.* **2022**. [CrossRef]
33. Bogatyreva, N.V.; Sokolov, A.Y.; Moiseeva, Y.M.; Gusev, Y.S.; Chumakov, M.I. Regulatory status of genome-editing plants: Perspectives for Russian Federation. *Ecol. Genet.* **2021**, *19*, 89–101. [CrossRef]
34. Hundleby, P.A.C.; Harwood, W.A. Impacts of the EU GMO regulatory framework for plant genome editing. *Food Energy Secur.* **2019**, *8*, e00161. [CrossRef]
35. Rosado, A.; Eriksson, D. Biosafety legislation and the regulatory status of the products of precision breeding in the Latin America and the Caribbean region. *Plants People Planet* **2021**, *4*, 214–231. [CrossRef]
36. Court of Justice of the European Union (CJEU). *Organisms Obtained by Mutagenesis are GMOs and Are, in Principle, Subject to the Obligations Laid down by the GMO Directive*; Court of Justice of the European Union: Luxembourg, 2018.
37. Max-Planck-Gesellschaft. Scientists Call for Modernization of EU Gene-Editing Legislation | Max-Planck-Gesellschaft (mpg.de)' Press Release 2019, 29 July 2019. Available online: https://www.mpg.de/13761643/scientists-call-for-modernization-of-the-european-genetic-engineering-law (accessed on 2 August 2021).
38. Van Der Meer, P.; Angenon, G.; Bergmans, H.; Buhk, H.J.; Callebaut, S.; Chamon, M.; Eriksson, D.; Gheysen, G.; Harwood, W.; Hundleby, P.; et al. The status under EU Law of organisms developed through Novel Genomic Techniques. *Eur. J. Risk Regul.* **2021**, 1–20. [CrossRef]
39. European Commission. *Study on the Status of New Genomic Techniques under Union Law and in Light of the Court of Justice Ruling in Case C-528/16, 2021*; European Commission: Brussels, Belgium, 2021.
40. Jorasch, P. The view of the European seed sector on genome editing in plant breeding. In *1st Plant Ed Conference Plant Genome Editing—State of the Art. COST ACTION CA 18111*; University of Novi Sad: Novi Sad, Serbia, 2019.
41. Graham, N.; Patil, G.B.; Bubeck, D.M.; Dobert, R.C.; Glenn, K.C.; Gutsche, A.T.; Kumar, S.; Lindbo, J.A.; Maas, L.; May, G.D.; et al. Plant genome editing and the relevance of off-target changes. *Plant Physiol.* **2020**, *183*, 1453–1471. [CrossRef]
42. EFSA Panel on Genetically Modified Organisms; Mullins, E.; Bresson, J.L.; Dalmay, T.; Dewhurst, I.C.; Epstein, M.M.; Firbank, L.G.; Guerche, P.; Hejatko, J.; Moreno, F.J. In vivo and in vitro random mutagenesis techniques in plants. *EFSA J.* **2021**, *19*, e06611. [CrossRef]
43. Young, J.; Zastrow-Hayes, G.; Deschamps, S.; Svitashev, S.; Zaremba, M.; Acharya, A.; Paulraj, S.; Peterson-Burch, B.; Schwartz, C.; Djukanovic, V.; et al. CRISPR-Cas9 editing in maize: Systematic evaluation of off-target activity and Its relevance in crop improvement. *Sci. Rep.* **2019**, *9*, 6729. [CrossRef]
44. Tang, X.; Liu, G.; Zhou, J.; Ren, Q.; You, Q.; Tian, L.; Xin, X.; Zhong, Z.; Liu, B.; Zheng, X.; et al. A large-scale whole-genome sequencing analysis reveals highly specific genome editing by both Cas9 and Cpf1 (Cas12a) nucleases in rice. *Genome Biol.* **2018**, *19*, 84. [CrossRef]

45. Li, J.; Manghwar, H.; Sun, L.; Wang, P.; Wang, G.; Sheng, H.; Zhang, J.; Liu, H.; Qin, L.; Rui, H.; et al. Whole genome sequencing reveals rare off-target mutations and considerable inherent genetic or/and somaclonal variations in CRISPR/Cas9-edited cotton plants. *Plant Biotechnol. J.* **2019**, *17*, 858–868. [CrossRef]
46. Lee, K.; Zhang, Y.; Kleinstiver, B.P.; Guo, J.A.; Aryee, M.J.; Miller, J.; Malzahn, A.; Zarecor, S.; Lawrence-Dill, C.J.; Joung, J.K.; et al. Activities and specificities of CRISPR/Cas9 and Cas12a nucleases for targeted mutagenesis in maize. *Plant Biotechnol. J.* **2019**, *17*, 362–372. [CrossRef]
47. Wang, Y.; Liu, Q.; Song, X.; Yang, X.; Han, L.; Romeis, J.; Li, Y. Unintended changes in transgenic maize cause no nontarget effects. *Plants People Planet* **2022**, *4*, 392–402. [CrossRef]
48. Turekian, V. The Evolution of Science Diplomacy. *Glob. Policy* **2018**, *9*, 5–7. [CrossRef]
49. Adeel, M. Food Security in a Post-COVID-19 World: Regulatory Perspectives for Agricultural Biotechnology; Science & Diplomacy: American Association for the Advancement of Science: 2021. Available online: https://www.sciencediplomacy.org/perspective/2021/food-security-in-post-covid-19-world-regulatory-perspectives-for-agricultural (accessed on 4 July 2022).
50. Rabitz, F. Institutional drift in international biotechnology regulation. *Glob. Policy* **2019**, *10*, 227–237. [CrossRef]
51. Mauduit, J.-C.; Gual Soler, M. Building a science diplomacy curriculum. *Front. Educ.* **2020**, *5*, 138. [CrossRef]
52. Arnaldi, S.; Lombardo, A.; Tessarolo, A. A preliminary study of science diplomacy networks in Central, Eastern and South-Eastern Europe. *Humanit. Soc. Sci. Commun.* **2021**, *8*, 1–9. [CrossRef]
53. Moedas, C. Science diplomacy in the European Union. *Sci. Dipl.* **2016**, *5*, 1–9.
54. Adeel, M. Towards a science diplomacy typology for regulation of agricultural biotechnology. *Sci. Dipl. Perspect.* **2022**, *1*, 74–77.
55. Hefferon, K.L.; Herring, R.J. The end of the GMO? Genome editing, gene drives and new frontiers of plant technology. *Rev. Agrar. Stud.* **2017**, *7*, 1–32. [CrossRef]

MDPI
St. Alban-Anlage 66
4052 Basel
Switzerland
www.mdpi.com

Plants Editorial Office
E-mail: plants@mdpi.com
www.mdpi.com/journal/plants

www.ingramcontent.com/pod-product-compliance
Lightning Source LLC
Chambersburg PA
CBHW061846210926
51850CB00006B/979
9783036585925